ENVIRONMENTAL DATA HANDLING
George B. Heaslip

THE MEASUREMENT OF AIRBORNE PARTICLES
Richard D. Cadle

ANALYSIS OF AIR POLLUTANTS
Peter O. Warner

ENVIRONMENTAL INDICES
Herbert Inhaber

URBAN COSTS OF CLIMATE MODIFICATION
Terry A. Ferrar, Editor

CHEMICAL CONTROL OF INSECT BEHAVIOR: THEORY AND APPLICATION
H. H. Shorey and John J. McKelvey, Jr.

MERCURY CONTAMINATION: A HUMAN TRAGEDY
Patricia A. D'Itri and Frank M. D'Itri

POLLUTANTS AND HIGH RISK GROUPS
Edward J. Calabrese

SULFUR IN THE ENVIRONMENT, Parts I and II
Jerome O. Nriagu

ENERGY UTILIZATION AND ENVIRONMENTAL HEALTH: METHODS FOR PREDICTION AND EVALUATION OF IMPACT ON HUMAN HEALTH
Richard A. Wadden, Editor

METHODOLOGICAL APPROACHES TO DERIVING ENVIRONMENTAL AND OCCUPATIONAL HEALTH STANDARDS
Edward J. Calabrese

FOOD, CLIMATE, AND MAN
Margaret R. Biswas and Asit K. Biswas, Editors

CHEMICAL CONCEPTS IN POLLUTANT BEHAVIOR
Ian J. Tinsley

RESOURCE RECOVERY AND RECYCLING
A. F. M. Barton

RESOURCE RECOVERY AND RECYCLING

ALLAN F. M. BARTON
MURDOCH UNIVERSITY
WESTERN AUSTRALIA

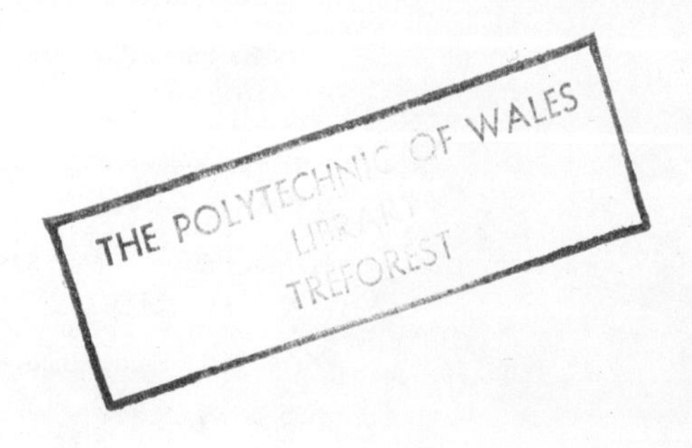

A WILEY-INTERSCIENCE PUBLICATION

JOHN WILEY & SONS

NEW YORK • CHICHESTER • BRISBANE • TORONTO

Library of Congress Cataloging in Publication Data:

Barton, Allan F. M.
Resource recovery and recycling.

(Environmental science and technology)
''A Wiley-Interscience publication.''
Includes index.
1. Recycling (Waste, etc.) 2. Waste products as fuel. I. Title

TD794.B37 628'.445 78-13601
ISBN 0-471-02773-1

Printed in the United States of America

10 9 8 7 6 5 4 3 2 1

The paper in this book was made from 65% recycled fiber by the Bergstrom Paper Company of Neenah, Wisconsin, a company specializing in recycled papers for books. This paper goes by the trade name of Thor Cote Medium 576 and is recycled from such groundwood-free materials as computer cards, used magazines, ''postconsumer wastes,'' and printer's trimmings.

SERIES PREFACE

Environmental Science and Technology

The Environmental Science and Technology Series of Monographs, Textbooks, and Advances is devoted to the study of the quality of the environment and to the technology of its conservation. Environmental science therefore relates to the chemical, physical, and biological changes in the environment through contamination or modification, to the physical nature and biological behavior of air, water, soil, food, and waste as they are affected by man's agricultural, industrial, and social activities, and to the application of science and technology to the control and improvement of environmental quality.

The deterioration of environmental quality, which began when man first collected into villages and utilized fire, has existed as a serious problem under the ever-increasing impacts of exponentially increasing population and of industrializing society. Environmental contamination of air, water, soil, and food has become a threat to the continued existence of many plant and animal communities of the ecosystem and may ultimately threaten the very survival of the human race.

It seems clear that if we are to preserve for future generations some semblance of the biological order of the world of the past and hope to improve on the deteriorating standards of urban public health, environmental science and technology must quickly come to play a dominant role in designing our social and industrial structure for tomorrow. Scientifically rigorous criteria of environmental quality must be developed. Based in part on these criteria, realistic standards must be established and our technological progress must be tailored to meet them. It is obvious that civilization will continue to require increasing amounts of fuel, transportation, industrial chemicals, fertilizers, pesticides, and countless other products; and that it will continue to produce waste products of all descriptions. What is urgently needed is a total systems approach to modern civilization through which the pooled talents of scientists and engineers, in cooperation with social scientists and the medical profession, can be focused on the development of order and equilibrium in the presently disparate segments of the human environment. Most of the skills and tools that are needed are already in existence. We surely have a right to hope a technology that has created such manifold environmental problems is also capable of solving them. It is our hope that this Series in Environmental Sciences and Technology will not only serve to make this challenge more explicit to the established professionals, but that it also will help to stimulate the student toward the career opportunities in this vital area.

Robert L. Metcalf
Werner Stumm

PREFACE

Recently there has been considerable concern expressed about the "crises" in energy, resources, and pollution. Energy demands, resource limitations, and environmental pollution are closely linked, and recycling can make a contribution to the solution of all three problems.

Ideas and techniques that have proved valuable in some situations have been collected, with the hope that they will be of use in other circumstances and will stimulate further recycling activity. The subject is too complex, and has received too little attention, for it to be tackled successfully by a single rationalized model, and I have taken instead a broad, unstructured view of the complex interrelationships; flexibility of thought and attitude are essential in this area.

The scientific bases and criteria for recycling are reviewed here in a simple manner; a minimum of technical knowledge has been assumed, and a glossary of scientific terms is provided. This is not primarily a recycling handbook, although it can be used as such by taking advantage of its numerous references, which permit easy access to bibliographies, reviews of the patent literature, and sources of technical data that have appeared up to mid-1976. These are mostly in the fields of chemistry and chemical engineering, but also included are some in physics and microbiology.

Good communication is critically important in the present state of development of a conserver society, and to facilitate direct exchange of information on recycling there are included in the bibliography the names and addresses of institutions with which the authors were affiliated.

My aim has been to present an unbiased survey of scientific and technological aspects of resources recycling and energy recovery, which is intended to be equally valuable to scientists and engineers, industrialists and environmentalists, teachers and students, professionals, and the concerned public. I have generally avoided making axiomatic statements in an area where only informed opinions are justified, but I consider that just by writing a book on recycling I have made my point of view clear.

I would like to express my appreciation to all those individuals and organizations whose ideas, information, and materials are incorporated here, and especially to members of Murdoch University who assisted me in many ways.

ALLAN F. M. BARTON

Perth, Western Australia
January 1979

CONTENTS

1. **Introduction** 1

Levels of Recycling, 2
Terminology, 3
The Broader Meaning of "Mining," 6
What Can We Recycle? 8
The Recycle Philosophy, 10

2. **Factors Influencing Recycling Feasibility** 13

Economics, 13
Availability and Value of Chemical Content, 14
Useful Physical Properties, 17
Collection and Transportation, 17
Separation, 18
Environmental Protection Pressures, 22
Energy, 24
Quality of Recycled Material, 26
Employment and Social Attitudes, 27
Public Relations and Advertizing Exercises, 28
Durability and Extended Life, 28
The Recycling Index, 29

3. **Physical Methods of Separation and Recovery** 30

Manual and Mechanical Sorting, 31
Electrical and Magnetic Properties, 32
Separation by Phase and Density Differences, 39
Throw, Bounce, and Slide Methods, 52
Surface and Membrane Processes, 54
Thermal Methods of Physical Separation, 60

4. **Chemical Separation and Conversion Processes** 65

Exchange Reactions, 66
Polymerization and Depolymerization, 73
Oxidation and Reduction Processes, 82
Hydrometallurgy and Pyrometallurgy, 93
Nuclear Fusion Torch, 94

5. **Microbiological Recycling** 95

Composting, 99
Controlled Fermentation, 102
Anaerobic Digestion, 104
Solar Energy Storage, 105
Recycling of Inorganic Materials, 107
Biodegradation of Plastics, 108

6. **Postconsumer Waste** 110

Composition of Urban Waste, 114
Packaging, 116
Beverage Containers, 118
Voluntary Separation at Source, 120
Collection of Urban Waste, 122
Terminal Sorting and Material Recovery, 122
Biological Treatment of Urban Waste, 126
Energy Recovery from Urban Waste, 132
Upgraded Fuels from Urban Waste, 138
Economic Factors in Urban Waste Treatment, 145
Sewage, 147
Scrapped Automobiles, 150
Discarded Tires, 154
Prototype Urban Waste Systems, 157

7. **Industrial and Agricultural Recycling Processes** 175

Building and Construction Materials from Wastes, 178
Solutions, Sludges, Dusts, and Slags, 187
Glass, 199
Gases, 201
Water Reuse, 207
Organic Liquids, 214
Metals, 215

Synthetic Polymers, 216
Textiles, 224
Cellulose, Carbohydrate, and Related Materials, 225
Hydrocarbons, 238
Radiation Sources, 240

8. Thermodynamics of Recycling **241**

Entropy, 241
Energy, 244
Energy Expenditures for Production and Recycling, 245
Efficient Thermal Energy Utilization, 248

Appendix 1 Recycling by Elements **250**

Carbon, 250
Helium, 252
Structural and Plating Metals, 252
Precious and Semiprecious Metals, 261
Less Common Metals, 261
Group Five Elements, 262
Sulfur, 262
Halogens, 263

Appendix 2 Recycling Review Publications **264**

Appendix 3 Addresses of Organizations **270**

Appendix 4 Glossary of Scientific Terms **287**

Appendix 5 Bibliography **300**

Index **379**

RESOURCE RECOVERY AND RECYCLING

1

INTRODUCTION

It was not very long ago that materials were recycled to a considerable extent because labor was much more readily available than materials. Garbage was fed to animals, containers were all reused, and tools were always repaired. Consideration of thermodynamic factors shows that waste-recycling problems are inevitable, and a compromise must be reached between material conservation and energy expenditure. There will always be difficulties associated with materials recycling, but the sooner we face them the less critical they will be.

Modern societies, in emphasizing disposal rather than recovery and recycling, have tended to use methods of waste treatment that are even more primitive than those practiced early in man's development: burning, burying, or dumping in rivers and oceans. Even now, considerable attention is directed toward advanced incineration and compaction techniques, with disposal still the primary aim. Some current trends in municipal waste disposal are even discouraging reclamation; air pollution control regulations are causing a move away from incineration (which provides the opportunity for energy recovery) toward landfilling, and public hygiene regulations prevent access for salvage to "sanitary landfills" without providing alternative reclamation opportunities.

It has been pointed out that modern society operates by using *energy* to turn *resources* into *junk* (Rose et al., 1972). Of this trio of problems, junk receives even less attention than energy and resources. Exponential growth of the waste problem in our affluent, disposal-oriented society now makes it obvious that recycling should be our goal, and it is necessary, therefore, for scientists and technologists to examine all aspects of fabrication and processing techniques to facilitate the reuse of materials. The philosophy of recycling provides a new set of priorities, and it is the aim here to consider various recycling processes in the light of the associated chemical and physical properties of materials.

The need for recycling is most apparent in a small, completely isolated system, demonstrated dramatically in the manned lunar expeditions. For example, hydrogen and oxygen after reaction together in fuel cells to provide electric power were "recycled" as drinking water. The space program has had the highly beneficial effect of causing us to look at our planet as a spaceship, emphasizing the importance of recycling at all levels. Nevertheless, the great technological achievement of moon landing was accompanied by astronauts dumping their refuse on the moon in the time-honored terrestrial manner. Another fact to keep in mind is that despite the prospects promised by science fiction, it is unlikely that we can import from outside the planet any appreciable quantity of materials.

LEVELS OF RECYCLING

In order to clarify and systematize recycling and polution problems, Breeling and Moore (1972) have considered a hierarchy of systems ranging from an atom to the Universe (Table 1.1). At all levels below number 16 there is a theoretical

Table 1.1. Systems Hierarchy

1	atom
2	molecule
3	pure gas, liquid, or solid
4	compound or alloy
5	simple component, e.g., shaft, differentiated cell
6	complex component, e.g., motor, kidney
7	unit, e.g., clock, automobile, man
8	home, factory, family
9	city, corporation
10	state, industry
11	nation
12	planet
13	solar system
14	galaxy
15	galactic cluster
16	universe

Source. Breeling and Moore, 1972. Reprinted from Western Hemisphere Nutrition Congress III, Symposia Specialists, P.O. Box 610397, Miami, Florida 33161, with permission.

upper limit to the effectiveness of recycling. It is possible that at the 16th level, the Universe is in fact a perpetual motion machine, the only system capable of complete containment of its energy and matter. (This possibility has been explored dramatically by Asimov (1956) in his science fiction story "The Last Question.") Certainly at all lower levels, material can never be recycled fully, and *any* recycling requires the expenditure of energy. When the recycling of a particular material is considered, it is necessary to decide on the level at which the problem should be tackled. For example, although glass bottles are discarded all over the world, the chemical constituents of glass are not in short supply. Indeed, it has been estimated (Rose et al., 1972) that if everyone on earth continually drank beer and continually discarded the empties, geologic processes would turn the empty bottles back into sand before we ran out of resources for making more bottles. Here the main problem exists at levels 8 and 9. Discarding tinned steel cans is another matter; many materials such as tin are in relatively short supply on the level 12 scale. Except for the loss of gases such as hydrogen and helium from the top of the atmosphere, material is conserved at level 12; but recycling involves not only conserving a particular element but also preventing it from becoming disseminated throughout the earth's crust and biosphere.

These examples illustrate the importance of *time* in recycling decisions. Another example may be of value here (Rose et al., 1972). In the construction of accommodation we can identify three distinct time scales:

1. That of the speculative builder, 1 or 2 years;
2. That of the average house owner, who looks forward 10 or 20 years and provides maintenance to protect the structure before "recycling" becomes necessary;
3. That of society as a whole, concerned with depletion of forests and energy reserves, and with optimum land use.

Short time scale considerations eventually involve greater energy and resource costs than long term planning, although the initial, apparent costs of the latter may be greater.

TERMINOLOGY

The term "recycling" in its narrowest, and probably original, sense denoted the return of a discarded material or article to the same product system, such as the return of waste paper to make new paper. Recently, however, many broader definitions have been proposed. For example, the concept of recycling applied to a forest tree may involve its use for recreation, shelter, furnishing, paper and board, food and chemicals, heat and electricity (Auchter, 1973; Carr, 1970). There are many examples of recycling in nature: the carbon, nitrogen, and oxygen cycles, the water cycle, the building up of sedimentary rocks (Chadwick and Goodman, 1975; Garrels et al., 1975; Hughes, 1975b; Moore and Moore, 1976; Stumm and Davis, 1974). This book is concerned with the processes for recycling discarded or waste materials however they arise (thus defining recycling in its broadest sense), but concentrates on "artificial" or man-made rather than natural processes.

If the fabrication of a product, "A," from an isolated material, "1," is depicted by step (1) in the series shown here, "recycle" involves either recovery of the original material (step (2)), or recovery of a new material (step (3)), or (usually least satisfactory) conversion to combustion products to recover energy (step (4)):

$$\text{isolated material 1} \rightarrow \text{fabricated product A} \tag{1}$$

$$\text{fabricated product A} \rightarrow \text{isolated material 1} \tag{2}$$

$$\text{fabricated product A} \rightarrow \text{isolated material 2} \tag{3}$$

$$\text{fabricated product A} \rightarrow \text{combustion products plus energy} \tag{4}$$

The "chain" of steps (1)-(3) may be extended:

$$\text{isolated material 2} \rightarrow \text{fabricated product B} \tag{5}$$

$$\text{fabricated product B} \rightarrow \text{isolated material 3} \tag{6}$$

and there is obviously an infinite number of possible schemes, which may be illustrated also in terms of flow diagrams (Bundi and Wasmer, 1976).

It is useful to distinguish between two types of recovery process:

i. Closed-loop, direct, or nonsacrificial recycling, or reuse, such as steps (1) and (2);
ii. Open-loop, indirect, or sacrificial recycling, or new use, such as steps (3) and (5).

In sacrificial recycling, there is a net loss of the original raw material when it is downgraded for its new use in other products. For example, although glass can undergo nonsacrificial recycling indefinitely by remelting, glass as an aggregate for road surfacing or building construction can be recycled "sacrificially" only once.

If elements are considered to be the components being recycled (Appendix 1), all pathways are cyclic except those involving nuclear reactions. More often, however, the aim is to preserve groupings of atoms with particular, useful characteristics or high internal energy values; there is little practical advantage in recycling carbon as carbon dioxide, but other carbon compounds are valuable fuels. This introduces the subject of energy, because in each of the steps (1)-(3) it is usual for energy to be expended. A balance must be achieved between expenditure of energy and conservation of material; if the material is extremely rare, no energy cost is too great. Extensive recycling occurs only when justified economically, and economic conditions depend on both supply and demand. The balance is shifting as supplies dwindle and demand grows; more and more materials now justify the expenditure of energy and effort necessary for conservation.

Table 1.2 lists the recycling possibilities, ranked in order from "most preferred" to "least preferred."

Table 1.2. Categories of Recycling

Type of Recycling	Examples
Reuse of article	Returnable, refillable bottle
Nonsacrificial recycling: reuse of material at similar quality level (primary recycling)	Color-sorted glass cullet; newspaper repulping for newsprint production
Nonsacrificial recycling: reuse of material at a lower quality level (secondary recycling)	Roofing felt from waste paper; fiberglass insulation from bottles
Sacrificial recycling: new use for material (tertiary recycling)	Road paving material from glass; composting of paper
Thermal recycling:	
Conversion to storable fuel	Pyrolysis of urban waste
Direct incineration to produce energy	Incineration of urban waste

Because of the ambiguity associated with many of the terms used in this field, it is necessary to state here the meanings implied when they are used in this book. The word *wastes* (and recently the equivalent term *residuals*) refers to materials discarded by community activities, and includes solids, liquids, and gases. *Refuse* normally means "solid waste," but this term is avoided here. *Garbage* is domestic or household food waste, *rubbish* is domestic nonfood waste, and *residential waste* or *domestic waste* is the combination of garbage and rubbish. *Urban waste* is taken as including all waste collected by a local authority other than sewage, that is residential waste, street waste, and other *municipal waste*, plus *commercial* (office and shop) *waste*. This usage broadly follows that of the

U.S. Office of Science and Technology (OST, 1969). In addition, the general term *postconsumer waste* is used here to include all the preceding categories, as well as abandoned cars and litter.

Flintoff (1974b) and Thomas (1974) have distinguished carefully between the terms *recycling* (the actual process for reusing materials, either direct, indirect, or energy utilization), *recovery* (collecting homogeneous wastes for reuse within a factory), *reclamation* (the collection by scrap merchants of homogeneous or mixed wastes to be sold for recycling), and *salvage* (the extraction of homogeneous wastes from a mixture of wastes).

Another distinction that has to be made is between *obsolete scrap (postconsumer scrap)* which is of most concern to society, and *home scrap* and *prompt scrap* which is generated within an industry and is relatively easily recycled. "Home scrap" or "in-house scrap" or "revert scrap" is the unavoidable nonproduct output of the industry manufacturing a material such as steel or plastic, and "prompt scrap" is the nonproduct output of fabrication operations in a subsequent stage of production. Recycling figures quoted for a material often refer to the total of home plus prompt plus obsolete scrap, and therefore do not necessarily give a true indication of the amount of material returned by society to the manufacturing industry. Even when these distinctions are made, the data from different sources are not always directly comparable (Sawyer, 1974).

THE BROADER MEANING OF "MINING"

In the United States, a lead has been taken by the Interior Department's Bureau of Mines in considering waste and scrap as resources. According to the Act of 1910 (amended in 1913 and 1915), the Bureau is charged with aiding in the conservation and orderly development of the nation's resources, and one way to conserve resources is to reuse or recycle metals and minerals already removed from the earth's crust. The Bureau's experience in this field* was recognized in the plans for the execution of the 1965 Solid Waste Disposal Act and the 1970 Resource Recovery Act. The Metals and Minerals Division of the Department of Commerce has been active also in projects concerned with reclaimed wastes (USDI, 1972).

Mention is made in Chapter 6 of the possibility that present-day rubbish dumps may be mined in the future; a practical demonstration that it is unnecessary and inappropriate to distinguish between secondary and primary sources of raw

*For example, Boyd, 1976; Cservenyak and Kenahan, 1970; EST, 1969a; Falkie, 1975; Kaplan, 1974; Kenahan, 1971; Kenahan et al., 1973; Mantell, 1975; MRW, 1976k; Spendlove, 1976; Sullivan and Makar, 1976; USDI, 1972.

Table 1.3. Types of Recyclable Materials

Material	Example	Recycle Potential	Recycle Rate, % of Potential
Manufacturing residues	Drosses, slags, skimmings	25-75% recoverable	Over 75
Manufacturing trimmings	Machining wastes, blanking and stamping trimmings, casting wastes	90% recoverable	Nearly 100
Manufacturing overruns	Obsolete new parts, extra parts	Variable compositions	Nearly 100
Manufacturing composite wastes	Galvanized trimmings, blended textile trimmings, coated paper wastes	Often not all constituents recovered	0-100
Flue dusts	Brass mill dust, steel furnace dust	Often not economical to recover	Under 25
Chemical wastes	Spent plating solutions; processing plant sludges, residues, and sewage	Often recoverable	Under 10
Old "pure" scrap	Cotton rags, copper tubing	Over 90% recoverable material	Over 75
Old composite scrap	Irony die castings, auto radiators, paper-base laminates	Often not economical to recover valuable materials	0-100
Old mixed scrap	Auto hulks, appliances, storage batteries	Not all materials recovered	Under 50
Solid wastes	Municipal refuse, industrial trash, demolition debris	Very low recovery rates now	Under 1

Source. Ness, 1972, with the permission of the National Association of Recycling Industries, Inc. (formerly NASMI).

materials. It is certain that mining tips will be reprocessed; coal slag heaps provide one of the largest accumulations of waste materials in all industrialized countries.

WHAT CAN WE RECYCLE?

The recycling industry is already well established, although it processes only a small proportion of all available wastes. An indication of the range of materials recycled in the United States has been provided by Ness (1972), together with an estimate of the recycle rate (Table 1.3). The utilization of manufacturing by-products is extensive, particularly in the first three categories of Table 1.3, and also in the slaughtering of animals for food, where there is little "waste."

There is no material that cannot be recycled if the incentive is sufficiently great. As an extreme example, jewellers and goldsmiths go to great lengths to recover gold dust because of its relatively high commercial value. On the other hand, supplies of metals such as zinc and tin with reserve lifetimes measured in a few tens of years are dissipated as surface treatments for steel. In the United States, for example, the ratios of obsolete scrap recycle to total consumption for some metals are shown in Table 1.4. Stated recycling ratios are

Table 1.4. Obsolete Scrap Metal Recovery as a Percentage of Consumption in the United States in 1971

Metal	Recycled Obsolete Scrap/Total Comsumption (%)
Antimony	61
Lead	35
Gold	30
Iron	28
Platinum group	25
Copper	24
Nickel	21
Silver	21
Mercury	21
Tin	19
Chromium	18
Tantalum	16
Zinc	5
Aluminum	4

Sources. Masters, 1974; USDI, 1972.

based often on the combined obsolete and prompt waste recovery, but information on the obsolete scrap alone as a percentage of total consumption, such as that in Table 1.4, may give a better indication of the present situation. It should be noted, however, that the useful life of a metal article is typically 20-30 years, so in the case of many metals the quantities recycled reflect the production at that time, which was far lower than current production, and this is particularly apparent for aluminum. As an example, the efficiency of obsolete scrap recovery (the percentage of obsolete scrap actually reused) in the United States is estimated to be 20% for aluminum and 45% for copper (50% and 30%, respectively, in the United Kingdom) (Chapman, 1974).

Solid wastes may be classified in many ways, but here the standard terminology of the U.S. Office of Science and Technology report (OST, 1969) is used, with the 1967 United States quantities as shown in Table 1.5. In 1971, the totals in millions of tonnes per year were estimated to be: urban, 209; industrial, 127; agricultural, 2161; mineral, 1544; totaling 4 billion tonnes (USEPA, 1974a).

Table 1.5. Generation of Solid Wastes from Five Major Sources in the United States in 1967, for a Population of 200 Million

Source	Pounds per Person per Day	Kilograms per Person per Day	Total (Millions of Metric Tonnes per Year)
Urban			
Domestic	3.5	1.6	116
Municipal	1.2	0.5	40
Commercial	2.3	1.0	76
Total	7.0	3.1	232
Industrial	3.0	1.4	99
Agricultural			
Vegetation	15.0	7	497
Animal	43.0	20	1425
Total	58.0	27	1922
Mineral	30.8	14	1020
Federal	1.2	0.5	40
Total	100	45	3313

Source. After OST, 1969.

These quantities are large, and some problems must be tackled on the broad scale, but local problems often have specific solutions (Rasmuson, 1973).

Recycling does not need to be on a large, industrial scale; all around us we see discarded items that are of value for other purposes. This extends to cultural activities; a book has been compiled of "Art from Recycled Materials" (Malcolm, 1974); instructions have been published for homemade recycled paper (Allaby, 1976); and there is a growing interest in museums of science and technology which provide a permanent role for some of the older obsolete mechanical and electrical equipment (Neal, 1971).

THE RECYCLE PHILOSOPHY

At the present time the subject of economics is all-important in recycling decisions, but the criteria used in deciding if recycling is worthwhile in a particular case should not be limited to economic factors; or if they are, economic costs should reflect all factors (*Ecologist,* 1972). Current economic costs do not include the "hidden" costs to society of manufacturing a product: pollution, deprivation of recreational facilities, dissipation of energy, and depletion of resources. Low-energy processes should replace high-energy processes wherever possible, and processes of mixing and dilution must be avoided. These are the thermodynamic criteria, which are discussed in Chapter 8. The main problem at present is that private and social optimal recycling rates do not necessarily coincide (Henstock, 1976).

Looking at our world around us, at first sight it might appear that nature itself is wasteful of energy and matter, but in fact within the confines of the solar energy budget a system has developed which recycles all essential materials. The important point to note is that the natural recycling periods are very long compared with the human lifespan, and society is interested in recycling on a timescale which is comparable with an individual's lifetime.

The term "recycling" now is established firmly in both conversational and scientific language, and is popular in newspaper headlines. Even more important than the examples of recycling reported in this book are the public attitudes to resources and conservation: The Canadian Trudeau government has emphasized the need for a *conserver society* to replace the present *consumer society;* Seaborg (1974, 1975a, 1975b) has called for a *recycle society;* and Boulding (1972) has contrasted the "closed" *spaceman economy* with the "open" *cowboy economy.* There are indications that this public awareness is developing, and recycling review articles have appeared recently in diverse publications.* In a recent

**American Metal Market* (1970), *American Youth* (McGough, 1972), *Brennpunkte* (Stumm and Davis, 1974), *The Bulletin* (Hoad, 1974a), *Chemical Engineer* (Pearson and Webb, 1973), *Chemistry* (Keller, 1973), *Chemistry and Industry* (SCI, 1976a, 1976b, 1976c), *Current Affairs Bulletin* (Peat, 1974), *Environmental Engineering* (Rasmuson, 1973), *Environmental Science and Technology* (EST, 1975a; Ganotis and Hopper, 1976; Kenahan, 1971; Ness, 1972), *ISWA Information Bulletin* (Bundi and Wasmer, 1976), *Journal of Environmental Planning and Pollution*

Table 1.6. Projects in Closed Cycle Living

Location	Project	Reference
Architecture students, Sydney University, Australia	Autonomous solar-powered house	Howden, 1974, 1975 Wheeler, 1975
Santee, California	Sewage water recycled for recreation lake	Stevens, 1967
New Alchemy Institute, Cape Cod, Massachusetts	"The Ark," a solar-heated wind-powered greenhouse and fishpond complex	*Futurist,* 1974
Architecture students, University of Minnesota	"Ouroboros," solar- and wind-powered house	Love, 1974
Day Chahroudi, Massachusetts Institute of Technology	"Biosphere," integrated house, greenhouse, solar heater, solar still	Love, 1974
"The Ecologist" community, Cornwall, U.K.	Near self-sufficient community of family farms	Milton, 1974
Biotechnic Research and Development community, South Wales, U.K.	"Soft," small-scale non-polluting technology	Brachi, 1974; Milton, 1974
Stockholm, Sweden	Clivus Multrum: aerobic digestion of human and garden wastes into compost	Chapter 6,"Sewage"; Milton, 1974
Reynolds Metal Co., Virginia, and other suppliers of recycled material	"Conventional" home built with recycled aluminum, glass, bricks (from mine tailings and glass)	EST, 1973a
Philips Research Laboratories, Aachen, West Germany	Solar house with scientific instrumentation	Sandscheper, 1975
Institute for Local Self Reliance, Washington, D.C. 20009.	Low technology waste recycling	Seldman, 1976
Casa del Sol, New Mexico State Universtiy	Solar demonstration house	San Martin, 1975

Control (Douglas and Jackson, 1973; Pearson, 1973), League of Women Voters (1972), *Metal Finishing Journal* (Jackson, 1972), *Nature* (1975), *New Civil Engineer* (Kinnersly, 1973), *New Scientist* (Coldrick, 1975; Hughes and Jones, 1975; Tudge, 1975), *Physics Today* (Rose et al., 1972), *Professionl Engineer* (Willson, 1974), *Public Works* (1976), *Readers' Digest* (Stevens, 1967), *Search* (Rosich, 1975a), *Science* (Abert et al., 1974), *Science and Public Affairs* (Berry, 1972), *Technology Review* (Wilson and Smith, 1972), and others collected in Appendix 2.

chemical engineering conference on "Treatment, Recycle, and Disposal of Wastes" (Johnston, 1975), more papers were concerned with recycling than were concerned with disposal. Instead of waste disposal, the more positive concept of a materials utilization industry has been proposed (Rosen and James, 1974); common products such as steel rods, glass bottles, and paperboard boxes being manufactured by an integrated factory from salvaged materials. The organic equivalent is the "Bioplex" concept (Forster and Jones, 1976). In the United States the federal government is now acting to provide a market for recycled materials (GSA, 1976). The recovery and utilization of wastes is being used as the central theme in university environmental chemistry courses, for example at DePaul University in Chicago (Melford, 1976) and at Murdoch University in Western Australia.

The philosophy of recycling is typified by utilizing two "wastes" to make a useful product, or designing a process so that all the "wastes" are useful by-products. It is also demonstrated admirably by an ice-cream cone or, if it can be devised, an edible cola bottle.

Associated with the recycle philosphy, and an extension of it, is the desire of some people for a lifestyle that conserves resources and energy (*Ecologist,* 1972). Usually an autonomous or self-contained house is the focal point of the enterprise, with one or more of the features: solar heating, solar cooling, methane-generating anaerobic digester, wind-driven electricity generator, greenhouse, fishpond, and solar water still (*Energy Primer,* 1976; Lucas, 1975; Vale and Vale, 1975). Table 1.6 lists some of these projects, which can be considered as small-scale "models" of a society in which conservation and recycling play dominant roles. Other models may be schematic or theoretical (Breeling and Moore, 1972; Rose et al., 1972), but they serve the same purpose of highlighting the need for recycling resources and avoiding waste (van Dam, 1975).

2

FACTORS INFLUENCING RECYCLNG FEASIBILITY

It is apparent from the fact that our society generates so many ''waste'' products that recycling is inherently more difficult than discarding. In general, it is difficult to recycle from economic, transportation, and chemical points of view. Recycling tends to be considered only when other courses of action are obviously unsatisfactory, either because of a shortage of ''natural'' raw materials, or because of environmental considerations. The aim of this chapter is to consider briefly the importance of several factors affecting decisions on recycling, some of which have been discussed by Bundi and Wasmer (1976). The remainder of the book concentrates on the scientific and technological factors.

ECONOMICS

Even if a process is scientifically possible it may not be technically feasible, and even then it may not be the economically preferred course of action. Economic considerations at present are the major criteria of recycle feasibility (Bridgwater, 1975; Clark, 1971; Henstock, 1976; Pearson, 1973; USEPA, 1974a, 1974b). This is typified by the description of recycling as ''utilization of the worthless'' (Müller and Schottelius, 1975; Ungewitter, 1938), and by the recommendation of the National Association of Recycling Industries: ''markets first; collections second.''

Most waste-utilizing processes produce intermediate products (those used for manufacturing other products) rather than consumer products, and most of these intermediate products are in competition with existing products. Buyers of intermediate products are profit-oriented, and act more rationally in an economic sense than ultimate consumers, so in the present economic system there is a limit to the extent of possible recycling (Manchester and Vertrees, 1973). The costs of producing useful materials from wastes can be reduced by improved technology, for example by the ''waste-plus-waste'' method. Thus obsolete blast furnaces may be used for high-temperature incineration of municipal waste (Davis, 1974), and waste alkaline electroplating solutions can be treated with acid wastes to precipitate the metals (Chapter 7).

Up to the present, direct economic factors, modified in some cases by taxation incentives or disincentives, have controlled the extent of recycling in any industry (U.S. Congress, 1972). Tax policies in many countries provide economic encouragement for continued and expanded use of primary products, to the direct economic disadvanage of recycled material. Mining subsidies and preferential freight rates for materials such as ores and woodchips may lower the prices of the virgin materials, encourage waste, and discourage recycling (Cutler and Goldman, 1973; Kakela, 1975). It should be noted, however, that freight rates when considered from the point of view of chemically equivalent amounts needed for the final product in fact may favor the secondary raw material (Albrecht, 1976). Often the market price of a commodity does not reflect resource depletion and environmental damage, so the long-term economic advantages of using recycled materials are not apparent. A tax on virgin materials might be appropriate, as proposed for plastics, for example (Milgrom, 1972). In the United States, Industrial Revenue Bonds now can be used for the purpose of building recycling plants, providing a source of lower-cost capital because of their tax free nature (Boyd, 1976). These were first used in Saugas, Massachusetts: see Chapter 6, "Prototype Systems."

In economic terms, waste recyclng is included in *residuals management* (Russell, 1971; Spofford, 1971). A residual is a nonproduct output with zero price in the current market, and may be either a material or low-grade heat. "Residuals management" is preferred to the approximately equivalent term "pollution control" because it conveys the information that there will always be some residuals, and that our main concern is keeping them to a minimum. Such studies have been made on: water in beet sugar production (Löf and Kneese, 1968); petroleum refining (Russell, 1973) and oil recycling (Sorrentino and Whinston, 1968); the steel industry (Russell and Vaughan, 1974; Sawyer, 1974); anaerobic fuel gas production from urban waste (Kispert et al., 1975); and the wastes handled by a disposal contractor (Bridgwater, 1975).

AVAILABILITY AND VALUE OF CHEMICAL CONTENT

The value of a chemical is determined by the availability of raw materials, difficulty of extraction, and demand. Wastes range from high-volume low-cost materials such as mine tailings, to low-volume, high-cost materials such as mercury or gold. If no particularly valuable chemical occurs in a waste material, recycling will not occur unless encouraged by subsidies or enforced by legislation. Thus in the United States the current potential value of resources in urban wastes is about $1000 million, but the present disposal cost is $1000 million (plus another $4000 million for collection)(CEN, 1974j). Obviously more than economic incentives are required to recover a significant proportion of these materials. However, the costs of some chemicals will increase dramatically in the near future as usable supplies decrease due to the present exponential increase in consumption (Lapp, 1973; Park and Freeman, 1970).

Consider a resource material, the usage of which is increasing at a rate directly proportional to the amount already in use. For example, in an unrestrained industrial society it is, perhaps, reasonable to expect the rate of iron ore usage at any time to be proportional to the amount of iron in use in that society at that time. Denoting the amount of material in use by m, this can be expressed mathematically as:

$$\frac{dm}{dt} = km$$

and it can be shown that on integration from a limit of $m = 0$ at time $t = 0$, it follows that

$$m = \exp(kt)$$

that is, there is an exponential relationship between the amount used m and the time t. This is sometimes described as a "J" curve (Miller, 1971a) because it exhibits a very rapid change after a long period with little perceptible change. Exponential curves are characterized by doubling times which are constant, and current doubling times for energy production, for example, are typically 20 years (corresponding to 3.5% increase per year). It is clear that such growth rates for energy and resource usage cannot continue indefinitely, even with maximum recycling of materials.

The history of mankind does not show a smooth increase in knowledge, resource utilization, and energy expenditure. Rather, new ideas and techniqiues have resulted in periods of relatively rapid change. For example, the Neolithic "revolution," although spanning millenia, was a dramatic change in contrast to the rate of previous development. The succeeding periods of rapid growth have been successively shorter; they may be described as the urban, industrial, and scientific revolutions (CRM, 1969). Together, they comprise a relatively brief transitional phase of exponential growth in population, technology, resource utilization and energy expenditure which will be followed inevitably by a leveling out or decay.

Figure 2.1 illustrates one estimate of reserve supplies of some resource materials. Numerical estimates of reserves and lifetimes of minerals are, of

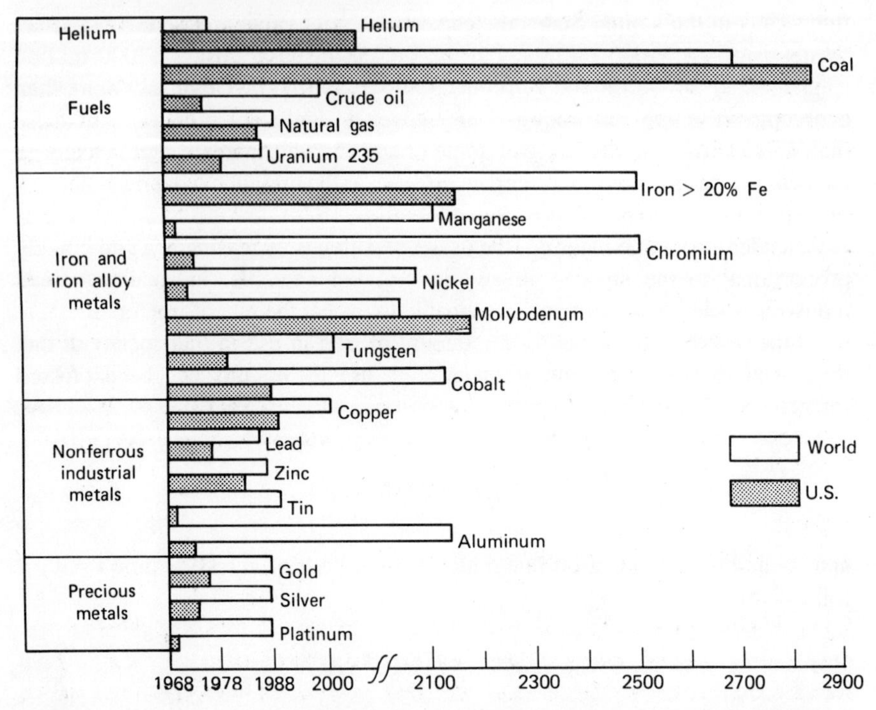

Figure 2.1. Estimated lifetimes of known recoverable reserves of mineral commodities at currently minable grades and at rates of consumption based on a stable world population of 3.5 × 10^9 and current rates of use. Note the time-scale changes at 2000 and 2100. (After Murdoch, 1971.)

course, unreliable, not only because of incomplete information on the geochemistry and configuration of the earth's crust, but also because of a variety of economic and technological factors. Despite this, exponential increase in usage means that a relatively large uncertainty in total reserves of a resource will cause only a minor uncertainty in the length of its lifetime.

As high-grade ores are depleted, the cost of production will rise rapidly. The only currently important major structural metals which will not be in short supply in the forseeable future are magnesium (from seawater), iron, and aluminum. Lower-grade ores in the case of the last two metals will require improved technology and greater energy, but there are no fundamental problems. The average aluminum content of most igneous rocks is about 8%, and basalts contain about the same percentage of iron. Such is not the case with copper, nor with titanium, which apart from major ores rutile (TiO_2) and ilmenite ($FeTiO_3$) is well dispersed (Bravard et al., 1972). Known and expected reserves of mercury, tin, tungsten, and helium will be nearly exhausted by early next century if present trends in use continue.

The important question is how long the exponential trend will continue. The development of new materials will modify demands, but it is difficult to evaluate their effects on the material systems (Swager, 1967). One view is that developing countries will be able to "leap-frog" the technology of present advanced countries, and achieve improved standards of living with lower per capita expenditure of energy and consumption of materials. Support for this idea is the leveling out in the use of some metals in countries with stable populations and relatively sophisticated technologies.

At the present time, one of the most critical resources is oil, both for fuel and as a chemical feedstock. Most countries have experienced problems since the oil "crisis" of 1973, but Japan is one of the worst affected of the industrial countries. One effect of this has been an intensification in recycling campaigns in Japan (CEN, 1974b; MRW 1976b).

USEFUL PHYSICAL PROPERTIES

Even if waste material has no very valuable chemical content, particular properties may make it worthwhile reclaiming. Refractory materials such as blast furnace slag, power station fly ash, and some iron and copper ore tailings can be used in the fabrication of construction materials (EST, 1970a). Blast furnace slag in particular is keenly sought after. The combination of binding properties and biodegradability of waste paper is utilized in grass seeding mixtures sprayed on road cuttings and other exposed earth surfaces. Some mining wastes can be used for ground stabilization, and coal ash slag may be suitable for mineral insulating wool manufacture. The resilience of waste rubber when it is incorporated into asphalt raises the softening point of road surfaces in hot weather and lowers the glass transition (brittle) point in cold weather; this has been used extensively in Phoenix, Arizona, which experiences a wide range of ambient temperatures. These and other examples are discussed in Chapter 7.

COLLECTION AND TRANSPORTATION

The relative location of sites of waste generation, processing plants, and markets are of great importance; processes that would be economically viable in one situation are not feasible in others. For example, a late 1960s United States car contains nearly the same percentage by weight of plastic materials as solid urban waste (2-3%), but while urban waste in the United States is collected in 20,000 sites, there are only about 100 automobile shredders. This makes the processing of plastic waste (by hydrolysis, for example) more feasible in the latter case (Mahoney et al., 1974). Mining wastes usually are produced far from potential markets, so blast furnace slag is preferred for construction materials, because steel works tend to be located near densely populated areas. Urban solid waste presents the most difficult collection and transport problem (Chapter 6). The problem of collection at central processing sites must be tackled; suggested solutions include low-altitude airships and pipelines for encapsulated wastes (Weiss, 1976).

SEPARATION

Separation problems may be chemical or physical. This is exemplified by copper; although copper has a relatively high value, less than half of the obsolete waste copper in the United Kingdom and the United States is recycled (Chapman, 1974). One reason for this is that in electric motors, copper windings are usually surrounded by a large mass of iron (presenting a physical separation problem); another reason is that a considerable proportion of the copper is used in brass, and chemical separation for recycling is often found to be uneconomic.

If physical separation is possible, this is preferable to chemical separation which usually requires large inputs of energy, chemicals, and technology. The chemical and technological complexity of separation procedures is illustrated by a system devised to extract molybdenum, cobalt, chromium, and nickel from superalloy waste (Brooks et al., 1969) (Figure 2.2). The large quantities of water, chlorine, and sodium carbonate involved should be noted; the balance that must be achieved in chemical, energetic, and economic factors determines the viability of recycling in a particular situation.

Some materials by their very nature are dissipated widely. Titanium recycle potential is limited, since 90% of the titanium produced at present is dispersed as titanium dioxide pigment (Bravard et al., 1972). Zinc (on galvanized iron) and tin (on tin-plated cans) are similar examples. In some cases, a new technological development makes an apparently difficult separation feasible; cryogenic separation of wire from rubber or plastic (in waste cable and tires) is a good example. Details of specific separation procedures are collected in the next chapter, and the problem of separation of urban waste is considered in Chapter 6.

Many of the separation problems encountered at present could be avoided by inprovements at the design stage. For example, incorporation of iron compounds in polymer beads used as filter aids (a substitute for diatomaceous earth) facilitate separation and washing; the magnetic characteristics permit the beads to be gathered up and washed, or they can be shaken in an alternating magnetic field so that the dirt is dislodged (Bolto et al., 1974; *Ecos,* 1974a). This technique has

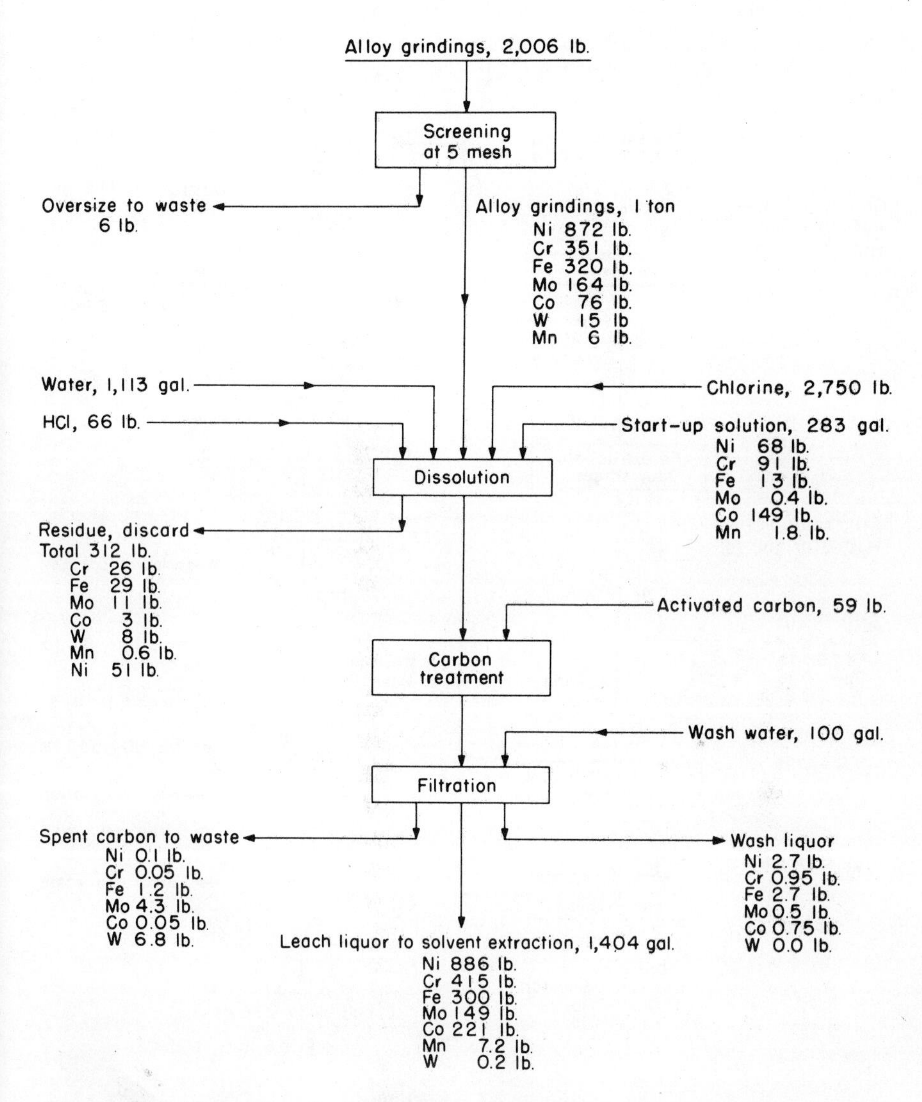

Figure 2.2 (a), (b), and (c). Process flowsheets and materials distributions for the recovery of molybdenum, cobalt, chromium, and nickel from superalloy waste. (From Brooks et al., 1969, with the permission of the Bureau of Mines, U.S. Department of the Interior.)

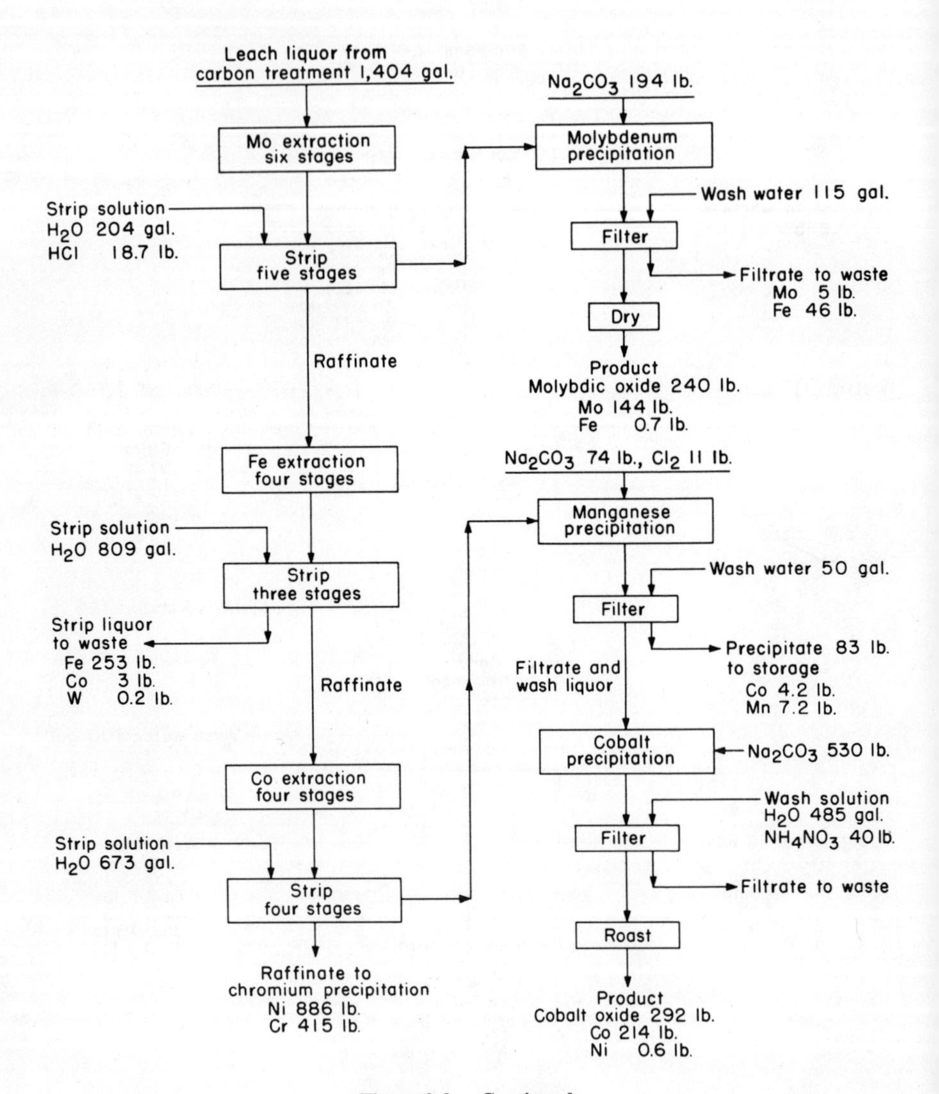

Figure 2.2. Continued.

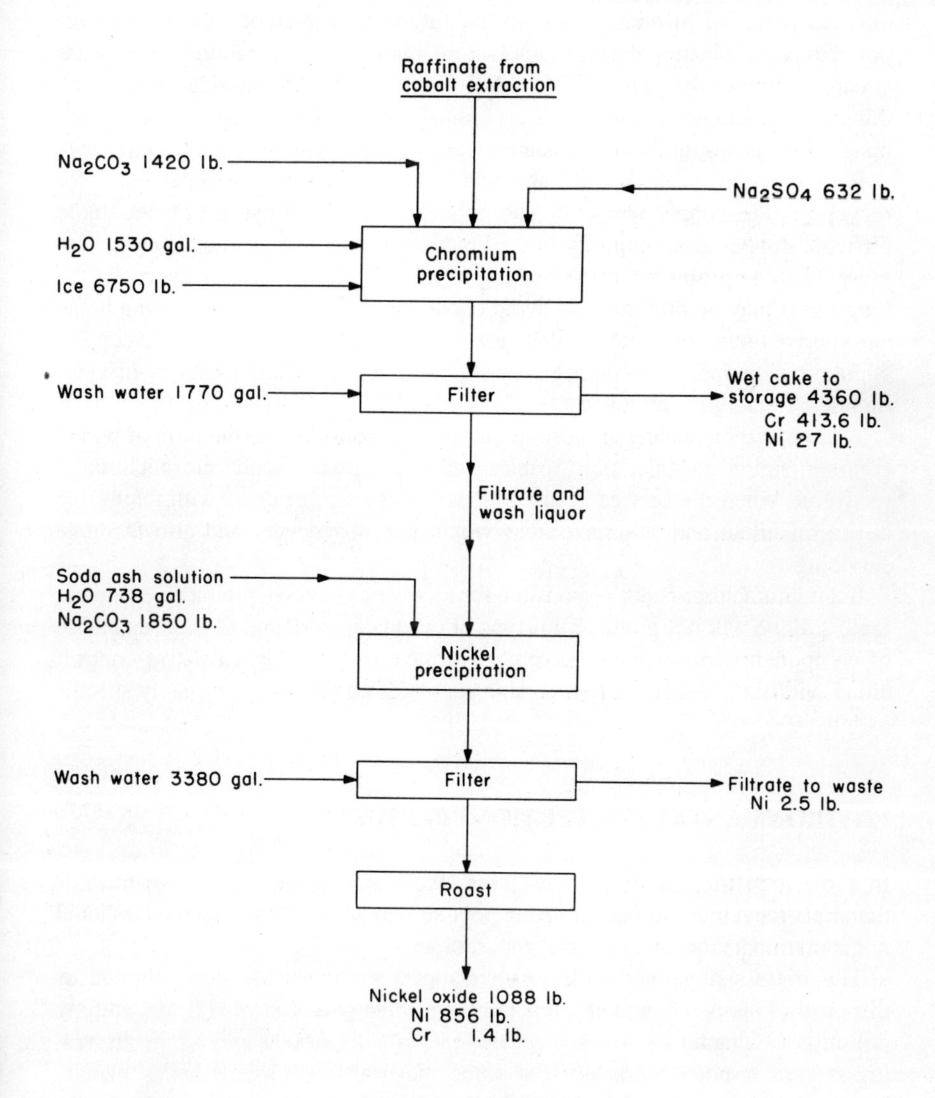

Figure 2.2. Continued.

obvious potential in other fields: in the future some method of coding incorporated in all plastics during manufacture may assist separation in the waste stream (Baum and Parker, 1974a; McRoberts, 1973). Miniaturization (in such things as calculators and radios), although reducing the initial demand for materials, has brought with it repair and recycling diffficulties (Henstock, 1976).

It may prove necessary to substitute materials to improve separability for recycling. The copper wiring in automobiles makes the recycling of steel more difficult; copper as an impurity in steel reduces formability. Although aluminum presents some problems in the making of reliable electrical connections, in the long run it may be preferable to avoid copper and to use aluminum wiring in the automotive industry (CRM, 1969; Rose et al., 1972). Alternatively, copper-containing components could be located in a few standard locations in automobiles, to facilitate removal.

A large and increasing proportion of copper is consumed in the form of buried communication and electricity cables and water pipes, which are not usually recovered when the service is discontinued. A system of ducts with removable covers in urban and suburban areas would permit recovery and also facilitate servicing.

If a manufacturer is not responsible for recovering his own products, economic factors alone will not promote this type of rationalization, and ease of separation of components for eventual recyclng will have to be under legislative control, along with the present purity or design standards and quantity or quality specifications.

ENVIRONMENTAL PROTECTION PRESSURES

In some industries, and in some places, the major factor in the drive towards materials recycling has been a strong popular demand that these materials should not contaminate the adjacent soil, water, or air.

The traditional solution for waste disposal problems has been dilution or dispersal. This is a "natural" process, and although it contributes to "entropy pollution" (Chapter 8), it is acceptable for waste disposal as long as the dispersing system is not overloaded. For some materials such as the heavy metals, however, the amount safely disposable by dispersive methods is very much less than that produced by our present industries, so they *must* be recovered, whatever the economic cost (Newell, 1975a; Rose et al., 1972). Chemical wastes are often less noticeable than consumer wastes, but they are potentially more dangerous (Chadwick and Goodman, 1975; Coleman, 1975; Rasmuson, 1973; Walker, 1971). In economic terms, environmental factors imply that wastes have a *negative* economic value. Many waste-recovery methods have been proposed, but then rejected on economic grounds: it is cheaper to dump the wastes and pollute waters and the atmosphere, than to process them. Recovery processes, however, will tend to become more "economic" when antipollution regulations prohibiting the discard of residuals are strengthened and enforced. Already in many countries environmental impact reports are necessary before an industry is established or land use is changed.

Some manufacturers are now finding that their waste recovery programs in fact are running at a profit, and this introduces the concept of waste streams as misplaced resources (Bennett and Lash, 1974). Examples are whey treatment in the dairy industry, animal feedstuffs as by-products from grain alcohol, building materials from power station fly ash, metal salts from metal pickling and electroplating industries, and reusable water in many industries. To take the example of dairy wastes quoted by Porges (1960), the average daily waste from a small dairy may contain the equivalent of 45 kilograms of dried skim milk. Complete aerobic biodisintegration of this would consume about 48 kilograms of oxygen, and at an oxygen solubility of 8.4 parts per million in water at 25°C, 5700 tones of aerated water would be required. If this amount of oxygen is not provided, the anaerobic processes lead to unpleasant, polluted conditions; yet this "waste" material is a well-balanced microbiological nutrient which can be utilized in a number of ways. A solution to the effluent water quality problem in many process industries is total water recycle, and this may be no more costly if introduced at the planning stage. Some wastes, such as mine tailings for example, probably will present disposal problems always, because there is no economically attractive use; environmental pressures will lead eventually to adequate, if unethusiastic, disposal measures.

It should not be overlooked that recycling processes can have their own environmental impacts, even to the extent that the net result is the reverse of that intended. Deinking of newsprint can produce materials that can pollute waterways, and full and detailed investigations are necessary whenever large-scale resource recovery is planned, along the lines of that carried out by Hunt and Franklin (1973) on the environmental effects of recycling paper. "Resource and environmental profile analyses" (Hunt and Franklin, 1975; Purcell and Smith, 1976) or "environmental impact matrices" (Bundi and Wasmer, 1976) are moves in the right direction.

Some of the side effects are more subtle. Thus reclaimed copper can contain cadmium, which is used extensively for plating steel, and there is the possibility of its leaching out of copper water pipes to reach dangerous levels in water supplies. The same considerations apply to impurities in the steel used for food cans, and selenium is concentrated in some plants used as cellulose sources, so constant quality control vigilance is required (Breidenbach and Floyd, 1970). The use of polychlorobiphenyls (PCBs) in carbonless papers can result in these chemicals finding their way into wastewater from repulping processes. If urban waste is used as a fuel source, it must be remembered that it contains some potentially hazardous elements in concentrations greater than those found in fossil fuels (Marr et al., 1976); for example, lead, iron, copper and zinc from inks and colored wrapping materials, and zinc from photosensitive copy paper. These can find their way into the atmosphere (Campbell, 1976). The risk of microbiological and chemical contamination of domestic water supplies drawn from wastewater sources is still significant, and it is generally considered that current scientific knowledge and technology are not sufficiently advanced to permit the direct use of treated wastewaters as sources of public supply (Pringle, 1974).

In the United States, the Environmental Protection Agency has promoted recycling extensively, by means of grants for demonstration plants (Chapter 6), sponsorship of research and publication, and compilation of bibliographies. Many of the publications cited in Appendices 2 and 5 result directly or indirectly from the activities of the USEPA. Under the authority of the 1965 Solid Waste Disposal Act and the 1970 Resource Recovery Act, guidelines have been drawn up with the intention of increasing the utilization by Federal Agencies of both recycled materials in purchased products (USEPA, 1976a) and source separation systems (USEPA, 1976b).

ENERGY

The processes of collecting glass bottles, moving them back to washing plants, washing them, and sorting them, require expenditures of energy, so although the manufacture of bottles from raw materials is also energy intensive, it is not immediately obvious that recycling glass bottles is a good idea. It turns out that it is energetically worthwhile to recycle whole bottles but not broken glass (Chapter 6). Similar factors apply for all materials, and a considerable amount of data is necessary in order to be able to make a valid decision in each case. It is certainly true, however, that energy consumption is an essential criterion in assessing the value of recycling. If, as appears to be the case, usable energy is to become scarcer and therefore more expensive, this factor will play a more dominant role in determining the place of recycling in society.

The Oak Ridge National Laboratory has carried out an evaluation of the total energy expenditures in the production and recycle of important structural metals (Bravard et al., 1972), with the results summarized in Table 2.1, which expresses the relative energy requirements of recycling and extraction from present and future primary ores. These figures show that in all cases there are significant energy savings in recycling, and as might be expected, these are greatest for aluminum and magnesium where the oxidation energy of the metal is high. More precise comparisons between processes and between strategies can be made by using "net energy analysis," which takes into account the direct and indirect

Table 2.1. Relative Energy Requirements of Recycling and Production from Ores for Several Metals, with Possible Future Ores in Parentheses

Metal	Source	Recycle energy/Production energy (%)
magnesium	sea-water	1.5
aluminum	50% bauxite	2.5-3.9
	(30% bauxite)	2.2-3.4
	(clays)	2.0-3.0
	(anorthosite)	1.8-2.8
iron	high-grade hematite	29
	magnetic taconite	27
	(specular hematite)	24
	(nonmagnetic taconites)	24
	(iron laterites)	20
copper	1% sulfide ore/98% scrap	4.7
	1% sulfide ore/impure scrap	11.5
	(0.3% sulfide ore/98% scrap)	2.6
	(0.3% sulfide/impure scrap)	6.2
titanium	high-grade rutile ore	31
	ilmenite rocks	26
	ilmenite beach sands	26
	ferruginous rocks	26
	(high titania clays)	25
	(high titania soils)	19

Source. Based on data from Bravard et al., 1972

energy that goes into every activity in the production of the finished material, and the energy value of the final material (Khazzoom et al., 1976; Purcell and Smith, 1976). Hirst (1973a, 1973b, 1975) has considered the energy savings in recycling steel, aluminum, and paper: If one-third of the United States production of these materials in 1970 had been from recycled scrap, the overall energy savings (from mine or waste, respectively, to primary products) would have been almost 0.7% of the total United States energy use. Energy required for sorting or separating the scrap from a mixed waste stream is not included in Table 2.1, but this would add only about 10% to the recycle energy unless an attempt was made to recycle the majority of the metal in circulation.

As the proportion of material recycled increases beyond a certain point, the cost of separation increases rapidly, and for any material under any particular set of conditions, there exists a point at which the energy expenditure on recycling equals the energy required to extract it from ''natural'' ores. This has been illustrated for copper by Stumm and Davis (1974) (Figure 2.3). The minimum in

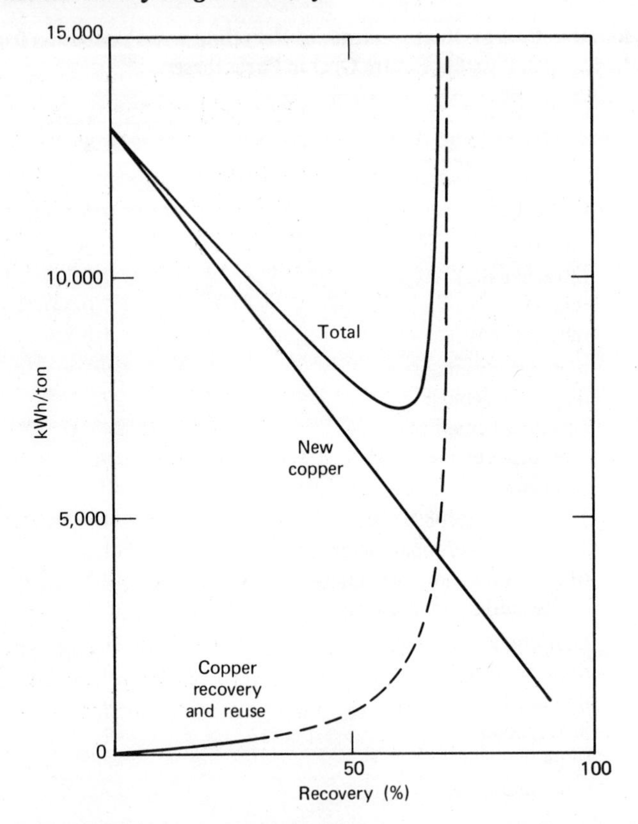

Figure 2.3. Optimization of energy consumption for the recycling of copper. (From Stumm and Davis, 1974, with permission.)

the "total" curve occurs at about 60% recovery, and the present recycle rate for copper approaches this figure in some countries; the steep rise in energy expenditure for recovery and reuse beyond this point should be noted. Chapter 8 includes further discussion in thermodynamic terms of mining, manufacture, recycling, and pollution.

QUALITY OF RECYCLED MATERIAL

There is frequently prejudice on the part of the public against recycled materials relative to natural ones. Sometimes this is justified, but usually it is not. It is certainly true that in the case of paper the secondary fiber can be no better than the lowest-grade fiber present in the scrap, and it must be used in applications that are compatible with its quality (Hunt and Franklin, 1973). In addition, the fibers do suffer some degradation (shortening) in processing, so 100% recycling is impossible for technical as well as for other practical reasons. Nevertheless, the quality of recycled paper fibers is perfectly adequate for many applications (Chapter 7). Similar arguments apply for any material that is not reduced to its simplest chemical form in the recycling operation; thermoplastics tend to undergo oxidation during reprocessing, but this can be overcome to a great extent, so thermoplastics can be molded into products with specifications which are adequate for many applications. (Unlike thermoplastics, thermosetting plastics cannot be recycled in their original form.)

When a material is reduced to its simplest molecular or atomic form during reprocessing, then no distinction can be made between primary and secondary materials. Impurities in recycled metals may differ from those in metals produced from ores, but they are not necessarily any more of a difficulty when the appropriate technology is developed and specifications are defined clearly (Alter and Reeves, 1975). Some of the properties of ferrous can scrap steels can be superior to those of primary materials (Cramer and Makar, 1976).

In principle, there is little difference between reprocessing sewage wastewater into drinking water, and drawing water from a heavily contaminated river for treatment; the end products can be biologically and chemically indistinguishable. In practice, however, we must ensure that the reclamation processes are fail-safe, and it is common sense that renovated wastewater should not be used as a drinking supply until there is no other practicable choice.

EMPLOYMENT AND SOCIAL ATTITUDES

Patterns of employment change as technology changes, and sudden alterations in technical processes in an attempt to conserve resources are, understandably, resisted by the work force. One example of this is the throw-away beverage container industry. The Oregon legislation banning ring-pull cans and imposing compulsory deposits on beverage bottles (Chapter 6) caused a canning plant to close. Although there is a redistribution rather than a reduction in labor, it is human nature to oppose sudden changes of this kind, so this forms another barrier to increased recycling of resources.

Social factors are also important in patterns of consumption. As well as the general convenience aspect of high-cost, throw-away packaging which would be difficult to withdraw, there is considerable variation in attitudes shown towards recovering the deposit on returnable bottles. In the United States it has been found that inner city dwellers buy beverages in smaller quantities and are reluctant to recover the deposits; suburbanites who buy in larger quantities and travel by car are more likely to return bottles (Hannon, 1972). Psychological aspects of recycling by consumers are now receiving attention (Geller, 1975; Geller et al., 1973, 1975). Middle-class suburban residents are more likely to assist in recycling programs by delivering material to collection centers, retaining newspaper for paper drives, or even in sorting household wastes. This topic is considered in Chapter 6.

PUBLIC RELATIONS AND ADVERTISING EXERCISES

As recycling becomes a more socially commendable practice, some manufacturers are mounting recycling campaigns with extensive publicity. Some of these, notably by aluminum manufacturers, seem efficient and successful, although they account for only a small percentage of the industry's output. As long as recycling occurs, it does not matter whether the incentive comes from tax legislation or public relations. What must be prevented, however, are public relations exercises which are only superficially recycling, such as collecting cans and dumping them elsewhere, or collecting refillable bottles and smashing them for remelting.

DURABILIY AND EXTENDED LIFE

One method of minimizing the recycling problem is to plan for longer useful product lifetime without increasing the difficulty of eventual disposal (OST, 1969; Rose et al., 1972; Thomas, 1974). This procedure is in conflict with our present economic and commercial bias toward a high productivity, throw-away lifestyle (Packard, 1960). It is, however, necessary to strike a balance because improved performance, safety, or economy in a particular appliance may lead to beneficial obsolescence. Certainly in those areas where rapid change is no longer occurring it is important that products do not have "built-in obsolescence," and that their components are repairable or replaceable on a modular basis. In the long run it will be economic pressure which prevails. When material costs increase sufficiently it will no longer be possible to squander them on nondurable items.

From the point of view of energy, planning for extended life may well be more valuable than planning for recycling. Berry (1972) has considered the energy costs in the manufacture, junking and recyling of automobiles, showing that extending their lifetimes by a factor of three would achieve greater energy savings than would be possible with the maximum amount of recycling. The main *energy* factors in the automotive industry are associated with iron and aluminum, whereas the *resource* problems are the other metals and here recycling is more important.

THE RECYCLING INDEX

Russell and Swartzlander (1976) have devised recently a recycling index (RI) to identify chemicals and products with good recycling prospects. The potentials for collection (in the form of a Collection Index, CI) and for processing (in terms of the Processing Index, PI), are multiplied together:

$$(RI) = (CI)(PI)$$

CI ranges from zero for dispersed materials such as fertilizers, detergents, pharmaceuticals, and surface coatings, to one for materials with established collection systems; and PI has a corresponding range of values from zero to one, depending on the technology available. A survey of the most important chemicals used in the United States showed that sodium sulfate had the highest RI value (0.67), reflecting its major use in the kraft pulping process. Hydrochloric acid has RI = 0.34, because it can be recovered readily from commercial chlorides and pickle liquors. The major organic chemicals have moderately high RI values (0.23, 0.16, and 0.14 for ethylene, benzene, and propylene, respectively) because of their high consumption rates in polymer manufacture and the growing interest in scrap plastic recovery.

Although such an index vastly oversimplifies the complex factors involved in recycling, it is a useful basic guide.

3

PHYSICAL METHODS OF SEPARATION AND RECOVERY

Before a material is processed for recycling it is usually necessary to separate it from a variety of other materials. This process of separation opposes the natural tendency to increasing disorder (increasing entropy within the system) and so always requires the expenditure of energy. Materials have many unique physical and chemical properties which may be exploited in separation processes, ranging from color, density, and texture which are utilized intuitively in manual sorting, to magnetic properties which require mechanical equipment.

In most cases the cost of sorting or separating is relatively high, whatever method is employed, and in the past salvage industries have been based primarily on reclaiming wastes from commercial and industrial operations that yield materials in relatively large, clean, homogenous batches. Inherently valuable materials in consumer wastes are not utilized to any great extent because their concentrations are low. To take another example, scrap dealers do not usually handle safes, which are made of a sandwich steel and concrete construction that cannot be separated economically.

If possible, "physical" separation methods are usually preferable, but in many cases "chemical" processes are necessary. (The distinction between physical and chemical processes is sometimes arbitrary and artificial, but it is a convenient basis for classification.) A recent publication by the American Chemical Society (ACS, 1969) highlights the importance of chemistry as the basis for action on many environmental problems, including the recycling of wastes. Sometimes it is possible to change physical characteristics to assist separation, for example drying to modify electrical conductivity, applying vacuum to porous materials to modify density, and chemical alteration of surface characteristics. Separation processes are well established in chemical engineering (King, 1971), so much of the technology required for resource recovery is already available.

MANUAL AND MECHANICAL SORTING

Hand-sorting, although labor-intensive and costly, still is used widely in salvage industries. It ranges from simple hand-picking of urban waste to sorting textiles, removing impurities from recycled paper, and determining the composition of alloys and plastics with the aid of physical and chemical tests. Hand-sorting is the "Maxwell demon" approach to recycling. (Maxwell in 1871 invented a mythical being of molecular size who could separate fast-moving molecules from slow-moving ones.) Entropy of mixing associated with disorder is overcome by human endeavor with a relatively small expenditure of energy. Thus in most factories and workshops with a little organization it would be possible to keep different kinds of scrap metals separate.

The emphasis in this chapter, however, is on mechanical separation, because it is only by the application of modern automated techniques that significant progress will be made in resource recycling in the kind of society we have at present. It is significant that the recovery rate for paper and paparboard in the United States has been falling since 1944, when 37% was recycled, to the present time (less than 20%), although it is estimated that 66% is potentially reclaimable (Myers, 1971a). Improved technology may enable this trend to be reversed.

Any sorting operation always involves a prior coding or identification process (Wilson and Smith, 1972). In some cases coding is simple, as in the magnetic separation of ferrous from nonferrous metal. More ofen, it is necessary to deal with multicomponent items (for example a beer can with tinned steel body, aluminum top, and a paper label). Here there are two possibilities. One is to use several sensors which observe and measure different properties, as a human sorter would, and so to come to a conclusion about the nature of each item. The second is to grind or shred everything so finely that most individual particles are homogeneous, and then separate them. This has the disadvantage that we give up the order and concentration of material that already exists, and then have to build it up again from a much less ordered level.

In a magnetic separation operation, coding and sorting take place at the same time. This is not usually the case, but the response to the physical properties (such as color) of a sensor (such as a beam of light) can operate a switch to divert items by conveyor belt or air blast to the appropriate hoppers.

The first stage in the development of automation from manual "picking" of waste is to use manual coding and machine switching; a human coder operates the controls of a mechanical separator. The next stage is complete automation, and the mechanical coder/separator which most closely approximates to manual sorting is an urban waste sorting machine developed for larger items of waste at the Massachusetts Institute of Technology (Senturia, 1974; Senturia et al., 1971; Wilson and Senturia, 1974). A computer examines data from several types of sensor (metal detector, impact sensor, and infrared sensor) and determines a category for each individual item. This technique is still in the development stage, but may in the future complement the shredding, sieving, winnowing, and slurrying processes now being introduced for recovery from urban waste. In conjunction with this sorting system, a detailed survey has been carried out of the size characteristics of urban waste (Winkler and Wilson, 1973). It appears that sorting by size alone affords a method of preliminary concentration of the various components. Metallic objects make up 40% of the total mass in the 3- to 5-inch (8-13 centimeters) range, glass objects account for 38% of the mass in the 5- to 10-inch (13-25 centimeters) range, and 67 mass % of objects over 12 inches (30 centimeters) are paper products.

In a few favorable cases materials have unique characteristics which permit a clear-cut separation, such as ferromagnetism in the case of ferrous-nonferrous metal separation. More commonly, however, materials exhibit a range of properties in any physical or chemical characteristic, and it is necessary to develop a separation process based on variation between the various components in some suitable property such as density or solubility. If materials are finely divided or intimately mixed, it is frequently necessary to use complex chemical processes to recover the components in reasonably pure states.

ELECTRICAL AND MAGNETIC PROPERTIES

Magnetism has been used for many years in drum and belt separators for mineral beneficiation, impurity removal, and scrap metal sorting. As indicated in Figures 3.1 and 3.2, the permanent magnet or electromagnet is usually stationary, and a steel drum or belt rotates past the magnet, so that magnetic metals are diverted from the waste stream (Engdahl, 1969; Hershaft, 1972; NCRR, 1974a; Sealy, 1975). The magnetic fields may be of alternate polarity around the periphery of the drum to agitate the magnetic items and dislodge the nonmagnetics, and the carry-over of nonmagnetics can be made negligible by the introduction of an additional magnetic drum. Colored glass can be separated from clear glass by differences in magnetic susceptibility due to varying iron content (Rose et al., 1972). Prior treatment is often necessary before magnetic separation, for example the crushing of incinerator clinker and cryogenically embrittled metals, or the shredding of automobile scrap. Electromagnetic separation of nonferromagnetics can be carried out if ferromagnetic material is incorporated, as in the recoverable filter aids in the form of granular magnetic polymers (Bolto et al., 1974) which may replace diatomaceous earth.

In the form just described, magnetic separation is appropriate only for the ferromagnetic materials (iron, nickel, and most ferrous alloys) but electromagnetic methods are not limited to ferromagnetics. Recently the *high gradient magnetic separation* (HGMS) technique has been developed (CEN, 1974c;

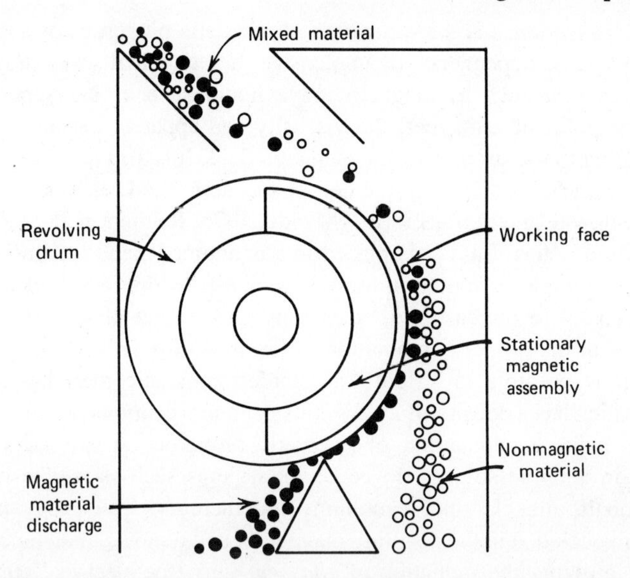

Figure 3.1. A magnetic drum separator. (From NCRR, 1974a, with the permission of the National Center for Resource Recovery, Inc., Washington, D.C., U.S.A.)

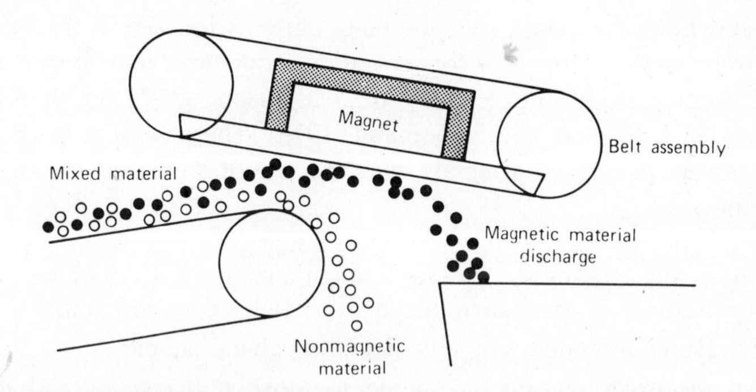

Figure 3.2. An overhead belt magnetic separator. (From NCRR, 1974a, with the permission of the National Center for Resource Recovery, Inc., Washington, D.C., U.S.A.)

WWT, 1976b). HGMS devices, which involve canisters packed with finely divided ferromagnetic material (such as stainless steel wool) in a magnetic field, are able to separate even very small and weakly magnetic (for example, paramagnetic) particles from a liquid slurry. High magnetic field gradients are created along the filament edges, trapping magnetic particles. To recover the magnetic material, the magnetic field is turned off and the material is flushed out.

Ferrohydrodynamics is the utilization of magnetic properties of a medium to control the spatial dependence of its density. If a colloidal solution of a ferromagnetic material such as magnetite in a liquid such as kerosene is placed between the poles of a magnet, the viscosity and apparent density of the fluid vary according to the strength of the magnetic field. It is then possible to separate particulate material fed through the cell of magnetic fluid; particles of different densities follow different trajectories (Mantle, 1976; Reimers et al., 1976). If the magnetic fluid (''ferrofluid'') is placed in a nonhomogeneous magnetic field of sufficient magnitude, a levitation force is generated within the fluid, causing its apparent density to increase, so ''canceling'' the effect of gravity on objects within it. With the proper combination of ferrofluid strength and magnetic field gradient, it is possible to ''float'' all nonferromagnetic materials, including platinum which has a density of 21.5 grams per cubic centimeter.

Another indirect application of magnetic properties is the use of ferrous hydroxide in alkaline solution to coprecipitate ions such as magnesium, manganese, cobalt, nickel, zinc, cadmium, and mercury. The heavy metal fine particles are occluded in the resulting magnetite, and removed magnetically.

Devices utilizing the induction of *eddy currents* (the electrical currents that flow in any conductor when the electromagnetic field through it is changing) may be used to separate nonferrous metals from nonconducting (nonmetallic) materials, in urban waste for example. As the mixed waste is fed over the device, the magnetic field associated with the eddy currents induced in the metals causes them to be attracted and held for separation, or diverted into hoppers (Campbell, 1974; Kenny, 1974; Preston, 1976; Schlömann, 1975, 1976; Sommer and Kenny, 1974; Spencer and Schlömann, 1975). There are four basic methods of eddy current induction in metals, and any one of them may be used for metal separation:

1. Physically moving the sample through a magnetic field which is generated, for example, by permanent magnets beneath a conveyor belt;
2. Physically moving a magnetic field through the sample;
3. An electrical phasing technique to move a magnetic field through the sample;
4. Varying with time the magnetic field through the sample.

In each case, the density of the conductor and its ability to sustain an eddy current flow and so generate its own magnetic field are the bases of selection properties. This information can be expressed in terms of the ratio of electrical conductivity to density (Table 3.1).

The simplest way of moving material through a magnetic field is to allow it to slide or roll down a ramp over an array of permanent magnets; different metals are deflected to different extents. A separator consisting of a rotary drum with a

Table 3.1. Electrical Conductivity (σ) to Density (ρ) Ratio for Some Nonferrous Metals

Metal	$\sigma/10^7 S\ m^{-1}$	$\rho/10^3 kg\ m^{-3}$	$\sigma\rho^{-1}/10^3 S\ m^2\ kg^{-1}$
Aluminum	3.5	2.7	13.1
Copper	5.9	8.9	6.6
Silver	6.3	10.5	6.0
Zinc	1.7	7.1	2.4
Brass	1.4	8.5	1.7
Tin	0.9	7.1	1.2
Lead	0.5	11.3	0.4

Source. Sommer and Kenny, 1974.

lining of permanent magnets on the inner surface covered with a suitably designed flexible stainless steel layer can split mixed materials into three fractions: Magnetic particles adhere to the drum surface until the flexible steel layer separates from the magnets at a particular position; nonmagnetic nonmetals fall out of the drum; and nonmagnetic metal particles are controlled by eddy-current induced forces when the drum rotates (Schloemann, 1976). In the so-called "aluminum magnet" a poly-phase, alternating electomagnetic field sweeps the aluminum (which has a conductivity/density ratio much higher than that of other metals commonly present in wastes) laterally off a moving belt (Campbell, 1974). These devices are still largely in the development and prototype stages.

The direct measurement of *electrical conductivity* may be utilized. One or more of a series of electrical probes maintained at a potential with respect to an earthing conductor makes contact with each item, and the current through the article codes it on the basis of conductivity and permits it to be diverted by an air blast or similar sorting device (Figure 3.3).

Electrostatic precipitation is based on the fact that a charged particle placed in an electric field will move towards a collecting electrode, and is widely used to extract grit and dust from gases, particularly from incineration combustion gases (Mantell, 1975; Ross, 1968; Skitt, 1972). A high voltage is applied between the discharge or ionizing wires (negative) and the collecting plates (positive). The ionized gas molecules so produced move toward the collecting plates, colliding with dust particles which thereby become negatively charged and move toward and adhere to the collectors (Figure 3.4). The electrostatic principle can also be used for the separation of plastics from other material. As the mixed waste is fed on to an earthed drum rotating in a high electric field gradient, the plastic is held on the drum and removed by a fixed wiper blade (Figure 3.5). From a feed of 90.7% paper and 9.3% plastic, a mixed plastic concentrate containing 99.4%

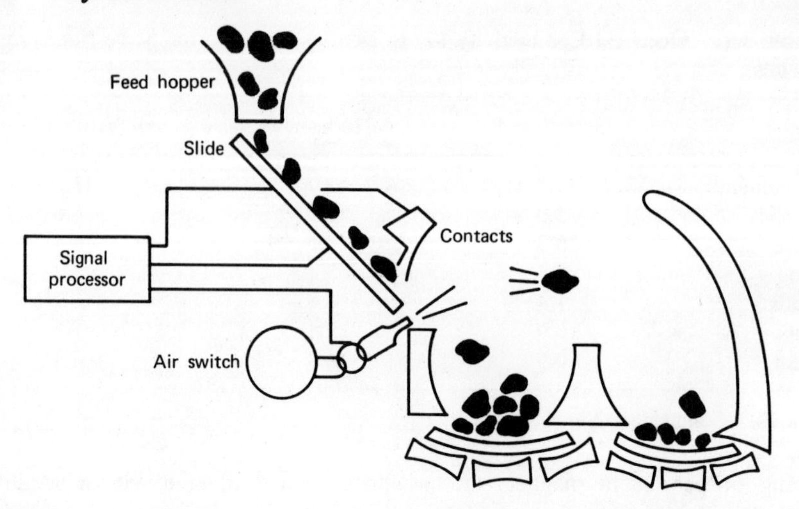

Figure 3.3. A conductivity sorter. (Sorter by Ore Sorters Ltd.; after Wilson and Smith, 1972, from *Technology Review,* copyright 1972 by the Alumni Association of the Massachusetts Institute of Technology.)

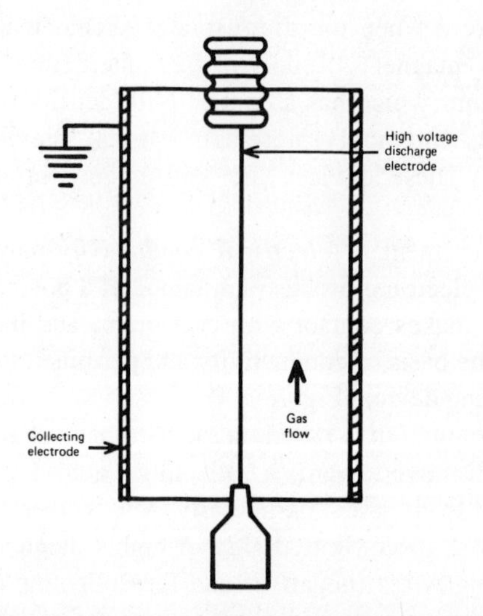

Figure 3.4. A simple electrostatic precipitator. (After Ross, 1968.)

plastic and a 99.9% pure paper residue were obtained (Sperber and Rosen, 1974). The technique can be used also for the separation of conductors such as aluminum from nonconductors in mixed solid wastes (Cummings, 1974, 1976).

Electromagnetic radiation ranging from microwave, through infrared, visible, ultraviolet, X-rays, and emission from radioactive sources, is available for

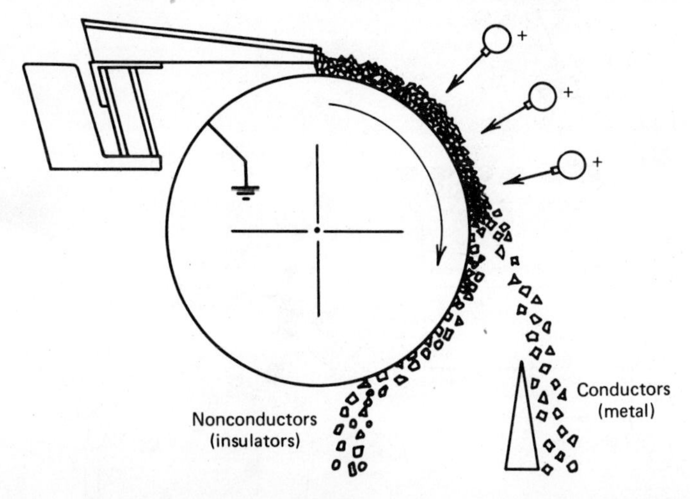

Figure 3.5. A simple electrostatic separator. (Separator by Carpco Laboratories; from Cummings, 1974.)

coding and separating items of waste material, and all these techniques are being investigated (Wilson and Smith, 1972). The simplest method is probably the optical scanner, using a broad band of *visible radiation* to separate optically transparent from opaque materials: for example, sorting glass from aluminum as these materials have nearly equal densities (2.7 grams per cubic centimeter) and are difficult to separate by density methods. Bottles or glass fragments can be sorted into green, amber, and clear categories as they fall from the end of a belt, by means of three photocells which compare the color of the glass to a reference background color, and operate rapid-acting air blasts to divert each item appropriately: Figure 3.6 (Cannon and Smith, 1974; Cummings, 1974, 1976; Wilson and Smith, 1972). Reflected light can also be used. The spectrum of *infrared radiation* diffusely reflected from a surface shows all the major infrared absorption lines characteristic of the reflector. Infrared radiation is superior to visible light for many sorting purposes because it is not misled by superficial colored dyes or inks, and because it can be used in normal room light. Reflection is more convenient than transmission in practical situations, and diffuse spectra from "dull" reflectors are more informative than specular (mirror) reflection from "shiny" surfaces. In the sorting of a limited range of items one wavelength is adequate, but in the Massachusetts Institute of Technology automated urban waste sorting system mentioned previously, a sensor measures the diffusely reflected light intensity at four infrared wavelengths produced in turn by a rotating filter wheel (Senturia, 1974; Senturia et al., 1971; Wilson and Senturia, 1974). Careful selection of these wavelengths allows the pattern of relative intensities to be made extremely sensitive to the nature of the reflecting material

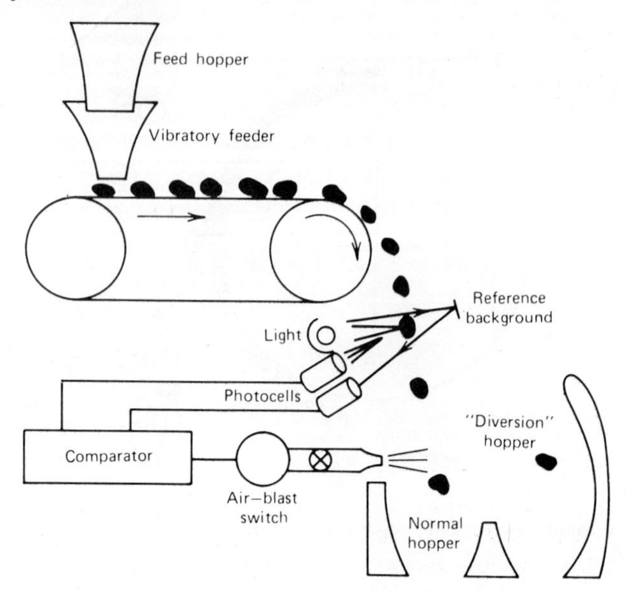

Figure 3.6. A color sorter. (Sorter by Sortex Co.; after Wilson and Smith, 1972, from *Technology Review*, copyright © 1972 by the Alumni Association of the Massachusetts Institute of Technology.)

(Figure 3.7). The four peak heights are measured electronically and computer-processed.

A proposed extension of the spectroscopic method is a device in which a high-power pulsed laser or a spark beam vaporizes a very small portion of an object, the vapor then being analyzed in a manner similar to the widely used flame photometry technique. In principle, this method could provide adequate information for precise classification of waste items. It has even been suggested, in connection with the recycling of poly(tetrafluoroethylene) (Arkles, 1973), that a high-energy laser beam could vaporize small, dark, light-absorbing objects to remove them completely from light-colored objects which would reflect rather than absorb the radiation.

An application of *ultraviolet* radiation is its use to precipitate silver salts from photographic fixing and bleach-fixing solutions, thus rejuvenating the processing solutions and recovering silver (Morrison and Lu, 1975).

The *thermoelectric* (Seebeck) effect has been utilized for sorting mixed metals. This is the effect in which an electrical potential is generated when the junctions in a circuit containing two dissimilar metals are at different temperatures. The magnitude and sign of the emf generated when a metal completes a circuit between probes at different temperatures depends on the nature of the metal and permits identification and separation (*Physics Education*, 1976).

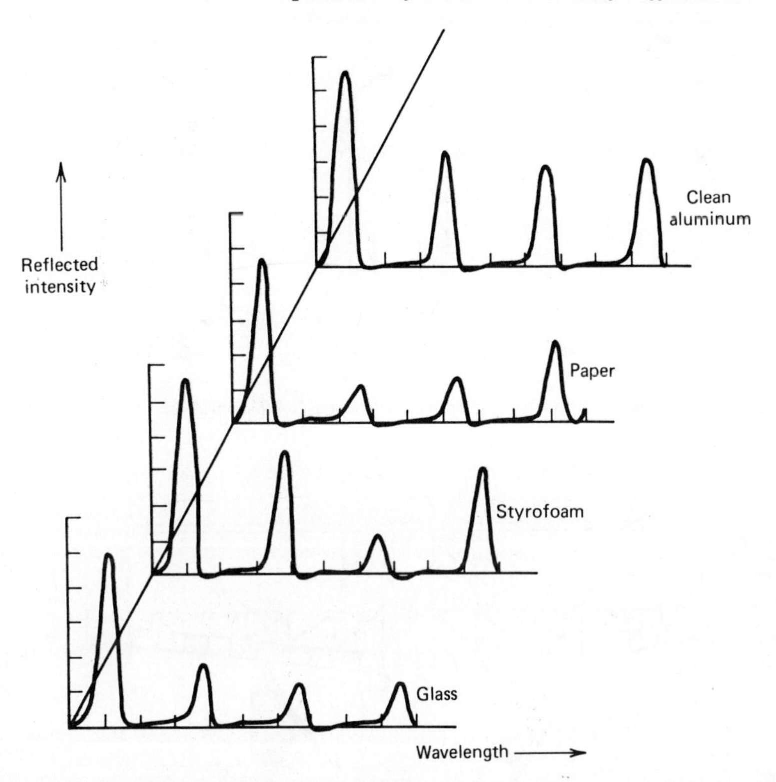

Figure 3.7. Reflection spectra from an infrared sensor, showing the different relative intensities at four wavelengths from various materials. (From Senturia, 1974; Wilson and Senturia, 1974. Reprinted with permission from "Computers and People," February 1974; copyright © 1974 by and published by Berkeley Enterprises, Inc., 815 Washington St., Newtonville, Mass. 02160.)

SEPARATION BY PHASE AND DENSITY DIFFERENCES

Filtration and Screening

Filtration and screening are obvious techniques for the separation of solids from liquids, or size-separating solids, with the filters or screens chosen for the particular application, and with an appropriate pressure drop between inlet and outlet. The subject has been dealt with exhaustively in a number of books and reviews on waste treatment.* The usual distinction between "screens" and "filters" is that the former generally contain regular holes of relatively large size

*For example, Besselievre and Schwartz, 1976; Hartinger, 1975; Koziorowski and Kucharski, 1972; Mantell, 1975; Nemerow, 1963; Rickles, 1965; Ross, 1968.

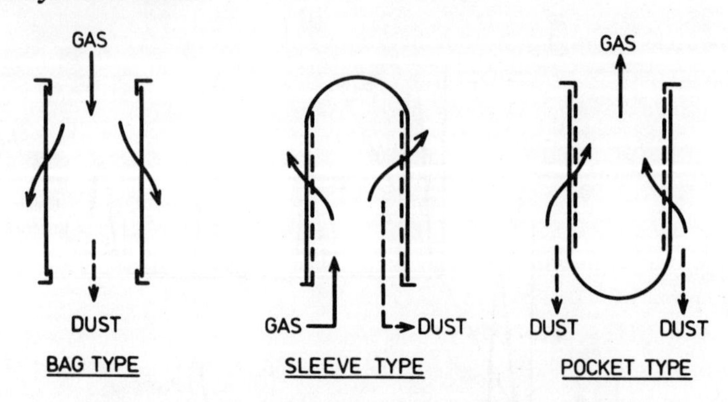

Figure 3.8. Three fabric filter configurations for gas/solid separation. (After Ross, 1968.)

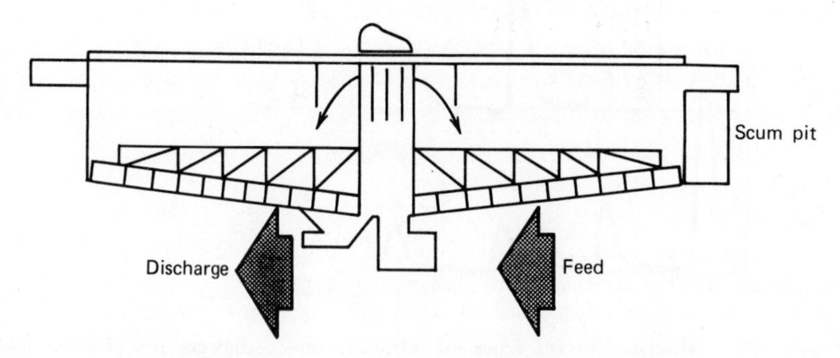

Figure 3.9. A settling tank or clarifier. (From Rickles, 1965, with the permission of *Chemical Engineering*.)

and operate with low pressure drops, while the latter usually have irregularly shaped holes of relatively small size (for example, a fibrous sheet) and operate with relatively high pressure differentials. Dust can be removed from gas streams by a fibrous or laminar filter material such as cotton, wool, synthetic fiber, wire wool, or paper, which may be used dry or wetted with a suitable liquid. Particles ranging from 0.01 micrometer to coarse screen sizes may be filtered successfully at better than 99% efficiency. Three fabric filter configurations for gas/solid separation are illustrated in Figure 3.8. A new development is the *band filter,* in which the material is spread in an extremely thin layer over a fast-moving cloth belt (WWT, 1975a).

By screening, cellulose fibers can be screened from hydropulped urban waste, and even dry screening can remove a large proportion of the dirt from many materials (Cannon and Smith, 1974; Dean et al., 1971). Dissolved solids occasionally are removed by evaporation of the solvent, but the value of the recovered residue must outweigh the fuel cost, and more commonly the solution is concentrated by evaporation to facilitate crystallization and subsequent filtration rather than evaporated to dryness. If a dissolved material can be induced to crystallize from a filtered solution it can be recovered in a particularly pure form, although the method is expensive if dilute solutions first must be concentrated by evaporation. When evaporation is necessary, submerged combustion (the direct contact of hot combustion products with the liquid) has the advantages of high thermal efficiency and low capital and maintenance costs (WWT, 1975e).

Filtration rates of fine solids from liquids are often a problem, with sludge particles plugging the filter pores. Sedimentation tanks may be used to allow settling (and ''aging'' or recrystallization in the case of crystalline solids) (Figure 3.9), and flocculation tanks may be incorporated, in which chemicals are added to improve the separation of colloidal solids (Rembaum, 1974). Elutriation (washing the solids) can be reduce the amount of coagulant required. Addition of fly ash (from incinerators or power stations) substantially improves the settling characteristics of slurries in some cases (CEN, 1974d; Moehle, 1967) and thus two waste problems to some extent eliminate each other. Electrical coalescence can be used to facilitate the separation of small particles dispersed in a dielectric (nonconducting) fluid, such as the separation of water-in-oil emulsions. When microstraining is reduced to pore sizes of molecular dimensions, the technique is known as ultrafiltration, and is considered as a membrane process, below. This is particularly important in advanced water treatment.

Hydraulic (Wet) Beneficiation

The direct application of density differences (the *sink-float* approach) may involve a single liquid, but often relies on the use of several liquids differing in composition and therefore differing in density. For example (Holman et al., 1972, 1974; Jensen et al., 1974; Kenahan et al., 1973), experimental work has been carried out on the separation of three types of plastic waste using only water (Figure 3.10). The mixed plastics were fed into a sink-float separator that floated off the polyolefins (polyethylene and polypropylene); the other two components, after sinking, were transported by airlift to an elutriation column (see below) which separated poly(vinyl chloride) from polystyrene. Also, the five major types of plastic (listed, with their densities, in Table 3.2) in priniciple can be separated using water, calcium chloride solution, and two alcohol-water combinations. Nagaya and Adichi (1972) have applied a water-steam mixture in plastics separation. For the high-value, heavy metals more costly liquids such as tetrabromoethane, methylene iodide, and thallium formate/thallium malonate may be used (Engdahl, 1969; Pearson and Webb, 1973). Alternatively, water slurries of finely ground, inert materials such as magnetite, ferrosilicon, or

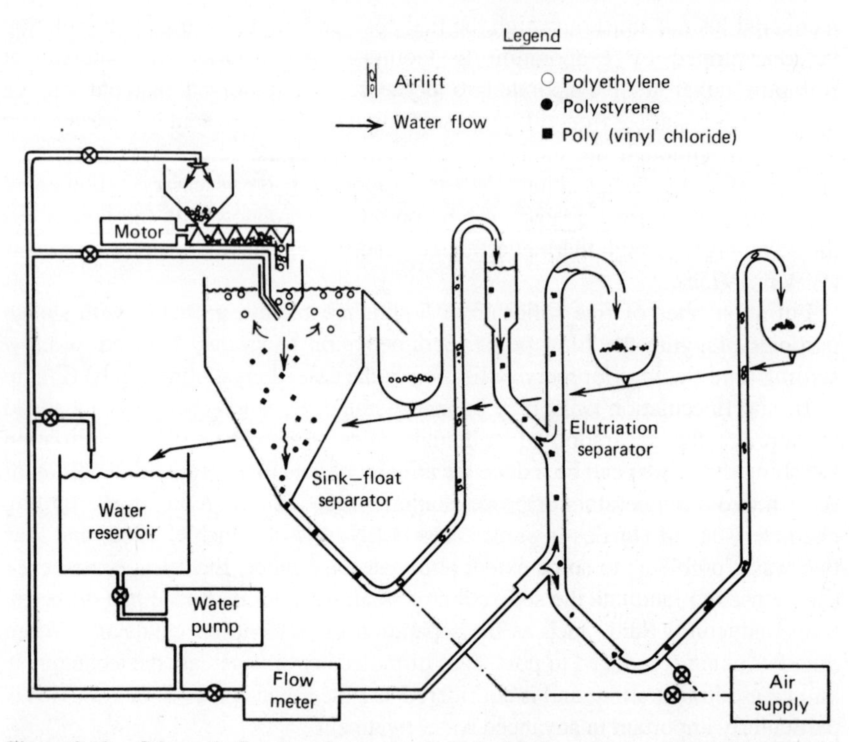

Figure 3.10. Schematic flowsheet for waste plastic hydraulic separator. (From Jensen et al., 1974; Kenahan, 1973; with the permission of the Bureau of Mines, U.S. Department of Interior.)

Table 3.2. Densities of Pelletized Plastics

Type	Density (g/cm^3)
Polypropylene	0.90
Low-density polyethylene	0.92
High-density polyethylene	0.94-0.96
Polystyrene	1.05-1.06
Poly(vinyl chloride)	1.22-1.38

Source. Jenson et al., in *Recycling and Disposal of Solid Wastes—Industrial, Agricultural, Domestic,* T.F. Yen, Editor, Ann Arbor Science Publishers, Inc., 1974, with the permission of the publishers.

galena can be used to provide "heavy media" with densities up to 4 grams per cubic centimeter.

Water classification, elutriation, or rising current (RC) separation (Figure 3.11) utilizes an upward-flowing water current to carry lighter particles while denser materials fall back through the column (Degner and McChesney, 1974). The two methods, water classification and sink-float with water-galena slurries, gave comparable results in separating nonferrous metals from nonmetals in automobile shredder rejects (Dean et al., 1972, 1975; Froisland et al., 1972; Valdez et al., 1975).

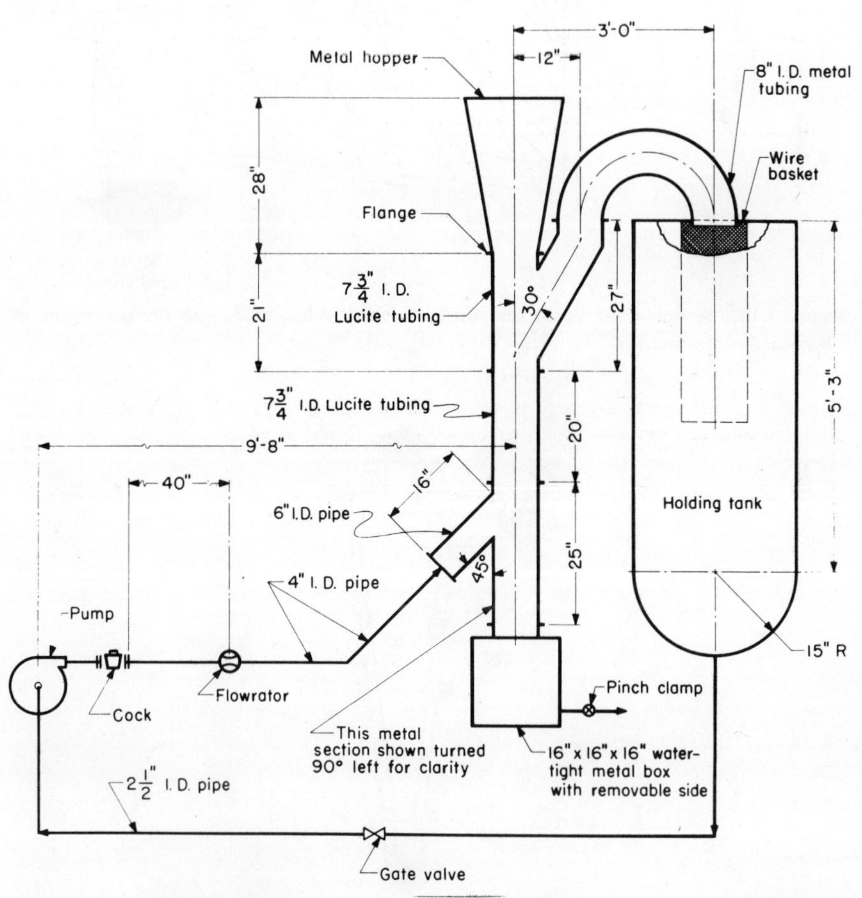

Figure 3.11. Water elutriating column designed for separating nonferrous shredded automobile scrap. (After Froisland et al., 1972, with the permission of the Bureau of Mines, U.S. Department of the Interior.)

An alternative method which has been tested for concentrating metal from wire-stripping wastes is *oscillating stratification* (the "*mineral jig*"). The material to be sorted is fed onto a liquid-solid bed in which the liquid is alternately forced through the gravel bed and drawn back. The materials are stratified in the order of their densities, the heavier material sinking through the

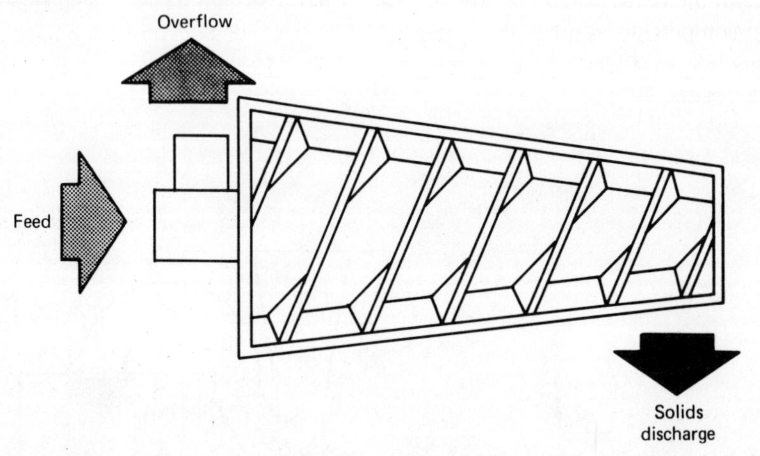

Figure 3.12. A solid-bowl liquid centrifuge. (From Rickles, 1965, with the permission of *Chemical Engineering*.)

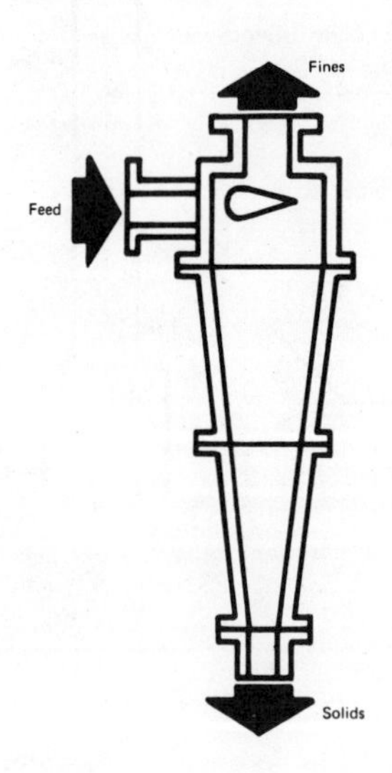

Figure 3.13. A liquid cyclone. (From Rickles, 1965, with the permission of *Chemical Engineering*.)

gravel bed and out through a screen at the bottom, while water carries off the lighter material in the overflow.

The beneficiation processes just described rely on gravitational forces; more efficient are the devices utilizing other forces (Mantell, 1975; Nemerow, 1963; Skinner, 1972). *Centrifuges* (Figure 3.12) produce a gravity-like force by acceleration of the feed, and in *cyclones* (Figure 3.13) the force is applied by introducing the pressurized stream tangentially to an inverted cone. In these ways dirt, glass, and metal objects can be separated from an organic slurry in urban waste treatment; and suspended matter can be separated from liquid used in food processing, allowing reuse of brines or syrups and eliminating some waste disposal problems. A combination of centrifugal forces and viscous drag is used in a *vortex classifier;* water or a "heavy medium" (a suspension of very fine solid particles) is used for separating inorganic materials, while air is suitable for the less-dense organic materials. In the inward-flowing air vortex the material is distributed in a continuous manner over a range of combinations of densities and viscous drag properties which characterize different objects.

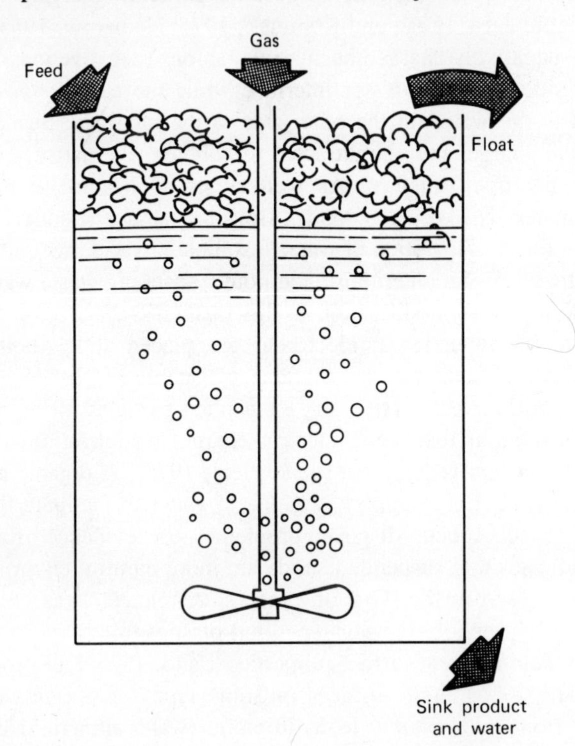

Figure 3.14. Separation by flotation. (From Rickles, 1965, with permission of *Chemical Engineering*.)

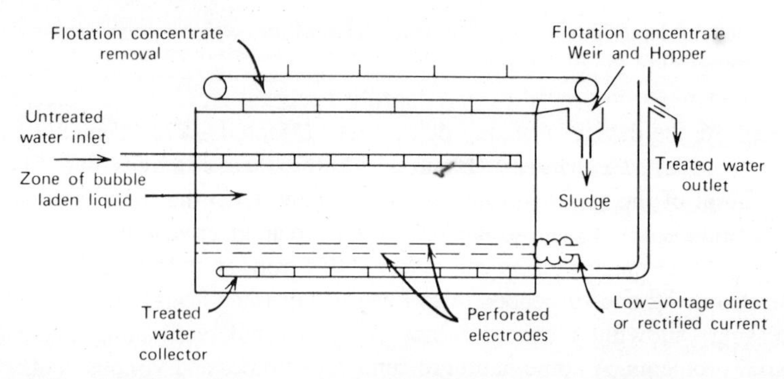

Figure 3.15. Schematic diagram of electroflotation with blade removal of floated solids. (Saint Gobain, Techniques Nouvelles; from Kuhn, 1971b, with the permission of the author.)

In the *froth flotation* process a gas (usually air) is pumped through a slurry, and the bubbles attach themselves to solid particles, buoying them up to the surface (Figure 3.14) (Lash and Kominek, 1975; Nemerow, 1963). The bubbles are usually negatively charged because the anions (negative ions) collect mainly on the air side of the air-water interface while the cations (positive ions) are distributed in the water on the other side of the interface. Suspended particles with positive charges may be attracted to bubbles; alternatively bubbles can be trapped in the floc structure. As well as its application to the recovery of inorganic materials like metal oxides from incinerator residues, froth flotation can be used for the separation of water-insoluble fats and precipitated protein, as in the treatment of slaughterhouse and poultry-packing waste waters (Hopwood and Rosen, 1972). A modification of this method is *electroflotation* or *electro-flocculation*. Two horizontal electrodes are placed at the bottom of a tank through which effluent passes. Streams of bubbles of hydrogen and oxygen resulting from the electrolysis carry suspended particles to the surface (Figure 3.15). It is thought that some electric charge neutralization of the colloidal particles also occurs (Kuhn, 1971b; Surfleet, 1970). If organic material (wood oil or mineral oil, for example) is added to water to facilitate flotation, it can contribute to subsequent disposal problems, but evidence from the mining industry indicates that suspended solids are more harmful environmentally than the chemicals used in the flotation process. A related "wet" technique is the separation of hydrophobic (water-repelling) plastics from mixed plastic waste on the basis of their different surface properties (Saito, 1975; Saito and Izumi, 1976; Saitoh et al., 1976). The surfaces of some types of plastic can be changed selectively from hydrophobic to hydrophilic (water-attracting) by means of a wetting agent. Air bubbles introduced into the separation cell adhere to the surface of hydrophobic plastic particles, buoying them up to the surface; the effective adhesion between air bubbles and plastic surfaces is such that relatively

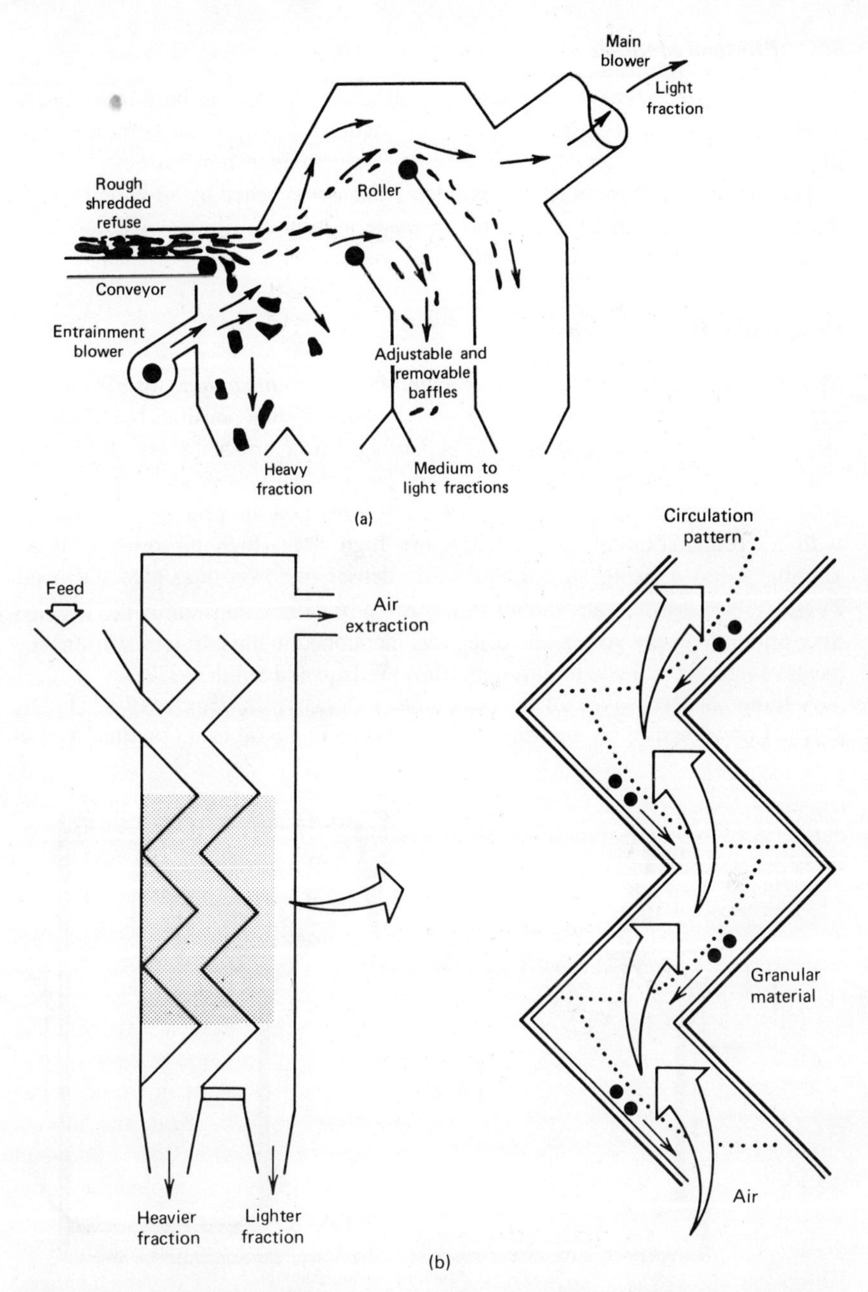

Figure 3.16. Schematic designs of air classifiers. (a) From Cannon and Smith, in *Recycling and Disposal of Solid Wastes—Industrial, Agricultural, Domestic*, T. F. Yen, Editor, Ann Arbor Science Publishers, Inc., 1974, with the permission of the publishers. (b) Zig-zag air classifier from NCRR, 1974c, with the permission of the National Center for Resource Recovery, Inc., Washington, D.C., U.S.A.

large plastic particles (at least several millimeters thick) can be floated easily. This technique can be used for the elimination of poly(vinyl chloride) from mixed plastic wastes, and the separation of foreign matter from plastics.

The various wet beneficiation methods often are preceded by air winnowing in the case of mixed wastes such as urban waste, to remove fibrous material which would otherwise absorb large quantities of water.

Pneumatic (Dry Air) Separation

The process alternatively known as *winnowing, air elutriation,* or *air classification* is useful for the separation of materials of different densities but of similar size. Examples are the separation of paper labels from crushed glass, differentiation of plastic wastes, resource recovery from urban waste, and separation of auto shredder material (Boettcher, 1972). It works on the principle that objects with a combination of low density and high drag (high air resistance) are entrained into a rising air stream, while denser or lower-drag pieces decend. Typical air classifiers are shown in Figure 3.16. The extension of the density-drag principle to the vortex classifier was mentioned in the previous section; this has given good results in preliminary trials (Wilson and Smith, 1972).

A rather different type of air separation process is the *fluidized bed* (Figure 3.17). Low-pressure air entering through the porous base of an inclined bed of

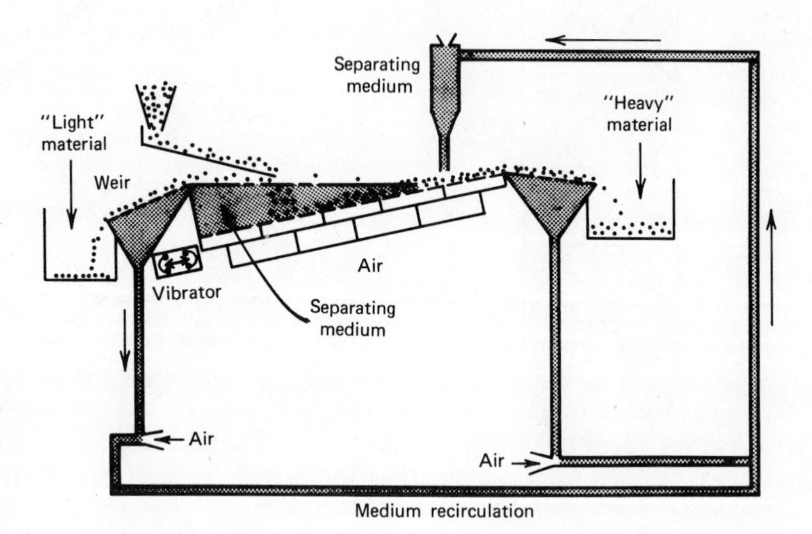

Figure 3.17. Schematic design of a fluidized bed separator. (From Pearson and Webb, 1973, with the permission of the Controller, Her Majesty's Stationery Office.)

particles causes them to act as a fluid, with an effective density that can range from less than that of water (1 gram per cubic centimeter) to more than 6 grams per cubic centimeter depending on the type of particles and the air flow rate. This is a ''dry'' version of the mineral jig described previously. Fluidized beds impart many of the desirable characteristics of a true fluid to solid particles (facilitating dispersion, improving flow characteristics, reducing viscosity, increasing sensitivity to gravitational forces, and introducing adaptability to control and metering) without the disadvantages of adding water or other liquid (Douglas and Sayles, 1970). The separation of a two-component mixture of solid particles of different densities is carried out by introducing it into a medium with an effective density intermediate between the two solid densities, so that one component ''floats'' while the other ''sinks.'' Powders such as those listed in Table 3.3 or blends of different densities are selected; the sizes and shapes of the particles have a secondary effect on the apparent density of the bed.

Table 3.3. Bulk Densities and Fluid Bed Densities for Various Materials, Particle Size between 150 and 200 Mesh B.S. (between 75 and 106 micrometers).

	Density (g/cm³)	
Material	Bulk Density	Apparent Fluid Bed Density
Silica sand	1.15	0.8
Barytes	1.9	1.7
Magnetite	2.25	2.0
Ferrosilicon(ordinary grade)	3.0	2.7
Ferrosilicon(special grade)	3.3	3.0
Copper	3.9	3.5
Lead	5.0	4.5
Tungsten carbide	7.0	6.3

Source. Table 1, Douglas, E. and Sayles, C.B., *AIChE Symposium Series,* **67**(116), 202 (1971), with the permission of the American Institute of Chemical Engineers.

The separation of nonferrous incinerator clinker into light (aluminum-rich) and heavy (copper-rich) fractions has been carried out at Warren Spring Laboratory on a fluidized bed of ferrosilicon, magnetite, or iron particles, the density of the medium being chosen to lie between the densities of the materials to be separated (Baum and Parker, 1974a; Skitt, 1972). Aluminum and copper from shredded automobile radiators can be separated cleanly in the same way. An extension of the method utilizes the *direct* air-fluidization of the dry materials to be processed; this has been called a pneumatic pinched sluice (Douglas and Sayles, 1970;

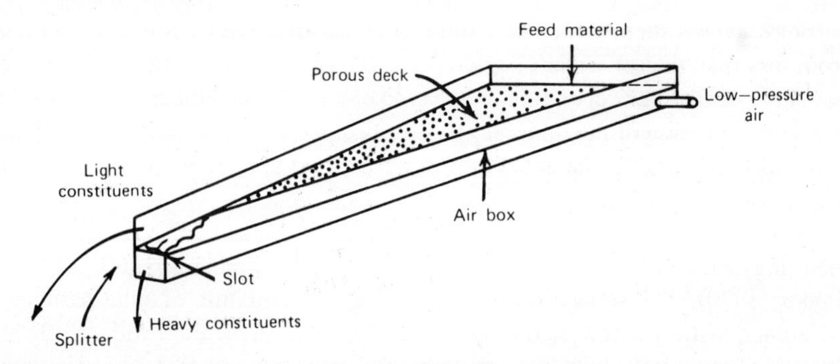

Figure 3.18. Schematic design of a pneumatic pinched sluide. (From Pearson and Webb, 1973, with the permission of the Controller, Her Majesty's Stationery Office.)

Pearson and Webb, 1973) and is illustrated in Figure 3.18. Low-pressure air is admitted to the airbox, and the dry material to be sorted is fed to the wide end of the sluice. During the fluidized flow, the heavier particles sink, and move at lower velocities because of frictional contact with the deck surface. The lighter layers move more rapidly and follow different discharge flight paths.

Solvent Extraction

A very effective way of separating materials is to dissolve selectively or extract one of them into a liquid phase, leaving the remainder in a solid, a gas, or another (immiscible) liquid phase. In principle, any mixture of gases, liquids, or solids can be separated into its components with the aid of one or more solvents under certain conditions of temperature and pressure, but in practice there are several problems. Selectivity between gases is not usually good unless a chemical reaction removes one gas, as in the reaction of carbon dioxide with limewater to precipitate calcium carbonate, or the formation of a formate from carbon monoxide and an alcohol (Sittig, 1975), for example. Ionic solids (salts) tend to dissolve in water and other ''ionizing'' solvents, while covalent materials (the majority of organic compounds) have greater solubilities in organic solvents like hydrocarbons, acetone, and carbon tetrachloride. A practical guide to solubilties is provided by solubility parameters (Barton, 1975) which are particularly useful for the solvent fractionation of plastics (Seymour and Stahl, 1976).

Although some soft scrap metals like iron and copper have been dissolved in molten zinc (which subsequently is removed by sublimation), metals usually are converted to their ionic forms, by reaction with acid, for example, before solvent separation methods can be used. Many metal ions are soluble in water, but the use of nonaqueous solvents such as acetonitrile permits selective dissolution of some metals (Parker, 1972a, 1973). Acetonitrile is particularly suitable for the extraction of copper. Metal ions may be extracted from water into organic solvents with the assistance of long chain amines (Clarke, 1974):

$$\underset{\text{amine salt in organic phase}}{2R_3NHX(org)} + \underbrace{M^{2+}(aq) + 2X^-(aq)}_{\text{metal salt in aqueous phase}} \rightleftharpoons \underset{\text{metal-amine complex in organic phase}}{(R_3NHX)_2MX_2(org)}$$

Solvent extraction is a well-established process for the recovery of metals from primary treatment of ores, particularly copper ores. There are commercial solvents available for iron, copper, nickel, cobalt, zinc, mercury, molybdenum, tungsten, uranium, vanadium, chromium, rare earths, and many other metals. Solvent extraction methods have been developed for various waste metal systems, with several types of solvent (Eccles et al., 1976; Fletcher, 1972, 1973; Harrison et al., 1976; Hughes, 1975; Lakshmanan and Lawson, 1973, 1975; Lawson, 1975; Reinhardt, 1975). Unless water is a suitable solvent, these methods involve the expense of a solvent, not all of which may be recovered in each cycle, as well as considerable chemical engineering resources. Consequently, solvent extraction is not used extensively at present in recycling, although this situation will change as the necessity for the recycling of certain materials becomes greater, wastewater metal content becomes more tightly controlled, and the extra cost becomes acceptable.

An example is the solvent extraction procedure developed on a laboratory scale for recovering zinc, iron, and trisodium phosphate from waste phosphate sludge (Powell et al., 1972b). The sludge was dissolved in hydrochloric acid, and iron was extracted (as ferric chloride) in isopropyl ether. Zinc was extracted by di-2-ethylhexylphosphoric acid in kerosene, and trisodium phosphate was recovered by crystallization. Solvent losses were reported to be low, and although the process was complex, it was considered commercially viable. Other applications include the recovery of nickel and cobalt from scrap (Fletcher, 1973), metals and acid from stainless steel pickling solutions, and zinc from drainage water in rayon manufacture (Reinhardt, 1975).

Work is in progress to find suitable solvents or solvent mixtures for separation or extraction of polymers. For example, polystyrene dissolves in *p*-xylene, and after filtration it can be precipitated as a pure powder by the addition of methanol (Yamazaki et al., 1974). Polyethylene can be recovered also with xylene as a solvent (Fukui et al., 1974). Solvents have been found which dissolve polystyrene, poly(vinyl chloride) and the polyolefins. For example, 15% cyclohexanone—85% xylene dissolves all these polymers and results in the formation of three phases at 115-125°C, with one polymer type predominating in each phase. At 15% total polymer in the polymer-solvent mixture, the separated phases appeared to contain the three types of polymer in better than 99% purity (Sperber, 1975; Sperber and Rosen, 1974, 1975). Polymers may be recovered also from aqueous latexes by freezing in thin films in the presence of a suitable solvent such as an alcohol (Ekiner, 1975).

Solvent extraction is not limited to low-temperature liquids. High-temperature extraction is possible in molten salt systems, for example between liquid slags (silicates) and liquid halides in metal recovery applications now being developed (Bogacz et al., 1975). Nor is it restricted to liquids. Solid particles of polymeric resins are now being developed for extracting metals from solution (*New Scientist,* 1976e), facilitating complete separation of the two phases.

Solvent extraction has been included here in the "physical methods" chapter, but in fact in many cases the interactions between solutes and solvents are sufficiently strong and specific that they should be classed as chemical interactions. A technique related to solvent extraction but relying on the formation of a dispersion rather than a solution is illustrated by the mist generated during a resin spray coating operation being collected in a solution, forming a stable pigment dispersion which is subsequently used as a thinner for the resin coating material (Tsuchiya, 1976).

THROW, BOUNCE, AND SLIDE METHODS

Hard materials may be separated from soft materials by dropping the mixture onto a moving belt. The hard items bounce off, the soft ones remain. This can be used to separate glass from compost manufactured from urban waste, for example (Figure 3.19). A much more precise use of this characteristic property of

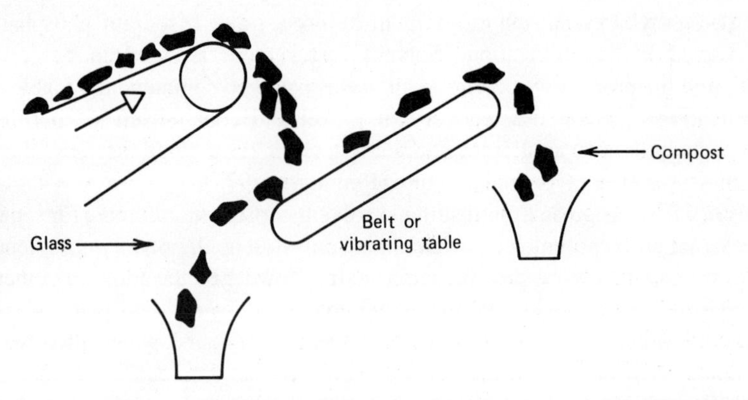

Figure 3.19. A bounce separator used to remove glass from compost. (Separator by MetroWaste; after Wilson and Smith, 1972, from *Technology Review,* copyright 1972 by the Alumni Association of the Massachusetts Institute of Technology.)

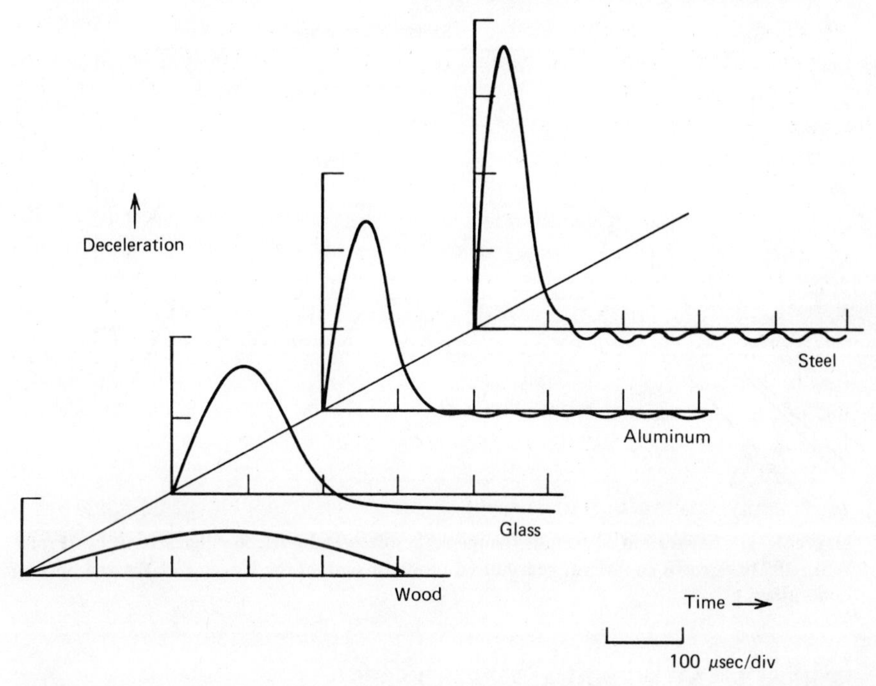

Figure 3.20. Deceleration data for samples of steel plate, aluminum plate, glass, and plywood. (From Senturia et al., 1971; Wilson and Senturia, 1974.)

materials is the impact sensor (Senturia, 1974; Senturia et al., 1971). Urban waste items are struck by a tool on which is mounted an accelerometer, and the deceleration of the tool depends on the material (Figure 3.20). This information can be computer analyzed to assist in the coding process.

The difference in frictional resistance between various components of a mixture and a striker/slider plate can be used to separate metallic from nonmetallic items (Figure 3.21). Solid-gas and solid-liquid friction is an important factor in the density-drag separation methods discussed in the previous section.

A simple type of ballistic or inertial separator has been devised (Douglas and Birch, 1974), based on the fact that dense objects leaving a feeding conveyor are projected further than less dense ones, and this is assisted by the winnowing action of an air stream. Copper and lead fragments from shredded armor-covered cables have been separated on a laboratory scale in an apparatus in which the material falls over successive units, each consisting of an inclined plate and deflector. Differences in the frictional and elastic properties of the constituents induce different trajectories. The apparatus, although restricted in application, has the advantages that it is simple to construct and has no moving parts (Bigg, 1970).

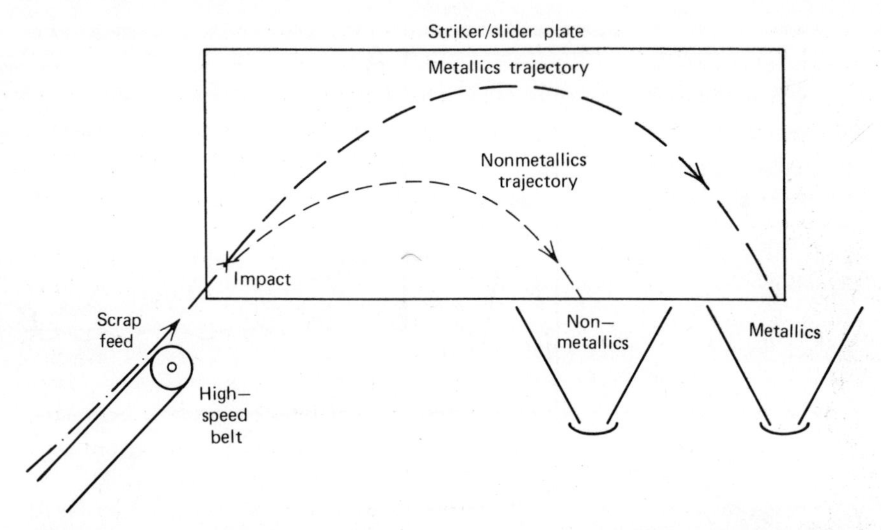

Figure 3.21. Separation of metallic/nonmetallic mixtures by friction characteristics. (From WSL, 1972b. Crown copyright; reproduced by permission of the Director of Warren Spring Laboratory.)

SURFACE AND MEMBRANE PROCESSES

Adsorbents and Filter Aids

Filter aids such as diatomaceous earth (a mineral composed of the fossilized silica skeletons of microscopic water creatures), fibrous asbestos, cellulose, and recently granular magnetic polymers (Bolto et al., 1974) are used to remove suspended solids at high filtration rates, but they do not have strong adsorptive properties. They depend for their action on the formation of a porous cake which traps suspended matter in narrow flow channels.

On the other hand, active or adsorbent materials with high surface areas (notably activated charcoal, coke, and alumina in slurries, packed beds or fluidized beds) are used to absorb a variety of soluble organic compounds from liquid and gaseous waste streams. This adsorbed material can be recovered by thermal or electrochemical ''reactivation'' of the adsorbent (Koziorowsi and Kucharski, 1972; Owen and Barry, 1972; Ross, 1968). A few inorganic compounds, for example iodine and molybdates, may be recovered also on carbon. As an example, nitrophenols at concentrations up to 1% in aqueous wastes are recovered in one chemical plant by acidification to a pH of 2.5, adsorption on granular carbon, followed by leaching with 5-10% sodium hydroxide solution to desorb them. A similar process can be used to recover phenol and acetic acid from brines (Fox, 1973), to reclaim alcohol from the vented air of plants using it as a solvent, and to recover methyl bromide from wheat silos after fumigation (Martin et al., 1975).

Active carbons are manufactured from carbon-rich material by controlled heating to obtain a carbonized solid (discussed further in Chapter 4) which is then treated with steam or oxidizing gases to develop the internal pore structure. It is also possible to make active filter aids capable of removing both insoluble suspended solids and some soluble impurities. Materials that are suitable for a range of filter aid and adsorption applications can be prepared from coal (Bowling, 1972) and from cellulosic wastes such as wood, textiles, and nut shells. Synthetic adsorbents (polymeric, porous, spherical beads with controlled surface polarities, surface areas, and pore sizes) are now available to facilitate the recovery of organic materials from wastewater. The capital costs are comparable with those of activated carbon, but operating costs for polymeric adsorbents appear to be more economical at high levels of dissolved organic adsorbates (Stevens and Kerner, 1975). Sometimes, simple and inexpensive adsorption media may be used. Thus the concentration of ionic constituents in sewage can be reduced in sand filters or by treatment involving lime and discarded rubber tires (Netzer et al., 1974; Russell, 1975), and newsprint is an economical filtration medium for wastewater from some industrial sources (Stauffer, 1975). Other inexpensive purification media are described in the "Ion Exchange" section of Chapter 4.

The surface properties of solids can assist separation both directly and indirectly. The special surface characteristics of the coalescing media in a new oil-water separator capture oil droplets and cause them to coalesce as they migrate through the media (*Industrial Recovery,* 1976f), and surface-wetting differences may be utilized in the separation of plastics and other materials by flotation, as discussed previously.

In the related liquid phase process of *foaming,* soluble surface-active agents concentrate at the solvent-air interfaces as a foam which may be removed. If the surface-active agents also complex or combine with inorganic metal ions, these ions may be removed by foaming, and the process is called *ion flotation* or *foam fractionation* (Clarke, 1974).

Membrane Processes

In membrane processes, an appropriate potential or driving force (concentration difference, voltage, or pressure differential) causes a flow of a component of a mixed waste stream through a membrane (Griffith et al., 1975; Meares, 1976).

Dialysis employs concentration differences as the driving force to separate solutes by means of their unequal diffusion rates through membranes (Nemerow, 1963). In the example of Figure 3.22, the energy required to separate partially the components in the mixture is provided by the dilution of the diffusate (the

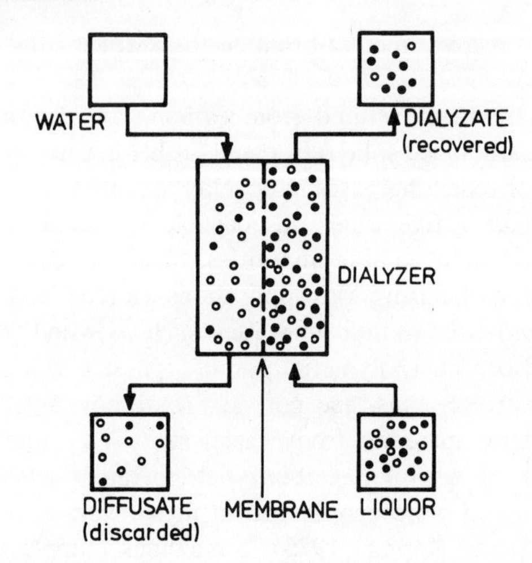

Figure 3.22. Block diagram of dialysis separation. (From Rickles, 1965, with the permission of *Chemical Engineering*.)

dilute, discarded solution) and of the dialyzate (the recovered solution, relatively richer in the required solute). Separations are made on the basis of either molecular size or of diffusion rates; dialysis is used usually to separate solute molecules which pass through a membrane from colloidal particles which are retained.

Electrodialysis uses electromotive driving force, with membranes able to pass selectively either cations or anions. The membrane bears positive or negative ionic charges, and so is able to act as an electric charge filter (Hartinger, 1975; Solt, 1976). Salts dissolve in water in the form of positively charged cations and negatively charged anions. If an electric field is applied, the cations move toward the cathode, while the anions migrate in the opposite direction, toward the anode. Cation-exchange membranes are permeable to cations but do not allow the passage of anions, while anion-exchange membranes pass only anions. Alternate layers of anion-exchange and cation-exchange membranes are placed between the electrodes as indicated in Figure 3.23, so that the solution between one pair of membranes becomes depleted in ions while the other solution becomes enriched. Electrodialysis is an ideal technique if a small volume at high concentration of a soluble salt is to be recovered from a large volume of dilute solution, for example ammonium fluoride waste from glass etching or hydrogen fluoride from quartz tube manufacture (Leitz, 1976). The process is still rather expensive for water desalination.

Ultrafiltration and reverse osmosis methods both use pressure gradients to force components of solutions through permeable and semipermeable membranes, which are becoming available in a variety of materials and designs

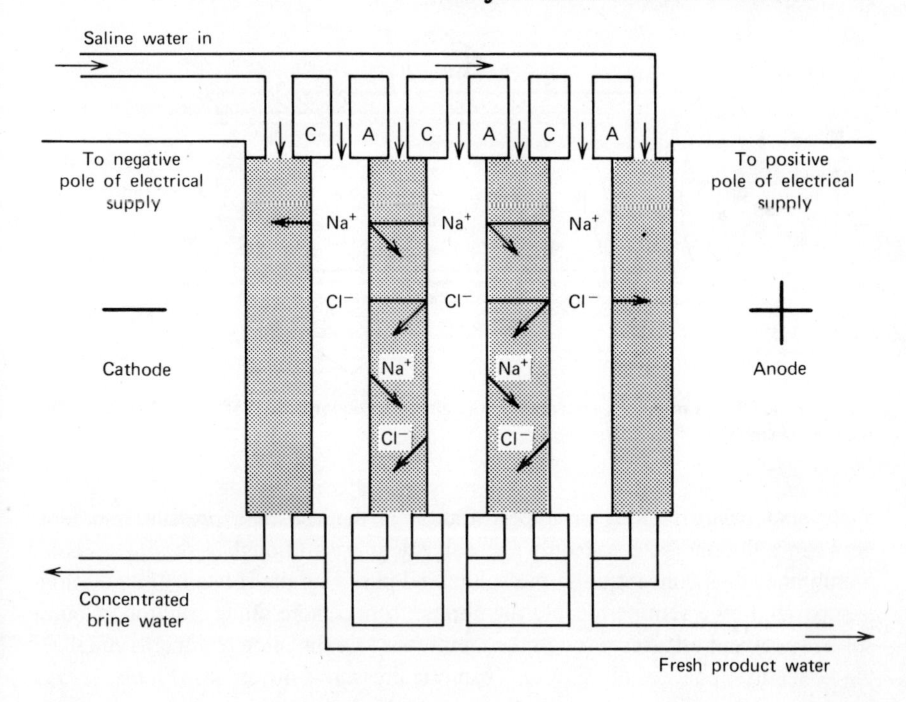

Figure 3.23. Block diagram of electrodialysis separation. (From Leitz, 1976, with permission. Copyright by the American Chemical Society.)

suitable for various purposes (Smith, 1975). In *ultrafiltration* (AERE, 1976; Bennett and Lash, 1974; Blatt, 1976; Hartinger, 1975; Nordstrom, 1974; WWT, 1976a) the waste stream is pumped through a porous tube on which a membrane or colloidal gel has been cast. The hydraulic pressure causes water and some dissolved low molecular mass materials (less than about 500 atomic mass units) to pass through the small pores, but emulsified oils or suspended protein particles are retained and concentrated within the tube (Figure 3.24). The "skinned" membranes of cellulose acetate or synthetic polymer now used are anisotropic, consisting of a thin high-density ultrafiltration layer (0.1-0.5 micrometer thick) supported on a strong porous matrix (of the order of 100 micrometers thick). Typically, the trans-membrane pressure is 200-700 kilopascals (30-100 pounds per square inch), with a membrane flux of 0.04 meter per hour (15 gallons per square foot per day). Ultrafiltration has been applied to the removal of: paint from solutions in the automobile industry; dispersed or dissolved oil; whey protein and food protein; ferric hydroxide from pickle liquors; lignin and lignin sulfonates in pulp and paper manufacture; and gluten (wheat protein) from starch factory effluent (Griffith et al., 1975).

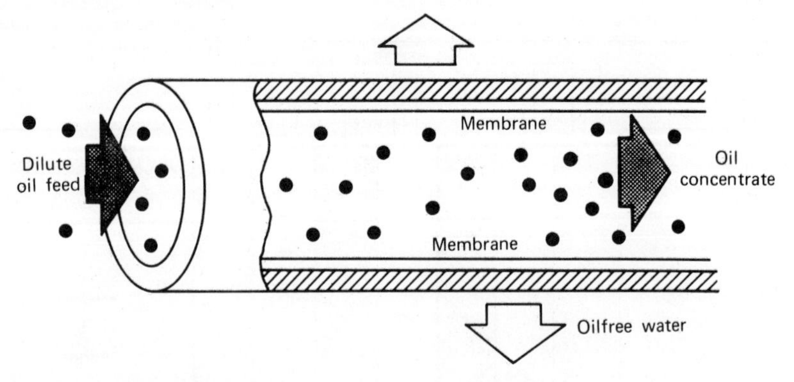

Figure 3.24. Diagrammatic illustration of the ultrafiltration process. (After Nordstrom, 1974, with the permission of *Pollution Engineering*.)

Reverse osmosis (RO) or hyperfiltration is the use of a pressure gradient between solutions to cause only the solvent to pass through a semipermeable membrane, thus concentrating the solute and purifying the solvent. If a solution is separated by a semipermeable membrane from a more dilute solution or pure solvent, solvent diffuses through the membrane in a direction tending to equalize the concentrations, resulting in an "osmotic pressure" difference (Figure 3.25). If pressure is applied to the more concentrated solution to exceed this value, the direction of solvent flow is reversed. Solvent passes through the membrane in the reverse direction: "reverse osmosis." Skinned membranes are used also in RO. A thin (usually less than 0.1 micrometer), dense, semipermeable skin of cellulose acetate (or, more recently, aromatic polyamide, polysulfone, polycarbonate, or cross-linked polyethenimine) is deposited uniformly on and anchored to a thicker (about 20 micrometers), strong and tough porous layer supported in some way to prevent rupture. Pressures must be higher for RO than for ultrafiltration: 4-7 megapascals (600 to 1000 pounds per square inch) pressure is required for fluxes of 0.04 meter per hour (15 gallons per square foot per day); the active layer is denser and the membrane stronger in RO.

Several types of RO system are now available commercially, each with particular advantages in certain situations (AERE, 1976; Griffith et al., 1975; Hartinger, 1975; Illner-Paine, 1970; McBain, 1976; Truby and Sleigh, 1974a, 1974b). Major applications are in wastewater treatment (Bailey et al., 1974; Cruver, 1975; Cruver and Nusbaum, 1974), including desalination of saline or brackish water (Harris et al., 1976; Kremen, 1975) and total recovery of water and dye in textile dyeing operations (Brandon, 1976; Porter and Brandon, 1976); in the sugar industry (Baloh, 1976); in the concentration of metal plating liquors (Clark and Fowler, 1974); and in conjunction with ultrafiltration to separate protein and lactose from cheese whey and similar food byproducts (Horton et al., 1972; Bennett and Lash, 1974). Wastewater streams containing 1-5% water-

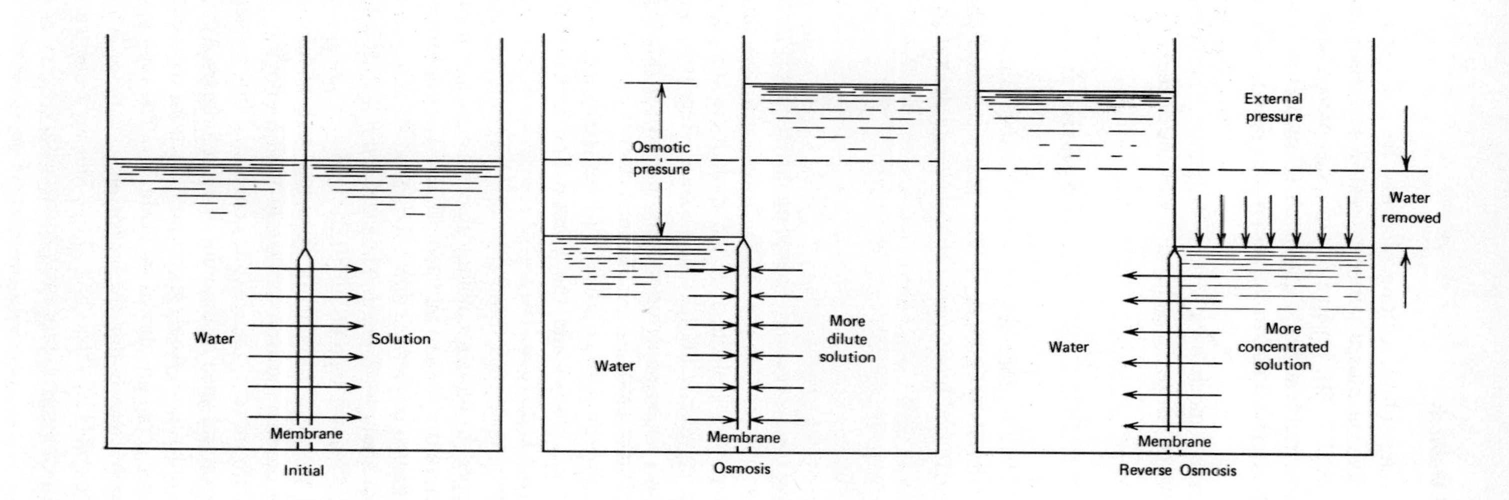

Figure 3.25. Illustration of the processes of osmosis and reverse osmosis.

emulsified or water-soluble cutting oils may be processed to produce liquids containing 30-60% oil plus a permeate with a low chemical oxygen demand (Zerlaut and Stake, 1974); reverse osmosis has proved satisfactory also in the recycling of papermill water and chemicals (Brown et al., 1976; Morris et al., 1972; Wiley et al., 1972) and in the removal of pesticides from water (Chian et al., 1975).

Because reverse osmosis uses surface ''filtration'' rather than deep filtration, fouling layers can be removed readily by hydraulic or chemical methods.

THERMAL METHODS OF PHYSICAL SEPARATION

This section includes processes in which heat treatment modifies physical properties of materials (Hartinger, 1975). Many chemical changes are brought about also by increased temperature, but these are dealt with separately, in the next chapter.

Melting

Selective melting (or ''sweating'') based on differences in melting points may be used to separate metals (Mantle, 1976; Sittig, 1975). A furnace fitted with a sloping hearth allows collection of the molten fraction, and some provision usually is made to vibrate the melting mixture. Recently, baths of molten salts have been shown to be superior for this purpose: oxidation of the metals is inhibited and cleaner products are obtained. Development of this technique has been carried out by the U.S. Bureau of Mines (Kenahan, 1971; Kenahan et al., 1973; Leak et al., 1974) and the Warren Spring Laboratory in the United Kingdom (Pearson and Webb, 1973).

Copper-containing ferrous scrap such as generators, alternators, and electric motors are dipped in molten calcium chloride (cheaper than barium chloride, which has been used also) and agitated briefly. The copper melts, and flows out of even inaccessible crevices to collect at the bottom of the container where it can be tapped off in about 99% yield. The steel scrap is little affected, and the salt can be reused. Cleanest separations (0.01% copper remaining in the iron) have been obtained by prior dipping or spraying with sodium sulfate or sodium silicate, which modify the iron surface to inhibit alloying with the copper. If several metals are present, the mixture, contained in a basket of inert and high melting point material such as titanium, may be immersed successively in baths of increasing temperature. Analyses of three fractions recovered from nonferrous auto fragmentizer scrap are shown in Table 3.4. The high proportion of zinc in the copper-rich fraction is due to the presence of brass in the scrap; selective melting is not capable of separating alloys into their elements. For this separation, chemical metallurgical processes are necessary.

Table 3.4. Products from the Selective Melting of Nonferrous Auto Fragmentizer Scrap (Percentages)

Metal	Zinc-rich Fraction (290°C)	Aluminum-rich Fraction (400°C)	Copper-rich Fraction
Zinc	85-94	4.8-17.0	28
Aluminum	4-13	71-85	1.4
Copper	0.03-0.7	2.0-3.2	64
Tin	0.7-1.0	0.8-2.3	0.2

Source. Pearson and Webb, 1973, with the permission of the Controller, Her Majesty's Stationery Office.

The technical feasibility of vaporizing metal impurities from recycled automotive scrap steels by various *vacuum melting* methods has been evaluated (Carlson and Schmidt, 1973; Carlson et al., 1974). Copper and tin levels are decreased to acceptable values, and chromium can be removed if the carbon concentration is reduced first. The high cost of this smelting method compared with an electric furnace is an economic barrier to its use, but it appears that a high grade stainless steel (type 304: 19 mass % chromium, 10 mass % nickel) can be prepared from automotive scrap by a single melting step in induction furnaces (utilizing the heating effect of induced eddy currents), and as some stainless steels are manufactured already by vacuum melting processes, the greater cost might be more acceptable here. *Electroslag melting,* in which the heating is carried out by the passage of an electric current through a molten salt or slag and impurities are removed from the molten steel by extraction into the slag layer, is successful for the removal of chromium, tin, and aluminum, but not copper (Carlson et al., 1974).

Distillation

Distillation involves heating a mixture of compounds so that the components of highest vapor pressure vaporize preferentially and are deposited as liquids (or occasionally as solids) on a cooled surface. Several components can be separated in batch processes by careful temperature control, but recovery of a single solvent frequently is carried out by an in-line, continuous process. In *steam distillation* superheated steam is introduced directly into the mixture, and the distillate which condenses contains a mixture of the more volatile components and water. If the products are immiscible with water, separation of the layers occurs automatically, but in the recovery of water-soluble materials, *fractionation* is required. This is the separation of liquid mixtures by the use of differences in boiling points and molecular masses. The vapors flow up a column, and if the heat input is adjusted, liquids of known composition can be removed at various heights up the column. It is a relatively expensive process, justified only for the recovery of more valuable materials. This is true also of *vacuum distillation,* where heat is applied to a system maintained under a vacuum in order to lower the boiling point. In *vacuum centrifugal distillation* a spinning evaporator cone enables a film to form. The vaporizable fraction evaporates rapidly, and with the help of the centrifugal force the unvaporized residue collects in a gutter around the edge of the cone (Rees, 1976). The vacuum methods are necessary for the recovery of heat-sensitive or oxidizable materials.

Freezing

The eutectic freezing technique involves holding a mixed system at a particular temperature such that a pure component freezes out. It is particularly appropriate for water purification: high-purity ice crystallizes, with up to 99.9% of the impurities remaining behind in a concentrated solution. (This effect may be seen in some homemade fruit juice ice-blocks. The water crystals tend to separate, leaving regions of concentrated fruit juices and sugar syrup.)

A variation on this technique is freeze-drying applied to aqueous solutions (Avila et al., 1973; Mantle, 1976). If a stream of droplets of the solution impinge on the inner surface of a vortex of a rapidly stirred, refrigerated, immiscible hexane phase, small frozen droplets are formed, and recovered by screening. Alternatively, in a continuous process liquid droplets are introduced into an immiscible refrigerant of higher density, where they freeze and float to the top of the column. In either case, the frozen droplets can be freeze dried by sublimation of the ice phase, leavng porous spheres of metal salts.

Cryogenic Embrittlement (Cryopulverizing)

Rubber tires cooled to −60°C in liquid nitrogen (boiling point −196°C) can be granulated by a hammer mill and the resulting rubber crumb (as well as fiber fluff and metal wire) used for a variety of purposes, for example the rubber for sports surfaces and the fiber and rubber for carpet underlay (Chapter 6). The same cryogenic technique (Wary and Davis, 1976) can be applied to cans (MRW, 1976d; *Solid Wastes,* 1976b), plastics (Braton and Koutsky, 1974; Jarvis, 1959), and to insulated copper and aluminum wire. After embrittlement, insulation and coatings separate easily from the wire (CIG, 1975b). At a suitable temperature, a roller system enables the embrittled insulation to be fractured and dislodged without damage to the wire, which remains ductile. The significant properties here are the ductile-brittle transition temperature of metals and the glass transition temperatures of polymers (−30°C for zinc-based die casting alloys, −40°C for poly(vinyl chloride)-neoprene insulation, −60°C for poly(vinyl chloride)-fabric insulation). Alternatives to liquid nitrogen are mixtures of acetone or methanol with dry ice (solid carbon dioxide), but liquid nitrogen has proved the most practicable and economical.

At a temperature of −120°C iron is very much more brittle than aluminum and copper. By precooling mixed metals in liquid nitrogen it is therefore possible to shred ferrous metals into flakes which are much smaller than nonferrous fragments accompanying them. Baled car bodies and other mixed metal scrap pulverized in this way can yield a high-quality scrap of approximately 99% ferrous material. The first selective cryogenic embrittlement plant of this kind was operating on a pilot scale in Liège, Belgium, in 1971. The technique is particularly suitable for the recovery of copper from small electric motors. Cryogenic techniques have been investigated by the U.S. Bureau of Mines for the separation of aluminum and copper from zinc die castings (automobile shredder nonmagnetic residue). At −65°C the zinc is brittle but the other two metals remain malleable, so that crushing at this temperature, followed by screening, allows good separation (Bilbrey, 1974; *Iron and Steel,* 1971).

A significant technical and economic factor in the future application of cryogenic embrittlement is that the vaporization of natural gas after production and shipping in liquefied form provides a large heat-absorbing capacity at a suitably low temperature for cryogenic embrittlement (CEN, 1974b).

Thermoplastics

Polymeric materials that become soft when heated and can be molded, extruded and shaped are called "thermoplastics" (in contrast to the thermosetting or chemically hardened plastics which do not soften on reheating and cannot be remolded). Thermoplastics, particularly polyethylene, polypropylene, polystyrene and poly(vinyl chloride), comprise nearly three-quarters of the total plastics produced in the United States, for example (Jensen et al., 1974), and because at least 2% of the total salvageable urban waste is plastic, the scope for remolding of thermoplastics is considerable. Most of the waste thermoplastic generated within manufacturing processes is recycled, but waste scrap of mixed and unknown properties usually is unacceptable in current processes (Chapter 7). There are examples of certain products being manufactured from particular items available in large quantities (high-density polyethylene milk bottles being recycled as drainage tiles, for example). Other products such as traffic markers, parking barriers, and paving blocks, where material specifications need not be so precise, utilize thermoplastics recovered by sink-float and air elutriation techniques into the classes indicated in Table 3.2.

Another application of the property of thermoplasticity is the separation of plastic film in waste from paper, either by the adherence of the film to a heated surface, or by thermal contraction of the film followed by air classification and/or screening (Laundrie and Klungness, 1973a, 1973b). The thermoplastic properties of wax can be utilized in the hot water treatment of wax-impregnated cardboard and corrugated board (Mohaupt and Koning, 1972), and there are many other examples.

Drying

This is one of the everyday applications of heat, and although there are numerous examples of such processes in recycling, only one is mentioned here, chosen because it illustrates the recycle philosophy applied to a simple yet important system. The Department of Chemical Engineering of Drexel University, Philadelphia, has developed a process for the manufacture of a stock feed ingredient by dehydrating wet cow manure extrusions in a packed bed, using superheated steam as the drying medium in a closed cycle system. The closed cycle eliminates odor problems; the superheated steam kills virtually all organisms and is more efficient than air for the purpose because it can be totally recycled, whereas air would require a purge stream to control moisture build-up (Grossman and Thygeson, 1974a; Thygeson and Grossman, 1975; Thygeson et al., 1971).

Even if other forms of fuel are expensive, solar heating can be used for the evaporation of water from aqueous solutions. For example, salt can be recovered from tanning brines by simple processes of settling and solar evaporation.

4

CHEMICAL SEPARATION AND CONVERSION PROCESSES

Most of the processes described in the previous chapter were concerned with physical separation, without the breaking of chemical bonds. The methods discussed in this chapter, on the other hand, involve the use of heat and/or chemical reagents to bring about separations accompanied by significant chemical changes, and to bring about conversion to more useful or more easily collected products (Barbour et al., 1974; Engdahl, 1969).

Chemical processes play an important part in the recovery of metals from waste solutions and solids, as illustrated for example, in Figure 2.2. Frequently, these are variations of existing commercial production methods, which form the basis of the whole of the metallurgical industry. Also, chemical reactions are often used for the removal of undesirable components from furnace gases and other waste gases. Thus, the removal of sulfur dioxide from furnace effluent gases is carried out usually by catalytic oxidation to sulfur trioxide, which is more soluble in water and produces the potentially useful sulfuric acid:

$$SO_2(g) \quad + \quad \tfrac{1}{2}O_2(g) \quad \xrightarrow{\text{catalyst}} \quad SO_3(g)$$

sulfur dioxide gas *oxygen gas* *sulfur trioxide gas*

$$SO_3(g) \quad + \quad H_2O(1) \quad \rightarrow \quad H_2SO_4(aq)$$

sulfur trioxide gas *liquid water* *sulfuric acid aqueous solution*

Chemical reactions show promise for the large-scale conversion to useful materials of waste polymers, both natural polymers such as cellulose and lignin, and synthetic polymers or plastics, equivalent in importance to the present-day oil refining. Many of the methods now being used in advanced water treatment are chemical rather than the traditional biological or physical methods, and these will become increasingly important as exploitation of natural water supplies reaches its inevitable limit and wastewater receives more attention. All these and other processes are described in this chapter and illustrated further in Chapter 7. The biochemical transformations carried out by microbiological systems are treated separately in Chapter 5.

EXCHANGE REACTIONS

Chemical Precipitation

A standard separation procedure for recovering a chemical from a mixed soluton is as follows:

1. Precipitate the material by adding another chemical, perhaps simply by changing the acidity in the case of metal ions (Clarke, 1974);
2. Separate the precipitate from the solution by filtration or centrifugation;
3. Dissolve the precipitate in a suitable solvent, and return it to the process or recrystallize to recover a pure product.

An example is provided by the method used for the recovery of iron cyanides in photographic bleaching baths. Ferrous salt is added to precipitate cyanide, and after separation and dissolution the ferrocyanide product is oxidized to ferricyanide, which can be returned for use in the photographic bleach process (Bard et al., 1975). Other examples may be found in Chapter 7.

A more recent development is the use of *precipitate flotation* instead of precipitate sedimentation (Flett and Pearson, 1975). Flotation agents such as amines and aliphatic acids may be used to make the precipitate hydrophobic and so assist the froth flotation process which was described in the previous chapter.

Ion Exchange

Ion exchange is the reversible exchange of ions between a liquid solution and an ion exchange resin (usually in the form of solid beads in a packed column or fluidized bed, but occasionally as a liquid). The technique is particularly suitable for the recovery of ions present in solution at low concentration (Clarke, 1974), and although satisfactory for brackish or hard water purification (Arden, 1976; WWT, 1976e), it is unsuitable for seawater desalination because of the high salt concentration. The resins used were originally inorganic solids (''zeolites'') but now most are three-dimensional organic networks to which are attached many ionizable groups. Polystyrene sulfonates, which are cationic (for exchange of positively charged groups), and poly(vinyl pyridine) and polyacrylics (anionic for exchange of negatively-charged ions) are now used extensively. Hydrogen ions (H^+) on the cationic resin, $(RCO_2)H$ or $(RSO_3)H_2$, can be replaced by metal (M) ions:

$$(RCO_2)H + M^+(aq) \rightarrow (RCO_2)M + H^+(aq)$$

acid form of resin | *metal ion in solution* | *metal bound to resin* | *acid in solution*

The bound metal ion is released subsequently (to form a concentrated solution for recovery) by means of the addition of strong acid:

$$(RCO_2)M + H^+(aq) \rightarrow (RCO_2)H + M^+(aq)$$

metal bound to resin | *free acid in solution* | *acid form of resin* | *free metal ion in solution*

Alternatively, one metal ion can replace another:

$$(RCO_2)Na + M^+(aq) \rightarrow (RCO_2)M + Na^+(aq)$$

sodium form of resin | *metal ion in solution* | *metal ion bound to resin* | *sodium ion in solution*

In this way, a pollutant ion may be removed from the waste stream and concentrated for reuse. In Australia, a process (Sirotherm®) was developed to utilize low-grade heat (under 90°C) instead of chemicals to regenerate resins (Blesing, 1971; Bolto, 1976; Calmon and Gold, 1976; Stephens, 1976; Weiss et al., 1966). The special mixed resin absorbs ions (salt from brackish water, for example) at 20°C, but when the temperature is raised to 80°C the position of equilibrium is altered and a concentrated salt effluent is produced:

$$(RCO_2)H + (R'N) + Na^+(aq) + Cl^-(aq) \underset{80^\circ C}{\overset{20^\circ C}{\rightleftharpoons}}$$

cationic resin | *anionic resin* | *sodium chloride solution*

$$(RCO_2)Na + (R'NH)Cl$$

sodium chloride bound to resin

In some situations, a toxic or valuable ion is present in small amounts together with relatively large amounts of other innocuous or low value ions, and for this purpose there have been developed specific ion exchangers (Calmon and Gold, 1976). These are particularly useful for mercury residues.

Resins are ideal for the concentration and recovery of the more valuable metal ions from such sources as electroplating baths: the noble metals, silver, copper, zinc, nickel, tin (as stannate) and chromium (as chromate), for example (Koziorowski and Kucharski, 1972; Robinson et al., 1974; Ross, 1968). Copper from cuprammonium waste liquors is absorbed as follows:

$$(RSO_3)H_2 + 2NH_4^+(aq) \xrightarrow[\text{neutralization}]{} (RSO_3)(NH_4)_2 + 2H^+(aq)$$

acid form of resin	*ammonium ion in solution*	*ammonium form of resin*	*hydrogen ion in solution*

$$(RSO_3)(NH_4)_2 + Cu(NH_4)_2^{2+}(aq) \xrightarrow[\text{absorption}]{} (RSO_3)Cu(NH_4)_2 + 2NH_4^+(aq)$$

ammonium form of resin	*cuprammonium ion in solution*	*copper bound to resin*	*ammonium ion in solution*

$$(RSO_3)Cu(NH_4)_2 + 2H^+(aq) \xrightarrow[\text{stripping}]{} Cu(NH_4)_2^{2+}(aq) + (RSO_3)H_2$$

copper bound to resin	*hydrogen ion (acid) solution*	*cuprammonium ion recovered*	*acid form of resin*

Uranium oxide may be recovered from the huge volumes of acid leach mine waters, containing typically 10 parts per million of uranium, by ion exchange.

Resins may be used also for removing or recovering any organic materials which can be converted to an ionized (charged) form, particularly acids such as tartaric acid and acetic acid. Protein can be removed from process effluents with cellulosic ion exchange resins (Grant, 1974, 1976; Jones, 1976; Sittig, 1975; WWT, 1975c). The protein is recovered from the resin by regenerating the bed with brine and/or alkali: the regenerant solution is neutralized to the optimum pH with sulfuric acid, and the protein coagulated by heating. Resins are, however, vulnerable to organic fouling, and pretreatment of many liquors is necessary. Macroporous (large-pored) resins now used in some sewage treatment systems (Bailey et al., 1974) have overcome many of these problems.

There are types of resin suitable for ion exclusion (preferential washing-out of ionized species), ion retardation (preferential washing-out of nonionized species), and chelation (removal of heavy metal ions from solution by binding them with complexing groups on the resin rather than exchanging them for hydrogen ions) (Rickles, 1965; Ross, 1968). The last technique is less well developed, but has the advantage of lowering the total ion content of a solution rather than replacing one ion by another. Liquid ion exchangers are also useful, particularly in solvent extraction (Fletcher, 1971); and an important new development is continuous, fluidized bed operation (EST, 1976d).

Although synthetic ion exchange resins are efficient, they are relatively expensive, and alternative cheap or free materials have been suggested for the removal of toxic heavy metal ions from mining and industrial waste streams

(Lightsey, 1975; Randall et al., 1974). Many agricultural and forestry by-products (peanut skins, walnut expeller meal, wood barks, coconut husks) contain tannin, and in laboratory tests have proved to be excellent scavengers for Cu^{2+} (cupric), Hg^{2+} (mercuric), Pb^{2+} (lead), Zn^{2+} (zinc), and Cd^{2+} (cadmium) ions, reducing their concentrations to levels below the usual effluent standards. Each sorbed metal ion displaces two hydrogen ions, so the position of equilibrium is acid dependent and the metals can be recovered in concentrated solution by acid treatment, as illustrated previously for the synthetic resins. Pine bark can reduce the biological oxygen demand of pulp and paper mill sludge by up to 95%, and as the nitrogen content of the bark is increased in this process, it becomes more valuable as a soil conditioner (Lightsey, 1975).

Electrochemical Processes

When an electrical potential difference is applied between two electrodes in an electrolyte solution (that is, in any solution containing ions and therefore electrically conducting), reactions occur at the electrodes at certain characteristic voltages (called electrode potentials). If the appropriate voltage is used in a suitable electrolyte, it is possible on the one hand to plate out selectively metals or to evolve liquids and gases, and on the other to dissolve metals selectively. The collective term for all such processes is *controlled potential electrolysis*.

The main problem in many recovery systems is the dilute nature of the solution, resulting in high electrical resistance in the solution and depletion of metal ion concentration in the vicinity of the cathode ("concentration polarization"). Much can be done to improve the efficiency of electrolyte flow cells by introducing vortices in the solutions near the electrodes, and by using extended surface electrolysis (ESE) cells (Keating and Williams, 1976). The ESE cell has a sandwich construction with a high area cathode, porous separator layers, and a screen anode rolled together into a spiral "jelly roll" configuration and inserted into a pipe. Great promise is shown by the increased surface areas and decreased interelectrode distances in fluidized bed electrode cells and in flow cells packed with carbon electrode materials (Chu et al., 1974; Clarke, 1974; Fleischmann and Pletcher, 1975; Flett, 1971; Hills, 1976; Surfleet, 1970). One form of packed bed cell is the dipolar cell, where the electrical potential gradient down the complete cell is made up of alternating polarities at metallic "electrodes" spaced at intervals and separated by porous carbon. The extent of electrochemical and chemical reactions can be controlled by the spacing of the "electrodes" and the rate of liquid flow through the cell.

A cylindrical electrode rotating in a cylindrical flow-through cell (Figure 4.1) permits higher mass transfer rates in dilute solutions, and the growing dendritic (needle-like) crystals are broken off in the turbulent liquid flow, forming a metal powder (MRW, 1976a, 1976c; WWT, 1976c). Related to this method but suit-

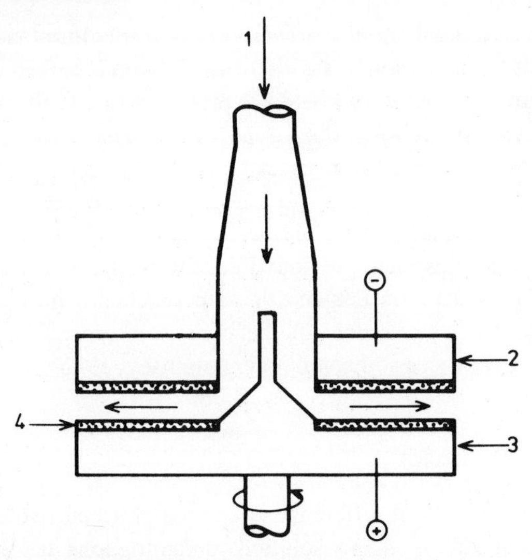

Figure 4.1. Schematic diagram of rotating electrode electrolysis cell: 1, electrolyte flow; 2, stator; 3, rotor; 4, electrodes. (Eco-Cell™; after MRW, 1976c, with the permission of Ecological Engineering, Ltd., Macclesfield, U.K.)

able for even more dilute solutions is the "pump cell" (Figure 4.2), which is based on the same principle as the "colloid mill". Flat, circular electrodes are rotated at high speeds relative to each other, and the liquid is introduced at the axis of rotation. The cell path-length is short (typically 0.1 millimeter) and the mass transfer rate is high, so dilute solutions can be electrolyzed with good electrical efficiency. The cell is self-pumping, but as an alternative, a pump may be incorporated to oppose the natural direction of flow to provide control over the residence time of reagents in the cell in order to match the electrochemical reaction rate with the rate of any subsequent chemical reaction in the flowing liquid (Ashworth et al., 1975). Dilute solutions of copper, zinc, and nickel ions can be converted by such methods into fine slurries of metal powders in water which can be separated subsequently by cyclone or similar liquid/solid separation techniques. High dilution is no longer a limiting factor in material recovery by electrolysis.

Metals such as copper, zinc, and aluminum usually are extracted and purified by electrodeposition in any case, so the extension of the process to waste recovery (preceded, if necessary, by chemical dissolution from mixed solids) has been relatively simple. Thus zinc metal can be recovered from smelter flue dusts (Higley and Fukubayashi, 1974; Powell et al., 1972a), and copper and other nonferrous metals can be separated from complex mixtures in the scrap from the electronics industry or from acid pickling wastes (Kuhn, 1971a). Electrolysis has

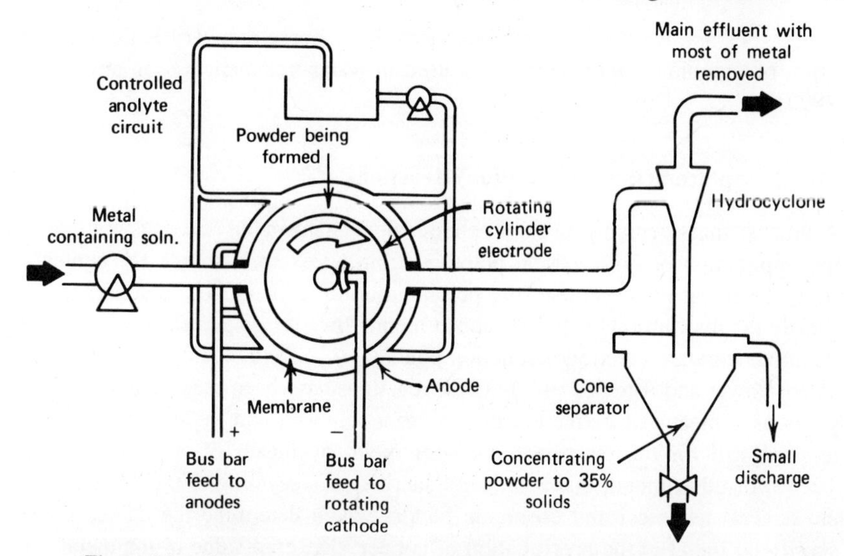

Figure 4.2. Schematic diagram of pump cell. (After Ashworth et al., 1975.)

been applied also to the recovery of gallium (Sleppy and Goheen, 1975), silver (Cooley and Dagon, 1976), nickel, cadmium, tin, arsenic, gold, and chromium (Kuhn, 1971b).

Some metal salts, such as nickel and cobalt salts, have standard electrode potentials which are very close in aqueous solution, making separation difficult by this method, but when dissolved in a molten salt such as a mixture of lithium chloride and potassium chloride at several hundred degrees Celsius, they can be recovered separately. Nickel can be plated out with less than 0.1% cobalt from starting alloys with up to 5.5% cobalt, and after all the nickel has been removed, the voltage can be adjusted and the cobalt may be plated out (CE, 1968). Fused salt electrolytic refining of scrap aluminum is an obvious extension of its method of manufacture.

Selective dissolution could in principle be used for the separate recovery of metals from articles like junk autos. The metallic assembly could be made an electrode (the anode) in a tank of electrolyte, and as the voltage increases the component metals would dissolve progressively and plate out on the other electrode (the cathode). With the appropriate electrochemical engineering techniques, each metal could be obtained in a relatively pure state. Iron would be recovered in a powder form, unsuited to conventional processing techniques, but powder metallurgy techniques are being developed. Future commercial viability depends on factors such as the relative costs of electricity and of conventional smelting fuels.

Another application of electrochemistry in resource recovery is the regeneration of granulated activated carbon used in water treatment (Owen and Barry, 1972).

Metal Displacement (Cementation) Reactions

A process that occurs by an electrochemical mechanism but that does not require the application of an electrical potential from an external source is ''cementation'' or metal displacement: the precipitation of a metal from solution by the sacrificial dissolution of a less noble metal, that is, of a metal with a more negative standard electrode reduction potential (Case, 1974; Fisher and Groves, 1976; Power and Ritchie, 1975). These reactions have been used for at least 2000 years as a means of extracting metals from solution, and the oldest process is applied still for the recovery of copper metal by means of scrap iron (EST, 1973b) or other metal; this is indeed ''sacrificial'' recycling of the metal. There are several technical and economic factors which determine the choice of the sacrificial metal in the precipitation of copper: the scrap value of the metal, the nature and weight of the resulting sludge which must be disposed of, and the rate of reaction (Coleman, 1975). On technical grounds, magnesium and aluminum are preferred, but frequently economics favor iron. ''Precipitate copper'' requires three times its weight in scrap iron, and tinfree light-gauge sheet is required.

$$\underset{\text{scrap iron}}{Fe(c)} + \underset{\substack{\text{copper ions}\\ \text{in solution}}}{Cu^{2+}(aq)} \rightarrow \underset{\substack{\text{ferrous ions}\\ \text{in solution}}}{Fe^{2+}(aq)} + \underset{\substack{\text{copper}\\ \text{metal}}}{Cu(c)}$$

Silver metal can be recovered from photographic wastes by bringing the waste solutions into contact with turnings of a light metal alloy containing aluminum and magnesium, or with steel wool (Cooley and Dagon, 1976).

Chemical Vapor Transport

If an element forms a suitable volatile compound, it is sometimes possible to distil the element from a mixture or alloy and condense it in a pure form. The halogen compounds of titanium ($TiCl_4$) and aluminum ($AlCl_3$ and $AlCl$) are good examples. A new method has been suggested for the recovery of aluminum from alloys by way of the volatile aluminum fluoride (Layne et al., 1974):

$$2Al(alloy) + MgF_2(1) \xrightarrow{1500\text{-}1650°C} 2AlF(g) + Mg(g)$$

aluminum in alloy — *molten magnesium fluoride* — *aluminum fluoride vapor* — *magnesium vapor*

$$2AlF(g) + Mg(g) \xrightarrow{<1350°C} 2Al(1) + MgF_2(1)$$

aluminum fluoride vapor — *magnesium vapor* — *aluminum liquid* — *magnesium fluoride liquid*

The alloy is reacted with molten magnesium fluoride at a temperature such that the reaction products have a substantial vapor pressure. In the absence of air, the vapors flow into an adjacent condenser and produce immiscible liquid layers of aluminum and of magnesium fluoride which can be tapped off. Containment of these materials at the high temperatures is a problem, but graphite, silicon carbide, aluminum nitride, and titanium carbide may be suitable.

POLYMERIZATION AND DEPOLYMERIZATION

In order to understand the way in which polymers may be degraded and depolymerized for reuse, it is necessary to consider their structures and the way they are synthesized. The word "polymer" comes from the Greek meaning "many parts," and polymers are materials whose molecules are made up of a large number of individual units, "monomers." Thus if natural rubber is heated, the hydrocarbon "isoprene" (2-methyl-1,3-butadiene) distils off, providing a good indication of the polymer structure:

```
       H CH3 H  H  H CH3 H  H  H CH3 H  H
       |  |  |  |  |  |  |  |  |  |  |  |
   ...-C--C==C--C--C--C==C--C--C--C==C--C-...  →
       |        |  |        |  |        |
       H        H  H        H  H        H

                 H     H  H
                 |     |  |
                 C==C--C==C
                 |  |     |
                 H CH3    H
```

Synthetic polymers can be produced either by *condensation* of monomer molecules (with the elimination of a simple molecule, often water), or by *addition* of monomer molecules containing double bonds (Jones and Chandy, 1974).

An example of a condensation polymerization is the formation of Nylon-66, a copolymer (a *polyamide*) of a diamine and a dicarboxylic acid each having 6 carbon atoms (hence "66"):

$$H_2N(CH_2)_6NH_2 \; + \; HOOC(CH_2)_4COOH \xrightarrow[-H_2O]{} -CONH(CH_2)_6NHCO(CH_2)_4CONH-$$

Polyesters are formed in a similar manner by condensation between organic acids or acid anhydrides with alcohols, accompanied by the elimination of water.

In the case of addition polymerization, reaction takes place after the initial formation of reactive species such as ions or free radicals, with subsequent addition of the reactive fragment in such a way that another reactive fragment is generated. (A free radical is a reactive molecular species with an unpaired electron, indicated by a dot in its molecular formula.) For example, the formation of poly(acrylonitrile) can be described as follows:

$$\underset{\textit{hydrogen peroxide}}{H_2O_2} \rightarrow \underset{\textit{hydroxyl radical}}{2HO\cdot}$$

$$\underset{\substack{\textit{hydroxyl}\\ \textit{radical}}}{HO\cdot} + \underset{\textit{acrylonitrile}}{\begin{array}{cc} H & H \\ | & | \\ C & = C \\ | & | \\ H & CN \end{array}} \rightarrow HO-\begin{array}{c} H \\ | \\ C \\ | \\ H \end{array}-\begin{array}{c} H \\ | \\ C\cdot \\ | \\ CN \end{array} \xrightarrow{H_2C=CHCN}$$

$$HO-\begin{array}{c} H \\ | \\ C \\ | \\ H \end{array}-\begin{array}{c} H \\ | \\ C \\ | \\ CN \end{array}-\begin{array}{c} H \\ | \\ C \\ | \\ H \end{array}-\begin{array}{c} H \\ | \\ C\cdot \\ | \\ CN \end{array} \rightarrow \text{etc.}$$

Four of the most common synthetic polymers are the addition polymers which can be represented chemically as

$$-(-CHX-CH_2)_n-$$

where X can be: hydrogen, in polyethylene;
phenyl, in polystyrene;
chlorine, in poly(vinyl chloride); and
methyl, in polypropylene.

Polyethylene, which is opaque and fairly tough, is probably the most common plastic, and is used in the majority of plastic bottles, as well as in the form of film. Polypropylene, which chemically is very similar to polyethylene, recovers its shape when deformed more readily than does polyethylene, and it is used for plumbing and to an increasing extent in packaging. Poly(vinyl chloride) is used widely in shoes, plumbing, and floor and wall surfaces, but its clarity and flexibility make it a popular substitute for glass in packaging. Polystyrene is also transparent, but is more rigid, and despite its brittleness has numerous applications in containers, toys, and home appliance components. These are all thermoplastics, so reuse by remolding is possible.

The significance of the distinction between condensation polymers and addition polymers is that the former are more likely to be reclaimed by solvolysis and the latter are more likely to be reclaimed by thermal depolymerization (see below).

All organisms are made up to a large extent of natural polymers, the main ones being: cellulose (the long-chain carbohydrates which are the primary structural material of plants: APS, 1976; Goldstein, 1976), lignin (a three-dimensional polymer of phenylpropane units that acts as a cement between the cellulose fibers), chitin (the protective shell of insects), and protein (condensations of amino acids). Natural rubber is based on the isoprene unit, as shown previously, and synthetic rubber is made by the copolymerization of butadiene and styrene, or of isoprene and butadiene, with various additives. As natural and synthetic polymers have many features in common, they are discussed together in this section. The properties and manufacture of synthetic polymers have been discussed by Milgrom (1972) and Lederman (1974) in connection with recycling and reuse.

Thermal Depolymerization

Thermal degradation of a polymer involves a series of chemical reactions, some of which may be the reverse of the addition polymerization steps, and involve *chain-breaking* with reactive intermediates such as free radicals. For example, thermal bond rupture of poly(methyl methacrylate) yields a polymeric radical which in turn splits into a smaller polymer radical and methyl methacrylate monomer:

poly(methyl methacrylate)

$$-(-CH_2-\underset{\underset{COOCH_3}{|}}{\overset{\overset{CH_3}{|}}{C}}-)_n-$$

$$\downarrow$$

polymer radical

$$...-CH_2-\underset{\underset{COOCH_3}{|}}{\overset{\overset{CH_3}{|}}{C}}-CH_2-\underset{\underset{CH_3}{|}}{\overset{\overset{COOCH_3}{|}}{C}}-CH_2-\underset{\underset{COOCH_3}{|}}{\overset{\overset{CH_3}{|}}{C}}\cdot$$

$$\downarrow$$

smaller polymer radical + *methyl methacrylate*

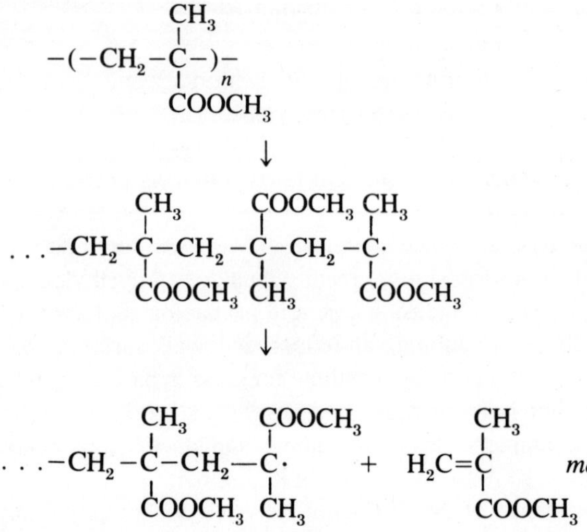

$$...-CH_2-\underset{\underset{COOCH_3}{|}}{\overset{\overset{CH_3}{|}}{C}}-CH_2-\underset{\underset{CH_3}{|}}{\overset{\overset{COOCH_3}{|}}{C}}\cdot \quad + \quad H_2C=\underset{\underset{COOCH_3}{|}}{\overset{\overset{CH_3}{|}}{C}}$$

An alternative possibility in thermal degradation is *cross-linking,* the formation of chemical bonds between neighboring chains by means of the side groups. In this case, the molecular mass increases, a rigid three-dimensional network forms, and if heating continues, eventually carbonization occurs. The two processes of chain-breaking and cross-linking may be summarized (Taylor, 1973):

$$\text{polymer} \xrightarrow[\text{breaking}]{\text{chain}} \begin{array}{c}\text{degraded polymer}\\ \text{(lower molecular mass)}\end{array} \rightarrow \text{volatile products}$$

$$\text{polymer} \xrightarrow[\text{linking}]{\text{cross}} \begin{array}{c}\text{degraded polymer}\\ \text{(higher molecular mass)}\end{array} \rightarrow \begin{array}{c}\text{network}\\ \text{polymer}\end{array} \rightarrow \begin{array}{c}\text{carbonized}\\ \text{residue}\end{array}$$

In some cases both processes are important and are in competition.

Poly(methyl methacrylate) and poly(methyl styrene) depolymerize to monomer in 100% yield, and poly(tetrafluoroethylene) gives greater than 95% yield of the monomer under the optimum conditions with superheated steam. Other addition polymers, such as polystyrene and polyethylene, tend to depolymerize but there are several competing reactions causing cross-linking of chains (Zerlaut and Stake, 1974). The polyethylene chain, for example, does not "unzip" during depolymerization, but is randomly broken and reformed, so the monomer yield at 200°C is only 0.1%. At high temperatures, fragmentation is more extensive. The low- and high-temperature mechanisms can be depicted as follows (Gutfreund, 1971):

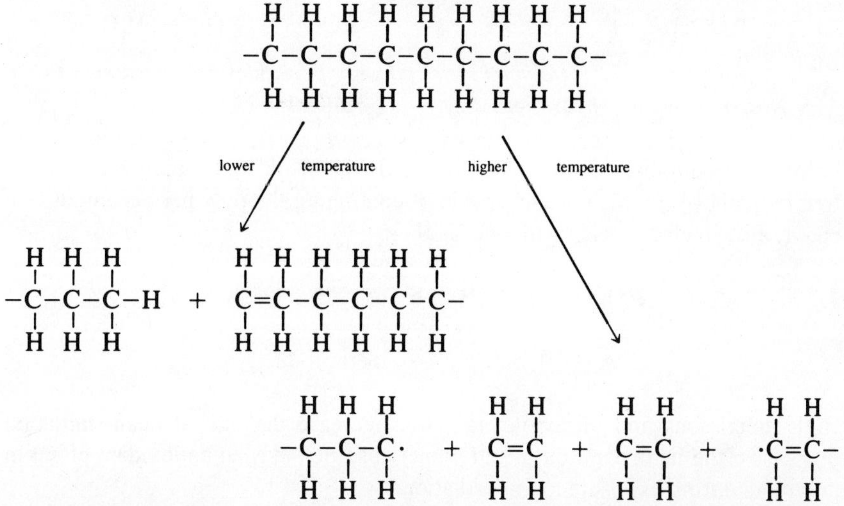

```
                       H  H  H  H  H  H  H  H  H
                       |  |  |  |  |  |  |  |  |
                      -C--C--C--C--C--C--C--C--C-
                       |  |  |  |  |  |  |  |  |
                       H  H  H  H  H  H  H  H  H

              lower     temperature          higher     temperature

 H  H  H                 H  H  H  H  H  H
 |  |  |                 |  |  |  |  |  |
-C--C--C--H      +       C=C--C--C--C--C-
 |  |  |                 |  |  |  |  |  |
 H  H  H                 H  H  H  H  H  H

                  H  H  H          H  H        H  H          H  H
                  |  |  |          |  |        |  |          |  |
                 -C--C--C·    +    C=C    +    C=C    +     ·C=C-
                  |  |  |          |  |        |  |          |  |
                  H  H  H          H  H        H  H          H  H
```

It has been found that it is possible to plasticize the surfaces of rubber particles by depolymerization with a hot gas flame (Sear, 1975); this introduces the opportunity to repolymerize and remold the rubber particles.

A general review of the reuse of synthetic polymers has been prepared by Sperber and Rosen (1974), and examples of the recovery of vinyl and cyclic monomers from addition polymers by thermal methods have been collected by Taylor (1973).

Low-Temperature Oxidation of Polymers

During the proces of converting polymers in powder or pellet form to fabricated products, oxygen is usually present and can react rapidly with polymers at the melt temperatures unless antioxidants are present (Scott, 1976). Chain propagation reactions which can take place if free radical initiators are present may be represented as follows, where R· denotes a polymer radical and RH is a polymer molecule:

$$R\cdot \quad + \quad O_2 \quad \rightarrow \quad ROO\cdot$$

polymer radical — *oxygen* — *polymer peroxide radical*

$$ROO\cdot \quad + \quad RH \quad \rightarrow \quad ROOH \quad + \quad R\cdot$$

polymer peroxide radical — *polymer molecule* — *polymer hydroperoxide* — *polymer radical*

$$\underset{\text{polymer hydroperoxide}}{ROOH} + \underset{\text{polymer hydroperoxide}}{ROOH} \rightarrow \underset{\text{polymer peroxide radical}}{ROO\cdot + RO\cdot} + \underset{\text{water}}{H_2O}$$

Peroxide decomposer antioxidants (denoted AH) work by forming less reactive free radicals (A·) which are incapable of continuing the polymer decomposition chain, but which can react with one another:

$$ROO\cdot + AH \rightarrow ROOH + A\cdot$$

$$A\cdot + A\cdot \rightarrow \text{inert products}$$

Both metal ions and ultraviolet radiation increase the rate of chain initiation processes; the incorporation of ultraviolet absorbers has an antioxidant effect in polymers during exposure to this radiation.

Oxidation and reduction reactions occurring at higher temperatures, in the presence of adequate oxygen, limited oxygen, or no oxygen, and resulting in more fundametal chemical changes, are discussed in subsequent sections.

Solvolysis of Polymers

"Hydrolysis" is a rather general chemical term applied to any reaction of a material with water, frequently in the presence of an alkali or acid, and sometimes at high temperatures and high pressures. In the same way that chain breaking is the reverse of addition polymerization, the hydrolytic degradation of polymers is formally the reverse of condensation polymerization. The term "solvolysis" extends the scope of "hydrolysis" to solvents other than water, such as alcohols.

The most interesting application of polymer hydrolysis in recycling is the suggestion that cellulose from paper and other organic waste materials can be converted economically into glucose (followed by fermentation to ethanol, for example). Cellulose, the major constituent of all trees and higher plants and therefore an extremely valuable potential source of carbon compounds, has the empirical formula $C_6H_{10}O_5$. When hydrolyzed with fuming hydrochloric acid, cellulose yields glucose in about 95% yield. This indicates that its structure is based on the glucose unit, but as it forms colloidal suspension in solvents in which glucose is soluble, cellulose is shown to be a very large polymeric molecule, and is usually written $(C_6H_{10}O_5)_n$:

$$\underset{\text{cellulose}}{(C_6H_{10}O_5)_n} + \underset{\text{water}}{nH_2O} \xrightarrow{\text{acid}} \underset{\text{glucose}}{nC_6H_{12}O_6}$$

Of course, some cellulosic materials are not chemically pure cellulose, but polymers of glucose with various amounts of other sugar units.

Hydrolysis of cellulose in acid solution has been used for cotton hulls and some woods, but the process of wood hydrolysis used in the United States during World War II was slow. Hydrolysis times in 0.5% sulfuric acid of the order of 3½ hours at 180°C were required, with an ultimate yield after fermentation of about 190 liters of ethanol per tonne of wood. Recent development of the idea is due to Porteous, who used a pressurized reactor to achieve a higher temperature, and suggested that the reaction be applied to organic waste disposal (Converse et al., 1973; Fagan et al., 1971a, 1971b; Grethlein, 1975; Porteous, 1969, 1975a). The sugar formed by hydrolysis undergoes further decomposition if the exposure to hot acid is continued, and Porteous, in optimizing the process, predicted that the best yield of sugar (55%) would be obtained in very short residence times (about a minute) at 230°C. Millett (1976) has suggested that rough separation of the complex lignocellullose (containing lignin, cellulose, and hemicellulose) could be carried out on the basis of different hydrolysis rates, thus facilitating the separation of various useful products, some of which are mentioned in Chapter 7.

An alternative way of breaking down cellulose molecules is to oxidize the cellulose first to dialdehyde cellulose and then to dicarboxyl cellulose before hydrolysis (Hearon et al., 1975). Suitable oxidation reagents are periodate (IO_4^-) for the first step and dinitrogen tetroxide (N_2O_4) for the second step; both reduction products of these reagents (iodate, IO^-_3, and nitric oxide, NO) can be oxidized and recycled.

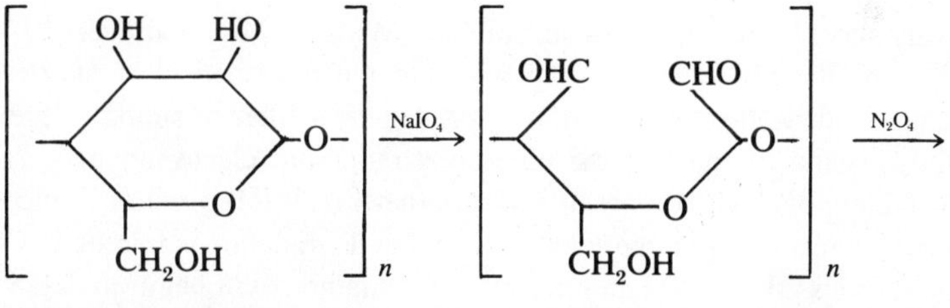

OH HO; -O-; O; CH_2OH; n $\xrightarrow{NaIO_4}$ OHC CHO; -O-; O; CH_2OH; n $\xrightarrow{N_2O_4}$

$(C_6H_{12}O_5)_n$, *cellulose*

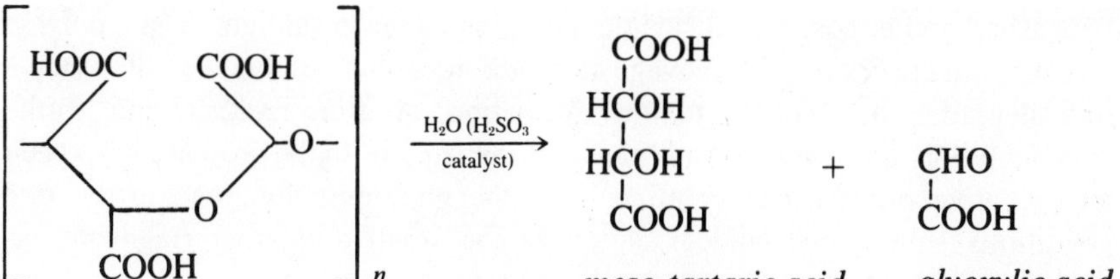

HOOC COOH; -O-; O; COOH; n $\xrightarrow[\text{catalyst)}]{H_2O\ (H_2SO_3}$ COOH–HCOH–HCOH–COOH + CHO–COOH

meso-tartaric acid *glyoxylic acid*

The glyoxylic acid can be converted into the more useful glycine by reaction with ammonia; with optimum conditions all reaction steps take place with nearly quantitative yields.

There is also a growing interest in the solvolysis of synthetic polymers, Taylor (1973) and Szilard (1973) having reviewed the reclamation of monomers and other products from condensation polymers by solvolysis. Poly(acrylonitrile) wastes hydrolyze to a water-soluble mixture of compounds containing amide and carbonyl groups (Verami and Mladenov, 1972), and the hydrolysis of polyurethane foam waste with steam at 200-350°C results in recovery of 2,4- and 2,6-toluene diamines, with a liquid residue consisting of approximately 98% polyproplyene oxide (Mahoney et al., 1974). It is probable that these products could be utilized economically (Campbell and Meluch, 1976; Sperber and Rosen, 1974). Polyurethane can be decomposed also by solvolysis with low molecular mass alcohols or amines (Campbell and Meluch, 1976). Considerable work has been carried out on the reclamation of the polyester poly(ethylene terephthalate), which has been depolymerized to low molecular mass fragments by dissolution in a mixture of the monomer and ethylene glycol at over 200°C for several hours. After separation, repolymerization can be carried out by standard methods.

Photolysis of Polymers

The study of photolytic degradation of polymers (chemical decomposition initiated by light) has been prompted by a desire on the part of a proportion of the public for degradable plastic containers, so that when they are littered they have relatively short lives and do not accumulate (Mark, 1974; Staudinger, 1974). The principle of the photochemical approach to the degradation of plastics is the application of the fact that window glass acts as a filter to sunlight, preventing the transmission of most of the ultraviolet radiation (electromagnetic radiation with wavelengths less than about 320 nanometers). If it is possible to incorporate chemical groups into polymer chains such that degradation is initiated by light of wavelength less than 320 nanometers, the containers will begin to degrade only when left outdoors (Evans, 1974; Gutfreund, 1971; Zerlaut and Stake, 1974). A variation on this approach is the use of a water-soluble ultraviolet-absorbing protective coating on ultraviolet-sensitive plastic. After use the container could be washed and exposed to ultraviolet radiation either in sunlight or in a polymer recovery plant. Photolytic cleavage of a carbon-carbon bond is the initial step in this degradation, leading to the formation of free radicals and further fragmentation: the reverse of addition polymerization. In the presence of oxygen, an oxidative process can continue after the photoinitiation, and in the right conditions subsequent biodegradation of the smaller polymer fragments can complete the process.

Other Methods of Depolymerization

There are several other methods of polymer degradation which may prove equally useful in the future, ranging from mastication (which is applied already to rubber reclamation) to various irradiation techniques with high energy radiation (which are at present little used) (Gutfreund, 1971; Taylor, 1973).

Bond breaking can be induced by *mechanical* means in mastication (''shear degradation''), and a related type of process occurs with *ultrasonic* degradation. Both processes result in degradation products with a narrow range of molecular masses. Mechanical milling is aided by *stress-cracking,* induced by some liquids that tend to promote brittleness (detergents had this effect on some of the early plastic kitchenware).

High-energy radiation (*ionizing radiation*), for example X-rays, or gamma rays from cobalt-60, can cause either bond cleavage and direct degradation, or cross-linking. In the latter case the rigid, brittle, three-dimensional framework formed by the radiolysis may be fractured easily by mechanical means. Poly(tetrafluoroethylene) is particularly susceptible to both shear wear and high-energy radiation, despite its chemical inertness (Arkles, 1973). The rapid and uniform heating of nonpolar plastics by *microwave* irradiation can be used in some circumstances for gasification (Tsutsumi, 1974).

The *biodegradation* of polymers is introduced in Chapter 5.

Synthetic Polymer Reutilization

Thermoplastic polymers may be re-extruded, in some cases afer degradation by one of the preceding methods. Sometimes the process is assisted by the addition of a small amount of an appropriate solvent. Short fibers can be prepared by solution spinning or melt spinning techniques applied to waste polymers or polymer mixtures, the short fibers being pressed into a fiberboard or paperlike nonwoven sheets. ''Graft copolymerization'' can be carried out. A catalyst or radiation is used to initiate radical formation on the existing polymer chains, resulting in chain cross-linking or copolymerization with groups introduced for special purposes. Thus wiping cloths could be manufactured by grafting cellulose or starch (to impart water absorptivity) or hydrocarbons (to enhance oil absorptivity) onto compressed fiber polymer sheet (IIT, 1971).

Inorganic Polymerization

There are several inorganic chemical systems that undergo reactions, either within the system itself or with other materials, to form polymers which by encapsulation or bonding impart useful properties to otherwise waste materials.

They include cements, mortars and pozzolan cements which solidify the material into a hardened mass; these are discussed in Chapter 7.

There are also newly developed processes based on sodium silicate. For example the Chemfix® process (Conner, 1974; Conner and Gowman) uses a two-part chemical system which reacts to form a stable solid, and also reacts with any additional polyvalent metal ions. The matrix is virtually a synthetic mineral, with tetrahedrally coordinated silicon atoms alternating with oxygen atoms in a linear "chain." The oxygen side groups on the chain bond strongly with polyvalent metal ions to form a cross-linked, three-dimensional polymer. These relatively insoluble metal silicates are resistant to leaching, and in fact there is evidence that the porous gel of silicate removes additional metal ions from solutions passing through it. The soluble silicate formulations are excellent bonding agents for various materials such as fly ash, papermill wastes, and sewage sludge (Falcone and Spencer, 1975).

OXIDATION AND REDUCTION PROCESSES

Originally, the term *oxidation* referred to the addition of oxygen to elements, and the term *reduction* meant reduction in the amount of oxygen associated with an element. Oxidation now has the broader meaning of *loss of electrons*. Thus the oxidation of the metal lithium is described by the equation

$$\underset{\textit{lithium}}{2Li} + \underset{\textit{oxygen}}{\tfrac{1}{2}O_2} \rightarrow \underset{\textit{lithium oxide}}{Li_2O}$$

If the formula for lithium oxide is written in its ionic form, it is clear that lithium has lost electrons during its oxidation:

$$\underset{\textit{lithium}}{2Li} + \underset{\textit{oxygen}}{\tfrac{1}{2}O_2} \rightarrow \underset{\textit{lithium ion}}{2Li^+} + \underset{\textit{oxide ion}}{O^{2-}}$$

Similarly, the reduction process in which the lithium ion, Li^+, is converted to lithium metal, Li, is associated wih a *gain of electrons* by lithium. The same principle applies to covalent (nonionic) compounds:

$$\underset{\textit{carbon}}{C(c)} + \underset{\textit{oxygen}}{O_2(g)} \underset{\text{reduction of carbon}}{\overset{\text{oxidation of carbon}}{\rightleftharpoons}} \underset{\textit{carbon dioxide}}{CO_2(g)}$$

(The symbols "c" and "g" denote "crystalline solid" and "gas.") Reduction of one material is usually carried out by another material which itself undergoes oxidation, and this material is often hydrogen or carbon monoxide:

$$\underset{\textit{hydrogen}}{2H_2(g)} + \underset{\textit{carbon}}{C(c)} \rightarrow \underset{\textit{methane}}{CH_4(g)} \qquad \text{(reduction of carbon)}$$

$$\underset{\textit{hydrogen}}{H_2(g)} + \underset{\textit{carbon monoxide}}{CO(g)} \rightarrow \underset{\textit{carbon}}{C(c)} + \underset{\textit{water}}{H_2O(g)} \qquad \text{(reduction of carbon)}$$

$$\underset{\textit{carbon monoxide}}{CO(g)} + \underset{\textit{water}}{H_2O(g)} \rightarrow \underset{\textit{carbon dioxide}}{CO_2(g)} + \underset{\textit{hydrogen}}{H_2(g)} \qquad \text{(reduction of hydrogen)}$$

The reversibility of the carbon-hydrogen-oxygen oxidation-reduction reactions enables one portion of a carbonaceous material to be oxidized in order to produce heat and reducing agents for reduction of the other portion. Various direct or indirect applications of these processes are summarized in this section.

Combustion (Incineration)

The simplest and most obvious way of recovering some of the values from any organic waste is by burning it for its energy content, and this may even be the most efficient utilization in many cases. It has an initial advantage in that the main reagent, oxygen, is available at low cost, and toxic organic contaminants, such as insecticides in crop wastes, are destroyed. Steam may be produced for heating and/or electricity generation, for desalinating water, and for drying the incoming fuel.

On a large scale, it has been suggested that coal refuse piles might be burned to provide both energy and a usable ash, but the main application has been in the disposal of urban waste. Heat recovery incineration techniques have been reviewed by many authors.* Problems arise in the incineration of synthetic polymers (plastics) which do not arise in the incineration of the natural carbohydrate polymers: the production of corrosive gases such as hydrogen chloride and chlorine from poly(vinyl chloride), the higher thermal output, and the

*For example, Baum and Parker, 1974a; Besselievre and Schwartz, 1976; Chamberlain and Müller, 1976; Chandler, 1976; Corey, 1969; Coulson, 1976; Cross, 1972; EPM, 1975c; NCRR, 1974c; Neal, 1971; Pavoni et al., 1975; Ross, 1968; Sebastian and Isheim, 1970; Skitt, 1972; Stephenson et al., 1975; Szpindler and Waters, 1976; Wilson, 1974a.

tendency of thermoplastics to clog grates. Recent innovations have included the use of fluidized beds and high-temperature molten salt and liquid metal reactors, which are discussed below.

Inorganic materials sometimes may be recovered by oxidation, for example the recovery of chlorine in the reaction

$$2FeCl_3(c) + \tfrac{3}{2}O_2(g) \rightarrow Fe_2O_3(c) + 3Cl_2(g)$$

ferric chloride *oxygen* *ferric oxide* *chlorine*

In the wood-pulp industry, a considerable proportion of the pulping chemicals can be recovered after incineration of the black liquor. Nonferrous coatings on sheet steel are often volatilized or burnt off, ideally with the dust then being collected. The lime coke left after the refining of sugar can be regenerated by calcination at 950-1000°C (Omori, 1975). "Destructive oxidation" of ferrous scrap involves the oxidation at 1000°C of low-value ferrous scrap from cans, appliances, and even cars, to high-grade iron oxide suitable for subsequent processing; alternatively the ferrous scrap can be used as a reducing agent in the upgrading of the nonmagnetic taconite ore to a magnetic, separable form of iron ore (USDI, 1972; Kenahan, 1971).

Pyrolysis

Pyrolysis or "destructive distillation" is chemical decomposition of a carbonaceous material by heat (usually 400-900°C) in the absence of oxygen, or in a controlled oxygen environment. The process has been used for many hundreds of years, notably to convert coal to coke for the iron-making industry and for town-gas production. A general schematic diagram for the process is shown in Figure 4.3.

Pyrolysis leads to the formation of four phases in relative amounts which depend on the pyrolysis conditions:

1. A *gas* component, including such low molecular mass compounds as hydrogen, methane, carbon monoxide, and carbon dioxide;
2. An *oil,* including compounds such as organic acids, alcohols, ketones, and furfural, with molecular masses in general lower than those of the starting material;
3. A *char,* containing carbon and any inert inorganic compounds that were present; and
4. An *aqueous* phase containing some water-soluble organic compounds.

In general, the more rapid the decomposition (high temperature, high heating rate) the higher the yield of gas; a long, slow heating process results in higher

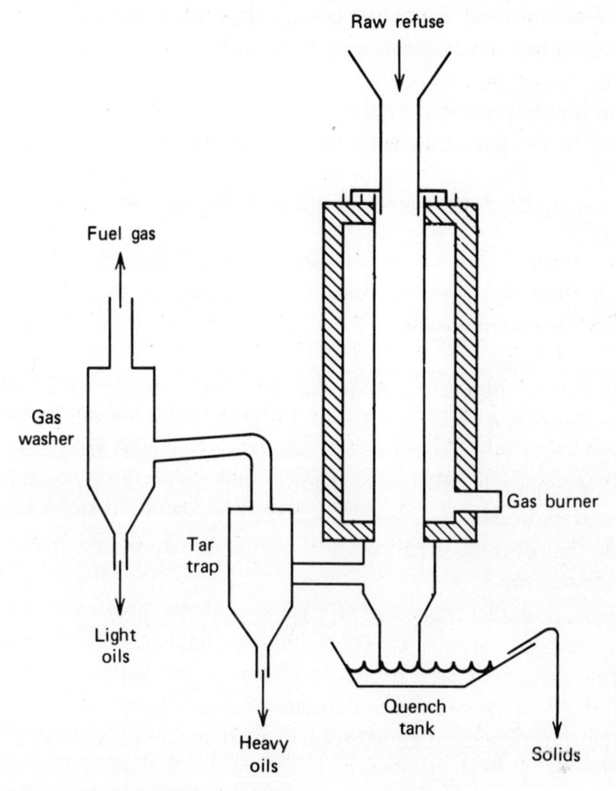

Figure 4.3. Schematic diagram of a pyrolysis reactor. (From WSL, 1972a. Crown copyright; reproduced by permission of the Director of Warren Spring Laboratory.)

proportions of oil and char. The mechanism of this process is complex, but a start can be made in understanding it by studying the thermal decomposition of cellulose; similar reactions occur in synthetic polymers.

On rapid heating in the absence of oxygen the cellulose molecule "explodes," and the fragments form many simple compounds including methane, carbon dioxide, hydrogen, carbon monoxide, and water (Burton and Bailie, 1974; Shafizadeh, 1971, 1975; Shafizadeh and Fu, 1973; Shafizadeh et al., 1973). At lower heating rates the molecules have sufficient time to reorganize into more thermally stable liquids, and eventually into solids that become increasingly difficult to break down.

As the temperature of wood is increased slowly in the absence of air, four distinct phenomena occur:

1. Up to 200°C the wood surface becomes dehydrated, and water with traces

of carbon dioxide and volatile organic compounds (formic acid, acetic acid, glyoxal) are evolved;

2. Up to 300°C these products are evolved in substantial quantities and the wood is converted into a char;
3. Up to 500°C the exothermic (heat liberating) reactions take place, and a variety of products including carbon monoxide, hydrogen, methane, formaldehyde, formic acid, acetic acid, ethanol, and hydrocarbon tars are formed;
4. Over 500°C some of these gaseous products (particularly water and carbon dioxide) can react with the residual char to yield further carbon monoxide and hydrogen.

The sequence of processes is similar for all carbonaceous material, and is indicated in Figure 4.4. The first part of this process (pyrolysis: the production of gas, oil, and char) is common to all methods of thermal treatment of organic materials. Subsequently, different reactions with steam, or hydrogen, or carbon monoxide and hydrogen are possible. Another possibility is that some oxygen is admitted, so that drying, pyrolysis, and partial oxidation take place in different regions of the reactor.

The various possible reactions of carbon with oxygen and the other gases are summarized in Table 4.1. The enthalpy change, ΔH, for each reaction indicates the extent to which that reaction is exothermic (producing heat: negative enthalpy change) or endothermic (absorbing heat: positive enthalpy change). The significance of enthalpy changes is discussed further in Chapter 8.

Table 4.1. Reactions Involving Carbon, Hydrogen, and Oxygen

		ΔH (kilojoules per mole)	
combustion	$C + \frac{1}{2}O_2(g) \rightleftharpoons CO(g)$	−111	(1)
	$C + O_2(g) \rightleftharpoons CO_2(g)$	−394	(2)
	$C + CO_2(g) \rightleftharpoons 2CO(g)$	172	(3)
carbon-steam gasification	$C + H_2O(g) \rightleftharpoons CO(g) + H_2(g)$	175	(4)
	$C + 2H_2O(g) \rightleftharpoons CO_2(g) + 2H_2(g)$	178	(5)
hydrogasification	$C + 2H_2(g) \rightleftharpoons CH_4(g)$	−75	(6)
	$CO(g) + \frac{1}{2}O_2(g) \rightleftharpoons CO_2(g)$	−283	(7)
water-gas shift	$CO(g) + H_2O(g) \rightleftharpoons CO_2(g) + H_2(g)$	3	(8)
methanation	$CO(g) + 3H_2(g) \rightleftharpoons CH_4(g) + H_2O(g)$	−250	(9)

Source. McIntyre and Papic, 1974.

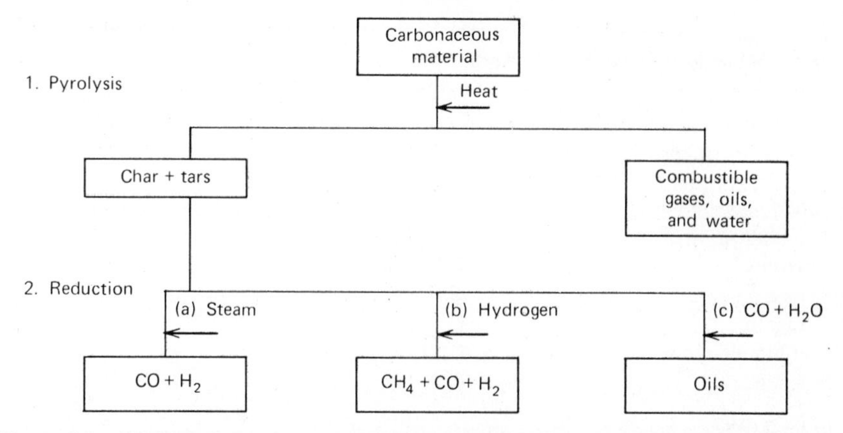

Figure 4.4. Methods for the thermal treatment of carbonaceous material. (After McIntyre and Papic, 1974. This figure was published originally in *The Canadian Journal of Chemical Engineering*, April 1974, pp. 268-272, "Pyrolysis of Municipal Solid Waste.")

Reactions (1) and (2), associated with combustion in an oxygen-deficient or oxygen-rich atmosphere, respectively, provide the thermal energy for subsequent endothermic reactions. The figures quoted are for 500°C; at higher temperatures the enthalpy change values differ. All reactions are reversible, and an increase in temperature shifts the equilibrium position of endothermic reactions (3)-(5) to the right. An increase in pressure causes a tendency to shift the reaction in the direction of decreasing number of moles of gas, so reactions (6) and (9) are shifted towards the right. At higher temperatures (800-1200°C) the reaction rates become very rapid, and are controlled by the mass transfer (diffusion) of reactant gases to interfaces between phases.

Treatment of the char with steam, reactions (4) and (5), increases the yield of gaseous products; reaction (4) produces *synthesis gas*. For many purposes, it is preferable for char and liquids to be converted as far as possible to gases so that only one fuel recovery or combustion method is required. This may be achieved more completely, either directly with hydrogen, equation (6), or with carbon monoxide and steam, equation (8), or by recycling the liquid product through the reactor. The key factor in producing a high-energy pyrolysis gas without additional hydrogenation, which is expensive, is a high hydrogen/carbon ratio in the fuel. The empirical carbon-hydrogen-oxygen formulas and hydrogen/carbon mole ratios for a number of fuels are listed in Table 4.2. It is apparent that, based on their potential to produce high hydrogen content fuel gases, the solid wastes are superior to the solid fuels in current use, and in fact they have hydrogen/carbon ratios approaching those of liquid hydrocarbons.

Some aspects of the thermal degradation of synthetic polymers were considered in the previous section. Subsequent processes resemble those just

Table 4.2. Compositions for Various Fuels and Solid Wastes Expressed as $C_6H_xO_y$ and as Hydrogen/Carbon Mole Ratio

	$C_6H_xO_y$	Hydrogen/Carbon Mole Ratio
Solid Fuels		
Cellulose	$C_6H_{10}O_5$	1.67
Wood	$C_6H_{8.6}O_4$	1.43
Peat	$C_6H_{7.2}O_{2.6}$	1.20
Lignite	$C_6H_{6.7}O_2$	1.10
Bituminous coal	$C_6H_4O_{0.53}$	0.67
Anthracite	$C_6H_{1.5}O_{0.07}$	0.25
Solid Wastes		
Urban waste	$C_6H_{9.6}O_{3.8}$	1.61
Newspaper	$C_6H_{9.1}O_{3.9}$	1.52
Plastic film	$C_6H_{10.4}O_{1.1}$	1.73
Domestic Waste	$C_6H_{9.9}O_{3.0}$	1.66

Source. Burton and Bailie, 1974

described for cellulose, but one difference is that inorganic acid gases from plastics can cause problems, particularly hydrogen chloride from poly(vinyl chloride).

The pyrolysis method has been applied to a variety of organic waste materials (Baum and Parker, 1974b), including tree bark (Finney and Sotter, 1975), sawdust and other cellulosics (Barrett, 1975), plastics (Menzel et al., 1973), scrap tires, feedlot manure (Engler et al., 1975), and urban waste. Urban waste pyrolysis systems are described further in Chapter 6. Pyrolysis seems particularly well suited to rubber tire disposal; one old tire can produce about enough carbon black for one new tire, as well as oil and gas (Collins et al., 1974).

Hydrogasification (Hydrogenation)

A common method of reduction of organic material is direct reaction with hydrogen in the presence of a catalyst at sufficiently high temperatures (500-700°C) and pressures (7-18 megapascals, 1000-2700 pounds per square inch) to produce methane, as in reaction (6) above, for example. Carbon monoxide produced during the decomposition of organics can be "shifted" to carbon dioxide and hydrogen, reaction (8), and the carbon dioxide then scrubbed out or removed by "methanation," reaction (9):

$$\underset{carbon}{C(c)} + \underset{hydrogen}{2H_2(g)} \rightleftharpoons \underset{methane}{CH_4(g)} \tag{6}$$

$$\underset{\text{carbon monoxide}}{CO(g)} + \underset{\text{water}}{H_2O(g)} \rightarrow \underset{\text{carbon dioxide}}{CO_2(g)} + \underset{\text{hydrogen}}{H_2(g)} \quad (8)$$

"water-gas shift"

$$\underset{\text{hydrogen}}{3H_2(g)} + \underset{\text{carbon monoxide}}{CO(g)} \rightarrow \underset{\text{methane}}{CH_4(g)} + \underset{\text{water}}{H_2O(g)} \quad (9)$$

"methanation"

This yields "substitute natural gas" (SNG) containing mainly hydrogen, methane, and ethane, and suitable for pipeline distribution, gas turbine operation, or (with a higher hydrogen content) in hydrogen fuel cells. In the United States, the Bureau of Mines has investigated hydrogasification processes for coal and various organic wastes (Steffgen, 1974; Wender et al., 1974; Wiser and Anderson, 1975). In this process, there is no liquid product, and the small amount of char formed (less than 5%) is added to a synthesis gasifier which forms synthesis gas from a portion of the waste, reaction (4):

$$\underset{\text{carbon}}{C(c)} + \underset{\text{water}}{H_2O(g)} \rightarrow \underset{\text{carbon monoxide}}{CO(g)} + \underset{\text{hydrogen}}{H_2(g)} \quad (4)$$

The carbon monoxide in this synthesis gas is then shifted to hydrogen for use in the hydrogasifier. The carbon conversion during hydrogasification cannot exceed about 50%, since some of the carbon is required for the production of hydrogen. Expensive high-pressure equipment is necessary, but the overall reaction is strongly exothermic so a high moisture content in the waste is acceptable without the addition of external heat, and a gas of high calorific value is produced.

Liquefaction with Steam and Carbon Monoxide

Cellulosic material as a fuel is greatly increased in value if it is treated in such a way that a low oxygen content liquid product is obtained, with the oxygen being removed as carbon dioxide (decarboxylation) rather than as water (dehydration), so that the hydrogen/carbon mole ratio remains high. The reduction of carbonaceous material in a waste or oil slurry with steam and carbon monoxide produces a high calorific value heavy fuel oil with a typical empirical molecular formula of $C_6H_{6.5}O_{0.5}$. The reaction occurs at temperatures of 250-400°C, which are significantly lower than those necessary for hydrogasification, but at pressures which are rather greater because of the steam pressure.

The U.S. Bureau of Mines Pittsburgh Energy Research Center (Appell and Miller, 1973; Appell et al., 1975; CE, 1971; CEN, 1971a; Franklin et al., 1973a, 1973b; Steffgen, 1974; Wender et al., 1974; Wiser and Anderson, 1975) and the

Worcester Poytechnic Institute (Kaufman and Weiss, 1975) have been investigating the process at a bench-scale level for processing sewage sludge, manure, wood wastes, and urban waste.

As in the hydrogasification process, the carbon monoxide may be produced from organic materials in another reactor, and need not be pure. Synthesis gas, a mixture of carbon monoxide and hydrogen, is satisfactory. An alkaline catalyst such as sodium carbonate or bicarbonate is necessary, although at higher temperatures the natural alkalinity of some organic waste is adequate. It is considered that there may be direct reaction of bicarbonate ion with carbon monoxide to yield formate ion:

$$\underset{\textit{bicarbonate}}{HCO_3^-} + \underset{\textit{carbon monoxide}}{CO} \rightarrow \underset{\textit{formate}}{HCOO^-} + \underset{\textit{carbon dioxide}}{CO_2}$$

Adjacent hydroxyl groups, in carbohydrates for example, undergo dehydration, followed by isomerization to ketones and reduction by formate ion:

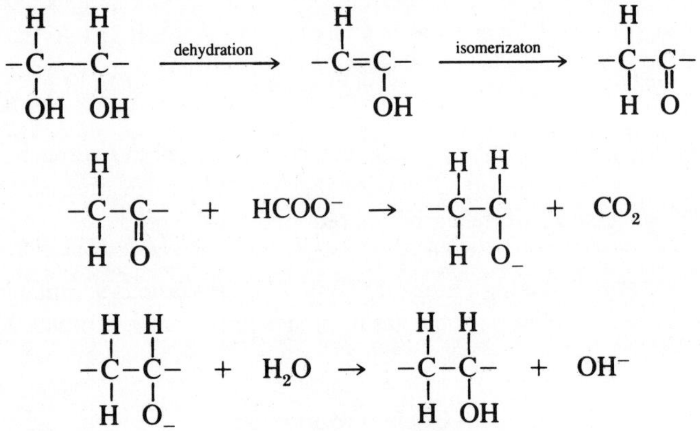

$$\begin{array}{cc} H & H \\ | & | \\ -C- & -C- \\ | & | \\ OH & OH \end{array} \xrightarrow{\text{dehydration}} \begin{array}{cc} H & \\ | & \\ -C= & C- \\ & | \\ & OH \end{array} \xrightarrow{\text{isomerizaton}} \begin{array}{cc} H & \\ | & \\ -C- & C- \\ | & \| \\ H & O \end{array}$$

$$\begin{array}{cc} H & \\ | & \\ -C- & C- \\ | & \| \\ H & O \end{array} + HCOO^- \rightarrow \begin{array}{cc} H & H \\ | & | \\ -C- & C \\ | & | \\ H & O^- \end{array} + CO_2$$

$$\begin{array}{cc} H & H \\ | & | \\ -C- & C- \\ | & | \\ H & O^- \end{array} + H_2O \rightarrow \begin{array}{cc} H & H \\ | & | \\ -C- & C- \\ | & | \\ H & OH \end{array} + OH^-$$

The overall effect is the replacement of a hydroxyl group by a hydrogen atom. Carbon monoxide then reacts with hydroxyl ion (possibly through bicarbonate ion as indicated previously) to regenerate formate:

$$OH^- + CO \rightarrow HCOO^-$$

This suggested mechanism explains why carbon monoxide is more effective than hydrogen in producing an oil rather than a gaseous reduction production, but is not yet conclusively established.

Gasification ("pyrolysis-incineration")

If carbonaceous material is introduced at the top of a fixed bed reactor, and some oxygen-containing gas (air, oxygen, steam) is introduced at the bottom, the organic material progressively undergoes the following processes as it moves down the reactor (Eggen and Kraatz, 1974a):

1. Drying;
2. First-stage decomposition, with the evolution of water, carbon dioxide, and carbon monoxide;
3. Tar formation;
4. Char formation, with evolution of hydrogen, methane, and ethane;
5. Char oxidation by the introduced oxygen or steam, or by carbon dioxide.

This, in effect, combines the two processes of pyrolysis and hydrogasification in the one reactor. Heat transfer and gas transport rates in the pyrolysis region are relatively low, so the pyrolysis process is essentially the same as that which occurs in an externally heated vessel under a slow heating rate. It has been suggested that producer gas could be generated in this way from coconut shells, thus improving the efficiency of some tropical island industries (Cruz, 1975); and the process is being investigated for the recovery of energy from forest residuals and municipal solid waste (Hammond et al., 1972, 1974).

If the organic material contains substantial amounts of water, clinkering (partial melting) of the ash does not occur, but otherwise in some cases the temperature reached may be high enough to melt any inorganic residue, and this feature is utilized in some high-temperature gasification systems. The output gas can be upgraded to a substitute natural gas (particularly if oxygen rather than air is used as oxidant) and used in the same way as the hydrogasification product.

Wet Combustion and Partial Oxidation

Wet combustion is a process in which organic material is pumped, together with water and air, into a pressurized reactor vessel (3.5-8 megapascals, 500-1200 pounds per square inch) at moderate temperatures (200-260°C). External heat is needed only to start the process; after that the rapid oxidation heats the water directly into steam, and also yields potentially valuable organic acids. Wet combustion has been applied to sewage sludge.

Another partial oxidation process involves bringing the finely divided solid waste into contact with an air-nitrogen mixture in a fluidized bed to produce gases, organic liquids, an aqueous phase, and a small tar fraction. A complex mixture of organic compounds results, the composition depending on the extent of oxidation, the limiting situation with no air used being that of pyrolysis (Skitt, 1972).

Hydrogen as a Chemical Feedstock

The hydrogen derived from pyrolysis or partial oxidation, after water gas shift, can be utilized as a chemical rather than as a fuel. The energy of waste materials is thus converted into products like methanol or an ammonia-based fertilizer by an enzymatic or chemical catalytic synthesis. As another example, it has been shown that sponge iron can be manufactured from iron oxide wastes by reduction with gases generated by the pyrolysis of shredded rubber tires (Dean and Valdez, 1972).

Reactors

Many thermal solid-gas reaction processes are carried out in *fixed-bed* reactors, of either batch or continuous type (Eggen and Kraatz, 1974a; McIntyre and Papic, 1974; Pavoni et al., 1975). In the case of continuous processes, which are becoming more popular, the fuel enters the top of the reactor and moves slowly down a vertical shaft or inclined bed through the reaction zones, while air or oxidizing gas and hot reaction products move upward.

Rotary kiln reactors overcome problems of fixed bed reactors which arise through channeling and uneven reaction, but are more expensive. The operating principles of *fluidized beds* were outlined in Chapter 3, where their use in mechanical separation was noted. The same technique is used in reaction processes, including incineration of coal (Piper, 1976) and high-temperature oxidation-reduction processing of solid wastes (Bailie and Burton, 1972; Bailie and Ishida, 1972; EPM, 1975c; Fisher, 1976; McIntyre and Papic, 1974; Szpindler and Waters, 1976). The beds are normally preheated by the burning of a gaseous fuel before the finely divided solid is introduced. Fluidized beds have the advantages of:

- High thermal capacity;
- High heat transfer rates by conduction-convection, leading to minimal formation of nitrogen oxides;
- Efficient gas-solid reaction, requiring less air or oxidant;
- Efficient agitation and mixing of the solid;
- Simple construction.

Disadvantages include:

- Temperature limitation due to softening of the fluidized solid (for example, 1100°C for a sand bed);
- Greater blower requirements than a fixed bed;
- High carry-over of solids with gas product;
- The need for size reduction before reaction.

Some applications of fluidized beds to the thermal treatment of urban waste are reviewed in Chapter 6. Fluidized bed incineration is one form of *suspension firing*. Another approach, *vortex incineration,* burns the fuel as it is blown tangentially into the combustion chamber, reducing residence time and improving efficiency relative to a fixed bed process.

A more direct approach to a high-temperature suspension "fluid" is the use of a high-temperature liquid. Thus, a *molten metal* such as lead may be used for pyrolysis (Brown, 1973; Crane et al., 1975). More commonly, an *ionic liquid (molten salt)* is used: various combinations of salts such as alkali metal halides, nitrates and carbonates (Kerridge, 1975). Examples include the oxidation or pyrolysis of wastes (Cover and Schreiner, 1975; Greenberg, 1972; Greenberg and Whitaker, 1972; Hammond and Mudge, 1975; Lessing, 1973); the removal of surface coatings (Klötgen, 1974); the pyrolysis of scrap tires (CEN, 1976b) and plastics (Menzel et al., 1973); and the electrorecovery of nickel and cobalt from their alloys (CE, 1968) and of gold from scrap electronic solders (Kleespies et al., 1970). In the case of oxidation, it is believed that a reactive oxidizing agent is present in small, equilibrium quantities in some melts. In other cases there are catalytic species present, and in electrorecovery the molten salt is acting as a highly conducting, mobile electrolytic liquid. As a final example of a high temperature liquid reactor, it is interesting to note that carbonaceous fuels and wastes in principle can be used for direct electric power generation by means of a liquid lead electrode and a molten carbonate salt electrolyte (Anbar et al., 1975).

HYDROMETALLURGY AND PYROMETALLURGY

In *hydrometallurgy,* which is widely used in the processing of copper and nickel ores for example, but which can be applied equally well to many wastes, the materials are dissolved in a suitable solvent and then separated by a variety of chemically based operations (Fletcher, 1976; Sittig, 1975). The solvent is often water, but may just as well be an organic liquid such as acetonitrile (Parker, 1972, 1973a) or a specially developed solvent (Fletcher, 1971, 1973; Lawson,

1975). The usual recovery scheme for metals is acid dissolution, separation of metal ions from solution, and recovery in the form of metal, oxide, or salt as appropriate (see Figure 2.3, for example). The method can be applied to complex alloy scrap such as nonferrous automobile shredder refuse (Coyle et al., 1976), bearing scrap, electrochemical machining sludges, and metal production and finishing wastes (De Cuyper, 1973). Solvent extraction is described in Chapter 3, and some examples of these processes are quoted in Chapter 7.

Pyrometallurgy (extraction of metals by processes requiring the application of heat) is a well established method of obtaining metals from ores, and in some fields is highly developed technologically, even if the scientific bases of the methods are not always fully understood. In some cases, only minor modifications to existing plants are necessary for pyrometallurgy to process waste materials from other industries or to more fully utilize their own by-products (De Cuyper, 1973; Sittig, 1975). Particular areas where progress is being made are electric furnace steel-making and foundry pigiron manufacture from cans and from automobile and large appliance scrap; and carbothermic reduction (heating with coke) of copper smelting slags to recover copper, tin, nickel, zinc, and iron. These and other examples are discussed in more detail in other chapters.

NUCLEAR FUSION TORCH

It has been suggested (CE, 1969; Eastlund and Gough, 1970; Gough and Eastlund, 1971) that a stream of exceedingly high-energy plasma from a nuclear fusion process could be used to ionize wastes and break them down to their elementary forms for collection. This idea has been criticized by Rose et al. (1972) for a number of reasons, including very low energy efficiency, and it appears to be a ''science fiction'' method at present.

5

MICROBIOLOGICAL RECYCLING

The breakdown and reformulation of organic materials by microorganisms may be either harmful or beneficial from the human point of view: Yen (1974b) distinguishes biodeterioration (the process of any undesirable change in materials caused by microbial activities) from biodisintegration (a useful or beneficial change when material is subjected to the actions of microorganisms). The relative roles played by chemical processes, biodeterioration, biodisintegration, and biosynthesis may be summarized in a cycle:

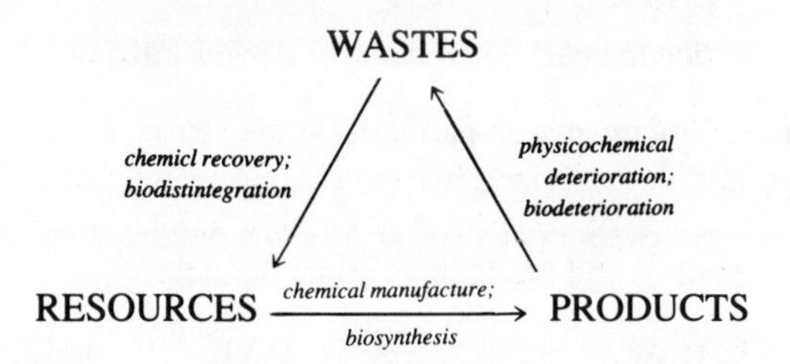

Microbial attack on materials is a continuing problem (Gilbert and Lovelock, 1975; Upsher, 1976), but biosynthesis is a means of meeting the increasing world demand for protein and energy, with waste materials providing a considerable proportion of the microbiological nutrients. The subject of microbiological chemical conversion is discussed in many reviews and books; one useful introductory account is the chapter in ''Environmental Chemistry'' by Manahan (1972); detailed mechanisms are considered by Stanier et al. (1971); and there are comprehensive works on industrial microbiology, for example those of Rainbow and Rose (1963), Casida (1968) and Beesch and Tanner (1974). Only an outline is presented here.

Bacterial cells are small in size (generally 0.3-3 micrometers), so their surface-to-volume ratios are high, and the insides of the cells are very accessible to the media in which they are located. Consequently, bacteria are capable of carrying out rapid chemical transformations. Fungi range downward in size from large, complex toadstools through molds to the microscopic, unicellular yeasts, but are generally much larger than bacteria. All organisms apart from those which utilize solar radiation directly by photosynthesis (algae and a few bacteria) are dependent on the availability of a suitable chemical energy source. A few bacteria (described as ''chemoautotrophs'') can obtain energy by the oxidation of reduced inorganic compounds, but most nonphotosynthetic organisms (bacteria and fungi) are ''chemoheterotrophs,'' requiring organic compounds as energy sources. They are classified as ''reducers,'' because they break down chemical compounds to more simple forms in order to extract energy. The number of microbial metabolites (chemical substances formed by microbiological action) already known is impressive (Miller, 1961), but only the commonest of these are mentioned here.

Microbial processes may be classified as aerobic (with oxygen) or anaerobic (without oxygen). Strict (or obligate) anaerobic microorganisms such as methanogenic (methane-producing) bacteria (Taylor, 1975b) are dependent on an oxygenfree environment in order to function, and they obtain energy from fermentation processes such as those shown here:

$$\underset{\textit{hydrogen}}{4H_2} + \underset{\textit{carbon dioxide}}{CO_2} \rightarrow \underset{\textit{methane}}{CH_4} + \underset{\textit{water}}{2H_2O}$$

$$\underset{\textit{acetic acid}}{CH_3COOH} \rightarrow \underset{\textit{methane}}{CH_4} + \underset{\textit{carbon dioxide}}{CO_2}$$

Many other microorganisms, in particular some yeasts, are called ''facultative anaerobes,'' and are characterized by two alternative mechanisms for obtaining energy. In the presence of oxygen they employ aerobic respiration, breaking down carbon compounds and producing carbon dioxide:

$$\underset{\textit{carbohydrate}}{C_6H_{12}O_6} + \underset{\textit{oxygen}}{6O_2} \rightarrow \underset{\textit{carbon dioxide}}{6CO_2} + \underset{\textit{water}}{6H_2O}$$

If no free oxygen is present, these microorganisms can carry out fermentations, such as the conversion of sugar to carbon dioxide and ethanol:

$$\underset{\textit{carbohydrate}}{C_6H_{12}O_6} \rightarrow \underset{\textit{carbon dioxide}}{2CO_2} + \underset{\textit{ethanol (ethyl alcohol)}}{2CH_3CH_2OH}$$

The yeast *Saccharomyces cerevisiae,* used (in different varieties) for both wine making and bread making, is the best-known example of facultative anaerobic microorganism (Harrison, 1963).

In fermentation processes, the ''average'' oxidation level of the end products is the same as that of the original material (substrate). Although some products are more oxidized and some less oxidized than the substrate, the total amount of carbon, hydrogen, and oxygen is unchanged. For example, in alcoholic fermentation the process is:

$$\underset{carbohydrate}{C_6H_{12}O_6} \rightarrow \underset{\substack{\text{carbon dioxide}\\ \text{(carbon in a more}\\ \text{oxidized form)}}}{2CO_2} + \underset{\substack{\text{ethanol (ethyl alcohol)}\\ \text{(carbon in a less}\\ \text{oxidized form)}}}{2CH_3CH_2OH}$$

In practice, this means that in fermentation we are sacrificing one portion of the substrate in order to obtain another portion in a more useful form. Fermentation of a solid carbonaceous waste can yield a smaller amount of either liquid ethanol or gaseous methane, both of which are more valuable fuels.

Organic wastes, whether of agricultural, industrial, or urban origin, are mixtures of carbohydrates and cellulose, as well as other compounds such as lignin, proteins, and hydrocarbons. Almost every carbohydrate and carbohydrate derivative can be used as a fermentable energy source by some microorganisms:

- *Hexoses,* sugars such as glucose and fructose;
- *Pentoses,* sugars such as arabinose and xylose;
- *Disaccharides,* sugars such as lactose, maltose, and sucrose;
- *Hemicelluloses,* polysaccharides made up of the order of 100 sugar units, which are fairly susceptible to attack by microorganisms;
- *Cellulose,* polysaccharides of 1000 to 10,000 sugar units, more resistant to microorganisms;
- *Sugar acids,* such as gluconic acid and glucuronic acid;
- *Polyalcohols,* such as mannitol and glycerol.

Glucose is fermented by almost all organisms capable of using any carbohydrate, and the two most widely distributed glucose fermentations are the alcoholic and homolactic fermentations. Both involve the intermediate formation of pyruvic acid by what is known as the Embden-Meyerhof pathway:

$$\underset{glucose}{C_6H_{12}O_6} \rightarrow \underset{pyruvic\ acid}{2CH_3\overset{}{\underset{\|}{C}}\underset{\|}{C}OH}$$

($2CH_3C(=O)C(=O)OH$)

Subsequent steps are:

$$\underset{pyruvic\ acid}{CH_3C(=O)C(=O)OH} \xrightarrow{\text{homolactic fermentation}} \underset{lactic\ acid}{CH_3CH(OH)C(=O)OH}$$

or:

$$\underset{pyruvic\ acid}{CH_3C(=O)C(=O)OH} \xrightarrow{\text{alcoholic fermentation}} \underset{carbon\ dioxide}{CO_2} + \underset{ethanol}{CH_3CH_2OH}$$

Various other products are possible; Table 5.1 lists some bacterial sugar fermentations that follow the pyruvic acid route.

Table 5.1. Bacterial Sugar Fermentations which Proceed through the Embden-Meyerhof Pathway and the Formation of Pyruvic Acid

	Class of Fermentation	Principle Products from Pyruvic Acid
1	Homolactic	$CH_3CHOHCOOH$, lactic acid
2	Mixed acid	$CH_3CHOHCOOH$, lactic acid
		CH_3COOH, acetic acid
		$HOOCCH_2CH_2COOH$, succinic acid
		$HCOOH$, formic acid (or $CO_2 + H_2$)
		CH_3CH_2OH, ethanol
2a	Butanediol	As in 2, but also
		$CH_3(CHOH)_2CH_3$, 2,3-butanediol
3	Butyric acid	$CH_3CH_2CH_2COOH$, butyric acid
		CH_3COOH, acetic acid
		$CO_2 + H_2$, carbon dioxide and hydrogen
3a	Butanol-acetone	As in 3, but also
		$CH_3CH_2CH_2CH_2OH$, butanol
		CH_3CH_2OH, ethanol
		CH_3COCH_3, acetone
		$CH_3CHOHCH_3$, isopropanol
4	Propionic acid	CH_3CH_2COOH, propionic acid
		CH_3COOH, acetic acid
		$HOOCCH_2CH_2COOH$, succinic acid
		CO_2, carbon dioxide

Source. Stanier, Doudoroff, and Adelberg, *The Microbial World,* 3rd ed., © 1970, p. 183. Reprinted by pemission of Prentice-Hall, Inc., Englewood Cliffs, New Jersey, U.S.A.

In some cases, more complex compounds such as cellulose are intially hydrolyzed (broken down) by an enzyme (cellulase in the case of cellulose) secreted by the bacteria. There are also highly specialized aerobic cellulose-decomposing bacteria, for example *Cellulomonas flavigena* (Callihan, 1975), in which the cellulase enzyme is associated with the cell surface and cannot diffuse away from it.

Proteins, the nitrogenous substances made up of amino acids which occur in the protoplasm of all plant and animal cells, are decomposed by aerobic micro-organisms with the liberation of ammonia, as well as organic acids, alcohols, and carbon dioxide, for example:

$$\underset{\displaystyle CH_2OH}{\overset{}{H_2NCHCOOH}} \xrightarrow{\text{hydrolysis and deamination}} \underset{\displaystyle CH_3}{O{=}CCOOH} + NH_3$$

Protein decomposition under anaerobic conditions (putrefaction) leads to the production of amines, mainly by anaerobic, spore-forming bacteria. Lignin, the very tough cell wall material in higher plants, is made up of aromatic units linked by aliphatic side chains and is extremely resistant to microbiological attack. Lipids (fats) yield alcohols and aliphatic (fatty) acids on aerobic fermentation:

$$\begin{array}{lcccl} CH_2OCR_1 \ (C{=}O) & & CH_2OH & & R_1COOH \\ | & & | & & \\ CHOCR_2 \ (C{=}O) & + 3H_2O \rightarrow & CHOH & + & R_2COOH \\ | & & | & & \\ CH_2OCR_3 \ (C{=}O) & & CH_2OH & & R_3COOH \\ \textit{fat} & & \textit{glycerol} & & \textit{aliphatic acids (fatty acids)} \end{array}$$

Even hydrocarbons are fermented by some bacteria. Oil spills on land and sea are removed eventually in this way, and rapid deterioration of some asphalt roads and roof sealers is attributed to the same mechanism.

In natural situations, aerobic composting processes in plant and animal wastes result in the evolution of carbon dioxide and ammonia (as well as other gases in trace amounts) and the ultimate formation of humus. Anaerobic processes, occurring under water in marshes (and in the gut of animals) produce methane, carbon dioxide, and peat (or manure). If the aim is simply to reduce the amount of organic waste matter, many processes are available. The products, with various possible degrees of subsequent treatment, are then potential soil fertilizers and conditioners. On the other hand, if specific degradation products (methane, or alcohol, for example) are to be recovered, much closer controls on the variables are required to guide and hasten the natural decay processes. Air-tight digesters can replace animal intestines for anaerobic processes, and open compost piles can be replaced by aerated bins or rotating drums. In both cases there is an optimum ratio of carbon to nitrogen of between 15 to 1 and 30 to 1 for most efficient performance of the microorganisms.

COMPOSTING

The conversion of biodegradable material into soil conditioners was one of the original methods of waste treatment: attractive because it combined disposal with reclamation in a "natural" manner which required no complex equipment or technical knowledge (Biddlestone and Gray, 1973). Although there have been successful large-scale plants in operation to convert urban waste into compost (Cardenas and Varro, 1973; Hays, 1973; Meyer, 1972; Prescott, 1967; Wiley et al., 1966), there are still practical problems which have not been solved completely (see Chapter 6). There appears to be only limited interest in the application of composting to organic industrial wastes, other than those of direct agricultural origin (Troth, 1973). In addition to its use in soils, compost can be used also as a fuel, and as an animal feed supplement.

The important parameters in the composting process are well established (Biddlestone and Gray, 1973; Gray and Biddlestone, 1973; Skitt, 1972). If the

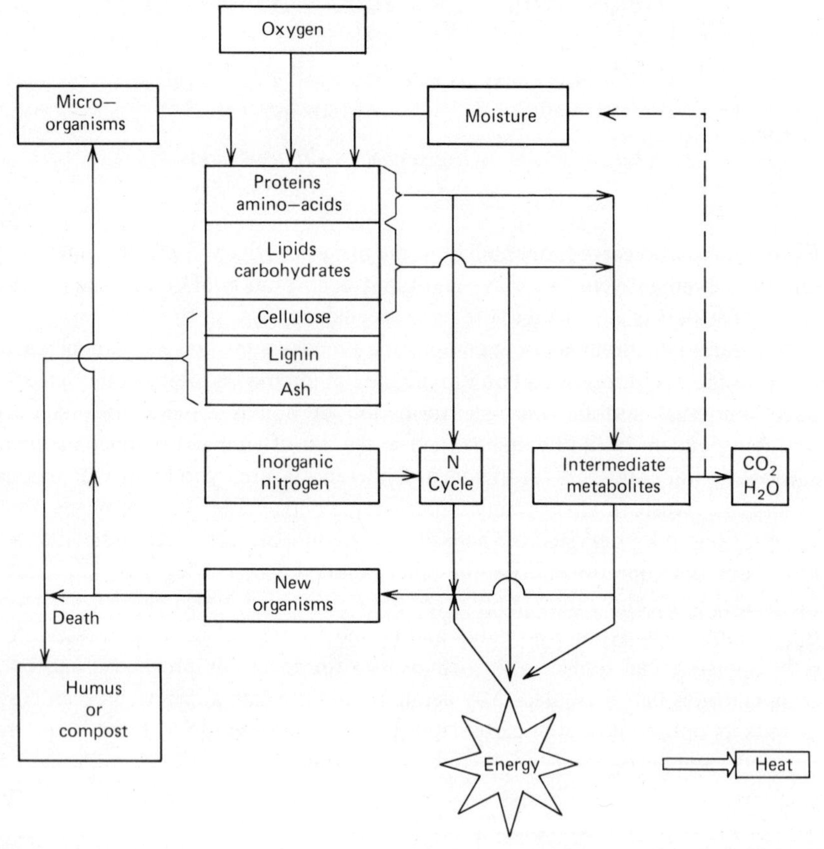

Figure 5.1. Overall flow sheet for the composting process. (From *Disposal of Refuse and Other Wastes,* by John Skitt, M. B. E., published by Charles Knight and Co. Ltd., London, England, 1972.)

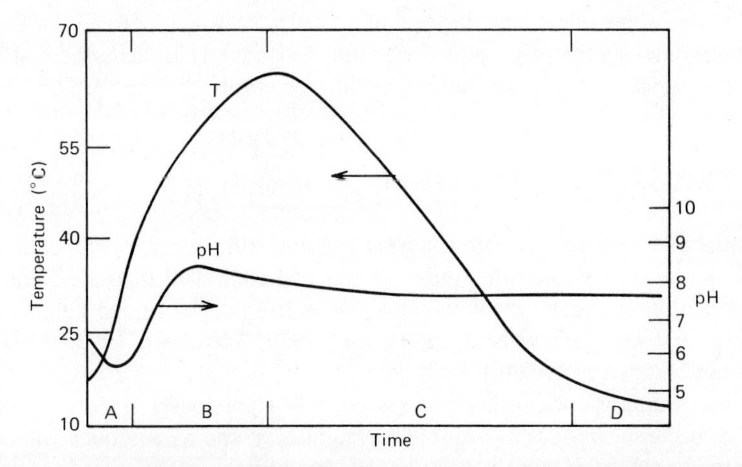

Figure 5.2. Temperature and acidity variation with time of a composting process. The acidity is expressed in pH units, where pH = $-\log_{10}$ (hydrogen ion concentration). The stages are: A, mesophilic; B, thermophilic; C, cooling; D, maturing. (From *Disposal of Refuse and Other Wastes,* by John Skitt, M. B. E., published by Charles Knight and Co. Ltd., London, England, 1972.)

moisture content of organic waste material is suitable (50-60%), and if it is aerated, the metabolism of the indigenous mixed microbial population speeds up. In addition to water, oxygen, and carbon compounds, the microorganisms require nitrogen, phosphorus, potassium, and some trace elements. A flowsheet for the process is shown in Figure 5.1. The temperature rises as biological oxidation of the carbon from the waste takes place, and at the same time the acidity increases (Figure 5.2). The first stage (A) is the multiplication of the indigenous mesophilic organisms (those organisms thriving between 25 and 45°C) which produce simple organic acids and cause a steep rise in temperature and an increase in acidity. The rise in temperature causes the activity of the mesophiles to fall, and the thermophiles take over (phase B). (Thermophilic or ''heat-loving'' organisms grow best between 45 and 75°C.) In phase C the reaction is continued by spore-forming bacteria and actinomycetes. The reaction rate slackens as the readily degradable material (waxes, proteins, hemicelluloses) is used up, and the temperature falls. In turn, thermophilic and mesophilic organisms invade and degrade the cellulose. Mesophilic bacteria hydrolyze the cellulose with the aid of an extra-cellular enzyme and assimilate the resulting low molecular mass, water-soluble fragments. Formation of antibiotics also occurs during the composting process and affects many of the organisms. Most pathogens are killed in less than three days when introduced into compost (Yen, 1974b). The first three phases are complete in a few days or weeks, but achievement of maturity in phase D requires months.

A related approach is land disposal of treated wastewater, sewage sludge, and other organic sludges by spray irrigation, but care is necessary as effluent irrigation of pastures has infected cattle with tapeworm.

CONTROLLED FERMENTATION

Considerable research is being carried out into the selective use of aerobic thermophilic microorganisms (pure cultures and controlled mixed cultures) for the formation of sugars, alcohols, and protein from cellulose and lignin wastes (Bellamy, 1973a, 1973b; Seal and Eggins, 1976; Trevelyan, 1975; Wilkinson and Rose, 1963; Yen, 1974b).

In the ordinary composting process described previously, there is a mixed flora of bacteria, mold, protozoa, and even insects, and a consequent variety of end products. Protein does not accumulate because the organisms consume one another. However, if thermophilic cellulolytic and lignolytic microorganism processes can be made to predominate, the composting rate is greater and the temperature is higher: 55°C for 24 hours results in almost complete destruction of human and animal pathogens, including viruses, bacteria and parasites (pasteurization). If the conditions are controlled so that they are the optimum for the desired organisms, and saprophytic and parasitic organisms are inhibited, the microbial single cell protein can accumulate at a maximum rate. Alternatively, particular fermentation products can be collected.

Research on cultures of some microorganisms (methanol-utilizing and methane-utilizing: Harrison, 1976) has shown them to be rather susceptible to contamination by other organisms growing on products of metabolism of the main culture, reducing efficiency, and producing substances that are possibly pathogenic to higher animals. Two techniques have been developed as alternatives to heat sterilization of media. Although formaldehyde at low concentrations is readily metabolized by organisms utilizing methanol and methane, it is a powerful sterilizing agent at concentrations of 0.1% and above. This means that the bulk medium can be sterilized with formaldehyde before it is fed to the continuous culture reactor, but after entry the formaldehyde is oxidized and contributes to the biomass. The second solution to the problem is the use of symbiotic mixed cultures, which are resistant to contamination because there is no residual carbon remaining in the solution if the organisms are carefully selected; a more complex ecological system has greater stability.

The common bakers' or brewers' yeast *Saccharomyces cerevisiae* can be used for anaerobic fermentation of waste liquids containing sugars, to produce ethanol at a conversion rate as high as 90% (Trevelyan, 1975). The yeast can be recovered by centrifuging, and recycled with no make-up required (Bennett and Lash, 1974). Microbiological conversion to other alcohols, various carboxylic acids, ketones, and (under anaerobic conditions) amines is also possible (Castor, 1974; Miller, 1961).

One of the important new sources of human food and animal feed protein is single-cell protein (SCP).* The idea is not new. There are isolated examples of algae and yeasts having been harvested and eaten centuries ago. Only recently, however, has it been recognized that protein from single-cell organisms offers the best opportunity for dramatic increases in food production. The reasons for this are that the microorganisms do not depend on agricultural land fertility or even on climatic conditions; genetic experiments for protein improvement can be undertaken readily; and SCP mass doubling times are of the order of minutes and hours, compared with days or weeks for plants and weeks or months for animals. SCP includes bacteria, yeasts, fungi and algae, and the ''crop'' is the powder of dead cells recovered by washing and drying. There are many problems still to be overcome, particularly in those products intended as human food, but the possibilities are almost unlimited.

Because carbohydrate wastes are produced at widely dispersed sites and often for only portions of the year, there is a need for low technology SCP fermentation processes (Rightelato et al., 1976). They must be capable of growth on a wide range of carbon sources, with high growth rates under a wide range of conditions, and with nontoxic high protein products which can be recovered easily. They must have also high conversion efficiencies, and this restricts the choice to aerobic processes (which have typical carbohydrate conversion efficiencies of 40%, considerably higher than those of anaerobic processes).

If the aim is to harvest the microorganisms as SCP, a symbiotic combination of organisms can be used: for example, one yeast that produces an enzyme capable of splitting starches into lower molecular mass carbohydrates, and a second yeast that grows rapidly on these carbohydrates. If desired, only the inital part of the fermentation process need be used; the enzyme cellulase released during the growth of fungi or bacteria in a cellulose pulp can be extracted and used to degrade waste cellulose such as newspaper to provide glucose. The commercial production of enzymes is a rapidly expanding industry (Meers, 1976). Some fungi such as *Trichoderma viride* (which was discovered during World War II in the South Pacific by its effect on United States cotton uniforms) are capable of utilizing cellulose and hemicellulose in nonsterile conditions. They are ideally suited for processing animal feedlot wastes, producing an essentially odorfree material with all the original nitrogen but reduced carbohydrate. *Trichoderma viride* growth characteristics on glucose and the rate of cellulase production are being determined (Mitra and Wilke. 1975; Spano et

*For example, Birch et al., 1976; Callihan and Dunlap, 1973; Callihan et al., 1975; Daly and Ruiz, 1975; Dunlap, 1973; Ghose, 1969; Kihlberg, 1972; MacLaren, 1975; MacLennan, 1974, 1975; Parker, 1973b; Skinner, 1975; Tannenbaum and Mateles, 1968; Vincent, 1969.

al., 1976), and this enzyme is regarded as being most suitable for waste cellulose hydrolysis.

ANAEROBIC DIGESTION

Anaerobic digestion results in the production of a valuable fuel gas (typically 70% methane, 30% carbon dioxide, and traces of oxygen and hydrogen sulfide). For cellulose, the formal chemical conversion and stoichiometry may be represented in a highly simplified manner as:

$$\underset{\textit{cellulose}}{C_6H_{10}O_5} + \underset{\textit{water}}{H_2O} \rightarrow \underset{\textit{carbon dioxide}}{3CO_2} + \underset{\textit{methane}}{3CH_4}$$

The carbon dioxide may be removed, for example by water carbonation tanks, and the methane stored in tanks or compressed into cylinders. Anaerobic organisms are unable to utilize lignin and lignin-cellulose complexes, but if the resulting sludge can be used as a fertilizer the process is feasible for agricultural wastes, although it tends to produce objectionable odors if not well managed and disease organisms are not killed as efficiently as in aerobic digestion.

Anaerobic digesters may be of the batch type or continuous, but in either case when the waste is first loaded there is an aerobic period while the oxygen initially present is removed and carbon dioxide is released. This is followed by a "liquefaction" phase in which the acid-producing bacteria by secreting enzymes break down fats, proteins, and most starches to simpler compounds. These include, in particular, the low molecular mass volatile organic acids such as acetic acid. During this period (about two weeks) the acidity increases (to a pH or 6 or less) and carbon dioxide is given off. The acid-producing bacteria are rather insensitive to changes in their environment, but these are followed in the "gasification" stage by methane-producing bacteria which reproduce more slowly and which are sensitive to the conditions. The acidity decreases slowly over the next 3 months or so, while ammonium compounds are formed (pH about 7) and the ratio of methane to carbon dioxide increases in the evolved gases. Finally, the mixture becomes well buffered (resistant to variation of acidity) at a pH between 7.5 and 8.5.

There are different sets of acid-producing and methane-producing bacteria which thrive in the mesophilic temperature range (optimum about 35°C) and the thermophilic range (optimum about 50°C). For a variety of practical reasons (efficiency of the mesophilic process, sensitivity of thermophilic bacteria to their environment, poorer fertilizer quality of thermophilic sludge, and difficulty of maintaining the higher temperature) the former temperature range is preferred usually, although thermophilic anaerobic digestion of solid waste for fuel gas production has been studied (Cooney and Wise, 1975). As the amount of methane produced in the mesophilic process increases markedly between 15 and 35°C, the digester is usually heated in cool and temperate climates. The "spent" charge or sludge is material having a far lower carbon:nitrogen ratio, and consequently it is a valuable agricultural fertilizer or animal feed supplement. It is composed of the dead bacterial cells as well as undigested organics such as lignin and inorganics such as grit.

One of the most publicized recent anaerobic digestion developments was the chicken manure-powered car (Grindrod, 1971). Although the application of sewage gas as a substitute for coal gas was well developed 50 years ago (Fulweiler, 1930), its use usually has been restricted to providing power for the operation of sewage treatment equipment. In Britain about two-thirds of the 5000 or so sewage installations use the heated digester process, producing more than enough methane to make them self-sufficient in energy (Coldrick, 1975). This sewage source is capable of producing only a small fraction of our total energy requirements, but the yield could be increased if all organic wastes were included (Hitte, 1976; Klein, 1972).

Practical reports of anaerobic digesters for agricultural wastes, with examples of low-cost digesters and bibliographies have been published, for example by Fry and Merrill (1973), but up to the present this process (unlike composting) is not widely used on either large or small scale. As the difficulties of disposing of animal feedlot wastes increase and as the demands expand for alternative fuel gas, anaerobic digestion should become more popular (Golueke and McGauhey, 1976). Anaerobic treatment of industrial wastes is receiving attention also (Scammell, 1975). The Bioenergy(TM) process is based on the development by Biomechanics Ltd in Kent, U.K. of an efficient method of preventing bacteria being washed out of the mixed and heated main tank, and gives a "zero cost" waste disposal operation under current economic conditions at a biological oxygen demand of 8000 milligrams per liter.

SOLAR ENERGY STORAGE

The concept of photosynthetic solar energy storage and nutrient reclamation ("energy farming" or "fuel farming") is now receiving considerable attention from both practical ecologists (*Futurist,* 1974; Imrie, 1975; *Mother Earth News,* 1972) and research scientists.*

*For example, Calvin, 1974; CEN, 1976a; Chedd, 1975; EST, 1976a; Golueke and Oswald, 1963; Golueke et al., 1973; IGT, 1976; Kemp and Szego, 1974; Kemp et al., 1975; Miller, 1975; Oswald and Golueke, 1960, 1964; Povich, 1976; Szego and Kemp, 1973.

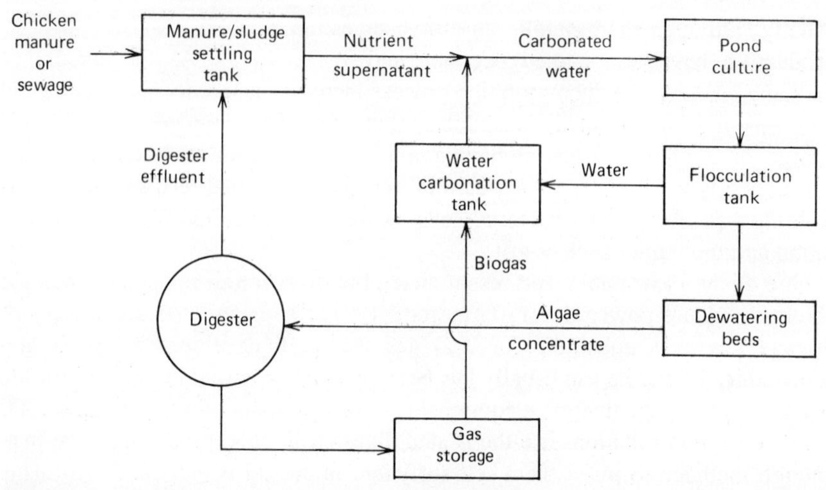

Figure 5.3. Flow diagram of sludge-algae-methane system. (Copyright © 1973, L. John Fry and Richard Merrill, from *Methane Digesters for Fuel Gas and Fertilizer* (New Alchemy Institute) Newsletter #3 is available for $4.00 from New Alchemy Institute West, P.O. Box 2206, Santa Cruz, CA 95063.)

In contrast to bacteria and fungi which consume fuels and are called "reducers," algae and plants (except in darkness when they metabolize organic matter) are "producers," utilizing light energy and storing it as chemical energy. A relatively efficient botanical solar energy-chemical energy converter uses the sludge-algae-methane cycle and closed nutritional and waste systems (Figure 5.3). The aerobic bacteria present in sewage break down the complex substances to forms usable by the algae, which are then harvested and anaerobically digested. The digester residues are recycled back into the ponds. Research is continuing into various aspects of this procedure, for example the effects of pretreatment of municipal wastewater with chlorine or cobalt-60 gamma irradiation on the subsequent hydroponic growth of plants. Hydroponics is the technique of growing plants directly in nutrient solutions rather than in soil. The nutrients may consist of fertilizers mixed from chemicals, but may just as well be derived from organic wastes such as sludge and effluent from anaerobic digestion processes (Fry and Merrill, 1973). Hydroponic ponds can range from open sewage lagoons, to sophisticated installations with transparent domes (Figure 5.4). Sludge is added slowly and continuously under a gravel bed, and it then percolates up through a layer of sand in which the plants are rooted. The plants may be water hyacinth (Wolverton and McDonald, 1976), pasture grasses such as rye and fescue, or green algae. Waste-grown algae now form a source of SCP for animal feed. Alternative methods are the oriental aquaculture system (sludge-algae-fish); "living filters" of grasses, reeds and other crops irrigated with

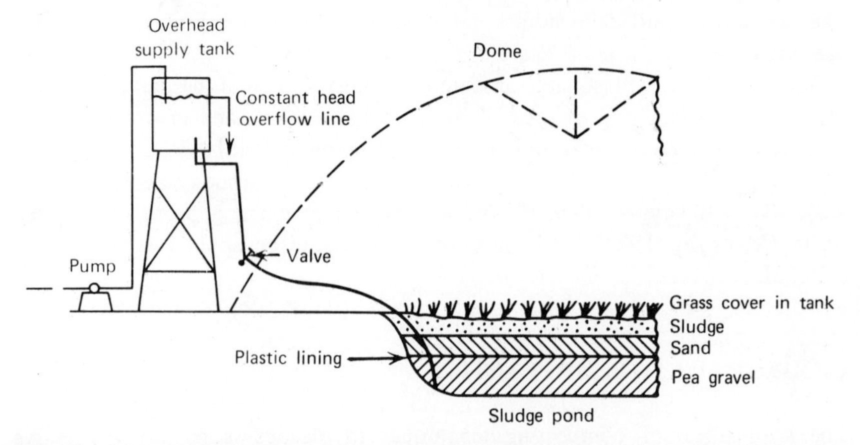

Figure 5.4. Hydroponic sludge culture of pasture grasses. (Copyright © 1973, L. John Fry and Richard Merrill, from *Methane Digesters for Fuel Gas and Fertilizer* (New Alchemy Institute).

nutrient solutions (CCEQ, 1973; Crites and Pound, 1976; EST, 1972a, 1975c; Mantell, 1975); and conventional forest fuel crops.

It should be noted, however, that photosynthesis is a relatively inefficient method of utilizing solar energy, if only because it uses only a part of the visible portion of the electromagnetic spectrum (Haneman, 1977; Madigan et al., 1971a). Thus cultivation of algae purely as a source of energy is probably not economically feasible, except when combined with the purification of wastewater and the production of food. There are also limitations imposed by supplies of water and phosphorus (Povich, 1976).

RECYCLING OF INORGANIC MATERIALS

Although the recycling of organic carbon and nitrogen compounds is the main result of microbial activity, sulfur, phosphorus, and metals such as calcium, iron, and copper are also involved (Garrels et al., 1975; WSL, 1975a). Some are important constituents of all living matter, but others are significant only in specific organisms. For example, iron bacteria form natural colonies, heavily encrusted with ferric oxide, in freshwater ponds and springs which contain reduced iron salts. The process of dissolving copper from mine wastes as the sulfate, by simply watering the surface and collecting the leachate which percolates through the dump, is assisted by the sulfur-oxidizing bacteria *Thiobacillus ferrooxidans*. They are plantlike in their metabolism; they utilize carbon dioxide and oxygen from the air and nutrients commonly found in soils.

The oxidation of sulfide to sulfate and of ferrous to ferric iron provides the source of energy.

There has been a reasonable amount of interest in several inorganic microbial processes: the leaching of ores of iron, copper, nickel, and zinc sulfides (by *Thiobacillus ferrooxidans*) and of gold and uranium (WSL, 1975b); the concentrating of aqueous solutions of metal ions such as iron, manganese, mercury, lead, zinc, and copper (Yen, 1974b); and the conversion of gypsum to elemental sulfur (Madigan, 1971b). All these processes could be applied extensively in materials recovery.

BIODEGRADATION OF PLASTICS

The application of composting techniques to plastics is usually ineffective because of their biological inertness. (Some polyesters, notably polyurethanes, are unusual among synthetic polymers in being susceptible to microbial attack (Upsher, 1976).) Vinyl plastics, particularly poly(vinyl chloride), show marked deterioraton after several years in the presence of soil microorganisms, but this is probably because the plasticizers such as glycol derivatives provide a source of nutrient. Most organic coatings undergo slow biodegradation of one type or another, and the deterioration of plastics would be beneficial from the point of view of solid wastes diposal (Elias, 1976; Yen, 1974b). The combination of photochemical and bacterial degradation processes has the potential of eventually destroying most plastics; ultraviolet irradiation of polyethylene results in free radical formation, followed by oxidation to hydroperoxides, polymer chain-breaking, and eventually microbiological degradation of the low molecular mass fragments. This degradation would be extremely slow in most cases, and it is noteworthy that poly(tetrafluoroethylene) is not susceptible to these processes. The carbon-fluorine bonds are not broken by the solar radiation that penetrates the atmosphere, and the absence of removable hydrogen atoms in the molecules prevents this material from being a substrate for enzymatic reactions (Arkles, 1973).

The biodegradability of polymers has received considerable attention, for example from the manufacturers of plastics in an attempt to improve the public image of plastic by preventing the accumulation of long-lasting plastic litter. Plasticizers are not incorporated in many packaging polymers, so the degradation rates are extremely slow, except in specially formulated "bio-cyclic" plastics (Guillet, 1973). Conventional polymers can be modified by the inclusion of chemical groups so that they are photobiodegradable. Alternatively, a proportion (typically 10-40% by volume) of plastics can be replaced by starch which is thus encapsulated within the polymer but which provides nutrition for microorganisms and leads to complete breakdown of the materials in soil or water (EPM,

1975b), or in more controlled fermentation conditions if desired. However, litter is a social problem, and may be solved more efficiently by education, legislation, or taxation (Chapter 6) than by the possibly hazardous method of making plastics biodegradable.

Biodegradability can be induced in some polymers (particularly polyethylene) by appropriate chemical pretreatment. After oxidation by nitric acid, various waste plastics were found to support microbial growth (Brown et al., 1974), providing a potential source of SCP food.

6

POSTCONSUMER WASTE

The present disposal practices for the large quantities of consumer wastes result in substantial losses of both resources and energy, and contribute to air and water pollution. The disposal problems in most countries are increasing dramatically, not only because urban populations are growing rapidly, but also because affluence and packaging techniques are increasing the per capita waste production. Estimates of postconsumer waste generation show great disparities, but the most recent U.S. Environmental Protection Agency estimate, based on material flow methods, of residential and commercial waste is an average of 3.75 pounds (1.7 kilograms) per person daily in 1973, totaling 144 million tons (130 megatonnes) for the United States population of 210 million (Smith, 1976). Projected totals for 1980 and 1990 are, respectively, 4.3 pounds (1.9 kilograms) and 5.0 pounds (2.3 kilograms) per person per day. (These figures are lower than earlier estimates (Dean et al., 1971; OST, 1969); it is emphasized that caution must be exercized in applying the national per capita rates to local level planning.)

The percentage of residential waste recovered for reuse of some kind is extremely small: 2% in Britain in 1967 (Skitt, 1972); 6-7% in the United States in 1973-1974 (Meyers, 1976; Smith, 1976); this is based mainly on paper. The local authority recycling which does take place is due often to the efforts of one enthusiastic person, as exemplified by Worthing in southern England (Bourne, 1972; Thomas, 1974).

Up to now the open dump, and more recently the "controlled tipping" or "sanitary landfill" methods have been, at least superficially, the most attractive from an economic point of view, but in some places the costs of land for such sites and of transport are becoming prohibitively high. Incineration prior to landfill is often necessary to reduce volume, but this contributes to air pollution if not carried out in efficient burners. Alternatives are pulverizaton or high-pressure baling before tipping, but these processes also entail additional expensive plant. Public action on material and visual pollution is increasing. Sometimes dumping at sea is an economically viable alternative, but this is only a limited and short-term solution. Figure 6.1 provides a flow chart of solid waste alternatives.

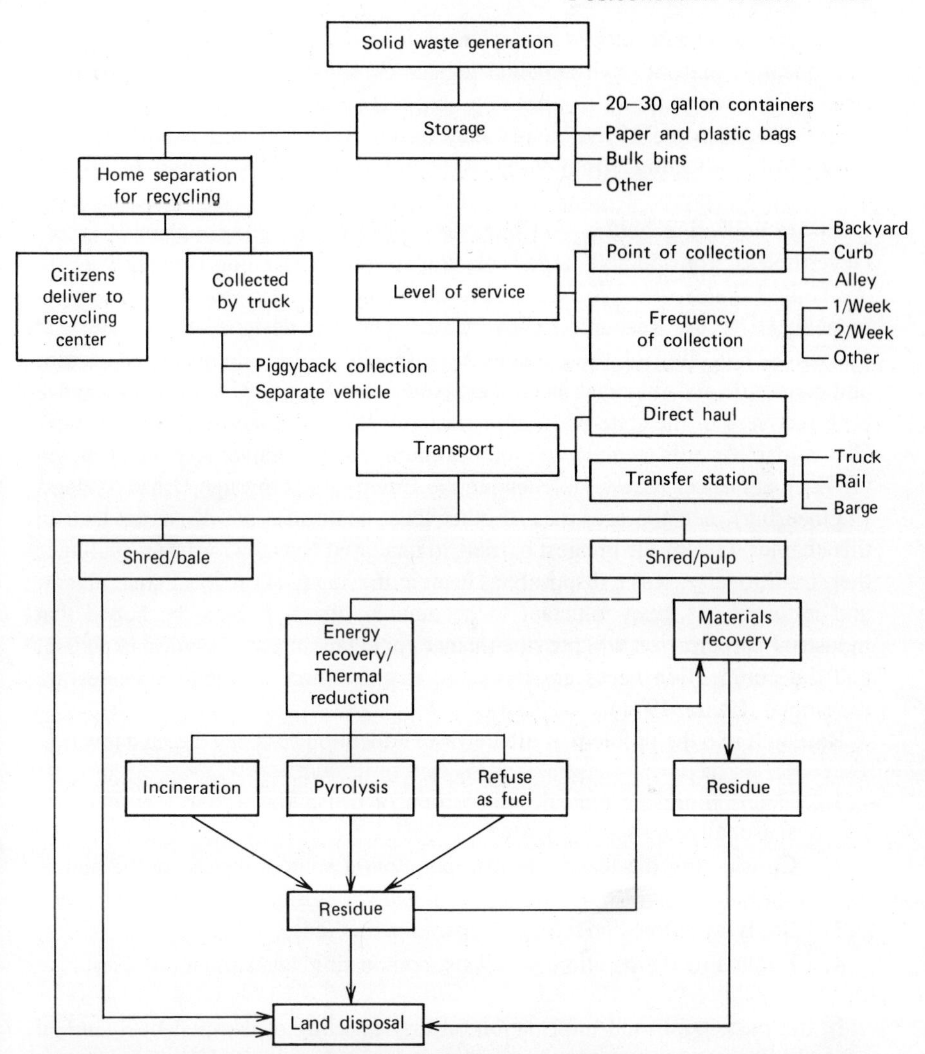

Figure 6.1. Flowchart of solid waste management alternatives. (After Willson, 1974. Reprinted with permission from *Professional Engineer* magazine.)

Already the average city dump contains some metals in concentrations greater than those of many minable ores, and our descendants may well look upon 20th century dumps as "man-made mines" (Kenahan et al., 1973; OST, 1969; *Technology Review,* 1971). Much of the present mineral recovery technology, particularly in the United States, is geared to high-grade mineral deposits. As the inevitable move occurs towards the utilization of lower-grade ores, the technology will be developed to recycle waste materials more efficiently.

Ironically, present environmental legislation sometimes hinders reclamation from waste. Air pollution control is prompting a move away from incineration, where at least some of the energy may be recovered, toward landfilling. The practice of "salvaging" from dumps is declining as access to sanitary landfills is controlled by public hygiene regulations, and some garbage collectors have given up collecting bottles "on the side" when the resulting income is taxed. Nevertheless, considerable large-scale and long-term planning is under way. In the United States, the 1965 Solid Waste Management Act received an amendment in 1970; the Resource Recovery Act, which is designed to promote and encourage recycling of solid wastes by research and development, education, and economic and financial incentives. New York State (1974) plans to achieve 60% recovery of the state-wide solid waste by 1984, and the trend in New York City is also towards resource recovery (Fenton, 1975). Denver Regional Council (1972) has Project REUSE (Renewing the Environment through Urban Systems Engineering), and the activities of other local authorities are discussed later in this chapter. One of the greatest barriers to increased consumer waste recycling is that, traditionally, waste disposal has been in the hands of the local government, and industry has been reluctant to become involved. It is to be hoped that industrial participation will provide finance for development, technical expertise, and the competition necessary to ensure that the more efficient processes are developed (Bailie, 1971).

Research into the problem of urban waste utilization has been directed toward

1. Determining the typical composition of urban waste, and how it varies seasonally and year by year;
2. Considering the feasibility of separation at source, that is, in the household;
3. Studying automated terminal separation methods;
4. Evaluating incineration, pyrolysis, composting, and similar proceses.

In the past, published information on the recycling of components of urban waste has been scattered, but the position is improving. In the United States, the National Center for Resource Recovery has published recently "Resource Recovery from Municipal Solid Waste" (NCRR, 1974c), and the Noyes Data Corporation has volumes in the Pollution Technology and Environmental Technology series on this subject (Jackson, 1974, 1975; Sittig, 1975). The increasing numbers of conferences, books and review papers on waste disposal now usually contain sections on recycling.*

*For example, Aleshin, 1974, 1976; Baum and Parker, 1974a, 1974b; Bond and Straub, 1973; Flintoff, 1974a; Johnston, 1975; Kirov, 1972; Mantell, 1975; Patrick, 1973; Pavoni et al., 1975; Ross, 1968; Stump, 1972; Yen, 1974a.

Table 6.1 summarizes potential recovery methods. The ultimate aim should be an integrated waste processing, materials recycling, and total energy utilization scheme, as advocated by Beal and Haselar (1972), for example. An important point that should be kept in mind is that if the economic and other incentives described in Chapter 2 have the effect of diverting the more valuable materials from the urban waste stream, the discarded residue will become even more difficult to utilize, and may well be useful only as energy or as a fuel (Jackson, 1975).

Table 6.1. Summary of Possible Types of Material and Energy Recovery from Urban Waste

Process	Material Recovery	Energy Recovery	Comments
Controlled tipping (optional pulverization and compression)	Landfill		Land availability and transportation problems
Shredding and air separation	Metals, glass, paper	Organic fuel	
Water pulping and screening	Cellulose fiber	Cellulose dried for fuel	
Incineration	Fly ash, glass, metals	Heat, electricity	High installation and maintenance costs; potential air emissions
Pyrolysis	Metals, glass	Fuel oil and gas	High installation costs
Liquefaction	Metals, glass	Fuel oil	High installation costs
Gasification	Metals, slag	Fuel gas	High installation costs
Composting	Compost, metals, glass	Compost dried as fuel	Expensive operation, limited demand for compost
Hydrolysis	Sugars, protein	Ethanol by fermentation	Under development

This chapter deals with the general techniques and problems of urban solid waste recycling, and specific details on glass, paper, ash, and metals are included in Chapter 7.

COMPOSITION OF URBAN WASTE

The nature of urban waste understandably varies with country, city, suburb, and season, and can be determined precisely only by analysis in each particular case (van den Broek and Kirov, 1972). It also varies with time as living patterns alter. Nevertheless, information on composition is necessary for planning disposal facilities (landfill capacities, incinerator design, and so on) and collection facilities (there has been a trend to lower overall density, for example). This information is required also for determining the feasibility of recycling various components of the waste.

Skitt (1972, 1973) has considered in detail the composition and analysis of domestic waste in Britain, where systematic analysis has been carried out since the 1930s. Table 6.2 summarizes the changes in domestic waste since that time, with figures predicted for 1980 (Gutt et al., 1974). The ash content decreased and the proportion of combustibles increased with the trend away from solid fuel fires and the change to gas, oil, and electrical home heating which was hastened by the Clean Air Act in 1956. Together with the increase in nonreturnable containers and packaging, these factors resulted in decreasing waste density.

Table 6.2. Average Composition of Domestic Waste in Britain since 1935

	Mass Percentage			
Component	1935	1963	1968	1980 (prediction)
Metal	4	8	9	9
Glass	3	8	9	9
Paper	14	23	37	43
Rag	2	3	2	3
Vegetable and putrescibles	14	14	18	17
Plastics	0	0	1	5
Dust and cinder	57	39	22	12
Unclassified	6	5	2	2

Sources. Gutt et al., 1974; Skitt, 1972, 1973.

In the United States, the National Center for Resource Recovery has carried out a comprehensive study of the composition of the urban solid waste stream, and of its resource recovery potential (NCRR, 1974c). Detailed information has been collected also by Barbour et al. (1974), Baum and Parker (1974a), Bond and Straub (1973), and Mantell (1975). This information is not reviewed here, but the details for one city will be presented. Madison, Wisconsin, may be considered representative of American cities in that its per capita discard rate is close to the national average (3.3 kilograms per day). Madison urban waste composition has been determined, together with the screen-sized distribution after shredding to less than 15 centimeters and air-drying (Dean et al., 1971), and this information is summarized in Table 6.3. The moisture content varied between 5 and 30%, and the average figure of 12.4% is used here. A separate survey of the nature and fiber grade of the paper products discarded in Madison in 1970 showed 47% newspapers, 13% magazines, 12% strong papers, and 28% other paper items (Myers, 1971a). The percentage ranges of components in United States urban waste have been reported (OST, 1969) as follows: garbage (food products) 5-15, paper 42-57, metals 1.5-8, glass 2-15, rags 0.6-2, garden debris 10-12, and ashes 5-19.

Table 6.3. Screen-sized Distribution of Urban Waste from Madison, Wisconsin in 1970 after Shredding to Less Than 15 cm and Air-drying to Remove 12.4% Water

		Mass Percentage of Shredded, Air-dried Waste			
Component	Mass Percentage of Undried Waste	−6 inches (15cm) +1 inch (2.5 cm)	−1 inch (2.5 cm) +4 mesh (0.5 cm)	−4 mesh (0.5 cm)	Total
Ferrous metal	5.7	4.7	1.8		6.5
Nonferrous metal	0.3	0.1	0.2		0.3
Paper	41.0	3.2	41.5	2.1	46.8
Cardboard	15.3	11.2	5.8	0.5	17.5
Cloth and leather	4.4	4.8	0.2		5.0
Wood and plants	1.5	0.5	1.1	0.1	1.7
Plastic	2.9	2.5	0.8		3.3
Rubber	0.6	0.7			0.7
Dirt and glass	15.9		2.3	15.9	18.2
Total		27.7	53.7	18.6	

Source. After Dean et al., 1971, with the permission of the Bureau of Mines, U.S. Department of the Interior.

Although the figures in Tables 6.2 and 6.3 cannot be compared directly (the former including domestic waste only, and the latter based on all urban waste) there is a broad similarity in the potentially valuable components (metal, paper, plastic). Differences are due to factors such as the greater use of undersink garbage grinders in the United States for food waste, and different packaging practices. In the less-developed countries, urban waste is characterized by a larger proportion of food organics, as illustrated in the results of a pilot recycling plant operated in Madrid, Spain (Cavanna et al., 1974, 1976). Varjavandi (1975) and Varjavandi and Fischof (1975) have reviewed urban waste trends in Australia.

Some very detailed analyses of composite urban waste have been made in certain places, and although these are of interest, information on how the composition varies with time is more valuable than very specialized isolated analyses. To take another point of view, the composition ranges of urban waste can be expressed in terms of its elements (Table 6.4). Neglecting the noncombustibles, this corresponds to the empirical chemical formula

$$C_{30}H_{48}O_{19}N_{0.5}S_{0.05}$$

This may be compared with the empirical formula of cellulose,

$$(C_6H_{10}O_5)_n \quad \text{or} \quad C_{30}H_{50}O_{25}$$

the similarity reflecting the high proportion of cellulose in the waste. The low sulfur value is significant when urban solid waste is considered as a potential fuel.

Table 6.4. Chemical Analysis Ranges of Urban Solid Waste

Component	Mass Percentage
Water	20-40
Carbon	15-30
Oxygen	12-24
Hydrogen	2-5
Nitrogen	0.2-1.0
Sulfur	0.02-1.0
Noncombustibles	15-25

Source. After Bailie, 1971.

PACKAGING

The question of packaging in urban waste has been discussed in recent years by numerous writers, including Abrahams (1971), Cannon and Smith (1974),

Darnay (1969), Darnay and Franklin (1969), Mantell (1975), New (1974), Nuss et al., (1975), and Sittig (1975). In the United States in 1971 between 5 and 8% of the typical household food bill was for packaging, and most of this material was not recycled. Packaging consumed about 47% of the United States' paper and paperboard, 75% of the glass, 29% of the plastics, 14% of the aluminum, more than 8% of the steel, and contributed 50 million tons to the 360 million tons of residential, commercial, and industrial solid waste (Heylin, 1971; USEPA, 1974b). In Britain it is estimated that £800 million was spent on packaging in 1968 (Skitt, 1972), and £1350 million in 1972 (*Industrial Recovery,* 1976d). The Australian Conservation Foundation estimated in 1974 that an average family spent $700 a year on packaging (Hoad, 1975b).

Containers have become more elaborate for food, confectionery and pharmaceutical supplies, and many garments are now sold in individual packs. Not only has the amount of packaging increased, but its nature is changing. Plastics and metals are replacing glass, and disposable bottles are replacing returnable ones. (The particular problem of nonreturnable containers is discussed in the next section.) The containers, especially aerosols, for many cosmetics, perfumes, pharmaceuticals and cleaners cost more than the contents. It is useful to distinguish between the process of *packing* (as an aid to handling, shipping, and warehousing) and *packaging* (to communicate a message about the product and assist its sale); and to consider what is necessary and what is unnecessary packaging (New, 1974). Table 6.5 presents one estimate of "unnecessary"

Table 6.5. Unnecessary Packaging in the U.K. as Estimated by New (1974)

Packaging Material	Percentage Unnecessary
Paper	40
Plastics	30
Metal	5
Glass	10
Total packaging	16

Source. Reprinted with permission from *The Ecologist* (Wadebridge, Cornwall, U.K.) **4**(6), July 1974.

packaging in Britain. A balance must be reached; certainly, individual packs have many advantages such as making self-service possible and reducing distribution costs (Honeysett, 1975), but expensive, competitive packaging can outweigh these benefits. It is obvious that the trend towards extravagant, disposable packaging should be discouraged, but if the consumer really wants to allocate a significant proportion of his budget for this purpose, more efficient means of container return and packaging recycling must be found. Manufacturers tend to claim that containers are designed to minimum specifications to contain, protect, and inform; that standardization would not lead to savings because it would result in some products being overpacked; and that the consumer does not want returnable packaging. It is often more efficient and economical to reuse packaging than to reprocess it, but even if the move to nonreturnable packaging is not reversible, it is essential that the waste be fed back into the materials cycle. Large-scale energy recovery from incineration or pyrolysis might appear to solve the problem, but it could have the effect of encouraging even more elaborate and wasteful packaging.

Plastic packaging presents special difficulties, particularly as it is necessary to make predictions of the rapidly changing plastics content and composition in urban waste when new disposal facilities are designed (Baum and Parker, 1974a; Lederman, 1974; Staudinger, 1970, 1974). The high stability of polyethylene, which makes it so suitable for packaging, causes recycling and disposal problems. Incinerators designed for materials with lower heats of combustion are often unsuitable for wastes with a high proportion of plastics. Poly(vinyl chloride) (PVC) causes special problems due to the formation of corrosive hydrogen chloride on incineration, and all thermoplastics tend to melt and clog furnace grates. In the future it may prove necessary to restrict the use of packaging materials that cause disposal problems.

The costs of packaging materials in energy terms have been considered by Berry and Makino (1974), Hannon (1972, 1973), Hunt and Franklin (1975), and Makino and Berry (1973) in the United States, and by New (1974) in the United Kingdom.

BEVERAGE CONTAINERS

Until the 1940s returnable glass containers were the rule, but at this time the steel and can companies saw the beer and softdrink market as a major expansion area. Even as recently as 1958, 98% of United States softdrinks and 58% of the beer was sold in returnable bottles (Hannan, 1972), but then aluminum companies moved in with an all-aluminum two-piece can. To compete against the entry of cans into their traditional market, glass bottle manufacturers introduced non-returnables, and with pressure from glass, steel, and aluminum container manufacturers, bottlers adapted their production lines to throwaway containers. This was not done because of bottle fragility or cost, but was in response to a change in consumer habits brought about by advertising and made possible by increasing public affluence. The glass industry did not develop a nonreturnable milk bottle, and so lost a large proportion of this market to wax- or polythene-impregnated cartons (Abrahams, 1971).

The traditional steel can is made in three pieces with a soldered or seamed side and separate top and bottom. Aluminum has been used in the tops of steel cans to facilitate opening, and the newer drawn-and-ironed two-piece steel can with the top sealed into a one-piece body is now coming into use. This development, together with tinfree steel (protected by a chemical surface treatment) and welding with organic cement rather than solder, has been prompted mainly by the success of the aluminum beverage can. Plastic bottles are now entering the market (CEN, 1975a; Wharton and Craver, 1975).

In response to environmental criticism, large aluminum producers have initiated programs to collect used aluminum containers (EST, 1969b). It is estimated that in the United States about 5% of all aluminum cans are recycled, and some collection schemes are reaching economic break-even point (Clark, 1971). In regions where aluminum beverage cans are marketed, two fractions of aluminum can be recovered from urban waste. Aluminum cans and formed containers are all of compatible alloys and can be reused for the same purpose with little addition of virgin metal. Other consumer durable and packaging aluminum alloys are suitable for the production of cast items such as lawn furniture (Blayden, 1974a). Some steel companies are collecting steel cans for detinning and steel recovery, and in certain circumstances they are being used for copper cementation, but the recovery of steel cans is not very attractive economically at present (Cannon, 1973; Hoad, 1974a; USEPA, 1974b). However, economic viability in volunteer recycling projects may be less important than the educational and social benefits (NCRR, 1974c), and such projects may be assisted by industries for their public relation and advertising benefits (Hoad, 1974a).

Even allowing for costs of handling, transport, and cleaning, the use of returnable bottles is cheaper (in both economic and energetic terms) than nonreturnable bottles or cans. It is estimated that the purchase price of softdrinks in throwaway glass is 30% greater than those in returnable containers (*Consumer,* 1976; Hannon, 1972, 1973) and the energy required to deliver beverages in throwaway containers is about three times that for returnable containers used, on average, 15 times. The marketing success of disposable containers is an example of consumer acceptance of greater convenience at the expense of increased economic and environmental cost.

A "resource and environmental profile analysis" (Hunt and Franklin, 1975) shows that the 10-trip returnable bottle is markedly superior to the present common alternatives (one-way glass, three-piece steel, and aluminum recycled to the extent of 15%), and is superior in some respects to new experimental containers (acrylonitrile-butadiene-styrene plastic and all-steel). It appears that collection, transport, chemical, and energy costs of recycling metal can materials are considerably higher than those involved in reusing glass containers, so despite public relations exercises of steel companies, the glass manufacturers may recover eventually the major share of the beverage container market.

It should be noted that the *energy* costs in retrieving glass from waste are higher than those in mining raw materials, so from an environmental and resources point of view it is important that bottles should be reused rather than remelted (Hannon, 1972, 1973). From a practical and *economic* point of view, recovery and cleaning difficulties often outweigh the energy considerations. If the glass can be collected cheaply enough, that is equivalent to new material costs (in the United States typically $20 per ton or 1 cent per pound), the process being ''subsidized'' by the efforts of volunteer individuals or community groups, the glass industry will accept old glass. In any case, about 5% of cullet (crushed glass) hastens the melting of the raw ingredients (sand, limestone, and soda-ash: see Chapter 7), but up to 30% or even 50% cullet can be used if it is of high grade. One large softdrink company in Britain announced that it favored reclaiming the glass from nonreturnable bottles for construction and road-surfacing materials as an energy conservation measure (MRW, 1974). It is unlikely that a detailed energy balance calculation would show this to be the preferred course of action for energy conservation.

Some American states (notably Oregon) and local authorities have legislated against nonreturnable drink containers (EST, 1975d; Loube, 1975; McCaull, 1974; Sittig, 1975; USEPA, 1974b; Waggoner, 1976). The Oregon ''Bottle Bill'' of 1972 required all softdrink and beer containers to carry minimum refund values and encouraged those containers used by more than one manufacturer. Ring-pull opening cans were banned. Surveys of roadside litter before and after the new law apparently indicated that not only had the number of beverage containers fallen dramatically, but also the volume of other litter had decreased. Such reversals of industrial and social practice entail other changes: new washing and refilling plants, new warehouses in supermarkets for bottle storage, and unemployment in can manufacture and canning plants; but there is increasing support for similar measures in local and state governments in the United States, Canada, and Australia (Bingham and Mulligan, 1972; Folk, 1973; Peat, 1974).

The recycling of glass and metal container materials in the United States has been summarized recently by Abrahams (1971), Bingham and Milligan (1972), Cannon and Smith (1974), Hannon (1972, 1973), Hunt and Franklin (1975), Nuss et al. (1975), and Sittig (1975); and Pausacker (1975) has discussed the situation in Australia.

VOLUNTARY SEPARATION AT SOURCE

The ideal situation would be the segregation of wastes into their components at source: the household, office, shop, restaurant, or public collection center (SCS, 1974). With a few exceptions, this practice does not occur, and strong incentives (financial, moral, or legislative) would have to be applied to obtain efficient sorting of a significant proportion of waste generated by a community. Some commercial establishments sell paperboard cartons to the secondary materials industry, and some householders store newspaper and bottles for volunteer collections. In some cases, local by-laws require collection of separate waste fractions, but this usually has the aim of expediting disposal rather than facilitating salvage. During war or national economic stress, recycling efforts by paper drives and scrap drives have proved valuable (Clark, 1971). Small communities have instituted voluntary recycling centers, where individual householders are asked to bring the separated waste (paper, glass, metal), and charitable organizations operate collections (Shelley, 1976; Taylor, 1975a). Some of these programs are successful (McGough, 1972; NCRR, 1974c), usually in relatively affluent suburban communities, such as the Ecology Action group in Berkeley, California (Clark, 1971). In the United States in 1972, 1000 aluminum centers collected between 1.3 and 1.4×10^9 cans; 8.8×10^8 glass containers were reclaimed; and at the 135 locations of three major steel companies, 6×10^7 cans were collected (NCRR, 1973a). In 1973 the aluminum can figures were 1.6×10^9 at 1200 centers (Dale, 1974). In large urban areas, successful voluntary recycling schemes are rare, although plastics are segregated successfully by householders in Japan, and remolded. Voluntary efforts frequently result in net economic and energy losses if full costings (including fuel) are carried out (Dudley et al., 1975).

Some detailed surveys have been conducted into public participation in recycling. A University of Wisconsin survey in Madison, Wisconsin, showed that only 23% of family heads believed that consumers themselves should attempt to recycle waste. Those that did tended to have had a better education and to hold a higher status job (Keller, 1973). On the other hand, it is encouraging that a survey of the attitudes of United States metropolitan housewives towards solid waste disposal (USEPA, 1972b) showed that a majority would cooperate in voluntary separation if given a lead by the authorities. Psychologists are studying individual and group reinforcement behavioral effects in the promotion of recycling (Geller, 1975; Geller et al., 1973, 1975). In another study, 35% of the staff of the U.S. Department of Agriculture Forest Products Laboratory in Madison volunteered to assist in a household paper separation project covering 14 consecutive days (Myers, 1971a). This study, while highlighting the difficulty in identification of high and low quality wood fiber products, concluded that if newspapers (low-grade fibers) were removed at source from domestic waste, the possibilities for successful recycle of other, higher grade wood fiber products from mixed urban refuse could be improved. One inexpensive method being investigated (Alter et al., 1976b) is asking householders to tie newspapers in bundles, allowing the bundles to mingle with other wastes during collection, then handpicking them from a conveyer belt.

Associated with voluntary separation and personal involvement is the problem of *litter,* on the beaches, in the countryside, along streets and highways, and internationally from seagoing vessels. The magnitude of the litter problem is a good indication that voluntary separation of waste is unlikely to succeed. In a study by the Highway Research Board of the U.S. National Research Council on roadside litter, quoted by Baum and Parker (1974a), along a one-mile (1.6 km) stretch of two-lane highway were found 770 paper cups, 730 empty cigarette packets, 590 beer cans, 130 softdrink bottles, 120 beer bottles, 110 whiskey bottles, and 90 beer cartons.

A possible solution that is being implemented or at least discussed in many parts of the world is a tax on goods that cause litter, to assist in the finance of litter control and recycling research. For example, all beverage containers that do not carry a certain minimum deposit might incur a tax. Although it seems unsatisfactory for people to pay for the right to litter, it might be necessary, and it is the pattern being followed in beverage container legislation (Bingham and Mulligan, 1972). The alternative is an idealistic solution: to change our life-styles (Pausaker, 1975).

COLLECTION OF URBAN WASTE

There is some conflict between the need to reclaim items of urban waste and the economic incentives to simplify the collection of waste by means of larger and more powerful packer trucks (Hershaft, 1972), and even by home compactors, which inevitably make the separation of reclaimable materials more difficult. This is the challenge faced by authorities responsible for the choice of an overall collection-reclamation-disposal system. Selection of collection methods is inevitably associated with the eventual fate of urban waste (Skitt, 1972). Domestic waste pipeline systems, which are sometimes advocated (Hershaft, 1972; Kinnersly, 1973; McGough, 1972; Zandi and Hayden, 1969) involve the transport of waste by air or water, and in the latter case the choice of terminal treatment for the resulting slurry is restricted (see below). The recent development of pipelines for transporting materials in capsules might solve this problem. Collection and transport at present contribute about 80% of the solid waste management costs (NCRR, 1973a), so there is considerable scope for improvement in this area.

TERMINAL SORTING AND MATERIAL RECOVERY

Ideally, a terminal waste treatment plant should be omnivorous, that is it should be capable of receiving all consumer wastes, recovering as many materials as possible, and converting the remainder to inoffensive products (such as carbon dioxide and water) with full utilization of any chemically stored energy. The major problem is sorting and separating the multitude of materials in consumer waste, and there are two distinct approaches to this. The usual aim has been to shred or pulp the waste to a more or less uniform size prior to classification, but recently at the Massachusetts Institute of Technology in the Unites States, and to some extent at the Warren Spring Laboratory in the United Kingdom, investigation has started into sorting by *items* into recyclable fractions. Most of the techniques discussed in this chapter fall into the former category.

Recovery from urban waste is basically a two-stage process: materials recovery (glass, metals, paper), and recovery of the total organic portion, usually as an energy source. The term "front end" is usually used for materials recovery with eventual conventional disposal of the residue and "back end" refers to the subsequent recovery of the organic material (Abert et al., 1974), or to inorganic recovery after incineration of the combustible material.

In principle, the front end separation techniques are those described in Chapter 3, with the relatively modest aims of recovering paper, ferrous metals (particularly tinned steel cans), nonferrous metals, and glass (Alter and Reeves, 1975; Douglas and Birch, 1976; Engdahl, 1969; Hansen, 1975; Henstock, 1975c; Hershaft, 1972; Jackson, 1975; Millard, 1976; Sittig, 1975). Ferrous metals are separated magnetically; the process for separating glass and paper is either dry, with air elutriation following milling and perhaps predrying, or wet, using hydropulping and screening. Most recovery systems in use, with the notable exception of the Black Clawson hydropulping system, are dry (Dale, 1974). Back end recovery of organics utilizes the chemical and biological process of Chapters 4 and 5.

Shredding

The process of dry urban solid waste size reduction (comminution) is carried out by impacting, shearing, grinding, milling, pulverizing, or flailing, but the term *shredding* is used most commonly, whether grinding or impacting actions are involved. Details are not included here, but are available from numerous sources.* Although shredders facilitate disposal of waste in landfilling by reducing the volume and assisting compaction, the most important application is for resource recovery. In order to achieve the most complete recovery of glass, metals, or energy, the waste should not exceed a certain top size because of the

*For example, Bond and Straub, 1973; Cannon and Smith, 1974; Diaz, 1975a; Engdahl, 1969; Hershaft, 1972; Hitchell, 1975; Ito and Hirayama, 1975a, 1975b; Jackson, 1975; NCRR, 1973b, 1974b, 1974c; Osell, 1973; Pavoni et al., 1975; Skitt, 1972; *Solid Wastes*, 1976c; Trezek and Savage, 1976; Wilson, 1974a; Wilson and Smith, 1972.

design of equipment such as air classifiers, which would otherwise have to be made significantly larger. After shredding, particles range up to about 15 centimeters (6 inches) in diameter, with an average of about 10 cemtimeters (4 inches). Shredders are subjected to heavy wear from glass and hardened steel objects such as ball bearings, and require frequent servicing.

For use as supplementary fuel in power stations, urban waste must be shredded to a top size of 4 centimeters (1½ inches). In the St. Louis pilot operation discussed below, 30 large metal hammers swing around a horizontal shaft and grind the solid waste against a steel grate, shredding the waste in one step. With a throughput of 650 tons per day, hammer retipping is required almost daily (Lowe, 1973). Two-stage shredding with air classification of the 10 to 20 centimeter (4 to 8 inch) particles between the two processes and fine shredding of only the light fraction is expected to reduce hammer wear.

Size reduction is expensive ($3-$4 per ton in the United States), but it is advantageous, whatever other treatment is used, even if this is restricted to landfilling (EST, 1973c; Ham et al., 1972; Hitchell, 1975; Marsden, 1973), and in the United States in 1974 there were 87 solid waste shredders in use (NCRR, 1974b). Magnetic separation may be used after shredding to remove ferrous metals. Together, shredding and magnetic separation form a basic system which can be used prior to landfill, incineration, pyrolysis, composting, or fiber recovery (Cannon and Smith, 1974). The relative advantages of subsequent processes (Hershaft, 1972) depend on many factors, including local demand for recycled materials.

A "semi-wet" pulverizing system was developed recently in Japan (Ito and Hirayama, 1975a, 1975b) to take advantage of the different mechanical properties of urban waste components and to incorporate both shredding and classification in one machine, with minimum energy expenditure. Glass, ceramics, and other brittle materials, together with food wastes, are pulverized in the first rotating drum stage and discharged through the first screen (Group I). The remaining waste is dampened, if necessary, to weaken the paper structure (hence "semi-wet"), and undergoes a process in which paper and materials with similar mechanical strength are selectively shredded and discharged through the second screen (Group II). All the remaining materials, which are ductile (plastics, textiles, leather, metals), are discharged from the open end of the drum (Group III). Test results showed that 47% of the paper (on a dry weight basis) found its way into Group II, where it made up 85% of the Group II content. The other classes of material were separated even more completely than this.

Air Classification

Air classification, winnowing, or air elutriation techniques have been used extensively in industrial processes, and are now being applied to the separation of low-density waste (paper, paperboard, plastics) from metals and glass (Fan, 1975). The City of Madison, Wisconsin, and the Forest Products Laboratory of the U.S. Department of Agriculture have operated jointly an air classification pilot plant in conjunction with a shredder, and satisfactorily separated paper, glass, metal, and plastic from raw urban waste (Auchter, 1973). Similar work has been carried out by the U.S. Bureau of Mines (Kenahan, 1971; Kenahan et al., 1973), the Stanford Research Institute (Boettcher, 1972; Hershaft, 1972), the Franklin Institute, Philadelphia (Franklin et al., 1973a), and Raytheon Service Company (Grubbs et al., 1976). The subject has been reviewed by the National Center for Resource Recovery (NCRR, 1974c). Air winnowing is sometimes used to remove light, fibrous materials which would absorb large quantities of water in subsequent wet processes.

Density Methods

''Float-sink'' separation (organics floating, inorganics sinking) can be carried out by a controlled water current method after shredding, air classification and magnetic separation (Alter et al., 1974; Degner and McChesney, 1974). ''Hydropulping'' (or ''hydrapulping'') reduces the organic components to fine particles to be carried in a water slurry. The dense materials are removed by gravity, and the ferrous metal, aluminum and glass may be recovered in good condition.

The Warren Spring Laboratory in Stevenage, United Kingdon, is developing an urban waste sorting system utilizing a ballistic inertial separator, as described in the later section on ''Prototype Systems.''

Screening

Dry screening on number 4 mesh screens (with 4.75-millimeter, 0.19-inch, openings) of one sample of urban waste (Dean et al., 1971) removed over 87% of the dirt and glass, but resulted in less than 4% loss of the total combustibles, making this simple process a potentially valuable separation method. Also, putrescibles have been concentrated in the undersize fraction resulting from screening with 65-millimeter (2.5-inch) round hole perforated plates (Cavanna et al., 1976). However, screening is usually difficult because of high water content, and is often preceded by air elutriation, which provides some drying. An alternative is *wet screening* after hydropulping and wet benefication. Paper and other organic fibers are collected on a screen to form a crude pulp which can be used for lower grade papers such as building paper, or as fuel. (The presence of grass and similar plant matter containing lignin reduces the quality of the paper produced.) The remaining organic matter in the slurry can be concentrated for incineration or composting. A prototype Black Clawson Hydrasposal Fibreclaim plant of this type has been built in Franklin, Ohio (see "Prototype Systems"). The alternative "Cal" Recovery System separates the paper dry by shredding, air classifying and screening before slurrying (Diaz, 1975b; Diaz et al., 1974; Golueke, 1975; Savage and Trezek, 1976; Savage et al., 1975; Trezek and Savage, 1976).

Methods for the Future

At the Massachusetts Institute of Technology a prototype recycling plant has been built which is designed to accept urban refuse from ordinary collection vehicles, and physically separate the waste into individual items. This *automated item identification* differs in principle from all the processes described previously (Senturia, 1974; Senturia et al., 1971; Wilson, 1974b; Wilson and Senturia, 1974).

After separation of loose paper and plastic film in the air stream, and the removal, by screening, of small material for shredding and air classification in the conventional way, the larger items are dropped by a conveyor into carts running on a continuous track. A metal detector, an impact sensor, and an infrared sensor (Chapter 3) interact with each item through a slit in the base of each cart. Data from the sensors are analyzed by a computer, and as a result of the classification decision the cart empties the item it contains at one of a series of dumping stations. If the various sensors do not agree on the composition of an item, it is sent by way of a hammermill to join the small particles. The process is illustrated in Figure 6.2.

The use of sensors of this type is not new. Color, opacity, or surface texture are often used with clean feedstocks for the separation of glass and ceramics, or for sorting glass by color. The novelty here is the attempt to separate the larger items of completely heterogeneous urban waste without prior shredding, incineration, or hydropulping. If successful, this technique will provide recovered products with less contamination than present methods. Initial trials indicated that the system was able to identify members of the categories glass, metal, plastic, and cellulose with better than 90% accuracy.

Finally, the most sophisticated method that has been proposed is the application of the *nuclear fusion torch,* but as pointed out in Chapter 4, this method does not appear to be energetically feasible, even given a viable nuclear fusion process (Rose et al., 1972).

BIOLOGICAL TREATMENT OF URBAN WASTE

The various types of biological processes outlined in Chapter 5 can be applied to the organic fraction of urban solid waste.

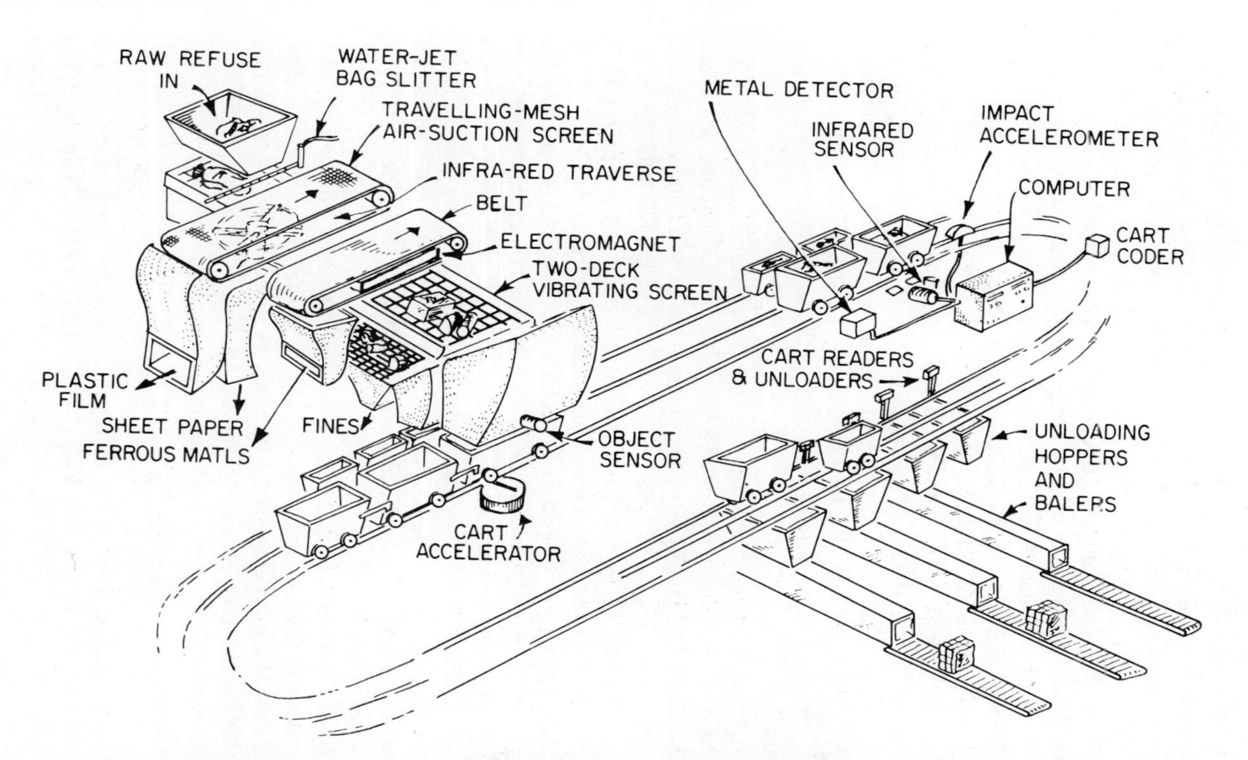

Figure 6.2. Diagram of the M.I.T. Presorter and large item sorter. (From Wilson, 1974; Wilson and Senturia, 1974. Reproduced by courtesy of David Gordon Wilson, Massachusetts Institute of Technology.)

Composting

Composting processes involve preparation (sorting, salvaging, shredding, perhaps adding water and extra nitrogen), stabilization (aerobic digestion for 5-7 days in aerated bins or for 2 weeks to 3 months in windrows), and upgrading (grinding, enriching, packaging). Air classification can be used to remove impurities from matured compost (Boettcher, 1972). Although there are no major technological problems in making compost from urban waste, and in some places it is a well-established method,* the process is expensive, and (particularly in Britain and the United States) there is often a problem in finding satisfactory markets for the product. In Britain in 1971, for example, only a dozen local authorities were known to be operating composting plants (Biddlestone and Gray, 1973). Many of those which began operating in the United States since 1951 have now closed (Bond and Straub, 1973; Hershaft, 1972; Meyer, 1972; NCRR, 1974c; OST, 1969; Pavoni et al., 1975). Composting projects have been most successful in Europe, particularly in Holland and Denmark.

Compost prepared from urban waste has little value as a fertilizer unless it is fortified, but it is particularly useful as a humus-forming agent and soil conditioner. It should be particularly attractive in areas where soil humus is depleted (eroded and marginal soils), or where the soil tends to pack solidly, as in the Bangkok area (Carter, 1975). The Libyan cities of Tripoli and Benghazi plan to convert their urban waste into compost to improve desert soils (ME, 1976a; MRW, 1976e). The City of Odessa in Texas plows its separated, shredded waste into the ground as a soil conditioner in this semi-arid region (Newell and Sandoval, 1976). Organic solid waste can be combined with sewage sludge in the composting process, since moisture must be added in any case, and the nutrients in raw sludge assist the decomposition of solid waste and augment the value of the compost (EST, 1968a; Flintoff and Millard, 1969; Prescott, 1967; Valdmaa, 1973; Wylie, 1955).

If the organic components in urban waste are treated by composting, it is always necessary to dispose of a proportion of the waste by some other means. However, if composting is combined with recovery processes for paper, metals, glass and plastics, this problem can be minimized. Plastics are normally biologically as well as chemically inert, but chemical pretreatment and selection of the appropriate microorganisms could make plastic composting feasible (Gutfreund, 1971).

In the United States, two plants in regular operation have used enclosed digester techniques (Cardenas and Varro, 1973; EST, 1971a, Franklin et al., 1973a, 1973b; Hershaft, 1972; Pavoni et al., 1975). In the Fairfield-Hardy

*Cardenas and Varro, 1973; Engdahl, 1969; Goldstein, 1969; Hays, 1973; Howard, 1972; Mantell, 1975; Meyer, 1972; Neal, 1971; Perna, 1971; Wiley et al., 1966; Wylie, 1955.

process at Altoona, Pennsylvania, the waste is ground in a wet pulper, dewatered in presses, and fed into a digester where it is stirred and aerated for 5 days. The Varro process, used by Ecology, Inc., in Brooklyn, New York, uses harrows operating in an enclosed digester with a 40-hour retention time. After the waste is shredded, up to 90% paper can be accommodated, and the noncompostibles need not be removed before digestion (only ferrous metals are removed in pretreatment). In the European Dano or "Bio-stabiliser" system, continuous rotation in large cyclinders for a few days at the optimum temperature for the inital thermophilic process is followed by milling and a maturing period in windrows (Howard, 1972; Mantell, 1975; Price, 1972). It is noteworthy that during the rotation very effective pulverization is achieved without the wear involved in other shredding and milling processes (*Solid Wastes,* 1976c).

An important practical problem is the possibility of excessive levels of some metal salts accumulating in soils which receive regular applications of this compost, due to the action of organic acids on metal waste components during the composting process (Mellanby, 1975; Neal, 1971; Skitt, 1972). Apparently, lead is not taken up to a significant extent by plants while the organic level of the soil remains high, but the potential danger exists. Small metal fragments which pass through the final sieve might be leached also, particularly in acid soils. This aspect is still receiving attention. Another potential contaminant is hormone herbicide or similar chemical introduced into the urban waste stream (Howard, 1972), although in large-capacity plants dilution usually overcomes any problems of this kind.

Skitt (1972) has considered the pros and cons of composting for urban waste disposal, reporting the conclusion of the Working Party on Refuse Disposal in Britain in 1971 that composting can make little contribution to the disposal of urban waste except in certain local circumstances. Pavoni et al. (1975) recently reviewed the situation in the United States, and came to a similar conclusion. Research is continuing to overcome the economic, toxicity, and distribution problems; one possible solution is the use of compost as a storable fuel (Wilson, 1974a).

Anaerobic Digestion

This microbial process occurs spontaneously in some urban waste landfills, and although the methane generated can cause problems, it can be utilized (EST, 1975e; Augenstein et al., 1976). Anaerobic digestion has been used in sewage treatment plants for at least 50 years, occasionally with the incorporation of urban solid waste, but scientific investigation of the anaerobic fermentation of solid waste began only recently (Klein, 1972). As natural gas reserves are limited, the methane generated by this method of urban waste disposal would be valuable for many communities (Bohn, 1971).

One of the problems is maintaining the optimum carbon:nitrogen ratio (a mole ratio no higher than 30:1). Solid urban waste, with a high cellulose content, typically has a carbon:nitrogen ratio of at least 60:1, so it is necessary to add nitrogen. Sewage sludge may well be used for this purpose as it has a high nitrogen content. Since the advent of the kitchen grinder disposal unit, sewage sludge has contained significant amounts of garbage. Taking this idea further, the organic waste could be passed as a slurry to the sewage system (Madigan, 1975), either by making kitchen grinders mandatory (that is, collecting separately only paper and inorganic waste), or by siting the solid waste collection centers adjacent to sewage treatment works equipped for methane production. However, these schemes do have the disadvantage of introducing all the pathogens present in the sewage, and these might not be destroyed completely in the anaerobic reactor, making subsequent disposal of the residue difficult.

It appears that anaerobic digestion is more applicable to rural than to urban waste disposal situations, but research and development are continuing (Pavoni et al., 1975; Yen, 1974b). The institutions involved include the Institute of Gas Technology with the "Biogas"® process (Ghosh and Klass, 1974; Klass and Ghosh, 1973; *Public Works,* 1976); the Dynatech Company (Kispert et al., 1975, 1976; Wise et al., 1975); the University of Illinois (Pfeffer and Liebman, 1976); the University of California (Diaz et al., 1974; Klein, 1972); and Massachusetts Institute of Technology (Cooney and Wise, 1975).

Direct Animal Feed Use

Food wastes from institutions and farmhouses are used widely for feeding animals, mainly pigs and chickens, all over the world. The feed should be cooked immediately before feeding to pigs, to minimize the risk of diseases such as swine fever and foot-and-mouth disease. This hazard is a significant one, but there was widespread opposition to the announcement by Australian agricultural authorties that swill feeding to pigs was to be banned (Pausacker, 1975), and the stringent sterilization regulations in Britain have caused a marked decrease in the use of pig swill (Mellanby, 1975). Obviously, central plants producing sterilized stock feed should be set up, and it would be most efficient to combine these with the main urban waste treatment facilities in some form of "urban farming" (Hughes and Jones, 1975).

Holloway and Scanlan (1975) have suggested that it is desirable to use ruminant animals (such as sheep) and nonruminants (such as chickens) directly in the recycling of cellulosic urban waste and in conjunction with an anaerobic reactor for other organic wastes, for the following reasons:

1. Ruminant animals can digest much of the cellulose fraction of the waste (Chapter 7);

2. The feces and urine of the animals provide enzymes to assist in the cellulose decomposition in the anaerobic digester;
3. SCP byproducts from the anaerobic reactor can be utilized as a non-ruminant animal feed supplement;
4. The animals provide food and fiber in the form of meat, hides, wool.

SCP Production

All the processes for single-cell protein production that were described in Chapter 5 in principle can be applied to urban solid waste: SCP harvested from algae fed on the sludge of anaerobic digesters can feed chickens and pigs, or cellulose can be converted enzymatically to sugars, followed by fermentation to produce a SCP such as yeast plus a stock feed residue. Because wastes arise even in microbiological conversions, the proposed integrated urban farm systems use animals, insects, fish, and microbes in coordinated systems producing food and

Table 6.6. Typical Heat Contents of Waste Materials and Fuels

Material	Btu per Pound	Gigajoules per Tonne or Megajoules per Kilogram
polyethylene	20,000	46
polypropylene	20,000	46
polystyrene	18,000	42
poly(vinyl chloride)	8,000	19
polyurethane	10,000	23
synthetic rubber (tires)	14,600	34
paper, paperboard	7,700	18
rag (cotton, linen)	6,400	15
typical urban waste	4,500-6,500	10-15
corncobs	8,000	19
rice hulks	5,000-6,500	12-15
bark	4,500-5,200	10-12
leaves	5,000	12
bagasse (sugar cane)	3,600-6,500	8-15
power station coal	11,000	26
oil	18,000	42
natural gas	23,000	53
brown coal	4,200	10
wood	9,000	21

Souces. CEN, 1974a; Sperber and Rosen, 1974.

fertilizer from urban waste and sewage (Holloway and Scanlan, 1975; Hughes and Jones, 1975).

ENERGY RECOVERY FROM URBAN WASTE

With its relatively high content of modern packaging materials, urban waste has a heat content between one-third and one-half that of power station coal, and better than that of a low-grade coal (Table 6.6). The relative merits of various methods of energy recovery have been discussed widely.* It is estimated (Hirst, 1973a, 1973b, 1975) that if 10% of the solid waste in the United States had been incinerated with energy recovery (in 1970), 25×10^9 kilowatt-hours of electricity could have been generated, that is 1.8% of total electricity consumption and 0.4% of total United States primary energy use.

Incineration

The disposal of urban waste by incineration offers advantages in volume reduction and the destruction of pathogenic microorganisms, and therefore has been used widely. Thus it is easy to see why energy recovery incineration is one of the two fully developed methods of utilizing the resources from mixed urban waste (the second being composting).

At its best, with the right equipment and in the right location, incineration can reduce the volume of organic matter by 90%, leaving sterile fly ash and residue. At its worst, the process creates air pollution and leaves a residue of 40% of the original volume. We are concerned here only with those installations which utilize the heat from waste combustion, that is, with recuperative incineration. Usually the waste is burned in an incinerator-boiler plant, the steam being used for electricity generation and/or space heating for buildings. Waterwall furnaces for this purpose, which were developed in Europe, permit operation at temperatures considerably higher than in furnaces with refractory walls, and incinerators of European design have been introduced into North America (ACS, 1969; EST, 1967). Information on incineration theory and design is widely available,† and problems in the economics and marketing of energy from waste

*For example, ANERAC, 1976; Bailie, 1971; Bailie and Doner, 1975; Blaustein and Tosh, 1975; Broussaud, 1976; Cross, 1972; Eggen and Kraatz, 1974b; EST, 1971b; Fife, 1973; Franklin et al., 1973a; Golueke and McGauhey, 1976; Hartley, 1975; Hiraoka, 1975; Jackson, 1974; Kasper, 1974; Kenward, 1975a; Kirov, 1975b; NCRR, 1974c; Pfeffer, 1974; Pfeffer and Liebman, 1976; Poole, 1975; Porteous, 1975b, 1975c; Rex, 1975; RRC, 1976; Sheng and Alter, 1975; Sheehan, 1975; Tillman, 1975; Varjavandi, 1975; Weinstein and Toro, 1976; Wilson, 1974a; Wilson and Freeman, 1976.

†For example, Baum and Parker, 1974a; Bond and Straub, 1973; Carpenter, 1971; Chamberlain, 1973; Corey, 1969; Coulson, 1976; Engdahl, 1969; EPM, 1975c; Essenhigh, 1970; Fisher, 1976; Franklin et al., 1973b; Harvey, 1973; Helliwell, 1972; Hershaft, 1972; Jackson, 1974; Mantell, 1975; NCRR, 1974c; Neal, 1971; Pavoni et al., 1975; Skitt, 1972; Stephenson et al., 1975.

have been discussed by Franklin et al. (1973a), Viscomi (1976), and Wilson and Swindle (1976).

European cities such as Amsterdam and Glasgow have utilized waste resources in this way for years; Paris has a large plant in operation at Issy-Les-Moulineaux; and the Greater London Council Edmonton installation provides electricity for the equivalent of 25,000 householders. The Norfolk Naval Station in Virginia, United States, has been using waste to generate steam since 1967; the Chicago Northwest Incinerator is designed to burn 1450 tonnes of waste per day and produce steam; and a steam-generating plant is being constructed at Saugas, Massachusetts (see the section "Prototype Systems" at the end of this chapter). The heat is used in some installations for generating electricity, drying sewage sludge, or the desalination of seawater; usually the ash is also utilized, for road construction, bricks, or concrete.

Air pollution problems must be solved, particularly as these incinerators are usually close to the urban areas they serve, and urban waste contains substantial amounts of toxic metals (Campbell, 1976). About 1% of the refuse is emitted as fly ash, and electrostatic precipitators and/or wet scrubbers are usually necessary. The majority of municipal incinerators in the United States, for example, probably do not meet air pollution standards in this respect, although suitable equipment is available (Jensen et al., 1974). Plastics in waste tend to cause problems because of their high heats of combustion, the tendency of thermoplastics to melt and clog grates, and the corrosion and air pollution resulting from release of gases, such as hydrogen chloride from poly(vinyl chloride) (EST, 1971c; Milgrom, 1972). At high tempratures, hydrogen cyanide can be emitted during polyurethane combustion. Although at the present plastics level of 1-2% in waste the difficulties are not great, some authorities believe that 3-4% will cause problems (Gutfreund, 1971).

Skitt (1972) has considered the factors involved in incineration heat utilization, with particular reference to Britain. Some United States observers have been pessimistic about the future of incineration (EST, 1972b), but recent increases in fuel costs may change this attitude. District space heating is attractive for local, densely populated areas, but electricity generation is economically feasible at present only when large plants are available to serve a high waste-producing population in a relatively small catchment area. From a total energy point of view it is, of course, most appropriate to obtain from incineration both electricity and low-temperature steam and hot water for house heating (Chapter 8). The economics of relatively small-scale incinerators with heat recovery (for example, in a hospital or apartment block) may be favorable in some circumstances (EST, 1972c). In this connection, other types of incinerator are being considered or are under development, such as fluidized bed furnaces (EST, 1968b), molten salt reactors, and molten metal reactors. All these techniques, discussed in Chapter 4, have potential advantages over conventional incinerators; they are more efficient on a small scale, and are less likely to produce air pollution. Another innovation is the development by the Combustion Power Company of a gas turbine driven by incineration gases generated by wastes in a fluidized bed (see "Prototype Systems").

As well as technical difficulties due to the nature and variability of urban waste, there is the problem of supply and demand in the energy produced. Energy recovery by incineration must be matched in place and time by consumption. A five-day week waste collection and incineration process is not capable of providing the sole source of heat or electricity on a continuous basis without extensive storage facilities. It is possible to store the fuel, either in solid form, or converted to gas or liquid, and some examples of this approach are considered in subsequent sections.

Substitute for Conventional Fuels

A modification of the incineration process is to use raw or separated waste as a substitute for a portion of the conventional solid fuel in industrial furnaces (EPM, 1976b; *Industrial Recovery,* 1976c: MRW, 1976g; *New Scientist,* 1976a; Young and Lisk, 1976); in kilns for the manufacture of cement and lightweight shale and slate aggregates (Chapter 7); and in electric power stations (Alter and Reeves, 1975; Tunnah et al., 1974). In a combined venture of the Union Electric Company, the city of St. Louis, and Horner and Shifrin, up to 25% of the coal in a 125-megawatt unit has been replaced by wastes. A similar process is being developed in England by the Warren Spring Laboratory and will be used by the Tyne and Wear County Council. These examples are included in the section, "Prototype Systems"; another example is the Ames, Iowa, Resource Recovery Plant (Funk and Russell, 1976). Some pretreatment of the RDF ("refuse-derived

Table 6.7. Comparison of Properties of Waste and Coal Used in St. Louis Power Station

Property	Waste	Southern Illinois Coal
Water content	20-31%	6-10%
Sulfur Content	0.2-0.3%	3.1-3.9%
Ash	9%	9.7-10.8%
Heating value		
Btu per pound	4100-5500	11,300-11,900
Megajoules per kilogram	9.5-13	26.3-27.6

Source. CEN, 1974a, with permission of the copyright owner, the American Chemical Society.

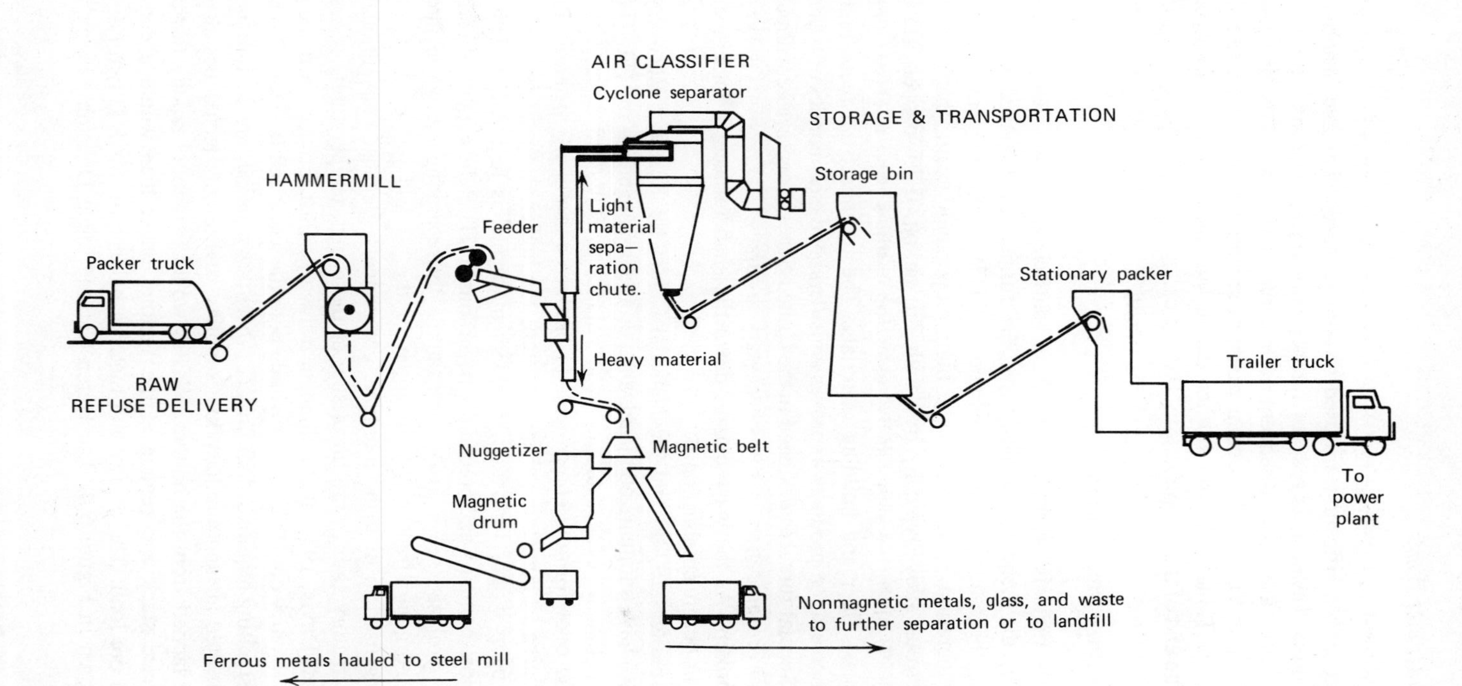

Figure 6.3. Diagrammatic representation of the St. Louis fuel preparation and resource recovery system. (After Lowe, 1973.)

fuel'') or WDF (''waste-derived fuel'') is necessary, and, typically, magnetic separation and air classification are used to remove noncombustible materials from the shredded waste. This reduces furnace component wear and improves the quality of the ash for subsequent utilization. The process is depicted diagrammatically in Figure 6.3, a comparison of some properties of waste and coal is made in Table 6.7, and further details are included in ''Prototype Systems'' at the end of this chapter.

High-Temperature Incineration

If the incineration is carried out at a sufficiently high temperature (1600-1700°C), the noncombustibles leave the furnace in a molten state, a process called ''high-temperature incineration'' (HTI), ''total incineration,'' or ''slagging incineration.'' As well as the high-temperature steam, this method produces a ferrous alloy (typically iron with 2% carbon, 0.7% copper, 1.5% silicon and 0.2% sulfur) which can be used for castings, and a glassy slag suitable for glass fiber and building materials. The revenue from these high temperature incineration products improves considerably the economics of waste incineration. Special furnaces designed in the United States accept waste without pretreatment or preseparation, but it is necessary to supply auxiliary fuel to keep the residue molten. As the waste passes down the shaft furnace, it undergoes partial pyrolysis before combustion.

Baum and Parker (1974a) and the National Center for Resource Recovery (NCRR, 1974c) have summarized the advantages and disadvantages of HTI. The advantages are that it reduces urban waste volume by 97% and destroys all putrescible and odor-forming components to produce a useful fill material. To offset this relative to conventional incineration, the higher temperature causes increased nitrogen oxides formation, supplementary fuel is required, and the operation is more complex and potentially hazardous. HTI can be combined with conventional pyrolysis (see below), for example by separating the waste into organic and inorganic fractions, with some of the oil and gas products of pyrolysis being used to melt the inorganic components. Because conventional incinerator refractory linings are unsuitable at these high temperatures, it has been suggested (Davis, 1974) that redundant ironworks blast furnaces in Britain could be used to carry out HTI, thus recycling both the waste and the plant. Both HTI and iron smelting require a shaft furnace where raw materials are fed into the top, with preheated air and auxiliary fuel supplies, and in both techniques slag and metal are tapped from the bottom. Wastes would pass more rapidly down the blast furnace stack and remain at the bottom longer than would a conventional iron ore load. The energy and material balance of this HTI-pyrolysis process is shown in Figure 6.4. The Union Carbide Linde Division ''Purox''

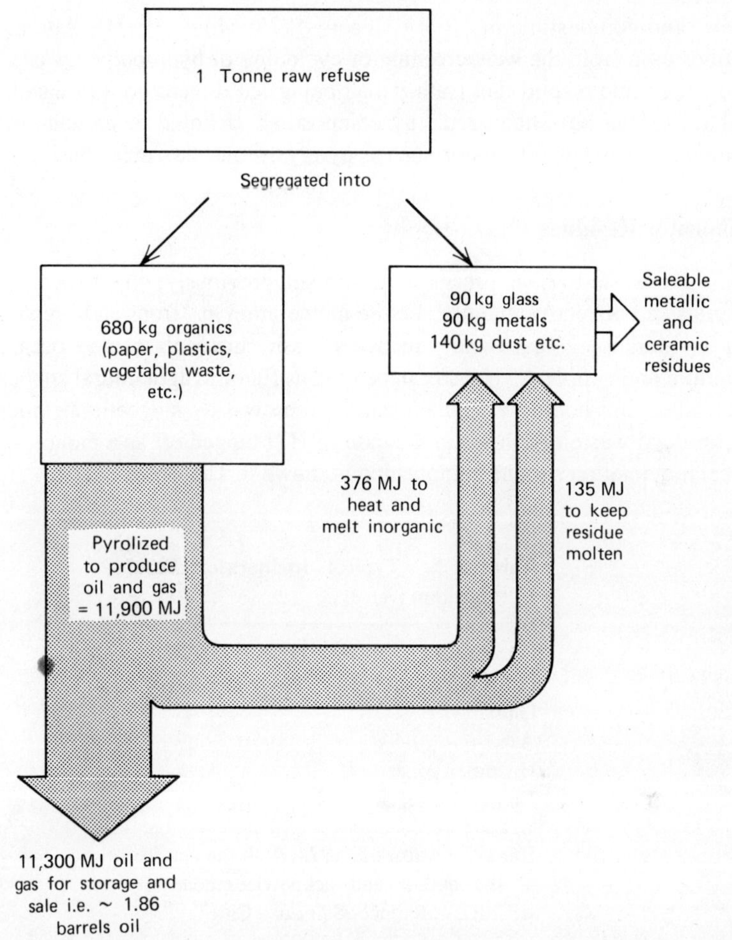

Figure 6.4. Energy and materials balance for a combined high temperature incineration—pyrolysis process. (From Davis, 1974. This first appeared in *New Scientist,* London, the weekly review of science and technology.)

slagging pyrolysis process uses pure oxygen to develop the high temperature, and the "Torrax" process preheats combustion air (see "Prototype Systems").

Storable Solid Fuel

Another approach to energy recovery is to produce a storable solid fuel: fuel recovery rather than direct energy recovery (Tillman, 1975). The urban waste is shredded, dried, air classified, and separated into ferrous metals, nonferrous metals, and combustible material (Allison, 1975; Morey, 1974). Alternatively, the fiber pulp from the wet screening or cycloning of hydropulped waste can be dried to provide a solid fuel (rather than being fed directly to a furnace); or the solid waste can be "stabilized" by composting. If liquid or gaseous fuels are required, a chemical conversion such as pyrolysis must be carried out.

Incinerator Residues

In all the waste-to-heat processes described previously, the noncombustible materials can be recovered either before incineration in "front-end" recovery, or after incineration ("back-end" recovery): ash, iron, glass, and occasionally aluminum and copper.* The ash can be used as fill or in agricultural applications. The ferrous and nonferrous metals can be removed by magnetic methods from the shredded waste or ash, or in the case of HTI tapped off as a molten alloy. A typical incineration residue composition is shown in Table 6.8.

Table 6.8. Typical Incinerator Residue Composition (%)

Ferrous scrap	30
(including cans	17)
Glass	43
Ash	17
Unburned paper	8
Nonferrous metal	2

Source. Ostrowski, 1974, with the permission of the author and acknowledgment to the AIChE Conference, Salt Lake City.

UPGRADED FUELS FROM URBAN WASTE

Analysis of urban waste (for example, Table 6.3) shows that there is up to 80% carbonaceous material, including 60% cellulose. If this organic material is chemically decomposed to lower molecular mass compounds by heating in the absence of oxygen or with limited oxygen, there are formed products that can be used as substitutes for fossil fuels and petrochemicals. This is in contrast to

*For example, Alter and Reeves, 1975; Cannon, 1973; Cramer and Makar, 1976; Cummings, 1974, 1976; Daellenbach et al., 1974, 1976; Dale, 1974; Dean et al., 1975; Henn, 1975; Jackson, 1975; Miles and Douglas, 1972; NCRR, 1974c; Nuss et al., 1975; Ostrowski, 1974, 1975; Sullivan and Makar, 1976; Sittig, 1975; Williams et al., 1975.

combustion or incineration of the organic material in an oxygen-rich atmosphere, where the carbon is reduced to its lowest energy level (carbon dioxide) and can be recycled only through plant photosynthesis. Discarded tires and plastics, in particular, are rich sources of fuel; 50% of the mass of a tire can be extracted as a low-sulfur, high-calorific value fuel (see Table 6.13).

Pyrolysis

The nature and proportions of pyrolysis products (gas, oil, char, and aqueous phase) depend on the temperature, pressure, heating rate, and heating duration, as discussed in Chapter 4, and in principle these factors can be varied to optimize production of the material which is more valuable in particular circumstances. Typical composition ranges of pyrolysis products are reported in Table 6.9. Considerable research into this process has been carried out by the United States Bureau of Mines (Schlesinger et al., 1973; Steffgen, 1974; Wender et al., 1974) and by other organizations. Analytical data on pyrolysis products from various solid waste components, information on thermodynamic and chemical engineering aspects, as well as details of prototype plants have been published.*

Because the temperatures in pyrolysis reactors are usually lower than those in incineration, and because the conditions are nonoxidizing, it is easier to extract clean glass and metal residues from the friable char. The pyrolysis can be carried out in a conventional closed retort, or in the more recently developed fluidized beds, molten salt reactors, and molten metal reactors. Alternatively, pyrolysis can be combined with incineration to bring about "gasification" (partial oxidation), or with high-temperature incineration ("slagging pyrolysis"), as discussed in Chapter 4. Advantages of pyrolysis are that the volume of the waste is reduced by 80-90%, and the plant may be located close to an urban center because there is minimal air pollution and little land area is required.

There is still a considerable amount of research and development to be done. Pyrolysis of poly(vinyl chloride) yields hydrogen chloride and a char material (Holman et al., 1974; Jensen et al., 1974), and the hydrogen chloride at high temperatures could bring about other, perhaps undesirable, reactions during the pyrolysis of urban waste; investigation of such reactions should be carried out. In domestic waste, the ammonia evolved from vegetable matter tends to neutralize the hydrogen chloride. It is estimated that the hydrogen chloride from up to 7% poly(vinyl chloride) in the waste stream could be neutralized in this way, and the present level is only about ½% (Pearson and Webb, 1973). Pyrolysis requires a

*For example, Alpert et al., 1972; Baum and Parker, 1974b; Bell and Varjavandi, 1975; Bond and Straub, 1973; Burton and Bailie, 1974; Crane et al., 1975; Douglas et al., 1974, 1976; EST, 1976b; Fife, 1973; Finney and Garrett, 1974, 1975; Franklin et al., 1973b; Hershaft, 1972; Ito and Hirayama, 1976; Jackson, 1974; Liebeskind, 1973; McIntyre and Papic, 1974; Morey et al., 1976; NCRR, 1974c; Pavoni et al., 1975; WSL, 1975d.

Table 6.9. Products of Pyrolysis of Domestic Waste

Product	Mass % of Waste	Calorific Value	Composition
Gas	15-50	13.5-12.5 MJ/m³ (350-550 Btu/ft³)	CO, H_2, CO_2, CH_4, C_2H_4, CH_6, and small amounts of other hydrocarbons
Oil	0-5	up to 23 MJ/kg (10,000 Btu/lb)	Complex mixture of hydrocarbons, aldehydes, ketones, etc.
Solids	20-40		
carbon char	4-20	14-21 MJ/kg (6000-9000 Btu/lb)	
ferrous	4-12		Mainly cans
glass	4-12		
nonferrous metal	0.2-1		
Aqueous	10-20		Contains dissolved organic compounds up to 5 mass %, B.O.D. up to 30 g/l

Source. WSL, 1975c. Crown copyright, reproduced by permission of the Director of Warren Spring Laboratory.

high capital investment, although there are available few detailed figures on capital costs and running costs (NCRR, 1973c).

Hydrogasification

When the organic fraction of urban waste is reacted with hydrogen at elevated temperatures, a gas is produced which is rich in methane. Wastes react at lower temperatures than does coal, and as a result the ethane yields are higher (ethane is formed in appreciable amounts only when the hydrogasification temperature is between 475 and 650°C). It is, therefore, possible to obtain a substitute natural gas (SNG) with higher heating value than the gas obtained from coal. A typical composition is: methane, 62%; hydrogen, 25%; ethane, 13%. As explained in Chapter 4, the hydrogen is produced from a portion of the waste to reduce the remainder, the complete system being as shown schematically in Figure 6.5. This process has reached the laboratory batch reactor stage in the Pittsburgh Energy Reseach Center of the U.S. Bureau of Mines (Steffgen, 1974; Wender et al., 1974).

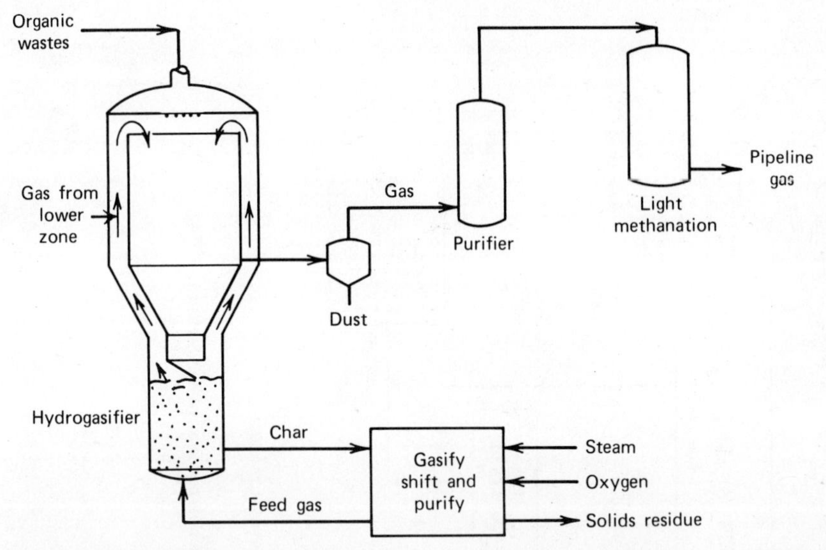

Figure 6.5. Diagrammatic representation of the waste hydrogasification process. (From Steffgen, 1974.)

Liquefaction

The Pittsburgh Energy Research Center has shown also that organic urban wastes react with carbon monoxide and water at temperatures above about 370°C to produce a heavy oil or low-melting bitumen (which can be used as a fuel oil) at a rate of approximately 2 barrels (320 liters) per ton of dry urban waste (Appell and Miller, 1973; Appell et al., 1971; CE, 1971; Steffgen, 1974; Wender et al., 1974). This work followed the observation that coal could be turned into a low-sulfur liquid fuel by reaction with carbon monoxide and water, and in fact carbohydrates are converted to oil at temperatures lower than those required for lignite. An alkaline catalyst is necessary, and although most wastes contain suitable salts, the reaction temperature may be reduced to 350°C or less if sodium carbonate or bicarbonate is added. Some of the oil produced in the reaction is recycled to slurry the incoming waste. The carbon monoxide may be introduced as synthesis gas made from a portion of the waste, as indicated in Figure 6.6. The process has reached the continuous bench stage after preliminary studies in laboratory batch reactors.

Partial Oxidation Processes

Partial oxidation of carbonaceous wastes at high temperatures (gasification)

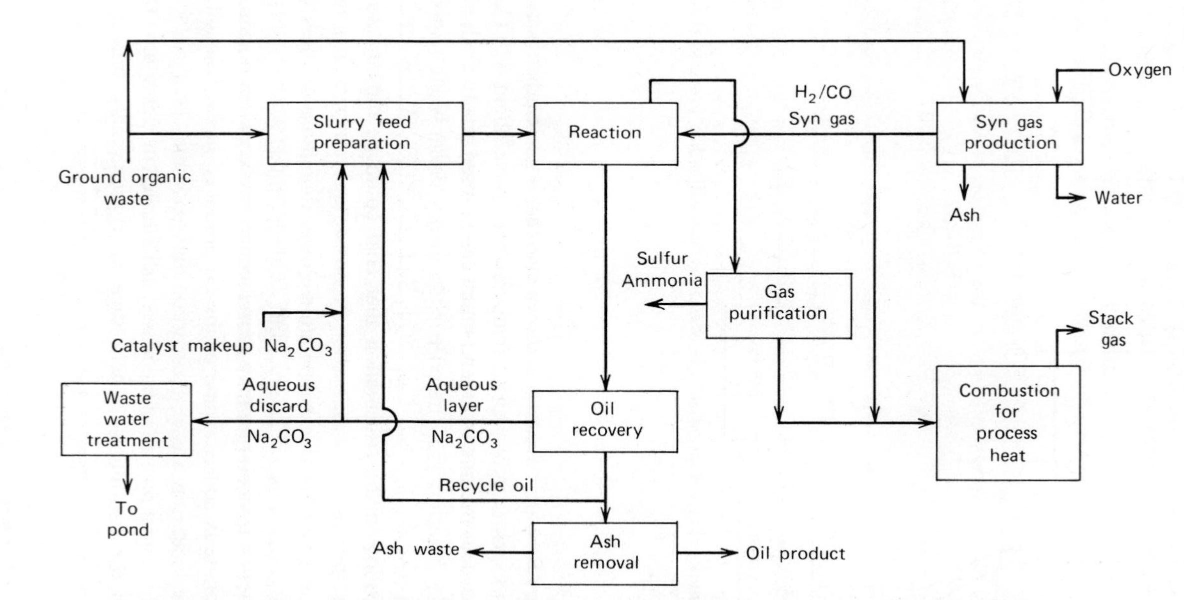

Figure 6.6. Diagrammatic representation of the waste liquefaction process. (After Steffgen, 1974.)

can produce carbon monoxide, hydrogen, and some hydrocarbons. In one form, air is injected directly into the lower portion of the waste, which is oxidized to generate the heat and reducing gases necessary for converting the upper portion into a fuel gas. Pure oxygen is used in the Linde unit (see "Prototype Systems") to reduce the volume of exhaust gases and minimize the generation of nitrogen oxides. If the temperature in the lower or oxidation zone is maintained below 700°C, the ash can be removed as a solid; alternatively, at a sufficiently high temperature (1000°C) "slagging pyrolysis" occurs; that is, a molten slag is obtained. If the temperature is in the 1200-1400°C range, the iron in the waste does not melt, but once oxidized it dissolves in the molten glass and decreases its viscosity so that drainage and tapping of the slag are facilitated. On the other hand, if the temperature is of the order of 1650°C the iron melts into a separate phase without oxidizing. This simplifies the recovery of the iron, but introduces new technical problems (Eggen and Kraatz, 1974a). The other partial oxidation processes that have been investigated for urban waste are wet combustion in an autoclave, and partial oxidation in a fluidized bed with an air-nitrogen atmosphere.

Chemical Hydrolysis

Chemical hydrolysis is suitable for processing wastes with a high cellulose content (or the fiber derived from urban waste) to produce glucose, which can be converted by fermentation to ethanol or SCP. Hydrolysis by hot dilute acid would be more rapid than enzymatic hydrolysis, but would require more expensive reactors; this technique has been proposed for urban waste (Converse et al., 1973; Fagan et al., 1971a, 1971b; Grethlein, 1975; Millett, 1975; Porteous, 1969, 1975a, 1975b, 1975c). In this process, which is in the exploratory stage of development, the cellulose is segregated from the rest of the waste, pulped, rapidly hydrolyzed at over 200°C in dilute sulfuric acid, cooled, neutralized, and fermented for about 20 hours at about 35°C. The resulting dilute aqueous ethanol solution is distilled to yield 95% ethanol. A disadvantage is that a high biological demand liquid waste stream must be treated before discharge.

Comparison of Oxidation/Reduction Processes

The methods that have been described show great ranges of product types and operating conditions. Some typical values of these variables are summarized in Table 6.10.

Comparison on the basis of product type, unit cost of waste conversion, cost of product, and cost of plant, are obviously difficult to make, particularly as there are few prototype plants in operation. Liquefaction, hydrogasification, and pyrolysis have been compared by the U.S. Bureau of Mines (Steffgen, 1974),

Table 6.10. Comparison of Oxidation/Reduction Processes for Fuel Recovery from Urban Wastes

	Gasification by Partial Oxidation	Hydrogasification	Chemical Hydrolysis	Liquefaction	Pyrolysis	Wet Partial Oxidation
Products	gas	gas	fermentable sugars	liquid, >95%	gas 50% oil 20% char 30%	liquid
Calorific value of gas	(fuel gas)	(SNG)			(fuel gas)	
Megajoules per cubic meter	~11	~37			~19	
Btu per cubic foot	~300	~1000			~500	
Pressure, megapascals	0.10	7-10	2.9	12-20	0.10	3-7
Pressure, psi	15	1000-1450	420	1800-2850	15	500-1000
Temperature, °C	700 (solid ash) 1650 (2 slag layers)	550-650	230	300-350	900	200-250
Gas added	O_2 or air	H_2(produced from waste)		H_2-CO (synthesis gas from waste)		air
Catalyst added			H_2SO_4	Na_2CO_3		
Stage of development	prototype	laboratory	laboratory	laboratory	several in operation	laboratory

showing that the capital cost, the operating cost, and the product fuel cost based on calorific value, are all substantially higher for liquefaction than for the other processes. The cost of the liquid fuel product, however, is estimated to be no higher than that of crude oil.

Other Chemical Products

The hydrogen produced from urban waste by partial oxidation followed by a ''shift'' of the carbon monoxide:hydrogen ratio (as described in Chapter 4) can be used as a feedstock for the catalytic synthesis of fuels such as methanol, or of ammonia (Cover and Schreiner, 1975; Seattle, 1975; Sheehan and Corlett, 1975). In the latter case, oxygen is separated from the air to carry out the combustion, and the nitrogen is subsequently reacted with the hydrogen:

$$C + H_2(g) \rightleftharpoons H_2(g) + CO(g)$$

$$C + \tfrac{1}{2}O_2(g) \rightleftharpoons CO(g)$$

$$CO(g) + H_2O(g) \rightleftharpoons CO_2(g) + H_2(g)$$

$$3H_2(g) + N_2(g) \rightleftharpoons 2NH_3(g)$$

Consideration has been given to the preparation of activated carbon from urban waste (Stevenson et al., 1973).

Another application of urban waste related to fuel production may be considered here. The nonmagnetic taconite ore contains 30% iron, but separation at present is uneconomic and it is rejected as waste. When roasted in a kiln furnace at 1000°C with urban waste, which contriubtes both fuel for heating and some metallic iron, the ''waste'' ore is converted to a 70% high-grade iron ore suitable for blast furnace production (USDI, 1972).

ECONOMIC FACTORS IN URBAN WASTE TREATMENT

The most obvious result of economic analyses of systems for resource recovery and energy recovery form urban waste is that they are not self-sustaining economic operations, but that when compared with traditional methods of waste disposal, they form viable alternatives in some situations. All methods require a large and localized source of waste, and easy access to industries utilizing the recovered materials. One review (Abert et al., 1974) found the economic ranking shown in Table 6.11 (although there is evidence that pyrolysis should be moved

Table 6.11. Urban Waste Treatment Economics Ranking in Order of Increasing Cost

Close sanitary landfill
Solid fuel recovery
Materials recovery
Composting with inorganic recovery
Pyrolysis with residue recovery
Remote sanitary landfill
Incineration with steam and residue recovery
Incineration with steam recovery
Conventional incineration
Incineration with electical energy recovery

Source. Abert et al., 1974.

toward the higher-cost end of the list), and projected 1980 economics show a similar order (Sheehan, 1975). More detailed analyses of the economics and systems planning are available in these and other reports referred to in this chapter. Because so few of the processes are fully developed with proven reliability, the relative merits of different processes are not completely clear, and when developed, some of the newer systems may prove to be superior to the "early starters." A recent opinion (Tillman, 1975) is that in the energy utilization area only three systems (Black Clawson "Fibreclaim," Union Carbide Linde "Purox," and Union Electric-Horner and Shifrin) have been operated at a sufficiently large scale over a sufficient period of time to permit careful analysis. In terms of cost effectiveness and environmental acceptability, another study (Schulz, 1975) found that the Union Carbide "Purox" and the Union Electric-Horner and Shifrin processes appeared superior.

We are in a transition period in urban waste treatment technology, and rapid changes are to be expected. In the task of detailed planning and comparison, computer simulation will be of assistance (Godfrey and Tupper, 1974). The effects of economies of scale in various systems are of critical importance (Mihelich, 1976). If a municipal authority is considering the installation of an energy or resource recovery urban waste treatment system, the most significant point to note at this stage is that there is a wide range of systems and potential products. Planning for waste *recovery* is more difficult than planning for waste *disposal*, and a manufacturing plant is more complex than a disposal plant. Local factors will influence strongly the decision:

- A nearby papermill may cause front end paper separaton to be favored;
- A local shortage of natural gas may be alleviated by a gasification process;
- Concurrent installation of a coal-burning electric power station may suggest solid waste fuel substitution;

- A nearby chemical industry may use ethanol produced by hydrolysis and fermentation;
- Nearby metal smelting and glass industries may justify recovery of noncombustibles as well as energy utilization.

This has been described as planning ''backwards'' (Kinnersly, 1973): determining which products are in demand, then deciding the systems that should be combined to provide them. Unless this is done, an efficient waste utilization process might be ''uneconomic'' relative to landfilling, for example, simply because there are no markets for the recovered material. If the urban waste supply is great enough, it could be possible to set up a self-contained ''material utilization center,'' depending on consumer markets rather than on raw material users (Rosen and James, 1974).

There is another alternative, an idea introduced earlier in this chapter. One school of thought considers that the high technology resource recovery plants are wasteful, and that local, low-technology urban waste collection and recycling systems can be both efficient and profitable (Pausacker, 1975; Seldman, 1976), just as village level technology may be more appropriate for agricultural waste utilization (Imrie, 1975; Imrie and Rightelato, 1976). Instead of building bigger and more complex systems which themselves use resources and energy and generate waste, we might modify our lifestyle to avoid mixing up the components of urban waste at the origin. Weekly municipal collections of food wastes only, and separate monthly collections of paper, metal, glass and plastics would be a simple method of introducting a far more economical recovery system if the public was prepared to cooperate. The alternative technology movement even extends to hand-crafted writing paper from waste paper (Allaby, 1976).

SEWAGE

As with urban solid waste, the usual approach to sewage in the ''developed'' countries since the introduction of the flush lavatory has been disposal rather than utilization. This policy is being reconsidered, both because ''disposal'' is becoming more difficult and because of a growing concern for conservation of natural ecological systems (Coldrick, 1975; Wood, 1975). Reclamation methods being investigated are mainly biological, but chemical systems may prove useful in some cases.

Biological Utilization

The principles of microbiological conversion of organic wastes were outlined in Chapter 5, and the processes have been described in detail for sewage by many authors (for example: ACS, 1969; Jenkins, 1963; Koziorowski and Kucharski, 1972; Masters, 1974; Nemerow, 1963). The first treatment for raw sewage is to allow it to settle in large sedimentation tanks. The deposited solid (primary sludge) retains 10-30 times its mass of water. This is then usually treated by an aerobic (oxidative) process which "stabilizes" it; this can be carried out by lagooning in oxidation ponds, or by activated sludge treatment. Biological treatment of sewage by means of activated sludge is an artificial intensification of the aerobic conversion process. The microorganisms form a suspension of finely flocculated material (activated sludge) which is added to the raw sewage, the mixture being agitated and aerated. The activated sludge has a large surface area, making it capable of adsorbing and oxidizing efficiently the fine colloids and suspended matter. As the aerobic processes continue, the quantity of activated sludge increases, and a portion is drawn off to treat further batches of primary sludge.

Both the sludge from primary sedimentation tanks, and the secondary sludge (after treatment with activated sludge, or on biological filters, which have a similar effect) may be processed by *anaerobic digestion*. In this process, various solid organic compounds are converted to liquid or gaseous products, particularly methane, which are being used to power many sewage treatment plants (Coldrick, 1975). The remaining solid material is called digested sludge; it has a much smaller volume than the primary sludge, does not putrefy, and dries easily. Wastewater, secondary sludge, and digested sludge have been used as fertilizers in conventional agriculture and horticulture and to assist in the reclamation of strip-mined land, in spray irrigation, and in ponds for hydroponics and fish farming (CCEQ, 1973; EST, 1972a; Pötschke, 1975; Sanks and Asano, 1976). The use of human as well as animal wastes as fertilizers has been reviewed comprehensively by Peterson et al. (1971).

Algae grown on oxidation ponds or in digested sludge ponds can be fed without processing to pigs, or dried for chicken feed (Priestley, 1976; WWE, 1973). *Yeasts* grown on sewage can be harvested and dried into a flaky concentrate containing 50% single-cell protein and suitable for human food. Over the last 30 or 40 years it has been suggested that digested sludge may be used directly as animal feed (Holloway and Scanlan, 1975; Pillai et al., 1967; Rao and Pillai, 1962). Its nutritive value is unquestioned, but care must be taken to monitor the final product, particularly with regard to accumulated metal salts and sterilization-resistant microorganisms. Nevertheless, the utilimate aim is total recovery of both organics and water from sewage (Greiser, 1971).

As noted previously in the section on the biological treatment of urban waste, sewage sludge is sometimes incorporated with urban waste before composting, to increase the nitrogen and phosphorus levels of the product. The same result is achieved on a domestic scale by devices such as Sweden's "Clivus Multrum," which may be installed in private dwellings, receiving toilet and kitchen wastes

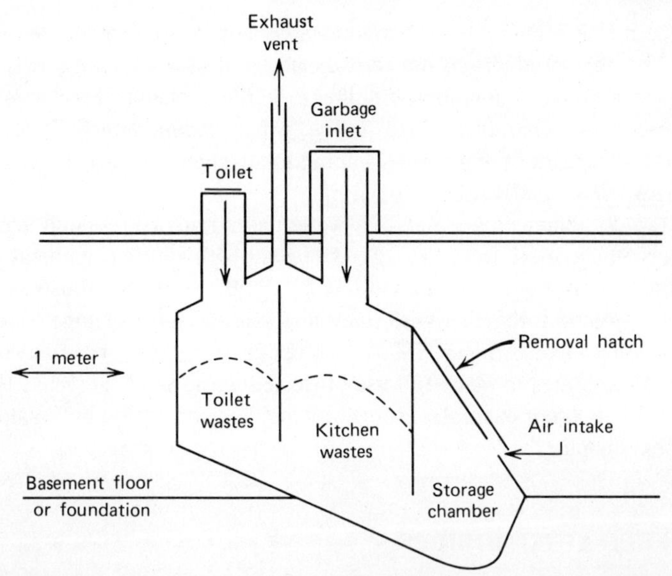

Figure 6.7. Diagram of the Clivus Multrum (After Milton, 1974, with the permission of the *The Futurist*.)

and producing compost (Lindstrom, 1975; Milton, 1974). It has no moving parts, uses no water, and overcomes the need for connection to main sewers (Figure 6.7). A layer of soil initially placed over the sloping bottom provides microorganisms which assist the decomposition. The Clivus Multrum is now available commercially in North America and Europe, and similar domestic sewage treatment systems in the United States have been described recently (*Compost Science*, 1976).

Chemical Reclamation

The wet oxidation and liquefaction methods for sewage sludge were referred to in Chapter 4. Sewage may be treated also with fly ash or soluble silicates. Fly ash, activated with acid and combined with small amounts of polyelectrolytes can be used in sewage treatment. It acts in the same way as activated carbon, removing most phosphate and organic matter to form a sludge which can be used as a fertilizer (Keller, 1973). Chemical fixation of activated sewage sludge with soluble silicates produces a solid material which when applied to soils has essentially the same inherent nutrients, water content and humus value as the wet, untreated sludge, but the potentially toxic heavy metals are converted to silicates which are available to plants only in quantities determined by the solubilities of the metal silicates (Conner and Gowman, 1975). It is not yet clear if the method is technically or economically suitable for large-scale fertilizer or soil conditioner production, but the silicate-fixed sewage sludge is suitable for land disposal in most situations. Another example of chemical reclamation is in a water recovery scheme at Lake Tahoe, where ammonium sulfate from the regeneration stream of ammonia-stripping ion exchange beds is being utilized as a fertilizer (*Chemical Week,* 1976).

The rapidly increasing demand for water has stimulated research into methods of water reclamation from sewage effluents, with physicochemical treatment supplementing biological treatment (ACS, 1969). Reverse osmosis is receiving particular attention, together with electrodialysis and ion exchange (Bailey et al., 1974; Sammon and Stringer, 1975). Santee, California was the first community in the United States to make full use of treated wastewater (Stevens, 1967), and there are many other examples, some of which are considered in "Water Reuse" in the next chapter.

SCRAPPED AUTOMOBILES

It is estimated (Cannon, 1973; Sawyer, 1974; USEPA, 1974b) that in the United States 7-8 million automobiles are discarded annually by the population of 220 million; that is, a disposal rate coefficient of 0.03 cars per person per year. The magnitude of the problem is indicated by the fact that in a recent year in the United States the automobile industry used 20% of the total steel consumed, 50% of the lead, 35% of the zinc, 7% of the copper, and over 10% of the nickel and aluminum (Lyons, 1972).

If the circumstances are favorable, a discarded automobile typically goes first to the auto wrecker or dismantler, who removes for sale those parts which are in demand, a valuable recycling operation that keeps millions of older-model vehicles on the road when they would otherwise become inoperable because of the lack of repair parts. The automobile then goes to the scrap processor, who removes by hand the more valuable scrap components: the battery, for its lead; the radiator, which is an excellent raw material for red brass or gunmetal (85% copper, 5% tin, 5% lead, 5% zinc); electric motor, generator and similar components for their copper; and the engine block and other cast iron components to be sent to foundries. This hand dismantling produces high-quality scrap, and although labor intensive, is still usually an economic operation. If properly stripped, the remaining hulk can produce high-grade steel scrap, but its disposal may still be a problem.

In reclaiming automobile bodies (after the removal of engine block, axles, and tires), there has been a progressive development in technology from cutting or guillotining, which is slow and labor intensive; to crushing them into bales, which makes subsequent separation of nonferrous metals difficult; to shredding

(Cannon, 1973; EST, 1970b); and recently to cryogenic fragmentation, which greatly facilitates separation and eliminates the danger of explosion from the petrol vapor in the fuel system when the hulk is burned or cut up with torches (Bilbrey, 1974; Braton and Koutsky, 1974; *Iron and Steel,* 1971). Steel can be removed magnetically from shredded bodies, and various physical separation methods such as liquid cyclone, selective melting, air separation, and hydrometallury can be applied to separate the nonferrous metals and nonmetals (Boettcher, 1972; Coyle et al., 1976; Dean et al., 1972, 1975; Dreisen and Basten, 1976; Froisland et al., 1972; Sittig, 1975; Valdez, 1976; Valdez et al., 1975).

If the automobiles are not to be shredded or fragmented, there is the problem of removing nonmetallics, and the usual pretreatment is burning out. Air quality controls now limit this practice in the open air, and "smokeless" incinerators have been developed for the purpose (Kaiser, 1969). The U.S. Bureau of Mines has developed recently an inexpensive smokeless incinerator with low operating costs (Cannon, 1973; Cservenyak and Kenahan, 1970).

Car shredders are extremely costly and require a large supply of disused vehicles, and auto scrap processors find it uneconomic to transport hulks to central locations. A solution to the problem of disposal in smaller cities and towns is the development of a mobile hydraulic press which crushes auto hulks to a thickness of 25 centimeters (10 inches). The press is mounted on a semi-trailer, which can carry also a fork-lift truck for feeding the bodies into the press. Up to 40, but usually 15-25 flattened car bodies can be carried on one semi-trailer for subsequent shredding at a central plant. However, transport costs may still exceed the steel value, and subsidies may be necessary. Also, there is evidence that in the Unitd Kingdom, car shredder capacity may soon exceed demand (MRW, 1976i).

The presence of an excessive amount of copper in steel adversely affects its ductility, and causes edge cracks during hot rolling, and its concentration is limited at present to 0.06% (Cannon, 1973). It is necessary, therefore, to develop metallurgical refining processes to remove this copper (Kenahan et al., 1973) or, as suggested in Chapter 2, to replace copper wiring in automobiles with aluminum. During subsequent smelting of the scrap steel, the residual aluminum would oxidize and enter the slag, perhaps replacing the aluminum now frequently added as a steel deoxidant. Copper is just one of the residual metals undesirable in quantities greater than some critical limit; others are nickel, tin, zinc, lead, aluminum, chromium, molybdenum, and vanadium. Also, sulfur accumulates in iron and steel being manufactured by normal refining processes if a considerable proportion of scrap is present. At the same time, standards for steel have become more stringent. Either new metallurgical processes must be developed (Daellenbach et al., 1976; Sittig, 1975), for example, the oxidation of contaminating metal (Grott, 1974); or vehicles must be planned and built for easy

Figure 6.8. The 1972 Mercury Montego, (a) before disassembly, and (b), sorted into its components: 1, light steel, less than ⅛-inch thickness; 2, heavy iron and steel, more than ⅛-inch thickness; 3, cast iron; 4, mineral wool; 5, glass and ceramic; 6, carbon, activated; 7, molded nylon; 8, Bakelite; 9, lead; 10, stainless steel; 11, asbestos; 12, copper and brass; 13, aluminum; 14, zinc die-cast; 15, mastic; 16, rubber; 17, polyurethane foam; 18, acrylic; 19, vinyl; 20, polyethylene styrene; 21, polypropylene; 22, nylon fabrics; 23, acrylonitrile-butadiene-styrene (ABS); 24, paper, fiberboard, and padding; 25, cotton, jute, textiles. (From Dean et al., 1974b. Photographs courtesy Bureau of Mines, U.S. Department of Interior.)

recycling (Dalyell, 1973); or the initial physical separation of the components before smelting must be improved. If the steel companies processed their own scrap, the quality could be improved (for example, by more hand dismantling) at little extra cost, but at present there is no economic incentive for a scrap dealer to remove more contaminants than necessary to achieve a particular grade of scrap, and as the steel company cannot depend on the quality, it will not pay a premium for ostensibly better scrap (Sawyer, 1974).

Considerable effort has gone into determining just what materials are available for recovery from scrapped automobiles, and also into studying the efficiency of the dismantling process (Sawyer, 1974). The U.S. Bureau of Mines in 1965 dismantled all components of 15 scrapped automobiles manufactured over a 10-year period, both to determine the fastest and most practical dismantling technique, and to obtain information on a "representative" scrap auto (Table 6.12). This composition, of course, is continually changing. In a 1972 extension of the project, the careful dismantling of a new car (Figure 6.8) showed a 17% decrease in cast iron and a 47% decrease in zinc, with a 69% increase in aluminum and a 31% increase in plastic up to 49 kilograms (17 pounds) (Dean et al., 1974b; Valdez et al., 1975). Comparable figures for Australia have bene given by Lyons and Tonkin (1975). It is predicted (CEN, 1974e) that there will be 180 kilograms (400 pounds) of plastic per car in the late 1980s. The effect of this will be to reduce the salvage value, unless suitable processes are devised for plastics reclamation (Campbell and Meloch, 1976; Mahoney et al., 1974; Valdez, 1976).

Table 6.12. Component Materials of a Representative 1954-1965 U.S. Automobile

Material	Kilograms	Pounds
Steel	1150	2530
Cast iron	230	510
Copper and brass	15	32
Zinc	25	54
Aluminum	23	51
Lead	9	20
Nonmetals	180	400
Total	1630	3600

Source. Kenahan, 1971.

One aspect of the obsolete automobile problem is technological, but there are also the social, environmental, economic and legal aspects, some of which have been considered by the U.S. Department of Commerce (USDI, 1972) and the U.S. Environmental Protection Agency (USEPA, 1972c). Because often there is nowhere to legally dispose of cars free of charge, it is estimated (USEPA, 1974b) that between 10 and 15% of the last owners succumb to the temptation of abandoning them in the coutryside or on a city street. Subsequent removal costs by the local authorities are then far greater than they would have been if a depot had been set up for this purpose. Twenty years ago the abandoned automobile was not a major problem. Open-hearth iron smelting methods could use almost 50% scrap, and the price of automotive scrap steel was high. The basic oxygen furnace produces steel in a much shorter processing time than does the open hearth furnace, so the proportion of scrap (containing considerable contaminants) must be lower because there is less time for oxidation of unwanted components. The change to the basic oxygen furnace, accepting only about 25-30% scrap, resulted in significantly lower prices for automotive scrap, which is less popular than other scrap sources because it has higher impurity levels. By 1967, over half a million cars were abandoned each year in the United States (OST, 1969). However, electric induction furnaces will accept up to 100% scrap, and demand for automotive scrap may improve if this method of smelting becomes more widespread.

Although from aesthetic and environmental points of view the removal of car hulks from the countryside and auto wreckers discard piles is highly desirable, the advantages in terms of energy consumption at present are significant but not phenomenal. More energy would be saved if automobiles were well maintained and their useful lives extended. Also, world iron ore reserves are still good, although some of the minor component metals soon may be in short supply. In terms of disposal economics, if the consumer has to pay for the disposal cost of the product, it is often cheaper to recycle it than to find somewhere to dump it legally. Because the final owner of an automobile is usually unwilling to pay for its disposal, some authorities are starting to make the first owner pay a levy in anticipation of its ultimate disposal, just as many consumers will have to pay in advance for the retrieval of their littered beverage containers.

A full analysis of automotive scrap recycling has been carried out recently by Sawyer (1974) for Resources for the Future, stressing economic as well as technological factors. This book contains references to numerous sources of information on the automobile scrap industry generally.

DISCARDED TIRES

Between 1 and 2 tires per vehicle are discarded each year, totalling over 250 million for the United States (Nuss et al., 1975; USEPA, 1974b). Tires are not biodegradable to any appreciable extent, whether dumped in the ocean or buried in landfill tips. The resilient, energy-absorbing structure of tires makes them difficult to fragment by normal techniques, and because of their shape they are expensive to transport when whole. Related to their resilience and shape is the tendency for tires to "float" up through other materials in landfills, working their way towards the surface (Braton and Koutsky, 1974; Search and Ctvrtnicek, 1976). The accumulation of tires on open tips is unsightly, a potential fire hazard, and a resting place for vermin. For these reasons, many landfill sites refuse to accept tires.

Open pit burning of tires is cheap and easy, but in many places atmospheric pollution controls have made this practice illegal. As well as natural and synthetic rubber, tires contain sulfur, carbon black, zinc oxide, and often nylon and fiberglass as well as wire. Combustion generates hydrocarbons, carbon particles, and oxides of sulfur and nitrogen; many municipal incinerators will not accept tires because of their sulfur content. Some tire companies pay more in freight than they receive from reclaiming centers, just to get rid of their used tires.

Generally, between 15 and 20% of discarded tires are retreaded or recapped, and the proportion is falling. Retreading postpones the disposal problem, but itself generates rubber waste when the remaining original tread is buffed off. There are numerous other uses for tires in their orginal form (Pincott, 1975; Search and Ctvrtnicek, 1976), but these applications do not go far toward solving the tire disposal problem, although they are interesting: floating breakwaters have been built to attenuate waves, by filling old tires with a buoyancy material such as foam polystyrene or polyurethane, and fastening them together in triangular arrays. Other uses include artificial reefs as fish havens in coastal waters where the ocean floor is sandy and flat for long distances, road crash barriers, embankment stabilizers, bumpers on docks, and for swamp reclamation and playground equipment. The use of tires for kindling forest wastes and burning out tree stumps is now illegal in many places. Ground rubber, probably because of its carbon black and sulfur content, under certain conditions can remove mercury and other metal ions from water (Netzer and Wilkinson, 1974; Netzer et al., 1974; Russell, 1975), and its use as a soil conditioner is now being studied (Sperber and Rosen, 1974).

Estimates of reclaimed rubber production as a proportion of the total rubber in tires for 1973 are 11.5% in the United Kingdom and 10% in the United States, having fallen from 27 and 26%, respectively, in 1963 (*New Scientist,* 1976d; RAPRA, 1976), but quoted reclamation figures vary widely according to the methods used in deriving them. Rubber reclaiming methods have been reviewed (Brothers, 1973; EPM, 1975a; *New Scientist,* 1976d; RAPRA, 1976; Sittig, 1975); in order of decreasing importance they are the wet digester process (heating chopped tires with steam and chemicals to char the fabric away from the rubber), the devulcanizing or dry process (autoclaving the finely ground rubber with oils), and the mechanical process (making use of the shearing heat generated within the rubber as it is torn apart in a cracker mill).

Considerable effort has gone into searches for uses for reclaimed rubber (Baum and Parker, 1974b; IRT, 1974; NCRR 1974c; Search and Ctvrtnicek, 1976; Sperber and Rosen, 1974), and a book has been published on the U.S. patents in this area from 1955 to 1972 (Szilard, 1973). Important uses at present are tire sidewalls and linings, sports and playground surfaces, various adhesives, and as an asphalt additive. The present demand for scrap tires is far less than the potential supply, and probably the only application that would absorb the necessary quantity would be road surfaces of rubberized asphalt (Brand, 1974; Braton and Koutsky, 1974; Bynum et al., 1972; EST, 1973d), and this use is subject to economic considerations. It is feasible only if it gives extended road life or other specific advantages such as an improved tire grip. The reclaimed fiber has been proposed for applications such as carpet underlay, or as a mulch for fruit trees (Teskey and Wilson, 1975), but frequently it is discarded (EPM, 1975a).

Pyrolysis of tires has been investigated, by the U.S. Bureau of Mines (Baum and Parker 1974; Collins et al., 1974; EST, 1969c; Wender et al., 1974) and in Japan (Inoue, 1974), for example. The range of pyrolysis products available from waste tires is shown in Table 6.13. Oil shale technology (pyrolyzing ground-up tires by direct contact with hot ceramic pellets, CEN, 1974f) and molten salt reactors (CEN, 1976b) are now under consideration. An alternative process is catalytic hydrogenation, (Wolk and Battista, 1973) which can provide heavy oils and carbon black. The economic feasibility depends on factors such as the cost of transportation, and a grid of 10 tire pyrolysis plants placed strategically across the United States has been suggested. The same problem of transport cost occurs in energy recovery by incineration, which is also being investigated (Moats, 1976).

Table 6.13. Products of Pyrolysis of Waste Rubber Tires

Product	Mass % of Rubber	Calorific Value	Composition
Gas	6-29	19-45 MJ/m^3 (500-1200 Btu/ft^3)	Mixed hydrocarbons; low sulfur content
Oil	31-66	42 MJ/kg (18,000 Btu/lb)	Less than 1% sulfur; low viscosity
Solid	35-62	28-33 MJ/kg (12,00-14,00 Btu/lb)	2-3% sulfur; about 4.5% zinc

Source. WSL, 1975c. Crown copyright; reproduced by permission of the Director of Warren Spring Laboratory.

The solution to many of the problems of tire disposal and utilization may be selective embrittlement, a process which is now being developed in many parts of the world. After cooling to about −60°C (−80°F) and hammer milling, tires are fragmented with almost complete separation of wire and fibers from the rubber. The degree of fineness of the rubber crumb and the nature of the fiber can be controlled by varying the process parameters. The advantages of this process include lower power requirements for the fragmentizer and the possibility of mobile units which can travel from one source of discarded tires to another. The rubber crumb may be used for sports, recreation and feedlot surfaces, soil improvement, cleaning oil spills, roofing, car underseal and other protective coatings, carpet underlay, and road surfaces (CEN, 1974g; CIG, 1975a; McMillan, 1975; Sperber and Rosen, 1974; WWT, 1975d). However, some investigators are not convinced of the overall superiority of the cryogenic technique.

PROTOTYPE URBAN WASTE SYSTEMS

In this section are found brief descriptions of some of the urban waste systems which have been suggested, planned, operated, or abandoned. All these systems have one thing in common: each results from a particular set of circumstances unique to a company or to a community. It should not be assumed that what works in one place will necessarily work somewhere else, but many of us shall have to consider in the near future what is best for our community, and the more alternatives we consider, the better.

Most of these are "high-technology" systems, although it is apparent that in the United States even large cities and large companies may be unable or unwilling to fund full-scale plants at the present state of technological development. Federal funds have been provided through the Environmental Protection Agency for a number of valuable "demonstration projects" (EST, 1971e, 1972h; Meyers, 1976), but inflation, unexpected technological problems, and lack of experience led to some failures, even of projects supported by the EPA (*Phoenix Quarterly*,1976). Several American states have acted to stimulate resource recovery (NCRR, 1976): since 1970, 21 states have passed some legislation, 12 having grant or loan authority and 6 having resource recovery operation authority.

There is still no clear indication of any one preferred direction which urban waste recovery should be taking: solid waste *collection* services appear to be optimized in population units of 50,000 or so, far smaller than the population usually considered necessary for "high-technology" central *recovery* facilities (Easton, 1976), and small-scale separation-at-source schemes are now being funded by the EPA (Meyers, 1976). It is noteworthy, however, that the city of Ames, Iowa, with a population of 40,000, is operating successfully a resource recovery plant which generates steam from municipal waste (Funk and Russell, 1976; Hopper, 1975; Mihelich, 1976).

Reports of prototype and demonstration projects have been collected by Baum and Parker (1974b), Boettcher (1972), Boyd (1976), Cheremisinoff (1976), Franklin et al., (1973b), Ganotis and Hopper (1976), Hartley (1975), Hershaft (1972), Hopper, 1975; Jackson (1974, 1975), Kuester and Lutes (1976), Lingle (1976), McEwen, 1976; Meyers (1976), NCRR (1974c), Pavoni et al. (1975), Stump (1972), Thomas, (1974), and USEPA (1974b). Technical information is not included here, but is available in many of the references.

Auckland City Council Compost ("ACCPOST")

Reference: Howard, 1972.

The City Council in Auckland, New Zealand, manufactures an organic soil conditioner ("ACCPOST") by the fermentation of urban waste in continuously rotating "Dano" bio-stabilizers at temperatures between 45 and 50°C for 5 days, followed by screening, and maturing by turning and watering in outdoor windrows for several months (Figure 6.9). Currently the plant accepts 40 tonnes per day, but this is to be increased threefold. A good market exists for the compost in home and commercial gardens in the Auckland area at a cost comparable to that of topsoil. As in any composting process, there is a residue to be disposed of (15% by volume, 40% by weight). The economic justification for the compost plant is not well established, but is influenced by limited landfill sites in the central Auckland City area. It is believed that at 120 tonnes per day the net cost will be equivalent to landfilling outside the city.

Battelle Northwest Gasification: Kennewick, Washington

References: Hammond et al., 1972, 1974; Pavoni et al., 1975.

In a pilot plant operated by Battelle Northwest for the U.S. Environmental Protection Agency and the city of Kennewick, Washington, residue from the pyrolysis of solid waste was reduced and gasified with an air-steam mixture as it progressed down through the reactor. As a result of the pilot operation, the Austin Company is constructing a 150 tons-per-day plant for "gasification" or "pyrolysis-incineration" of urban waste.

Black Clawson "Hydrasposal" Wet Pulping and "Fibreclaim" Recovery Process, Franklin, Ohio

References: Arella, 1974; Baum and Parker, 1974b; Cummings, 1974, 1976;

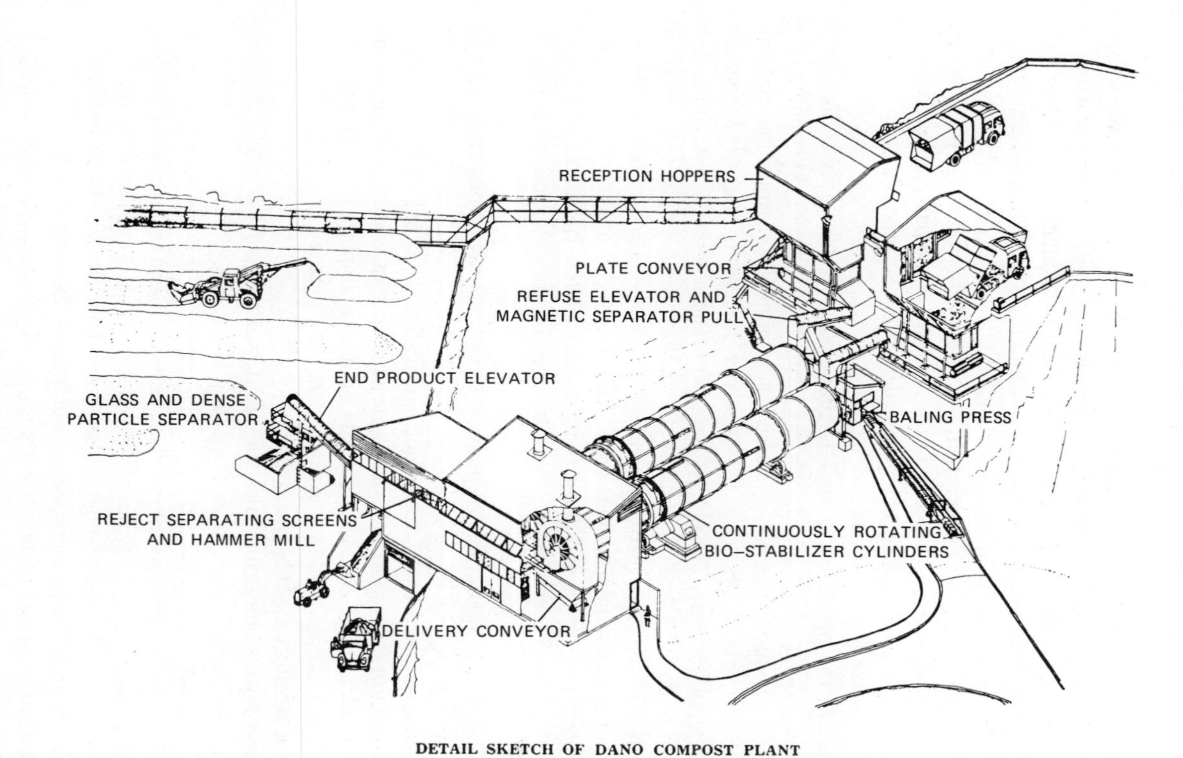

Figure 6.9. Sketch of the "Dano" compost plant used by the Auckland City Council. (From Howard, in *Solid Waste Treatment and Disposal,* N. Y. Kirov, Editor, Ann Arbor Science Publishers, Inc., 1972, with the permission of the publishers.)

Eichholz, 1972; Franklin et al., 1973a, 1973b; Hershaft, 1972; Hopper, 1975; Jackson, 1975; Kinney, 1974; Lingle, 1976; NCRR, 1974c; Pavoni et al., 1975; Tillman, 1975.

The Black Clawson Company, with a grant from the U.S. Environmental Protection Agency Office of Solid Waste Management Programs, has built one of the most complete recovery plants in the United States, at Franklin, Ohio (Figure 6.10). The Black Clawson Company is engaged in the manufacture of papermaking machinery in Middleton, Ohio, and fortuitously an engineer with the company was a councillor of the city of Franklin at the time Franklin was on the point of exhausting its solid waste landfill sites. The consulting engineers were A. M. Kinney, Inc., of Cincinnati, who were instrumental in the development of the "Hydrasposal" process which converts all pulpable materials to an aqueous slurry, a process also being considered for the production of a waste cellulose fiber cake for boiler fuel.

The heart of the operation is the "Hydrapulper," which receives the urban waste, shreds it with a high speed rotor and forms it into a water slurry containing 3-4% solids. Dense objects are removed continuously from the bottom, magnetically separated, and then the glass-rich fraction is recovered from the slurry by a liquid cyclone. The Glass Container Manufacturers' Association, also with the aid of an EPA grant, constructed a glass reclamation plant to operate in conjunction with the fiber recovery process. The dense residue from the Hydrapulper is dried and air classifed into three fractions; light (aluminum and plastics), medium (glass), and dense (nonferrous metals). The aluminum is purified by electrostatic separation. A "Sortex" optical system separates the glass by color.

A "defibering" screen removes plastics, twigs, and stringy material from the cellulose fiber pulp, which is then dewatered to approximately 40% solid content (the "Fibreclaim" process). Additional purification steps are necessary: treatment with sodium hydroxide and washing to remove grease and microorganisms, and perhaps bleaching. A market for this product is important; in Franklin the cellulose fiber has been sold to a nearby roofing felt manufacturer, but sales of all recovered materials cannot be guaranteed.

Screening rejects, including plastics, are dewatered and fed into a fluidized bed reactor for incineration, with the potential for energy recovery. Although complete combustion is claimed, and wet scrubbers remove particulates and sulfur oxides, the water vapor plume is an aesthetic disadvantage of the process. Much of the water is recycled, but waste water eventually must be removed from the system (a serious disadvantage of the process) and transferred to a secondary wastewater treatment plant incorporating an aeration lagoon.

The products recovered from 100 tonnes of urban waste are aproximately 18 tonnes (on a dry weight basis) of paper fiber, 6.5 tonnes of metal, and 4 tonnes of glass (Table 6.14).

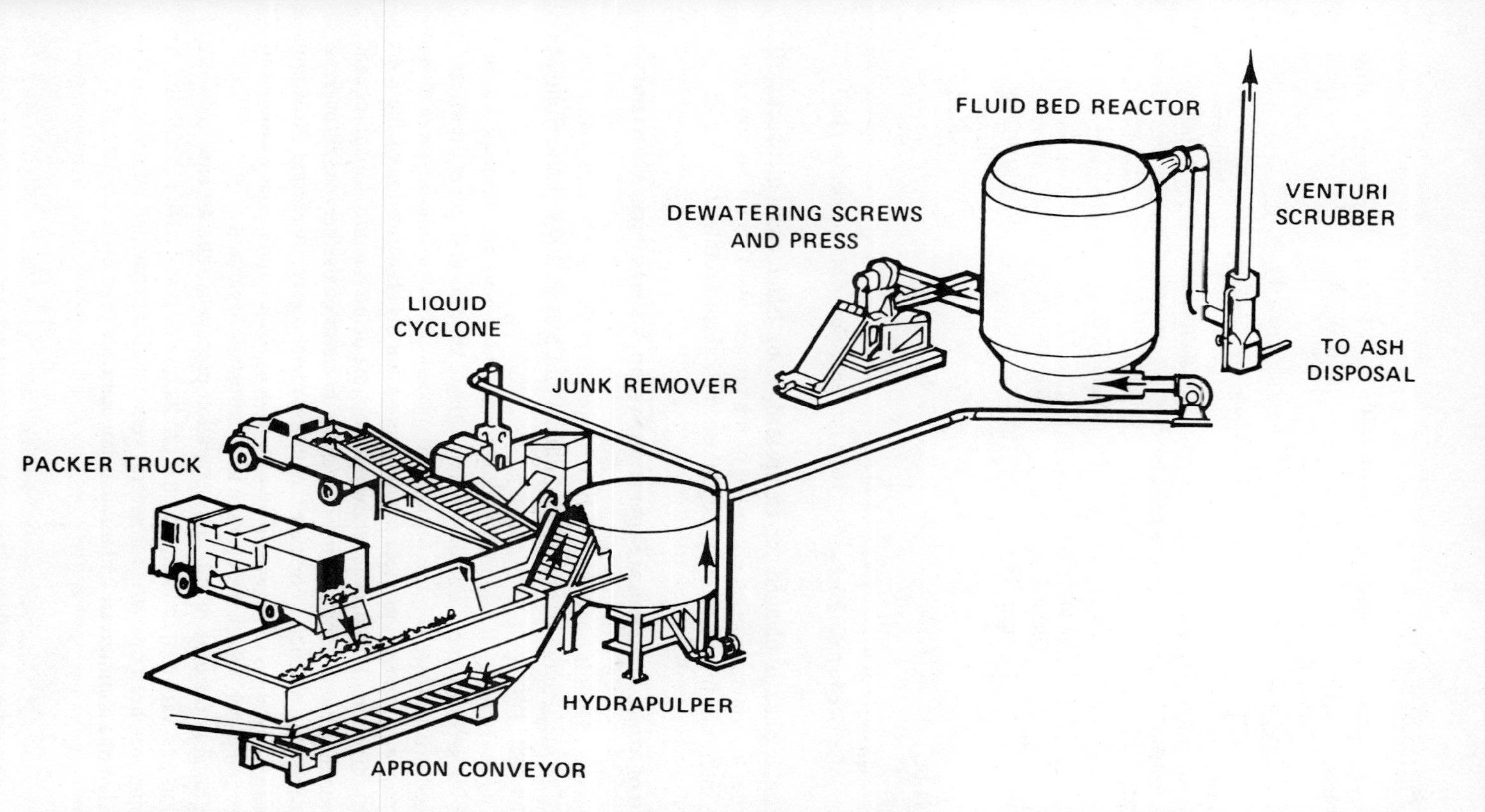

Figure 6.10. Diagrammatic illustration of the Black Clawson recovery plant at Franklin, Ohio. (From NCRR, 1974c, with the permission of the National Center for Resource Recovery, Inc., Washington, D.C., U.S.A.)

Table 6.14. Material Balance for the Black Clawson Process, Expressed as Mass Percentages

	Products (mass %)		
Waste Input (mass %)	Recycled	Landfill	Incinerated for Energy
Water, 25			
Metal, 10	Iron, 6 Aluminum, 0.5	Metal, 3.5	
Glass, 8	Glass, 4	Glass, 4	
Paper, 40	Paper, 18		Paper, food, plastic, textiles, 37
Food, plastic, textiles, 17		Food, plastic, textiles, 2	

Source. After Sperber and Rosen, 1974, with the permission of Marcel Dekker, Inc.

A similar plant is planned for Hempstead, Long Island, with a capacity 10 times that at Franklin. In an extension of the process, the Black Clawson company is interested in converting the cellulose into industrial ethanol.

Bureau of Mines Mechanical Separation System, College Park, Maryland

References: Douglas and Birch, 1976; Franklin et al., 1973b; Jackson, 1975; NCRR, 1974c; Pavoni et al., 1975; Sullivan and Makar, 1974, 1976; Sullivan and Stanczyk, 1971.

The U.S. Bureau of Mines at College Park, Maryland has developed a dry system to separate mechanically metals, glass, plastic, and paper from raw, unburned urban waste (Figure 6.11). The process involves shredding in a chain mill, magnetic removal of steel for detinning, and air classification of the nonmagnetic fraction. The glass is recovered from the dense fraction (for subsequent color sorting) by screening. This is feasible because of the very small average size of the glass particles compared with the metal pieces. A second shredding and air classification of the less-dense fraction separates nonferrous metals from plastics and paper, which then pass to an electrostatic separator.

The notable features of this dry separation process are the absence of water pollution and vapor emissions, and the production of many usable materials. A pilot plant has been operated for several years, and has provided information for the design of a commercial 250 ton-per-day plant.

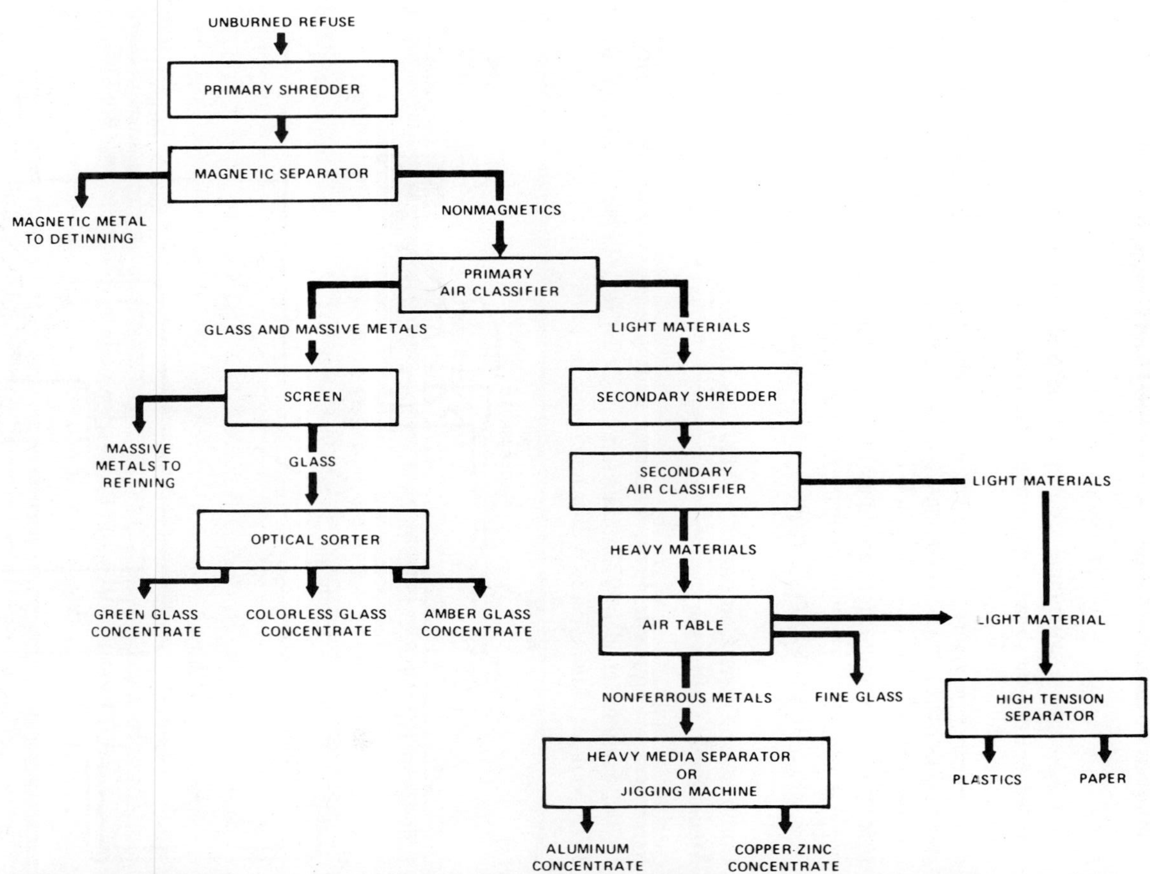

Figure 6.11. Flowchart for the U.S. Bureau of Mines dry urban waste separation system. (From NCRR, 1974c, with the permission of the National Center for Resource Recovery, Inc., Washington, D.C., U.S.A.)

Canterbury Municipal Council and the Steel Can Group, Australia

References: Pausacker, 1975; Spittle, 1975.

The Canterbury Municipal Council in the Sydney metropolitan area, with assistance from the Steel Can Group, has installed a ferrous metal magnetic separation unit which will be the first test in Australia of the viability of large-scale resource recovery from urban waste. This follows the installation of a shredder to reduce the waste volume and the cost of landfill disposal. The plant was opened in 1975, and although it is premature at present to comment on the success of the venture (influenced by the availability of relatively low-cost iron ore in Australia) it is significant that a country with very large mineral reserves is now considering this form of resource conservation.

Combustion Power Company CPU-400 Gas Turbine, Menlo Park, California

References: Campbell, 1974; Chapman and Wocasek, 1974; EST, 1970c, 1970d; Franklin et al., 1973b; Jackson, 1975; NCRR, 1974c; Pavoni et al., 1975; Van der Molen, 1973; Wilson and Freeman, 1976.

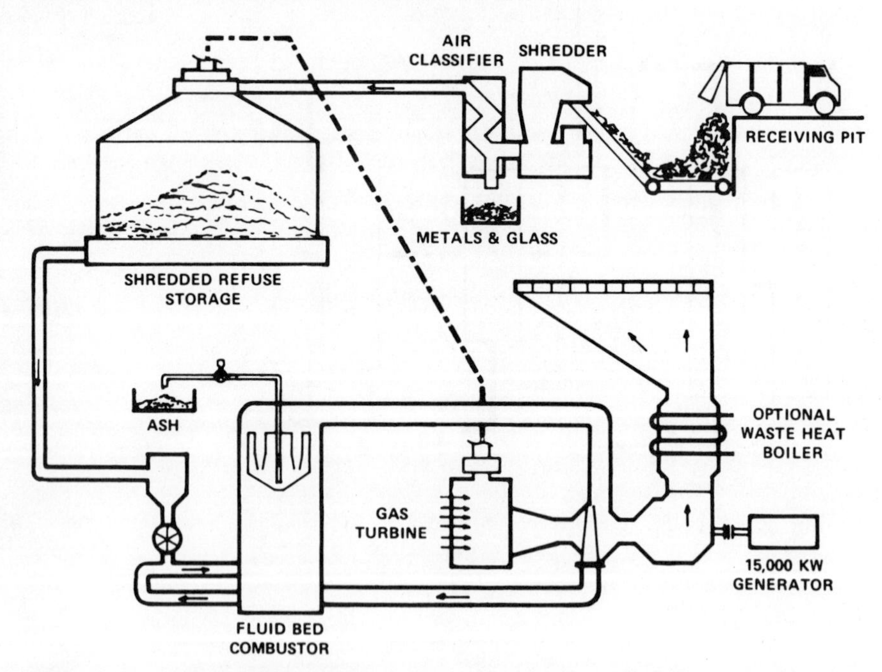

Figure 6.12. Diagramatic illustration of the Combustion Power Company CPU-400 system. (From NCRR, 1974c, with the permission of the National Center for Resource Recovery, Inc., Washington, D.C., U.S.A.)

In this dry process (Figure 6.12), which is being tested in Menlo Park, California by the Combustion Power Company with the assistance of the Environmental Protection Agency, urban waste is shredded and air classified, with the lighter, organic fraction being burned in a high-pressure fluidized bed reactor. The pressurized products of combustion are cleaned and used to power a gas turbine electricity generator. A turbine has the advantage that it can be put on or off line rapidly, which is preferable in view of the variable rate of urban waste production, but the disadvantage that the gases must be well cleaned before they are fed into the rather delicate turbine blades. A full-scale plant would be capable of generating about 5% of the electrical power needs of the community producing the waste, and as the turbine utilizes only about a quarter of the energy, excess heat could be used for supplementary community thermal energy requirements. Of course, the exhaust gas stream is subject to the same scrutiny by air pollution agencies as the gases leaving conventional incinerators.

The denser fraction of the air-classified material can be processed to yield glass, ferrous metal by magnetic separation, and aluminum and other nonferrous metals by eddy current separation.

Ebara Resource Recovery Project, Japan

References: Ito and Hirayama, 1975a, 1975b, 1976.

The Ebara Manufacturing Company, supported by the Office of Research and Development Programs of the Japanese government, has carried out a 3-year feasibility study of resource recovery systems for urban waste treatment, and is planning to install a 100 ton-per-day demonstration plant. This system features semi-wet selective pulverization (described earlier in this chapter), composting, and two-bed pyrolysis in which the fine solids are transferred between a fluidized pyrolysis reactor and a fluidized incinerator which oxidizes the char.

Edmonton Refuse Incineration Plant, Great London Council

References: *Energy Digest,* 1976; Pavoni et al., 1975; Porteous, 1975b.

This plant accepts 1800 tons of refuse in 600-700 truckloads between 10 a.m. and 4 p.m. on working days. This is transferred from bunkers to four of the five rolling grate boiler units driving three of the four 12.5-megawatt turbogenerators, which give the plant a net output of 30 megawatts of electricity. Although there were significant early teething troubles, the facility is now operating efficiently.

Groveton Paper Mill Incinerator, New Hampshire

Reference: Young and Lisk, 1976.

Until recently, the waste from the Groveton paper mill in Northumberland, New Hampshire was burned in the town's dump, but when the mill switched to complete tree utilization (*in situ* chipping) the space made available in the mill was used for the installation of an incinerator with steam recovery, and the company also accepted the town's solid waste. When some waste items caused problems in the incineration, the company ran a public education campaign, and now garbage is collected in plastic bags which are burned without opening or shredding.

This example illustrates how much can be achieved as a result of cooperation between industry, communities, and local authorities.

Hercules System, State of Delaware

References: EST, 1970e,1972a; Pavoni et al., 1975; USEPA, 1974b.

The Hercules system, constructed for the state of Delaware and supported by the Enviromental Protection Agency, is designed to accept domestic and industrial wastes and sewage sludge (Figure 6.13). Domestic waste is shredded and the ferrous metals are recovered. Clean paper and plastics are recovered from selected industrial solid waste before the remainder is combined with domestic waste and sewage solids for composting into a saleable soil conditioner. Plastics

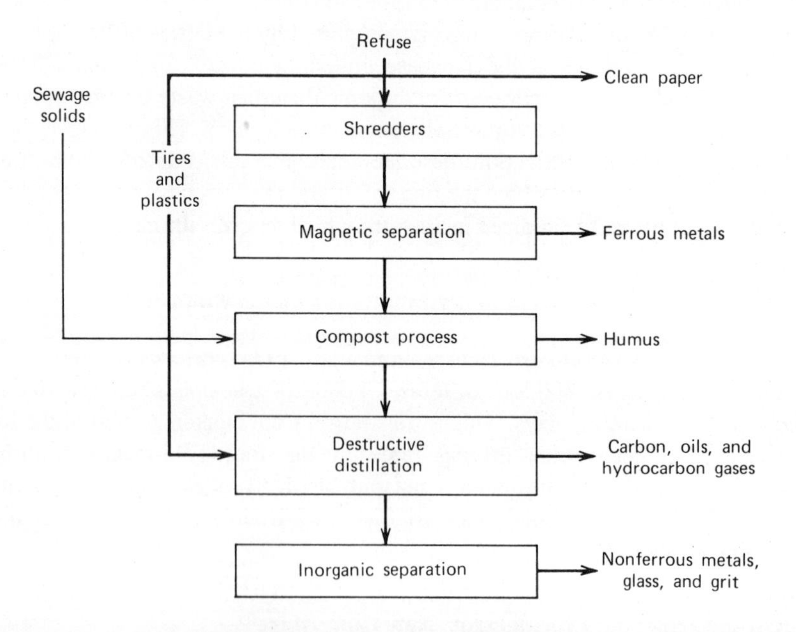

Figure 6.13. Simplified flowchart for the Hercules system. (After Bailie and Burton, 1972, with the permission of the American Institute of Chemical Engineers.)

and tires, together with rejects from the compost process, are pyrolyzed in a fluidized bed to yield hydrocarbons for fuel and carbon, together with a residue of nonferrous metal, glass, and unusable material totalling 20% of the domestic waste.

This system demonstrates that it may be preferable to employ both biological and chemical processing of wastes, and to combine the processing of several types of wastes.

Linde Division, Union Carbide, "Purox" Process, Charleston, West Virginia

References: Fisher et al., 1976; Franklin et al., 1973b; Jackson, 1975; Schulz, 1975; Tillman, 1975; Wilson and Freeman, 1976.

In the oxygen refuse converter system developed by the Linde Division of Union Carbide and tested in a 200 ton-per-day pilot plant in Charleston, West Virginia, pure oxygen rather than air (hence "Purox") is injected into the mixed solid waste. Before pyrolysis, the solid waste is coarse-shredded, the magnetic metal is removed, and the waste passes down through a drying zone in the vertical shaft furnace. In the lower portion of the furnace, oxidation generates the heat necessary to carry out pyrolysis in the upper portion and to produce a fuel gas which is scrubbed and delivered to the user. The char and inorganics fall from the pyrolysis zone to the oxidation zone, melting to a slag which flows out continuously and is quenched to a frit. The carbon monoxide resulting from char combustion passes upwards to enrich the fuel gas.

This process is primarily a fuel-generating waste disposal system rather than a resource recovery system, but it is compatible with any degree of front-end separation for material recovery. Any rejected inorganic components are incorporated into the frit; all organics are converted without generation of air or water pollutants to a saleable fuel gas, typically 47% carbon monoxide, 33% hydrogen, 14% carbon dioxide, 5% hydrocarbons and 1% nitrogen.

The requirement of oxygen is less of an economic problem if an adjacent plant is a big oxygen consumer and large-scale generation is possible. Some steel mills produce very large amounts of oxygen, and the "Purox" fuel gas, in fact, could be used as a reducing agent in steel production. The gas is also a potential source of methanol or ammonia, a process being considered for the city of Seattle, Washington (Seattle, 1975; Sheehan and Corlett, 1975).

Lowell, Massachusetts

References: *Phoenix Quarterly*, 1976; USEPA, 1974b.

The Environmental Protection Agency and the state of Massachusetts planned to finance a demonstration project in Lowell to recover metals and glass from the residue of an existing incinerator. Both the EPA and the city spent considerable sums before it was realized that it would be uneconomic to upgrade the air emission controls on the incinerator (a requirement stipulated by both State and Federal governments in their grant agreements) and to control leachate and other pollution problems in the adjacent lagoon and landfill. The project was abandoned.

Madrid, Spain

Reference: Cavanna et al., 1974, 1976.

A pilot plant for the mechanical separation and recovery of a range of products from urban waste has been operated by Empresa Nacional ADARO de Investigaciones Mineras, in conjunction with the U.S. Bureau of Mines.

Monsanto Enviro-Chem "Landgard" Pyrolysis Process, Baltimore, Maryland

References: Bowen, 1975; EST, 1971d, 1975f; Franklin et al., 1973b; Hopper, 1975; Jackson, 1975; Lingle, 1976; Pavoni et al., 1975; *Phoenix Quarterly*, 1976; Sussman, 1975; USEPA, 1974b; Wilson and Freeman, 1976.

A full-scale 1000 ton-per day plant is being built by Monsanto Enviro-Chem under an Environmental Protection Agency grant in Baltimore, Maryland, utilizing the "Landgard" system which has been demonstrated in 30-35 ton-per-day prototype plants in St. Louis, Missouri, and in Kobe, Japan (Figure 6.14). Shredded, unsegregated waste is heated in an oxygen-deficient atmosphere in a rotating kiln with the aid of auxiliary fuel oil, so that first drying, then pyrolysis, take place. The generated gases are burned with air in a gas purifier, and the hot combustion gases are heat-exchanged in boilers to provide steam for heating and air-conditioning. It is claimed that air pollution in this method should be far less than that with an equivalent incinerator; the hot residue is discharged from the kiln, water-cooled, and separated magnetically and by water flotation and

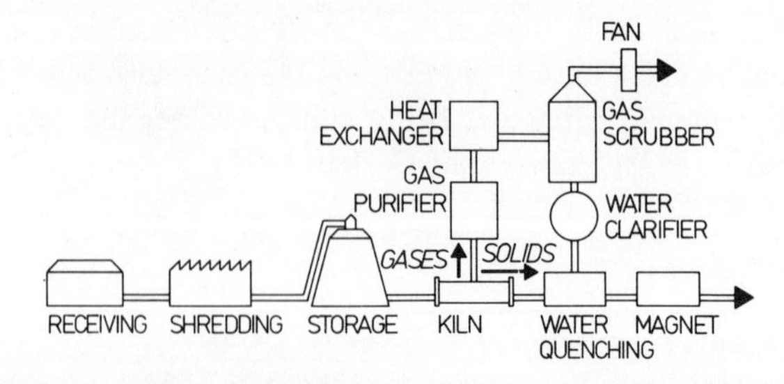

Figure 6.14. Diagrammatic illustration of the Monsanto Landgard process.

screening into ferrous metal, a glassy aggregate suitable for landfill, and a char. All process water is continuously recirculated.

Scaling up to 1000 tons per day from the pilot plant has been more difficult than expected. Solids remain in the kiln for a longer period than in the prototype, producing residues which differ in quantity and composition from those expected. Metallic salts vaporize and condense to extremely fine particulate matter. The plant was not in full operation 2 years after the scheduled completion date.

Although in the system as described the pyrolysis gases are burned on site to generate steam, it is considered that they could be burned together with coal, oil, or natural gas in nearby power utilities.

Nashville Thermal Transfer Corporation, Tennessee

References: EST, 1972b, 1972d, 1975e; Kenward, 1975c; NCRR, 1974c; Palmer, 1975; Pavoni et al., 1975.

A combination of environmental considerations and heating and cooling requirements in a city reconstruction project prompted the city of Nashville to establish the Nashville Thermal Transfer Corporation and to construct a steam-generating incinerator. It was associated with an extensive piped central heating and chilled water system; the steady demand for both summer and winter energy providing a high utilization rate for the equipment and "fuel." As in many other similar plants, there have been problems with the scrubbers in the incinerator chimneys.

National Center for Resource Recovery System

References: Alter et al., 1974, 1976a; Kinnersly, 1973; NCRR, 1974c; Pavoni et al., 1975.

The National Center for Resource Recovery, an agency funded by a number of United States users and manufacturers of packaging, is financing at New Orleans the first of a planned series of recovery plants located with markets for the recycled materials in mind. The emphasis is on front-end recovery of metals, color-sorted glass, and paper, with a proportion of the waste stream passing on to landfills or to an existng furnace, as in the Union Electric-Horner and Shifrin project. The design is based on considerable pilot-scale experience in the District of Columbia.

Occidental (formerly Garrett) Flash Pyrolysis, San Diego County, California

References: Finney and Garrett, 1974, 1975; Finney and Sotter, 1975; Franklin

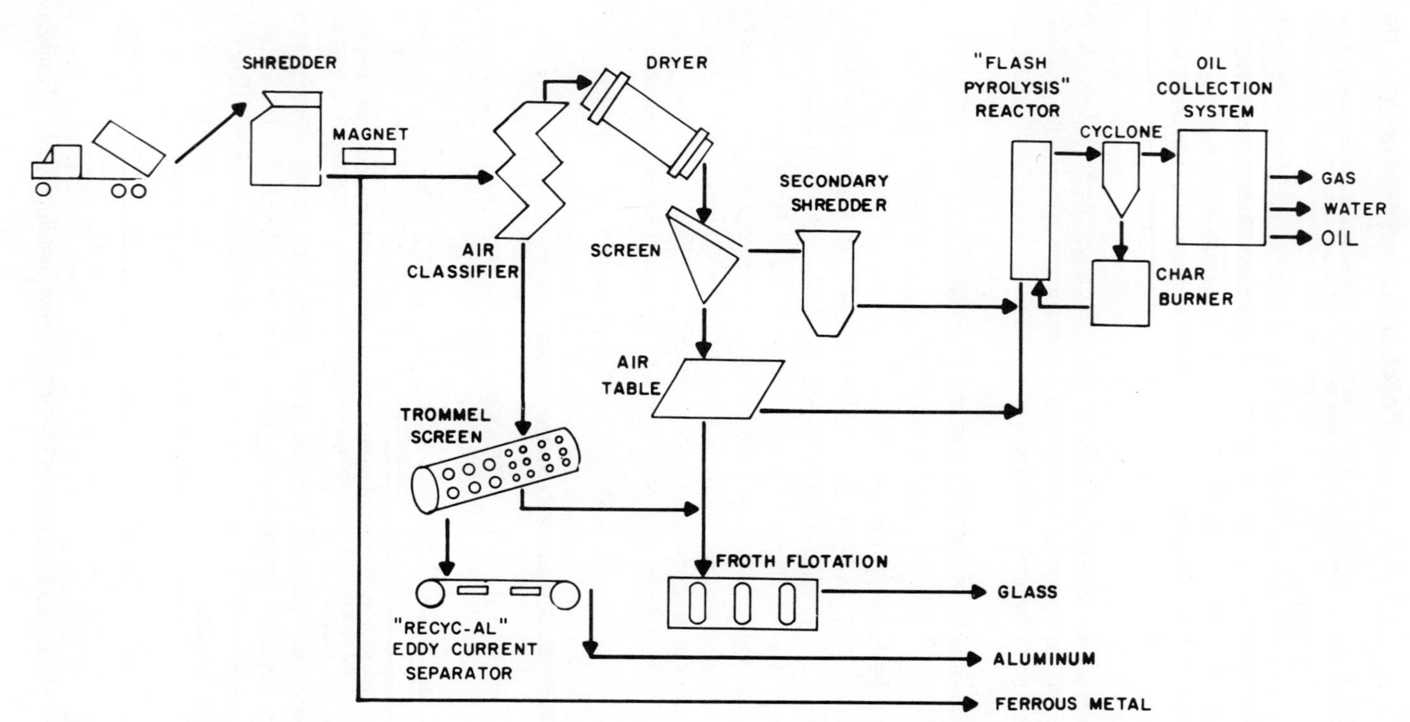

Figure 6.15. Simplified flow scheme of the Occidental Resource Recovery System. (From Preston, 1976, with the permission of the Occidental Research Corporation.)

et al., 1973b; Hopper, 1975; Jackson, 1975; Levy, 1975; Lingle, 1976; Mallan and Titlow, 1976; Morey, 1974; Morey et al., 1976; Pavoni et al., 1975; Preston, 1976; USEPA, 1974b; Wilson and Freeman, 1976.

A new 200 ton-per-day capacity plan in San Diego County, California, is being built using a flash pyrolysis process developed by Garrett Research and Development (now Occidental Research Corporation, a subsidiary of Occidental Petroleum). Substantial financial support is coming from the U.S. Environmental Protection Agency. The waste is shredded, dried, and the organic material is removed by air classification and screening before rapid flash pyrolysis of the finely ground organic fraction at 500°C (Figure 6.15). The resulting gas is used for heating within the plant, and on the basis of pilot plant results, the oil yield is 1 barrel (159 liters) per ton of waste. The low-sulfur oil is suitable for low-grade fuel (to be used by the San Diego Gas and Electric Company), and the char is a potential solid fuel or source of activated carbon. Metals and glass are separated from the inorganic fraction. The Occidental Energy and Resource Recovery process includes froth flotation for glass recovery, and an eddy current method for nonferrous metals. The flash photolysis process also shows promise for oil production from other organic wastes such as tree bark.

Refuse Energy Systems Co., Saugus, Massachusetts

References: CEN, 1974a; Hopper, 1975; Jackson, 1974; *Phoenix Quarterly,* 1976.

The General Electric Company is using steam generated in a 1200 ton-per-day waste burning plant operated by Refuse Energy Systems Company (RESCO) in Saugus, Massachusetts. This joint venture of Wheelabrator-Frye, Inc. (the U.S. licensee of the Swiss Company, Von Roll Ltd.), and De Matteo Construction Company, accepts waste from a number of Boston suburbs at a cost competitive with existing alternatives such as landfilling. Although metals and glass recovery is included in the process, and research is being conducted into the use of residues as road base and cement substitute, energy recovery is the main aim.

The approach of communities pooling their wastes, concentrating responsibility in a single private organization, and using well-tested technology (the Von Roll process is used in many facilities, worldwide), has resulted in a successful operation with minimal start-up difficulties.

Torrax Solid Waste Conversion System, Erie County, New York

References: Franklin et al., 1973; Pavoni et al., 1975; Stoia and Chatterjee, 1972; Szekely et al., 1975; Wilson and Freeman, 1976.

The 75 ton-per-day Andco-Torrax plant in Erie County near Buffalo, New York (Figure 6.16) was funded by the Environmental Protection Agency and also

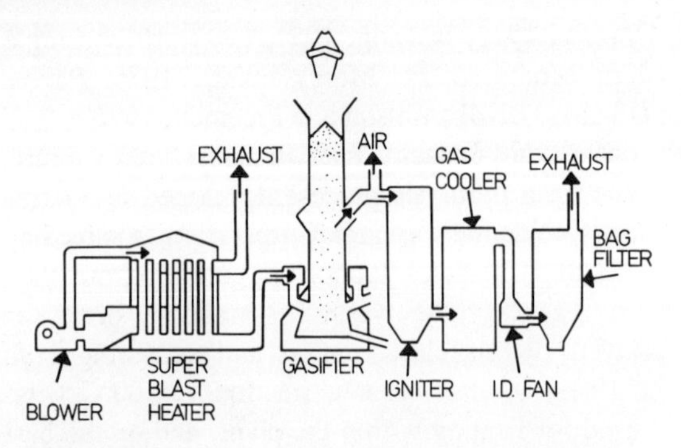

Figure 6.16. Schematic diagram of the Torrax solid waste conversion system. (After Pavoni et al., 1975; Stoia and Chatterjee, 1972.)

involved the Andco-Torrax parent companies (The Carborundum Company of Niagara Falls and Andco, Inc., of Buffalo), the American Gas Association, and the New York State Department of Environmental Conservtion. A commercial 200 ton-per-day unit based on the demonstration plant design is being constructed in Luxembourg.

The as-received waste is periodically charged into the top of the reactor, and slowly settles, the organic material being pyrolyzed before it reaches the high-temperature bottom zone. High-temperature, preheated air combines in the base of the fixed bed reactor with the carbon produced by the pyrolysis, to raise the temperature sufficiently high to melt the inorganics into a slag. This is tapped and fritted in a water chamber to produce as the only residue a low-volume, glassy aggregate that is inert and easy to handle.

The gases evolved during the pyrolysis are drawn off (by low pressure developed by an exhaust fan rather than a chimney stack) and mixed with a small excess of air in a torrential pattern in the secondary combustion chamber. The gaseous combustion products then are cooled (for example, by passing them through a steam-generating plant) and cleaned before exhausting to the atmosphere. The system has good energy recovery (up to 70% of the heating value of the waste) if the steam is produced on site.

The Torrax process has much in common with the Union Carbide ''Purox'' process. Both involve slagging pyrolysis gasification, and both are more concerned with waste disposal and energy recovery than with resource recovery.

Union Electric-Horner and Shifrin Supplementary Fuel Project, St. Louis, Missouri

References: CEN, 1974a; Franklin et al., 1973; Horner and Shifrin, 1973; Jackson, 1974; Klumb, 1976a, 1976b; Lingle, 1976; Lowe, 1973; Pavoni et al.,

1975; Schulz, 1975; Sutterfield, 1974; Sutterfield and Wisely, 1972; Sutterfield et al., 1972, 1974; Tillman, 1975; USEPA, 1974b; Wilson and Freeman, 1976; Wisely et al., 1974.

The first investor-owned utility in the United States to burn municipal solid waste as a supplementary fuel in the production of electric power was Union Electric, using waste from St. Louis, Missouri, with Horner and Shifrin as consulting engineers and with assistance from the Environmental Protection Agency.

The waste is shredded to a particle size of less than 4 centimeters (1½ inches), air classified, magnetically separated, and introduced to the extent of 5-27% by heat value via separate nozzles to pulverized coal-fired furnaces. Up to 650 tons per 16-hour day can be burned in this way. Some of the details were included earlier in this chapter, particularly Figure 6.3. The project has attracted considerable interest, because the capacity of existing furnaces in many cities would be adequate to allow the process to become the principal solid waste disposal method. The main reason for the refuse-derived-fuel process being well advanced compared with other disposal systems is that all equipment used is commercially available. Of course, other methods of disposal must be available for the noncombustible waste, and for the ash. Another restriction is that waste can be used as a supplementary fuel on a large scale only when the power station runs at high load levels throughout the year; that is, with a high summer load for air conditioning, for example, to match the winter domestic heating load. Equipment wear, corrosion, and air emissions, which have been studied thoroughly at St. Louis, do not appear to present undue problems.

A 7300 tonne-per-day system for the utilization of all the solid waste in the St. Louis metropolitan region (with 2.5 million people) is planned.

Warren Spring Laboratory Urban Waste Sorting Program, Stevenage, U.K.

References: Douglas and Birch, 1974, 1976; Kenward, 1975b; MRW, 1976h; Newell, 1975b; *New Scientist*,1974a, 1975b, 1976b, 1976c; *Solid Wastes*, 1976a; WSL, 1975e.

A pilot plant has been set up at the U.K. Department of Industry Warren Spring Laboratory to investigate the physical separation of urban waste, using mainly dry methods. At the head of the process circuit, a "bag-burster" has been designed to lacerate the bags with minimum effect on the contents, so that most bottles remain unbroken and cans are not squashed. (Milling is normally carried out near the end of the process, so that components can be separated prior to deformation and contamination.)

After liberation from its containers, the waste is sized on a large rotating cylindrical screen (trommel) to yield the following fractions: less than 13 millimeters (dust, dirt, ashes, grit, and cinders, discarded without further treatment); between 50 and 13 millimeters; between 200 and 50 millimeters; and greater than 200 millimeters (immediately milled and returned to the head of the circuit). The intermediate fractions are separated using the electromagnetic, ballistic/inertial, air classification and sink/float techniques, described in Chapter 3, to yield cans, rags, paper in three grades, nonferrous metals, glass, and organic material (potentially useful as compost or animal feed).

This plant is only experimental, but the U.K. Department of the Environment is supporting the design and construction of 10 tonne-per-hour mechanical waste-sorting plants in conjunction with the Tyne and Wear and the South Yorkshire County Councils, using the technology developed by the Warren Spring Laboratory.

Another recycling facility provided at the Warren Spring Laboratory in Stevenage is the Waste Materials Exchange, where lists are compiled of available waste materials and circulated to participating companies (see Figure 7.1 in the next chapter).

Worthing Corporation, England

References: Bourne, 1972; Thomas, 1974.

Bourne (1972) quotes a House of Lords debate in 1970, in which the success in recycling of Worthing Corporation was described. Waste paper, ferrous and nonferrous metals, fat, bones, bottles, and rags are recovered with the assistance of householders. Much of the success was attributed to the efforts of the chief engineer, who carried out a personal publicity campaign to instruct in waste disposal methods. Methane gas is obtained from sewage to power the disposal plant, and compost is made from sewage sludge.

This is an example where progress in recycling has been made by the unilateral action of one public body or one individual, and emphasizes what could be achieved by concerted action on a national scale.

7

INDUSTRIAL AND AGRICULTURAL RECYCLING PROCESES

Wastes are commonly dvided into the following categories, as indicated in Table 1.5:

- Urban (domestic, commercial, municipal)
- Industrial
- Agricultural
- Mineral

Solid waste is extremely varied in nature, and is usually classified according to the industry generating it, as in the Standard Industrial Classification in the United States for example, and the nature of these wastes has been reviewed by many authors in several countries.* In contrast, in this chapter the emphasis is placed on classes of materials and compounds with recycling potential, by whatever means they arise in industrial, mineral, agricultural, or urban waste processes.

Although problems of agricultural waste have increased recently because of intensive farming practices and large food processing complexes (Peterson et al., 1971), organic waste material other than urban waste always has been recycled or reclaimed to a considerable extent, and agriculture is the oldest form of controlled recycling. The larger scale of both primary and secondary operations is not necessarily a disadvantage; one feedlot of 150,000 head of cattle has received a substantial proportion of its feed from the waste derived from three potato chip factories (Grames and Kueneman, 1968).

Neither has recycling in industry been neglected: the utilization of by-products in certain areas is extensive (Alexander, 1975; Baum and Parker, 1974b; Sittig,

*ACI, 1973; ACS, 1969; Beretka, 1973; Bond and Straub, 1973; Coleman, 1975; Gutt et al., 1974; Koziorowski and Kucharski, 1972; Mantell, 1975; Neal, 1971; Nemerow, 1963; OST, 1969; Rosich, 1975a, 1975b; Sittig, 1975.

1975) and the theoretical bases of cyclic material flows in industrial processes are well developed (Nagiev, 1964). The glass industry has been an efficient consumer of its own waste cullet and also of slag and fuel ash from other industries since the middle of the 18th century (Scholes, 1974). However, in general the industrial recycling rates could be improved readily, although they are now falling in some cases (MRW, 1976j).

The major problem is that industrialists have looked upon waste recovery and recycling as sidelines to be handled by waste merchants or waste treatment consultants rather than as an integral part of the manufacturing operation (Coleman 1975; Dureau, 1975; Howard, 1976; Kelley, 1976; Neal, 1971). Small factories are often less inclined to introduce recycling procedures than are large plants, because economy of scale is important in this area, and in the future recycling and waste disposal considerations may well determine plant size in some industries. Central waste processing complexes, such as the prototype scheme operated in the United Kingdom by Re-Chem International (Coleman, 1975; WSL, 1975f), and waste exchange schemes (Müller and Schottelius, 1975) may be alternative solutions. For example, the U.K. Department of Industry is operating a waste materials information exchange scheme as a free, trial service (Allen and Poll, 1976; *Chemisty in Britain,* 1975), and Figure 7.1 illustrates a sample page from a U.K. Waste Materials Exchange Bulletin.

Because of their very large scale, mining operations have particular waste problems which are often of an engineering rather than a physicochemical nature, and engineering problems are not considered here. An indication of the progress in resource recovery in the mining industry is provided by the wide range of subjects covered in recent mineral waste utilization symposia (Aleshin, 1974, 1976). Urban waste, which presents special difficulties due to the complexity and variability of its composition and its dispersed nature, is considered separately in Chapter 6.

A common pattern for industrial pollution control in the past has been as follows:

1. Growth in the plant throughput increases the effluent discharged into atmosphere or waterway;
2. Pollution becomes obvious and public pressure results in an enforced examination of the process;
3. The nature and quantity of the waste components are determined;
4. Waste processing equipment is installed, with considerable capital expense and running costs.

Much more satisfactory than this terminal waste treatment is the following alternative method of profitable *pollution control at source:*

REF R	QUANTITY		MISCELLANEOUS
AA423C	30	TE/W	DRY 200 MESH POWDER, ASBESTOS/RUBBER/METAL (AL, CU)/FILLERS
AA433C	20	TN/W	FILTER CAKE, 24% OIL, 8% SILICA, 23% MIXED METAL HYDROXIDES
AA450D	1	TN/M	ENAMEL PAINT SLUDGE FROM WATER WASHED SPRAY BOOTHS
AA473C	3	TE/D	CHIPBOARD SAWDUST
AA474C	3	TE/D	CHIPBOARD OFFCUTS
AA481C	80	KG/W	WASTE INK
AA484D	2000	KG	POLYESTER GLUE
AA502E	30	CM/W	HARDWOOD OFFCUTS
AA516C	5	CM/W	WHITEWOOD OFFCUTS, TRIANGLES UP TO 18×6×1.5 (INCHES)
AA522C	3750	LT	GELLED ACID CATALYSED ALKYD/MELAMINE/FORMALDEHYDE RESIN
AA526F	20	TE/M	OFF-CUT TRIM BOARD/ALUMINIUM FOIL
AA530A	100	KG	WISPOFLOC 20 FILTER AID
AA543C	100	TE/W	FAILED CATHODE LININGS FROM ALUMINIUM ELECTROLYSIS CELLS
AA547C	2	TE/M	PRINTING INK (RETURNS FROM MACHINES) VARIOUS COLOURS
AA591C	5000	LT/M	GREY GLOSS PAINT, MANFTD FROM MIXED PAINT RESIDUES
AA643F	4	TE/M	UNSORTED DAMAGED WOODEN PALLETS
AA647F	900000		CAPS TO FIT 6 FL OZ BOTTLES NOZZLE TYPE 20 MM R4 WHITE
AA161D	30	SM/M	3MM MELAMINE LAMINATE OFFCUTS, TEAK & ROSEWOOD PATTERNS
AA724F	4	TE/M	SCRAP WOOD
AA759C	200	GL/W	SOLVENT BASED ADHESIVE RESIDUES
AA770F	500	LT	PAINT STOVING RED OXIDE IRON/ZINC/CHROME, TO DEF.80-13/1
AA771F	500	LT	PAINT UNDERCOAT STOVING DARK GREY, BS632, TO DEF.1045A
AA772F	800	LT	PAINT FINISH STOVING BRONZE GREEN, BS224, TO DEF.1045A
AA775E	287	KG	PHOSGUARD FLAME RETARDANT
AA777E	2088	KG	ELTASOL PX (25)
AA793C	30	TE/W	WOOD FLOUR BEST WHITE, ALL GRADES, 60 MESH-120 MESH
AA848G	6	TN/W	SPENT OAK BARK TAN
AA850C	5	TN/W	CHIPBOARD OFF CUTS
AA902G	100	TE	PACKING SHAVINGS BALED, APPROX 60KG EACH
AA912F	18	TE	GUM RESIN (IN 120 WOOD BARRELS)
AA915F	87	KG	IRGASTAB BC30 (SUPPLIERS CIBA-GIEGYLTD)
AA916F	1000	KG	CARIFLEX TR1102 (SUPPLIER SHELL CHEMICALS LTD)
AA921F	200	LB	ESTANE 5703 (SUPPLIERS B.F GOODRICH LTD)
AA922F	172	KG	TYRIL (SUPPLIERS COLE PLASTICS LTD)
AA923F	200	KG	QUALITEX (SUPPLIERS WM SYMINGTON & SONS LTD)
AA928F	86	KG	BITRAN 'H' (SUPPLIER GLOVER CHEMICAL CO LTD)
AA935A	600	KG/M	USED FURNACE GRADE CARBON BLACK
AA937A	1	TE/W	SILICON CARBIDE RODS-VARIOUS FORMS-PROCESS REJECTS
AB014F	250	TN	STRAMIT STRAW BOARD VARIOUS GRADES
AB069D	10	TN/W	DRY BALTIC REDWOOD SHAVINGS (PROBABLY BALED)
AB045C	100	/W	REEL, HARDBD, STEEL CTRS,97/8"OD × 63/4"WIDE 11/16"HOLE
AB073D	1	TN/W	SOFTWOOD OFFCUTS, SHORT 8" BOARD TO 8FT.2×1/2" STRIPS
AB075D	800	GL/M	PRINTING INK SLUDGE
AB124F	70	TN	EXTERIOR GRADE OIL TEMPERED HARDBOARD (1/4" × ASS SIZES)
AB130D	2	TE/W	SHOT BLAST GRIT
AB131D	100	/W	USED HARDWOOD PALLETS—VARIOUS SIZES
AB161D	60	TE/M	SPRUCE BARK CHIPS
AB162D	30	CM/W	SOFTWOOD SAWDUST
AB163D	3	CM/D	SOFTWOOD CHIPS AND CEMENT
AB183D	7000		SOLID FIBREBOARD CARTONS
AB186C	70	/W	EX CHEMICAL PALLETS 1.2 × 1M AND 1M SQUARE
AB214F	1	KG	CIBRACRON TURQUOISE BLUE GFP
AB221C	100	/W	WOODEN PALLETS EX BAGGED CHINA CLAY & STARCH (1200 NOW)
AB248F	7	TE/W	BEECH, LARCH WOOD FLOUR, 2% MOISTURE, FINER THAN 100 MESH
AB251F	40	CM/W	WOODCHIPS MIXED WITH ANIMAL DROPPINGS AND URINE
AB264F	2	TN/W	BROKEN AND LIGHTWEIGHT WOODEN PALLETS
AB265D	15	CM/W	SOFTWOOD OFFCUTS UP TO 12×3−4−5×13/8 IN

Figure 7.1. A page from a U.K. *Waste Materials Exchange Bulletin.* (Crown copyright; reproduced by permission of the Director of Warren Spring Laboratory.)

1. Determination of the nature and quantities of wastes expected from a planned plant or plant extension;
2. Investigation of potential markets for products derived from reprocessed waste, opportunities for recycle within the plant, and waste segregation procedures to prevent unnecessary mixing;
3. Inclusion of the necessary equipment for recycle and for by-product manufacture as an integral part of the plant;
4. Recovery of extra running costs (and perhaps even a profit) from the recycled material.

This chapter is concerned with those recycling processes that have proved successful environmentally and economically, and also with suggestions and ideas that still require development. It is to be hoped that success in some areas will encourage experimentation in others. Examples of "industrial waste disposal made profitable" have been collected by Bennett and Lash (1974), Fox (1973), McRoberts (1975) and Nemerow (1963), and other general reports and publications are included in Appendix 2.

BUILDING AND CONSTRUCTION MATERIALS FROM WASTES

The construction industries require large quantities of low-cost raw materials, and so have always provided opportunities for waste reuse. It is estimated that 7% of the 180 megatonnes (1.8×10^8 tonnes, 1.8×10^{11} kilograms) of aggregates used in the United Kingdom in 1968 were derived from waste materials. The extent of this utilization could be increased considerably. At a conservative estimate over 100 megatonnes of potentially useful quarry and mine by-products are currently produced each year (Blunden et al., 1974; Gutt, 1974). If the material produced in the form of slurries, including dredge spoil, is included, the figure is 10 times higher than this (Miller and Collins, 1974). The Building Research Establishment has published a detailed survery (Gutt et al., 1974, 1976) of the locations, disposal methods, and prospective uses of the major waste materials of potential value to the building industry in the United Kingdom.

In the United States, a survey of ash utilization up to 1969, conducted by the Bureau of Mines (Brackett, 1970) and reported by Baum and Parker (1974a) lists a large number of applications, but aggregate for buildings, road construction and fill material accounts for the bulk of the ash used. (This is, however, only 7% of the 22 megatonnes collected annually.) Similarly, only a small proportion of the ash and coal waste generated in Australia is utilized (Beretka, 1973; *Ecos*, 1976). Efforts are now being made to increase the amount of waste used (Emery, 1976a). For example, a computerized data bank has been set up by the Highways

Materials Laboratory of McMaster University, Canada to handle information on the utilization of wastes for highway construction (Emery, 1975; Emery and Kim, 1974), and an inventory for the same purpose has been compiled in the United States (Miller and Collins, 1974). Various types of slag are very valuable construction materials (Emery, 1976b). The mineral waste utilization symposia sponsored by the U.S. Bureau of Mines and the IIT Research Institute in Chicago contain numerous examples of waste material reuse in the construction industries (Aleshin, 1974, 1976).

Waste materials usually contain a variety of components in proportions that can fluctuate widely, and it is therefore necessary to determine their physical properties and to evaluate tolerance limits for potentially harmful components. The Building Research Establishment in Britain (Collins, 1976; Gutt, 1972, 1974; Gutt and Smith, 1976) and the CSIRO Division of Building Research in Australia (Beretka and Brown, 1974; *Ecos*, 1976; Tauber, 1966; Tauber and Crook, 1965; Tauber and Murray, 1968, 1969, 1971; Tauber et al., 1970, 1971) have played significant roles in this area. For example, in the crystallization of blast furnace slag, one component (dicalcium silicate) can crystallize in a form that may subsequently undergo transformation accompanied by volume reduction and possible spontaneous "failure." Slag in which this might be a problem is therefore excluded from use, and in the United Kingdom satisfactory material is described by a British Standards Specification (Gutt, 1972, 1974). The mechanical properties of mixtures of power station fly ash with soils and clays are now being investigated, so that stable formulations can be developed for road embankments and similar applications (Joshi et al., 1975).

An important question which should be receiving more attention than at present is: What happens to the materials from demolished buildings? (Wilson, 1975; Wilson et al., 1976.) Much of the resulting wood, bricks, and concrete is dumped, because construction methods are not designed for the recycle of building components.

Dense Aggregates and Structural Materials

Many industrial by-products are suitable, in original or modified form, for high-density structural materials. Waste from the iron and steel industry is already widely utilized. Blast furnace slag, resulting from the fusion of limestone (calcium carbonate), ash from the firing coke, and siliceous and aluminous residue from iron ore, has the range of compositions indicated in Table 7.1. This slag is tapped off from the top of the molten iron and crystallized by slow cooling in air. Ninety percent of the 9 megatonnes produced annually in the United Kingdom is processed to produce a material suitable for structural reinforced concrete, roadstone, or railway ballast (Gutt, 1972, 1974; Gutt and Smith, 1976; Gutt et al., 1974). A similar percentage of the United States production of 25

Table 7.1. Range of Compositions of Blast Furnace Slag

Substance	Range (%)
Calcium oxide, CaO	30-50
Silicon dioxide, SiO_2	28-38
Aluminum oxide, Al_2O_3	8-24
Magnesium oxide, MgO	1-18
Iron oxide, Fe_2O_3	1-3
Sulfur, S	1-2.5

Source. After Gutt, 1972, 1974, by permission of the Director, Building Research Establishment. British Crown copyright, Controller HMSO.

megatonnes in 1972 was utilized (Drake and Shelton, 1974). Slag from the conversion of pigiron to steel by means of a flux of limestone (calcium carbonate, $CaCO_3$) or dolomite (magnesium carbonate, $MgCO_3$) has a more variable composition and less uses. After weathering it may be used as a road aggregate (with good resistance to tire polishing) but it is unsuitable for structural concrete. Waste molding sands form another raw material for concrete.

In the U.S.S.R., alumino-silicate waste products from a range of industrial processes, particularly from iron blast-furnaces, form the basis of glass-ceramic or ''sitall'' production. (A glass-ceramic is a glass which has been devitrified or crystallized in a controlled manner so that it consists of a very large number of very small crystals. ''Sitall'' is a coined Russian term which is equivalent to ''glass-ceramic.'') It can be produced in several colors and is used for flooring and wall tiles as well as being pressed and cast into a variety of other forms (Scholes, 1974).

Some of the largest accumulations of waste materials are the coal-mine slag or shale tips (''culm banks'') which originate from the sedimentary rocks with which the coal seams are associated; and which amount to at least 3000 megatonnes in Britain alone (McElroy, 1975; Gutt, 1974). Coal waste utilization in Britain is increasing, and there is renewed interest in this in the United States (Maneval, 1974). In coal preparation, the mixed material is crushed to liberate the trapped coal and separated by float-sink, jigging, or froth flotation methods. In the ''unburnt'' form, the slag contains a considerable amount of coal, and some of this usually remains in the ''burnt'' material resulting from slow combustion in the tips. Only a minute proportion of the annual production is used at present as landfill material, although ambitious proposals have been made for land reclamation which would involve 400 megatonnes of shale (Kinnersly,

1973). In this case it was suggested that it be moved by rail and sea, although shale can be conveyed long distances also by pipeline. In the United States the annual production is of the order of 100 megatonnes, and is increasing. Possible uses being investigated include road base material (Maneval, 1974). Some coal-washing debris in Japan is used as a plasticizer in the manufacture of floor tiles from hard pottery stone waste. This contains ferrous sulfide as an impurity, which reacts with carbonaceous matter in the coal waste to prevent foaming during firing (Toyabe, 1974). The microscopic particles of coal dust in the coal washing wastes tend to be needle-shaped, and studies are being made with the aim of incorporating this dust into concrete and gypsum, so improving their strength.

Other mining operations may yield useful "wastes." The U.S. Bureau of Mines (Cservenyak and Kenahan, 1970; Kenahan, 1971) and the Canadian Mines Branch of the Department of Energy, Mines and Resources (Collings et al, 1974) have investigated a number of materials including gold-mine, copper-mine, and iron ore tailings for the manufacture of structural blocks, bricks, and tiles. The Bureau of Mines in the United States (Kenahan, 1971) and the CSIRO Division of Building Research in Australia (Tauber et al., 1971) have considered the use of "red mud," a by-product in the manufacture of alumina from bauxite, as a resource for the heavy clay industry and in other applications (Moodie and Hansen, 1975; Sittig, 1975). Low-grade ilmenite may also be of interest for the manufacture of ceramic products (Tauber et al., 1970). Oil shale yields only about 1 barrel (159 liters) of oil per ton of shale (Klass, 1975), and it is difficult to envisage a use for the massive amounts of waste to be generated by large-scale processing of shale.

A compacted mixture of waste calcium sulfate, limestone, ash and coarse aggregate has been evaluated for the construction of parking lots (Brink, 1974; EST, 1972e), utilizing both the ash and the waste lime sludges from the sulfur oxide scrubbers of coal furnaces. The clinker, ash or boiler slag residue from urban waste incineration is usually produced at too low a temerature for it to make a suitable concrete aggregate unless it is reprocessed, as it may contain residual organic matter, iron which could cause staining, and lead, zinc, and aluminum salts which could interfere with the setting of cement, but it may be used as landfill. High-temperature incineration produces a more useful, fused material, and a similar result can be achieved by remelting incinerator residue (together with ash and precipitated muds from water treatment plants if desired). Although urban wastes contain high percentages of sodium and potassium salts because of the presence of glass, their oxides tend to be volatilized in high-temperature incineration, and the resulting slag resembles industrial slags. Because of the vitreous (glassy) nature of the granulated waste incinerator slag, the thermal conductivity of concrete manufactured with it is much lower than that of conventional concrete made with crystalline aggregate, and this product is superior even to bricks as a thermal insulator (Maydl et al., 1974).

Waste glass in road surfaces of tarmacadam (asphalt, bituminous concrete) is being evaluated in laboratories and in actual road use (Abrahams, 1973; Cummings, 1973; Dickson, 1973; Jackson, 1975; Malisch et al., 1970, 1973a, 1973b, 1975; NCRR, 1974c). Although some reports claim good skid resistance, there is some doubt as to its economic and practical viability. The surface is hard, but some investigators claim that glass is likely to have a low polish resistance. It is possible that burnt colliery spoil will be superior in this respect (Gutt et al., 1974). One suggestion is that the glass-containing "Glasphalt" product be reserved for maintenance of city streets, near the source of glass supply. An advantage of including glass is that it has a higher heat capacity and lower thermal conductivity than conventional aggregates (Dickson, 1973; EST, 1973e) and so remains softer in cold weather for a longer time, facilitating repair operations. A considerable proportion (up to about one-sixth) of foreign material can be tolerated in the glass used for this purpose, so unrefined glass waste direct from collection centers is suitable. A further suggestion is that a mixture of waste glass, ground rubber and asphalt might make a satisfactory paving material: "scrap tire-beer bottle concrete" (Bynum et al., 1972). Related to this is glass-polymer composite (GPC), made by mixing crushed waste glass with methyl methacrylate, polyester-styrene, or similar monomer, and polymerizing. This has been used successfully for sewer pipe construction (Beller and Steinberg, 1973; Steinberg and Beller, 1974). A thermoplastic paving material has been prepared by blending a copolymer of ethylene and vinyl acetate with asphalt and a petroleum distillate or used oil (Anyos et al., 1973). To this basic formulation have been added for testing several waste materials as fillers: ground rubber tires, sulfur, and lignins from papermaking. If suitable waste plastics were available, this technique could form the basis of a range of useful products. Secondary sulfur has a wide range of potential applications in the construction industry (McBee and Sullivan,1976).

Interest has been shown in using waste glass as an aggregate in Portland cement concrete (Phillips, 1973; Stearns, 1973); this has been termed "Glascrete." Preliminary results (Johnston, 1974) show that a typical commercial glass could be used satisfactorily as aggregate in concrete, but only if certain conditions of alkalinity and composition were fulfilled. Replacement of the cement with fly ash to the extent of 25-30% by mass appears to assist development of normal strength and dimensional stability in glass-containing concrete, but it is recommended that the acceptability of such concrete be based on tests with the actual materials to be used.

Large fired construction panels containing glass and other inorganic wastes can be fabricated using a vibratory casting technique before firing (Howard and Shutt, 1973). Alternatively, building bricks can be made from a variety of solid wastes using an admixture with cement, water, and a proprietary chemical (EST,

1972f). (Such binding agents may well include sodium silicate.) The moist mixture is packed in a high-pressure mold, compressed, discharged, and allowed to cure for at least 24 hours. This process avoids kiln heating, with its associated pollution problems, which is required for clay bricks.

In connection with urban waste, it should be remembered that many cities owe much of their valuable real estate to landfills, and if public nuisance can be minimized this process does have merit, although it is an unfortunate end for potentially valuable resources.

Lightweight Construction Materials

A growing market for waste products lies in the lightweight aggregate field (EST, 1970f; Gutt, 1972, 1974; Gutt et al., 1974; Kunze, 1975; Minnick, 1970; Snyder, 1964). Hot blast furnace slag can be foamed for this purpose by the steam generated when the slag is cooled with a limited amount of water; this is followed by crushing and screening to produce irregularly shaped, porous particles. Another source of lightweight aggregate is power station waste: furnace clinker and boiler slag from traveling-grate stoked furnaces (which contains up to 25% carbon) and, more recently, from pulverized coal furnaces the bottom ash and the very fine flue ash ("fly ash") collected in cyclones and electrostatic precipitators (Faber et al., 1967; Meikle, 1975). The ash generated by pulverized fuel furnaces is sometimes called "pfa" (pulverized fuel ash), and it has the chemical composition range indicate in Table 7.2. Residual, unburnt carbon

Table 7.2. Ranges of Chemical Composition of Fly Ash in the United States and the United Kingdom

Constituent	U.S. Range (mass %)	U.K. Range (mass %)
Silica (as SiO_2)	34-52	38-58
Alumina (as Al_2O_3)	13-31	20-40
Iron oxides (as Fe_2O_3)	6-26	6-16
Calcium oxide (as CaO)	1-12	2-10
Magnesium oxide (as MgO)	0.5-3	1-3.5
Sulfate (as SO_3)	0.2-4	0.5-2.5

Sources. United States: Barton, 1967; Snyder, 1964 (courtesy of Battelle-Columbus Laboratories, Columbus, Ohio). United Kingdom: Gutt, 1974; Gutt et al., 1974. (by permission of the Director, Building Research Estiablishment. British Crown copyright, Controller HMSO)

content can range from less than 1% to up to 20%, depending on the burning efficiency (Stewart and Farrier, 1967). Under microscopic examination, fly ash particles show various forms, including hollow spheres of fused or partially fused silicates. Fly ash can be pelletized with water and sintered into a granular material suitable as lightweight aggregate, its performance depending on the type of coal generating the ash (including its iron content), the ash particle size, and carbon content (Brackett, 1967; Minnick, 1970; Snyder, 1964, 1967). Fly ash readily absorbs water, and this is considered to be of advantage during the concrete curing process. Concrete made with a suitable proportion of fly ash aggregate is stronger and less likely to crack on exposure to wide temperature fluctuations. This is particularly important in climates such as that in Canada (Collings et al., 1974), although care must be taken as the inclusion of fly ash in the cement as a pozzolan (see below) can inhibit the entrainment of air in concrete and make it more susceptible to frost damage. Fly ash also has beneficial effects on brick, and sodium silicate can be used as a binder in the manufacture of fly ash brick. As a power station by-product, fly ash has the advantage that it is available in just those areas where lightweight aggregates are in demand. Britain and France are among the countries using a large proportion of their power station ash, and use of this construction material in the United States is growing (Zimmer, 1967).

Colliery spoil, the clay or shale waste containing some coal, also can be converted into a lightweight aggregate (sometimes called "sintered pumice") suitable for concrete. The pelletized material is heated rapidly by igniting the fuel it contains; gas evolved is trapped within the softening pellet, causing it to expand (Gutt et al., 1974; Kunze, 1975). The waste from slate quarries can be treated in the same way, expansion being caused by gases resulting from decomposition. Similarly, waste glass, clay, and sodium silicate (typically in proportions 78:20:2) may be pelletized and expanded to form a lightweight aggregate (Liles and Tyrrell, 1976); and low-density ceramic products may be produced by bonding lightweight shale aggregates with crushed waste glass and firing at temperatures below those at which the component glass was originally prepared (Tauber and Crook, 1965). Lignin extracts from the wood pulping industry may be used as organic binders in these and similar concrete products; alkali lignins when the liquor used for treating the woodchips is alkaline, and lignosulfonates when the liquor is acidic.

Lightweight "bubble aggregates" can be produced by coating pellets of organic waste with clay, powdered shale, or powdered slate, and then firing them. The heating value of the organic material is utilized, and the resulting particles have densities and sizes which can be controlled for particular applications (Aïtcin and Poulin, 1974). Extremely lightweight concrete may be formed by incorporating granules of reprocessed waste plastic (Kinnersly, 1973; Sperber and Rosen, 1974; USEPA, 1974b). In a recent development, the Australian CSIRO has succeeded in making a low-density brick from a mixture by volume of half clay and half sawdust, providing a potential market for the three million cubic feet (90,000 cubic meters) of sawdust generated annually in Australia.

Cements and Pozzolans

The manufacture of Portland cement can utilize the waste products of some other industries, but at the same time it generates wastes itself. Required for Portland cement are materials containing four compounds: lime (calcium oxide), silica (silicon dioxide), alumina (aluminum oxide), and iron oxide. They may bc virgin materials such as limestonc (calcium carbonate), shale, clay, sand, shell, and iron ore, or by-products such as ash, slag, and tailings, combined in suitable proportions (Barton, 1967; Brown et al., 1976b; Kondo et al., 1976). For example, cement can be made from a 50:50 mixture of fly ash and limestone (CEN, 1976c). After mixing they are fired, with the result that water, carbon dioxide, the volatile alkaline oxides (sodium oxide and potassium oxide) and cement dust are driven off (see below: ''Solutions, Sludges, Dusts, and Slags''). The resulting intimate, sintered mixture of calcium oxide, silica, alumina, and iron oxides is then finely ground. When mixed with water and aggregate to form concrete, hydration reactions occur during the ''curing'' or ''setting,'' leading to strong chemical bonding of the components. During this process, calcium hydroxide is released.

Some finely-divided siliceous materials react chemically with calcium hydroxide and water to form compounds with cementitious propeties, and are called *pozzolans*. (A pozzolan is a material that by itself does not have cementing value, but when it reacts with metal hydroxides, particularly calcium hydroxide, in the presence of water it forms cementitious compounds.) Thus fly ash reacts wih the metallic hydroxides produced during the setting of cement, prevents them from leaching out, and decreases the susceptibility of the cement to attack by sulfate aerosols in air. The extent of pozzolanic activity of various types of fly ash is being studied (Joshi et al., 1975; Raask and Bhaskar, 1975), and finely ground glass is also a pozzolan (Pattengill and Shutt, 1973). If blast furnace slag is quenched (cooled rapidly) it does not crystallize but instead forms glassy granules which have setting properties when mixed with water. This can be used in the manufacture of *slag cements*.

There are two main incentives for the replacement of a proportion of the cement in concrete by pozzolans such as fly ash. One is economic, but there is also a technical reason. The hydration of cement is exothermic (heat generating) and in ''mass concrete'' (large concrete structures such as dams intended to resist applied loads by virtue of their mass) the associated temperature rise results in tensile stresses within the final structure, and reduction in cement content is one method of minimizing this effect (Philleo, 1967).

Utilization of urban waste as a cement kiln fuel has been developed by Associated Portland Cement Manufacturers in the United Kingdom, replacing a proportion of the pulverized coal in the ignition of chalk and clay. The residual ash is incorporated in the finely ground cement, so the waste is used totally, and acid gases from the incineration serve to neutralize the alkaline oxides formed during the cement manufacture (EPM, 1976a; *Industrial Recovery,* 1976a; Knights, 1976; ME, 1976b; MRW, 1976f; *New Scientist,* 1976a; *Surveyor,* 1976).

An integrated process for making both alumina and cement has been described (Grzymek, 1974): the fine clinker produced by burning crushed limestone and clay-bearing material, when treated with mild alkali, forms a metal-aluminate from which alumina can be extracted. The remaining slurry, when mixed with limestone and gypsum, is used in the manufacture of Portland cement. This is another illustration of the advantages that may be obtained by investigating in the design stage of any process all possible uses for ''waste'' materials.

Decorative Surface Treatments

Innovations in the building industry are more sought after in surface treatments than in structural materials, and there are many examples of the application of waste materials to surfaces. The light-colored waste sands from the china clay industry in the south-west of England are utilized to some extent as decorative construction material, even in areas where transport makes the cost higher than that of local sand. This does not, however, use a significant proportion of the 10 megatonnes produced each year (Gutt, 1972, 1974; Gutt et al., 1974). The use of slag glass-ceramics for wall tiles in the Soviet Union (Scholes, 1974) has been mentioned already. Copper mine tailings are being considered for the manufacture of surface treatments such as ceramic tiles and facing bricks, providing a wide range of possible colors (Pigott et al., 1971), and in Australia it has been shown that manganese-containing industrial wastes can be made to replace industrial manganese dioxide in brown bricks (Beretka and Brown, 1974). Waste basalt dust can be fused at 1200°C into a decorative stoneware, and if clay is incorporated a new stoneware product with better heat shock resistance is formed (Tauber, 1966).

A facing brick can be manufactured from a mixture of fly ash and bottom ash or slag, with a small amount of liquid sodium silicate as a binder (EST, 1970f). It is lighter than clay brick, and competitive in strength, appearance, and cost. Waste glass can be incorporated into terrazzo floors, and as chips (fused or unfused) in ceramic pieces in wall panels (Jackson, 1975; Keller, 1973; Scott, 1973; Tauber and Murray, 1969, 1971). Broken glass can be one of the ingredients of the surface layer of white facing bricks, and wire-brushed masonry blocks containing glass present an attractive facade. Glass used in decorative materials can command a better price than it would as an aggregate substiute, but the economics of glass recovery from urban waste probably will still be unfavorable, except as a contribution toward the cost of separation and disposal.

Insulation

Coal ash slag, asbestos mine tailings, and glass from urban waste have been made into mineral wools, suitable for thermal insulation, by melting and spinning into fibers. Also, by adding calcium carbonate to urban waste glass (which need not be color-sorted) and autoclaving, there are formed large pieces of foamed glass which can be machined and nailed. The foaming reaction is

$$\underset{\textit{calcium carbonate}}{CaCO_3} + \underset{\textit{silica}}{SiO_2} \rightarrow \underset{\textit{calcium silicate}}{CaSiO_3} + \underset{\textit{carbon dioxide}}{CO_2(g)}$$

and sintering of the glass particles prevents escape of the carbon dioxide gas.

None of these processes is using a significant proportion of the wastes available, but these are steps in the right direction (Goode et al., 1973; Jackson, 1975; Mackenzie, 1973; Marchant and Cutler, 1973).

SOLUTIONS, SLUDGES, DUSTS, AND SLAGS

Insoluble material resulting from chemical processes, washing, or wet scrubbing of effluent gases can be removed from water by one or more of the physical methods described in Chapter 3. If the sludge is to be dewatered for disposal, fly ash may be used as a filtering aid. Frequently, however, it is necessary to dissolve the sludge before one of the chemical techniques of Chapter 4 is applied; so it may not be advisable to increase the bulk of sludge.

In some processes, a reagent can be recycled by means of the relatively simple provision of wringers or drainage facilities to remove reagents before the product is washed, together with a settling tank. This can reduce both the amount of effluent and the cost of reagent. Recycling of unhairing liquors and chrome liquors in the hide-tanning industry provides a good example, where simple modifications to the conventional process are able to decrease the tannery effluent and save 20-25% of the reagent costs (Balas, 1974; Davis and Scroggie, 1973; *Ecos,* 1974c; France, 1975; Money and Adminis, 1974). The flux used in the resmelting of aluminum (a mixture of sodium chloride, potassium chloride, and cryolite) can be recycled by leaching, clarifying the brine, and evaporating; thus at the same time recovering the flux, recovering metallic aluminum, and preventing ground water contamination which might result from the leaching of dumped flux (WWT, 1976d).

The manufacture of plastics and related synthetic organic materials results in effluents containing alcohols, phenols, formaldehyde, and organic acids

(Koziorowski and Kucharski, 1972; Fox et al., 1973). Reduction in the levels of these compounds is often possible in conjunction with the recycling of cooling water. Sometimes the organics can be recovered by distillation, but concentrations are often low, and the usual aim in the past has been effluent treatment rather than recovery, frequently by adsorption on carbon or by microbiological methods.

The chemical methods of Chapter 4 are required for the recovery of metal ions from aqueous solutions and sludges (Clarke, 1974; Fletcher, 1976; Flett and Pearson, 1975; Sittig, 1975). Metal ions may be recovered from the effluent resulting from metal finishing operations or electrochemical machining, by precipitation with alkali. This can be summarized very simply as

$$\underset{\text{metal ion in aqueous solution}}{M^{n+}(aq)} + \underset{\text{hydroxide ion}}{nOH^-(aq)} \rightarrow \underset{\text{insoluble metal hydroxide}}{M(OH)_n(c)}$$

where M may denote perhaps a mixture of nickel, cobalt, copper, chromium, iron, and zinc. (The symbols "aq" and "c" denote "aqueous solution" and "crystalline solid.") Lime (calcium hydroxide) is often used as a source of hydroxide ions, but almost invariably hydroxide sludge coats the lime particles, resulting in inefficient utilization and dilution of the precipitated hydroxide which is being recovered. Sodium hydroxide may well be superior in the long run, although it is more expensive. The next step in the process is one of the usual hydrometallurgical processes: dissolution in sulfuric acid, filtration, and a series of reactions to remove successively the metal ions (Chapter 4).

Hydroxide precipitates are difficult to filter, and because of coprecipitation and entrainment, separations are often inefficient. An alternative approach is to precipitate a mixed metal salt and to make use of differences in the temperature dependence of solubilities (Fletcher, 1971). For example, nickel can be precipitated as $NiSO_4 \cdot (NH_4)_2SO_4 \cdot 6H_2O$. Aqueous slurries of some metal hydroxides can be reduced directly to the metal by hydrogen gas at elevated temperatures and pressures while other hydroxides are inert, thus providing another potential separation procedure.

Copper is frequently recovered by cementation with scrap iron or zinc:

$$\underset{\text{cupric ion in aqueous solution}}{Cu^{2+}(aq)} + \underset{\text{iron metal}}{Fe\,(c)} \rightarrow \underset{\text{copper metal}}{Cu\,(c)} + \underset{\text{ferrous ion in aqueous solution}}{Fe^{2+}(aq)}$$

Alternatively, controlled potential electrolysis can remove copper ions from solutions containing chromium and zinc ions. Oxidation with chlorine converts ferrous ions to ferric ions, so that all the iron can be precipitated as ferric hydroxide at a pH of 3 (still slightly acidic) with lime; most other metal hydroxides are not precipitated at this pH.

$$2Fe^{2+}(aq) + Cl_2(g) \rightarrow 2Cl^-(aq) + 2Fe^{3+}(aq)$$

ferrous ion in aqueous solution	*chlorine gas*	*chloride ion in aqueous solution*	*ferric ion in aqueous solution*

$$Fe^{3+}(aq) + 3OH^-(aq) \rightarrow Fe(OH)_3(c)$$

ferric ion in aqueous solution	*hydroxide ion in solution*	*ferric hydroxide precipitate*

Zinc and chromium can be precipitated with alkali, as in zinc recovery from viscose rayon waste water (De Jong, 1976; USEPA, 1971), or alternatively zinc can be separated by liquid-liquid extraction. Cobalt may be precipitated at a pH of 4, and nickel metal is recovered by electrolysis, or by reduction with hydrogen under pressure, or by crystallization of nickel sulfate. This kind of reclamation is economically viable only if large amounts of metal-bearing waste sludges or liquors are generated within the collection area (for example, 70 tonnes of nickel per annum) (Jackson, 1972).

The reclamation problem is more difficult in waste metal pickling solutions (mixtures of metal salts with relatively concentrated acids) and plating solutions (mixtures of metal ions with anions such as cyanide), and these are discussed below in more detail. Pickle liquors and other sources of metal ions are being investigated as alternative chemicals to replace alum and lime for phosphate removal from municipal waste waters, notably in the Lake Erie watershed (Scott and Horlings, 1974; Shannon and Fowlie, 1974). The advantages of this procedure are twofold: the provision of a precipitant at minimal cost, and the solution of a waste metal disposal problem.

Another ''waste-treats-waste'' technique is the selective removal of heavy metal cations from mining and industrial waste streams by contact with solid agricultural by-products such as peanut skins, redwood bark, and coconut husks (Randall et al., 1974). The metal ions may be released subsequently and the purifying columns regenerated by treatment with dilute acid and washing. Examples of solutions tested on a laboratory scale are copper mine runoff water, lead battery wastes, and mercury battery wastes. Fly ash can be used for the same purpose; on contact with water, the calcium oxide in fly ash generates hydroxide ions and heavy metal ions are precipitated as hydroxides or adsorbed. Aluminum, chromium, manganese, ferric, nickel, copper, zinc, cadmium, strontium, and lead ions can be removed in this way, but not mercury (CEN, 1975b).

The fumes or aerosols (with particles below 1 micrometer in average diameter) resulting from pyrometallurgy frequently contain significant quantities of recoverable metal (De Cuyper, 1973). For example, the melting of scrap brass results in fumes containing zinc oxide (as well as a solid residue of lead and copper oxides). Flue dusts from steel furnaces contain iron oxide as well as oxides of zinc, copper, lead, and other metals originating from any recycled scrap added to the steel-making furnaces (Cochran and George, 1976a; Dean and Valdez, 1972; Sittig, 1975). Electric arc furnaces, in which up to 100% scrap charges can be used, often yield flue dusts with more zinc than iron; urban waste incinerator flue dust can contain in excess of 0.5% tin (Jackson, 1973). A common technique for reclaiming zinc and copper from dusts is dissolution in sulfuric acid, followed by copper cementation and recovery of zinc by solvent extraction, electrolysis, or as crystalline zinc sulfate for agricultural use. Alternatively, an example of a pyrometallurgical reclamation process is the recovery of zinc from an electric furnace steel-making dust containing 25% zinc by mixing with carbon and heating to 1100°C in an oxygenfree atmosphere. Up to 99% of the zinc is recovered, and preroasting of the dust in air recovers as lead oxide 90% of the small amount of lead present in the dust (Cochran and George, 1976a; Hickley and Fukubayashi, 1974).

During the manufacture of cement, a considerable amount of dust (typically 10-20% of the raw material) is collected from the exhaust gases of cement kilns. Much of it contains alkali (sodium oxide and potassium oxide) which cannot be recycled within the cement process because low-alkali cement is required for some purposes. High potassium and lime content make the dust suitable for use as a fertilizer, and its alkaline nature makes it useful for the neutralization of industrial acid wastes. Despite these and many other possible applications, this dust is not utilized widely at present (Davis and Hooks, 1974).

One of the main sources of loss in the pyrometallurgy of nonferrous metals is the discard of slags (De Cuyper, 1973; Sittig, 1975). It has been found that a carbothermic reduction technique (heating to 1350°C in the presence of coke in an electric arc furnace) can recover a considerable proportion of the copper, tin, nickel, and iron present in copper blast furnace slag. A hydrometallurgical process can be used to recover aluminum, aluminum oxide, and fluxing salts (sodium and potassium chlorides) from aluminum dross furnace slag (Caldwell et al., 1974). The lead and zinc industries generate large quantities of waste, and the toxicity of lead is a particular problem (Ratter, 1975).

Calcium Sulfate

Calcium sulfate dihydrate, $CaSO_4 \cdot 2H_2O$, is a by-product in the manufacture of phosphoric acid by digestion of rock phosphate with sulfuric acid, a reaction which can be summarized as

$$Ca_5F(PO_4)_3(c) + 5H_2SO_4(aq) + 10H_2O \rightarrow 3H_3PO_4(aq) + HF(aq) + 5CaSO_4 \cdot 2H_2O$$

rock phosphate — *sulfuric acid* — *phosphoric acid* — *hydrogen fluoride* — *gypsum*

Five tonnes of this byproduct ''phosphogypsum'' are produced per tonne of phosphoric acid, totaling 2 megatonnes of gypsum annually in the United Kingdom (Gutt and Smith, 1973; Gutt et al., 1974). (For comparison, 2.9 megatonnes of natural gypsum and 1.4 megatonnes of natural anhydrite, $CaSO_4$, were produced in the United Kingdom in 1970.) Chemically derived gypsum is a potential raw material for the manufacture of plaster-of-Paris (calcium sulfate hemihydrate, $CaSO_4 \cdot ½H_2O$) and gypsum wallboard. For use as plaster-of-Paris, it must be calcined to eliminate some of the water, and some problems arise due to the approximately 1% of residual phosphate which it contains. The Australian CSIRO Division of Building Research has developed a method of producing from media rich in phosphoric acid a calcium sulfate dihydrate product which is almost phosphatefree (Adami and Ridge, 1968a; Ridge, 1975), and has studied the properties of the derived hemihydrate (Adami and Ridge, 1968b). It appears that cast gypsum derived from ''phosphogypsum'' produced in media rich in phosphoric acid can be stronger than that derived from mineral gypsum. However, impurities still hinder its general use in plaster products and in another potential application, as a retarder to delay the setting of cement. In the United Kingdom, only in the production of plaster-of-Paris does the use of ''phosphogypsum'' show a clear cost advantage over natural gypsum: the economic manufacture of plasterboard requires a large-scale operation and natural gypsum still dominates the field. Gutt and Smith (1973) have considered the extent of utilization of byproduct gypsum and anhydrite, with particular reference to the United Kingdom. One problem is that the level of radiation (due to radium) in gypsum derived from sedimentary phosphate ores is higher than that in material derived from igneous products.

The second significant source of secondary calcium sulfate is the anhydrite generated during the manufacture of hydrofluoric acid from calcium fluoride and sulfuric acid (Gutt et al., 1974):

$$CaF_2(c) + H_2SO_4(aq) \rightarrow CaSO_4(c) + 2HF(aq)$$

calcium fluoride — *sulfuric acid* — *calcium sulfate* — *hydrofluoric acid*

The production of this so-called ''fluoroanhydrite'' is small compared with that of ''phosphogypsum'', and its utilization is limited.

Gypsum in a relatively pure form is also produced in some coal furnace sulfur removal systems (EST, 1975g; Rasmuson, 1973), and although only relatively small quantities are available (compared with ''phosphogypsum'') it may be utilized more readily. By-product calcium sulfate is an important reserve of sulfur for the future. Although there is adequate sulfur produced at present, the surplus may disappear when fossil fuels with their sulfur ''impurity'' are exhausted.

Pulp and Paper Mill Liquors

Pulp and paper mill liquors are complex mixtures of suspended solids and solutes, with high organic content (for example, lignosulfonic acid, organic acids, and sugars), and consequently it has a high biological oxygen demand. Chemical losses occur principally in the pulping process. The major pulp digestion methods are groundwood (mechanical, that is essentially nonchemical); soda (with a mixture of soda ash, Na_2CO_3, and lime, $Ca(OH)_2$); kraft or sulfate (sodium sulfide, hydroxide, sulfate, and carbonate); sulfite (calcium or ammonium bisulfite and sulfurous acid at elevated temperatures and pressures); and semichemical (sodium sulfite followed by mechanical treatment).

During the sulfite process, the lignin that binds together the cellulose fibers is dissolved as the lignosulfonate salt, and the hemicelluloses are hydrolyzed to soluble sugars, leaving approximately one-half of the dry weight of the original wood as cellulose fibers. The spent sulfite liquor has approximately 10% solids, made up of about one-half calcium lignosulfonate and one-quarter carbohydrate, as well as the pulping chemicals. Calcium bisulfite cannot be recovered by burning because of the formation of insoluble calcium salts, but sulfur dioxide can be recovered from incineration of the waste liquors of the soluble base (sodium, magnesium, or ammonium) sulfite processes, and in the case of the ammonia system reacted with ammonia and water to regenerate ammonium bisulfite (EST, 1976c).

In the alkaline pulping process the chemicals are recovered from the spent black liquors by evaporation and incineration. Sodium carbonate, Na_2CO_3, and sodium sulfide, Na_2S, are formed on burning; treatment with calcium hydroxide, $Ca(OH)_2$, forms sodium hydroxide. Additional sodium sulfide is added, and the solution reused as a pulping liquor.

If incineration can be avoided, the carbohydrate is a potential source of ethanol or single-cell protein (by fermentation), and the lignin offers a possible source of synthetic organic chemicals which could rival coal tar and petroleum, as discussed later in this chapter. The sulfite waste liquors therefore have been investigated widely as sources of by-products (fuel, road binder, animal feed, fertilizer, building material, alcohol, turpentine, resin, aliphatic acids, and yeast), but the main problem is that the markets cannot absorb the large quantities of rather low-grade materials produced. One field in which there is a large potential market is the slow release fertilizer made by treatment of lignin sulfonates with oxygen and ammonia, providing "fixed" (organically bonded) nitrogen as well as being a good source of humus (Flaig, 1973). Apart from the recovery of the maximum proportion of sulfur pulping chemicals, the main effort has gone into producing an effluent with acceptably low biological oxygen demand and color. Soluble silicates are proving valuable aids to coagulation (Falcone and Spencer, 1975).

The wastes situation in the pulp and paper industry has been reviewed by Nemerow (1963), Pearl (1968), Timpe et al. (1973), and Wiley et al. (1972) in the United States, by Dewhirst (1976) in the United Kingdom, and by Koziorowsky and Kucharski (1972) in Europe. Although processing with almost complete water recycle and low chemical loss is possible (Cornell, 1975; Haynes, 1974; Rapson, 1976; WWT, 1976f), the associated costs are still unacceptably high in most cases. There is an alternative approach to the paper mill effluent problem: extruded mineral papers (with a high percentage of mineral fillers) and "synpulps" (papermaking fibers made from synthetic resins) may enable paper to be manufactured with lower waste output (Hoge, 1976).

Acids

Acid wastes, particularly sulfuric and hydrochloric acid solutions, are common in many industries, and recovery has been an accepted practice for a long time (Theilig et al., 1975). Dilute solutions may be utilized in other parts of the process, or concentrated if impurity levels are low. Decontamination and concentration are carried out by such methods as dialysis, electrodialysis, adsorption, or ion exchange. Submerged combustion (combustion of a fuel within the solution being heated) may be used for concentrating sulfuric acid if the contaminants are organic, with oxides of sulfur being removed from the effluent gases. Spray burning of waste sulfuric acid converts it to sulfur dioxide along with carbon dioxide and other gases. The sulfur dioxide is oxidized to sulfur trioxide and absorbed to make new sulfuric acid (Nemerow, 1963; Theilig et al., 1975). Waste concentrated sulfuric acid from the petrochemical industry is acceptable to fertilizer manufacturers for digestion of phosphate rock; the hydrocarbon impurities are tolerated because biodegradation occurs when the fertilizer is spread, but the seasonal nature of the demand for fertilizer causes sulfuric acid storage problems (Coleman, 1975).

Gaseous hydrogen chloride can be recovered on graphite or a similar adsorber and is often used for the production of chlorine and chlorination of hydrocarbons (Rickles, 1965; Theilig et al., 1975). Direct recycling of hydrogen chloride is possible in the production of vinyl chloride. Dichloroethane is synthesized from chlorine and ethylene, and then thermally decomposed. Hydrogen chloride from this step can be combined with acetylene to form further vinyl chloride:

$$\underset{\text{ethylene gas}}{CH_2{=}CH_2(g)} + \underset{\text{chlorine gas}}{Cl_2(g)} \xrightarrow{45°C} \underset{\text{dichloroethane}}{CH_2ClCH_2Cl(l)}$$

$$\underset{\text{dichloroethane}}{CH_2ClCH_2Cl(g)} \xrightarrow{500°C} \underset{\text{vinyl chloride}}{CH_2{=}CHCl(g)} + \underset{\text{hydrogen chloride}}{HCl(g)}$$

$$\underset{\text{acetylene}}{CH{\equiv}CH(g)} + \underset{\text{hydrogen chloride}}{HCl(g)} \xrightarrow{150°} \underset{\text{vinyl chloride}}{CH_2{=}CHCl(g)}$$

The combined process results in complete use of chlorine and avoids hydrogen chloride production, but is limited by the availability of acetylene. Chlorine can be produced from hydrogen chloride by electrolysis, or by reaction with oxygen (the Deacon process):

$$\underset{\text{hydrogen chloride}}{2HCl(g)} + \underset{\text{oxygen}}{\tfrac{1}{2}O_2(g)} \rightleftharpoons \underset{\text{chlorine}}{Cl_2(g)} + \underset{\text{water}}{H_2O(g)}$$

It is usually easier to recycle hydrochloric acid than sulfuric acid, but the former reagent is often avoided if possible, because of its corrosive effect on metals.

Ferrous Pickling Wastes and Similar Solutions

Because of the values of both the dissolved metals and the acids, the recovery of chemicals from pickle liquors in the metallurgical industries is fairly well developed. Large volumes of dilute sulfuric acid and metal salts arise during the removal of oxide scale from steel (Besselievre and Schwartz, 1976; Koziorowski and Kucharski, 1972; Nemerow, 1963), a reaction which can be summarized:

$$\underset{\text{ferrous oxide}}{FeO(c)} + \underset{\text{sulfuric acid}}{H_2SO_4(aq)} \rightarrow \underset{\text{ferrous sulfate}}{FeSO_4(aq)} + \underset{\text{water}}{H_2O}$$

It was estimated (IUPAC, 1963) that the total amount of metal to be treated annually throughout the world was 120 megatonnes, 12×10^9 square meters of surface, requiring 2.75 megatonnes of sulfuric acid. (It should be noted that alternative techniques can be used to minimize the volume of waste: the use of shot blast or other abrasive treatment in place of pickling; and the inhibition of rust chemically rather than by a film of oil or grease (Nemerow, 1963).) The sulfuric acid process for the conversion of ilmenite, $FeTiO_3$, to titanium dioxide, TiO_2, results in a similar effluent: a mixture of iron salts and sulfuric acid (Czerwonko, 1974).

Processes have been developed that can recover, among other products, the following from iron/sulfuric acid mixtures (Rickles, 1965):

- Ferrous sulfate heptahydrate, $FeSO_4 \cdot 7H_2O$, and/or sulfuric acid
- Ferrous sulfate monohydrate, $FeSO_4 \cdot H_2O$, and/or sulfuric acid
- Ferric sulfate and sulfuric acid
- Iron oxide and sulfuric acid
- Electrolytic iron metal and sulfuric acid
- Iron oxide and ammonium sulfate

Commonly, ferrous sulfate is crystallized out by adjusting the sulfuric acid concentration and the temperature, and/or by evaporation. Gaseous hydrogen chloride may be used also in the regeneration of sulfuric acid:

$$\underset{\textit{ferrous sulfate in aqueous solution}}{FeSO_4(aq)} + \underset{\textit{hydrogen chloride gas}}{2HCl(g)} \rightarrow \underset{\textit{crystalline ferrous chloride}}{FeCl_2(c)} + \underset{\textit{sulfuric acid}}{H_2SO_4(aq)}$$

The ferrous chloride is filtered from the sulfuric acid and heated in the presence of water to regenerate hydrogen chloride for treating further ferrous sulfate solution.

$$\underset{\textit{ferrous chloride}}{FeCl_2(c)} + \underset{\textit{water}}{H_2O} \rightarrow \underset{\textit{ferrous oxide}}{FeO(c)} + \underset{\textit{hydrogen chloride}}{2HCl(g)}$$

If oxygen is present, oxidation of the iron to the ferric state also occurs:

$$\underset{\textit{ferrous chloride}}{2FeCl_2(c)} + \underset{\textit{water}}{2H_2O} + \underset{\textit{oxygen gas}}{\tfrac{1}{2}O_2(g)} \rightarrow \underset{\textit{ferric oxide}}{Fe_2O_3(c)} + \underset{\textit{hydrogen chloride}}{4HCl(g)}$$

Valuable pigment grade iron oxides can be produced by precipitation of iron as a salt or hydroxide, followed by conversion to oxide (DeWitt et al., 1952).

The heavy metal ions in mixtures with sulfuric and phosphoric acids can be removed by dialysis, ion exchange, electrolysis, and other methods (Rickles, 1965); reference has been made previously to the utilization of these metal ions in pickle liquors to remove phosphate from sewage waters (Scott and Horlings, 1974; Shannon and Fowlie, 1974).

Acid recovery is simplified if hydrochloric acid rather than sulfuric acid is used for cleaning metal surfaces, and this is the current trend (Coleman, 1975; Jackson, 1972; Taylor, 1975c). The acid ferrous chloride liquors are concentrated by evaporation, and thermally decomposed to produce iron oxide as indicated in the preceding equations. Hydrochloric acid is recovered by condensation as constant boiling point acid. Alternatively, treatment with lime precipitates ferric hydroxide, and hydrochloric acid can be regenerated from the calcium chloride solution with sulfuric acid:

$$2FeCl_3(aq) + 3Ca(OH)_2(aq) \rightarrow Fe(OH)_3(c) + 3CaCl_2(aq)$$

ferric chloride solution | *calcium hydroxide solution* | *ferric hydroxide* | *calcium chloride solution*

$$CaCl_2(aq) + H_2SO_4(aq) \rightarrow CaSO_4(c) + 2HCl(aq)$$

calcium chloride solution | *sulfuric acid* | *calcium sulfate precipitate* | *hydrochloric acid*

Phosphate coating of metals is used widely in industry for corrosion protection and to provide a painting or enamelling base. When steel is treated with the phosphating solution, iron is dissolved and hydrogen is evolved, forming products which bond strongly to the metal surface:

$$3Zn(H_2PO_4)_2(aq) + Fe + \tfrac{1}{2}O_2 \rightarrow$$

zinc phosphate solution | *iron* | *oxygen*

$$\underbrace{Zn_3(PO_4)_2(c) + FeHPO_4(c)}_{\text{zinc and iron phosphates bound to steel}} + H_2O + 3H_3PO_4(aq)$$

zinc and iron phosphates bound to steel

The waste phosphate solution contains a sludge of zinc and iron phosphates; and phosphoric acid, zinc, and iron can be recovered efficiently (Powell et al., 1972b).

Nonferrous Metal Finishing Wastes

This type of industrial waste is generated in relatively small quantities at a great number of locations: for example, there are approximately 20,000 electroplating plants in the United States (George and Cochran, 1970). Also, there is a wide variety of materials in the wastes: metal salts, alkalis, acids, and mixed proprietary electroplating formulations (Boden, 1975; Koziorowski and Kucharski, 1972; Nemerow, 1963). For these reasons, application of many of the recovery methods is not economic, and the alternative course of action often is to change the nature of the process or materials. One way or another, however, considerable progress is being made (EST, 1974a; O'Connor, 1971).

In general, copper is readily recovered from all sources by electrolysis in acid solution, and nickel is obtained with rather more difficulty in most cases by acid dissolution followed by alkaline precipitation and chemical separation. It is difficult to electrowin selectively or to solvent extract nickel from mixed solutions. The recovery of metals other than copper, nickel, and chromium is usually incidental.

The annealing of copper and brass must be followed by removal of the oxide scale. Used liquors from the pickling of copper in sulfuric acid are usually treated electrolytically. Copper is plated out on the cathode, acid is regenerated at the anode, and the solution is returned to the pickling bath:

CATHODIC REACTION: $$\underset{\text{cupric ion}}{Cu^{2+}} + 2e^- \rightarrow \underset{\text{copper metal}}{Cu(c)}$$

ANODIC REACTION: $$2H_2O - 4e^- \rightarrow \underset{\text{hydrogen ion}}{4H^+} + \underset{\text{oxygen gas}}{O_2(g)}$$

Total removal of copper is unnecessary in this case, as it does not interfere with the pickling reaction. For copper alloys (in particular with zinc and tin) the process is more complex (Jackson, 1972; Kuhn, 1971a).

Etching solutions for brass frequently are composed of sulfuric acid and chromium trioxide (CrO_3) or sodium dichromate ($Na_2Cr_2O_7$). The resulting waste solution contains dissolved copper and zinc, together with chromium which has been reduced to its trivalent form during etching:

$$\underset{\text{hexavalent chromium ion}}{2Cr^{6+}(aq)} + \underset{\text{metallic copper}}{3Cu(c)} \rightarrow \underset{\text{cupric ion}}{3Cu^{2+}(aq)} + \underset{\text{trivalent chromium ion}}{2Cr^{3+}(aq)}$$

The recovery methods for chromium, copper, and zinc salts mentioned previously must be applied: the U.S. Bureau of Mines has been active in investigating these problems (Cochran and George, 1976a, 1976b; George and Cochran,

1974; George et al., 1976). Solvent extraction into an organic phase such as a tertiary amine or organic phosphate can extract relatively pure hexavalent chromium (Cr^{6+}). The electrowinning of copper and other metals from dilute solutions such as electrolysis rinse water is facilitated by fluidized bed electrodes and flow cells, in which the short inter-electrode distances overcome the problem of the high electrical resistance of dilute electrolyte solutions while allowing high flow rates and large surface areas, as discussed in Chapter 4.

Solutions used in the pickling of the wide range of nickel alloys include sulfuric, nitric, hydrochloric, and hydrofluoric acids, alkalis, and, occasionally, molten alkalis. Direct recovery of nickel by solvent extraction is possible if the other components are first removed, but it is more usual to find all the metals precipitated with hydroxide, followed by recovery of all the metals in turn from the sludge.

Electroplating wastes often contain metals in the form of complex metal cyanides such as $Cu(CN)_4{}^{3-}$, so it is necessary to segregate the cyanide wastes (which on acidification would release the highly poisonous hydrogen cyanide) from the chromium wastes which must be acidified and reduced to the metallic state. Chrome, nickel, and cyanidefree copper plating rinse waters may be concentrated by evaporation and returned to the plating system, although capital and running costs for this form of recovery may be high. Chromic acid solution can be recovered by reverse osmosis or ion exchange for return to the plating bath, but in the case of nickel and copper plating solutions both the contaminating metal ions (for example, ferrous ions) and the valuable plating metals are positively charged, so concentration but not separation can be achieved. (The resulting deionized water may be suitable for reuse without further treatment.)

Copper, cadmium, and zinc solutions containing cyanide may be treated electrochemically to recover the metal and destroy the cyanide in one operation:

CATHODIC REACTION:

$$\underset{\textit{cuprous ion}}{Cu^+} + \underset{\textit{electron}}{e^-} \rightarrow \underset{\textit{copper metal}}{Cu(c)}$$

ANODIC REACTION:

$$\underset{\textit{cyanide ion}}{CN^-} + \underset{\textit{hydroxyl ion}}{2OH^-} - \underset{\textit{electron}}{2e^-} \rightarrow \underset{\textit{cyanate ion}}{CNO^-} + H_2O$$

SUBSEQUENT REACTIONS:

$$\underset{\textit{cyanate ion}}{CNO^-} + 2H_2O \rightarrow \underset{\textit{ammonium ion}}{NH_4^+} + \underset{\textit{carbonate ion}}{CO_3^{2-}}$$

$$\underset{\textit{cyanate ion}}{CNO^-} + \underset{\textit{ammonium ion}}{NH_4^+} \rightarrow \underset{\textit{urea}}{NH_2CONH_2}$$

U.S. Bureau of Mines research (Cochran and George, 1976b; George and Cochran, 1970) has shown that combination of waste alkaline cyanide plating solutions with acid plating wastes results in the precipitation of metallic cyanides which can be treated to recover metals. *If carried out carefully under the correct conditions,* this operation does not result in the evolution of significant quantities of hydrogen cyanide. This "waste-plus-waste" method is attractive because it requires no additional reagents, concentrates the metals, and results in a relatively harmless effluent. Some commercial experience has been obtained (George et al., 1976).

In electrochemical machining (ECM) the workpiece is made an anode and the forming tool is the cathode in an electrolytic cell. Electrolyte is pumped at high velocity through the narrow gap between anode and cathode both for cooling and to remove the eroded material. ECM sludges thus contain finely divided metals, metal oxides, and hydroxides in an aqueous electrolyte (usually sodium chloride or sodium nitrate) (Boden, 1975). Filtration may be slow unless there is pretreatment. One method (suitable only for valuable metals) is evaporation to dryness followed by dilution (George and Cochran, 1974). The Bureau of Mines has studied the smelting of filtered ECM sludges (Cochran and George, 1976a), and the Warren Spring Laboratory in Britian has developed a hydrometallurgical process involving dissolution of the sludge in hydrochloric acid, precipitation of hydroxides, and solvent extraction of cobalt and nickel (Pearson and Webb, 1973). Minor amounts of iron, chromium, cobalt, molybdenum, niobium, titanium, and aluminum may be present also in ECM wastes. Similar types of solution result from electrical discharge machining (EDM) and from the chemical reclamation of nonferrous or low-iron nickel alloys (such as those used in jet engine turbine blades) when direct resmelting is not possible because of contamination with oil or abrasives or because identification of the composition of warn or broken parts is difficult. In these cases, similar processing methods are used (Brooks et al., 1969); see Figure 2.2. The use of electrochemical and solvent extraction processes for the recovery of this type of metal waste in Sweden has been reviewed recently by Reinhardt (1975).

GLASS

Glass was one of the first manufactured products; probably it was produced first by the Phoenicians, 3500 years ago. Its main use throughout its long history has been for containers, which today account for three-quarters of the total production and for much of the discussion and activity in connection with recycling (Cannon and Smith 1974; Hannon, 1972,1973; Jackson, 1975; Nuss et al., 1975; Sittig, 1975; Thomas, 1974; Weeden, 1975); the role of glass in beverage containers is discussed in Chapter 6. Although glass is not a scarce or valuable resource, probably more effort has gone into the recycling of glass than into the reclamation of any other material, except for water and perhaps metals. The scope for waste glass utilization in secondary products may be seen from the range of papers contributed to a symposium held in New Mexico (Albuquerque, 1973).

Manufacture and Reuse

Glass is made by heating together about 50% sand (silica, SiO_2), about 16% soda ash (sodium carbonate, Na_2CO_3), and about 12% lime or limestone (calcium carbonate, $CaCO_3$), together with usually 15-20% cullet (crushed glass) which facilitates the mixing and melting of the raw materials. Silica is inexhaustible, supplies of limestone are ample, and soda ash can be produced from a naturally occurring mineral or by the chemical processing of salt or brines (particularly the Solvay or ammonia-soda process). These raw materials are well distributed and readily available, so their costs are low and stable. In addition, manufacturers readily produce their own "in house" cullet, so purchased cullet is preferred only if its price is low and if it is of good quality (uncontaminated with colored glass and foreign material, especially metals).

The usual sources of cullet at present are bottling plants, dairies, breweries, etc., rather than urban wastes where glass recovery for this purpose is less economic (*New Scientist,* 1974b). Transportation costs can raise the price of cullet above that of new raw materials, as the plants are usually situated close to their raw material sources; but on the other hand, increased cullet usage results in reduced discharge of residuals (USEPA, 1974a).

If urban waste derived glass is used (Alter and Reeves, 1975), the usual separation methods are float-sink, and froth flotation using an organic liquid that adheres to the glass particles. The latter is more efficient, producing a glass purity of 99.9%, but is more costly. Efficient optical color sorting of the larger glass fragments is possible, and an electromagnetic method can be used to separate clear and colored glass granules. The cullet of lower color quality is suitable for products with less severe color specifications, and if the public would accept slight color imperfections in recycled bottles, new glass could contain up to 50% reclaimed glass. The production of fiberglass requires a high degree of homogeneity and purity, and this cannot be assumed to be an important market for scrap glass just because the color is unimportant.

New Uses for Glass

As the demand for low-grade glass, from urban waste or litter collection for example, is not great, and as the amount of recycled glass in new products has the 50% limit, many imaginative attempts have been made to find other "sacrificial" recycling uses (Albuquerque, 1973; *Industrial Recovery,* 1976b; NCRR, 1974c). The main applications, incorporation into wall, floor, and road surfaces, as construction aggregates, glass ceramics and insulation, were discussed earlier in this chapter, and there are numerous minor uses such as goldfish tank "sand," glass abrasive paper, and tiny beads for use in reflective highway paints.

In large, densely populated cities such as New York and Chicago the mass of glass discarded in urban waste is greater than the mass of road construction aggregate used; but in general there would be no problem in disposing of all waste glass in road building and surfacing. There would be no benefit in energy or in direct economic terms in doing this; replacement of crushed stone by glass in road paving means that the total energy expended would be increased about 60 times (Hannon, 1972). On the other hand, the glass might just as well be used in road construction as buried in landfills. The advantages are environmental, psychological, and economic to the extent that it would offset at least part of the urban waste collection and separation costs, and so encourage the recovery of more critical resources. Park and Bendersky (1973) have analyzed the economics and markets for secondary glass products.

Disposal of Glass

It is often said (Abrahams, 1971; Robsons, 1972) that glass is the ideal packaging material. It is made of raw materials that are plentiful; it is inexpensive (at least at present energy costs); it does not "pollute" the environment in a chemical sense; ground glass improves the soil conditioning value of compost (unless it is to be used in sandy soils); glass bottles actually assist incineration by providing air pockets, without contributing to air pollution (although melted glass can be troublesome when it deposits on walls and grates, its presence is an advantage in slagging incineration); and glass is an ideal material for landfilling. Nevertheless, discarded bottles on beaches, in streets and in the countryside present local hazards. As discussed in the previous chapter, a deposit system to encourage return of bottles for refilling reduces overall energy expenditure, and even if returned bottles are broken up for cullet with a small net loss of energy, there are environmental and social benefits.

GASES

Treatment of waste gases (discussed, for example, by Moore and Moore, 1976; Rickles, 1965; Ross, 1968; Sittig, 1975) usually involves removal of particulate matter by a physical process (filtration or electrostatic precipitation) and removal of undesirable or valuable gases, such as halogen compounds, oxides of sulfur and nitrogen, and ammonia, by a chemical process (absorption or "scrubbing"). Carbon dioxide is not usually removed, but it is possible to speculate that the time might come when large carbon-burning plants are required to remove this gas as well, to prevent further increases in its global atmospheric concentration.

The ideal treatment for waste or effluent gases is exemplified by a recent development in the aluminum industry (Cochran, 1974). Calcined alumina is used at 120°C to adsorb the hydrogen fluoride being evolved from molten fluoride baths at 1000°C. The source of hydrogen fluoride is, in fact, the reverse of the recovery reaction, and the cycle is closed by returning the fluoride-containing alumina to the smelting cell.

$$\underset{\textit{alumina}}{Al_2O_3} + \underset{\substack{\textit{hydrogen}\\\textit{fluoride}}}{6HF(g)} \underset{\substack{1000°C\\\text{HF generation}}}{\overset{\substack{\text{HF recovery}\\120°C}}{\rightleftharpoons}} \underset{\substack{\textit{aluminum}\\\textit{fluoride}}}{2AlF_3} + \underset{\textit{water}}{3H_2O(g)}$$

One of the commonest effluent gases is sulfur dioxide, which is being considered now not just as a major pollutant from fossil fuel power stations, metal smelters and oil refineries, but also as a potential source of cheap sulfur and sulfur compounds. There is great diversity in the reclamation processes (Table 7.3), but one problem is that worldwide reduction of sulfur emissions to an environmentally acceptable level cannot be achieved without oversupplying the sulfur market (Rasmuson, 1973); so unless other applications are found, such as in the construction industry (McBee and Sullivan, 1976), some of the more sophisticated recovery methods may be abandoned in favor of cheaper disposal methods. Recovered sulfur already accounts for more than 30% of the total world supply (CEN, 1975c). The production of sulfuric acid is an important method for removing sulfur oxides from copper, lead, and zinc smelter gases. Smelters producing more acid than they can sell may use it to neutralize another waste product, smelter slag, and at the same time extract metal values (Twidwell et al., 1976).

The U.S. Bureau of Mines is developing a process to remove sulfur dioxide at levels below 1% and produce elemental sulfur. A citric acid/sodium citrate absorption solution is used, and the sulfur dioxide is reacted with hydrogen sulfide generated from steam, methane, and some of the product sulfur (Kenahan et al., 1973).

$$\underset{\textit{sulfur}}{4S} + \underset{\textit{water}}{2H_2O} + \underset{\textit{methane}}{CH_4} \rightarrow \underset{\substack{\textit{carbon}\\\textit{dioxide}}}{CO_2} + \underset{\substack{\textit{hydrogen}\\\textit{sulfide}}}{4H_2S}$$

Table 7.3. Methods of Removal of Sulfur Oxides from Effluent Gases

Type of Process	Agent	Products	Comments
Absorption	ammonia	ammonium sulfate	
Absorption	sodium sulfite	sodium bisulfite	
Absorption	buffered citrate (citric acid/sodium citrate) or buffered phosphate	reacted with H_2S to form sulfur, or steam stripped to recover SO_2	Kenahan et al., 1973
Absorption	magnesia slurry	sulfur dioxide/sulfuric acid	
Absorption	lime or limestone slurry	calcium sulfite/calcium sulfate	Gogineni et al., 1973
Absorption	soda ash, regenerated with zinc nitrate and lime	sulfur dioxide	
Absorption	sodium hydroxide (possible electrolytic regeneration)	(sulfur dioxide, hydrogen, oxygen)	
Absorption	molten carbonate	hydrogen sulfide	Yosim et al., 1973 (see text)
Adsorption	copper oxide or manganese dioxide	sulfur dioxide	
Adsorption	limestone powder	calcium sulfite/calcium sulfate	
Oxidation	air/catalyst	sulfuric acid	economic only for high SO_2 levels
Adsorption/ oxidation	activated carbon in fluidized bed	internal reaction with H_2S to form sulfur	
Wet oxidation	ozone/manganese sulfate solution	sulfuric acid	
Vapor phase reaction	ammonia/water	ammonium sulfite; ammonia regenerated with calcium sulfite disposal	Shale, 1973

Sources. EST, 1975g; Rasmuson, 1973; Rickles, 1965; Sittig, 1975.

$$2H_2S + SO_2 \rightarrow 3S + 2H_2O$$

hydrogen sulfide — *sulfur dioxide* — *sulfur* — *water*

Sulfur is also formed in a dry, fluidized activated carbon process (Ball et al., 1973), which can be summarized by the following chemical reactions in which "g" denotes "gas" and "s" indicates that these molecules are sorbed on to the carbon surface:

$$SO_2(g) + \tfrac{1}{2}O_2(g) + H_2O(g) \xrightarrow[\text{carbon}]{\text{activated}} H_2SO_4(s)$$

sulfur dioxide — *oxygen* — *water* — *sulfuric acid*

$$3H_2(g) + 3S(s) \xrightarrow[\text{carbon}]{\text{activated}} 3H_2S(g)$$

hydrogen — *sulfur* — *hydrogen sulfide*

$$H_2SO_4(s) + 3H_2S(g) \xrightarrow[\text{carbon}]{\text{activated}} 4S(s) + 4H_2O$$

sulfuric acid — *hydrogen sulfide* — *sulfur* — *water*

$$S(s) \xrightarrow{\text{heat}} S(g)$$

OVERALL REACTION:

$$3H_2(g) + SO_2(g) + \tfrac{1}{2}O_2(g) \rightarrow S(g) + 3H_2O(g)$$

hydrogen — *sulfur dioxide* — *oxygen* — *sulfur* — *water*

A less valuable product is obtained (although the process is often more convenient) by the full oxidation of calcium sulfite derived from sulfur dioxide scrubbers. The resulting nearly pure gypsum slurry settles rapidly, and can be utilized for building products. The calcium sulfite is formed either by direct reaction with lime or limestone, or indirectly as in the ammonia injection process (Shale, 1973). During the scrubbing operation, some sulfites are oxidized to sulfates.

SCRUBBING:

$$2NH_3(g) + SO_2(g) + H_2O(g) \rightarrow (NH_4)_2SO_3(c)$$

ammonia — *sulfur dioxide* — *water* — *ammonium sulfite*

$$NH_3(g) + SO_2(g) + H_2O(g) \rightarrow NH_4HSO_3(c)$$

ammonia — *sulfur dioxide* — *water* — *ammonium bisulfite*

REGENERATION OF AMMONIA:

$(NH_4)_2SO_3(aq)$	+	$2NaOH(aq)$	$\rightarrow$	$Na_2SO_3(aq)$	+	$2NH_3(g)$	+	$2H_2O$
ammonium sulfite		*sodium hydroxide*		*sodium sulfite*		*ammonia*		*water*

$NH_4HSO_3(aq)$	+	$NaOH(aq)$	$\rightarrow$	$NaHSO_3(aq)$	+	$NH_3(g)$	+	H_2O
ammonium bisulfite		*sodium hydroxide*		*sodium bisulfite*		*ammonia*		*water*

REGENERATION OF SODIUM HYDROXIDE:

$Na_2SO_3(aq)$	+	$Ca(OH)_2(c)$	$\rightarrow$	$CaSO_3(c)$	+	$2NaOH(aq)$
sodium sulfite		*calcium hydroxide*		*calcium sulfite*		*sodium hydroxide*

$Na_2SO_4(aq)$	+	$Ca(OH)_2(c)$	$\rightarrow$	$CaSO_4(c)$	+	$2NaOH(aq)$
sodium sulfate		*calcium hydroxide*		*calcium sulfate*		*sodium hydroxide*

A process that reminds us of the need to look beyond conventional liquid systems is the use of a mixture of molten alkali carbonates at 450-850°C (Yosim et al., 1973). In the following reactions, M denotes a mixture of 32 mass % Li, 33 mass % Na, 35 mass % K:

SCRUBBING:

$$SO_2(g) + M_2CO_3(l) \rightarrow M_2SO_3(l) + CO_2(g)$$

$$SO_3(g) + M_2CO_3(l) \rightarrow M_2SO_4(l) + CO_2(g)$$

$$M_2SO_3(l) + \tfrac{1}{2}O_2(g) \rightarrow M_2SO_4(l)$$

DISPROPORTIONATION: $4M_2SO_3(l) \rightarrow 3M_2SO_4(l) + M_2S(l)$

REDUCTION:

$$2M_2SO_3(l) + 3C(s) \rightarrow 3CO_2(g) + 2M_2S(l)$$

$$M_2SO_4(l) + 2C(s) \rightarrow 2CO_2(g) + M_2S(l)$$

REGENERATION:

$$M_2S(l) + CO_2(g) + H_2O(g) \rightarrow M_2CO_3(l) + H_2S(g)$$

As lithium is costly, it is recovered from the M_2CO_3 filter cake by an aqueous process. The hydrogen sulfide is converted to sulfur for recovery.

Hydrogen chloride in gaseous form, or its aqueous solution hydrochloric acid, is formed as a by-product in numerous processes, particularly in the manufacture of chlorinated organic compounds (EST, 1975h; Rickles, 1965). Examples are isocyanates (used in polyurethane manufacture), chlorofluorocarbon refrigerants and aerosol propellants, and silicones.

$$COCl_2 + RNH_2 \rightarrow RNCO + 2HCl$$

carbonyl chloride *amine* *isocyanate* *hydrogen chloride*

$$RCl + HF \rightarrow RF + HCl$$

alkyl chloride *hydrogen fluoride* *alkyl fluoride* *hydrogen chloride*

$$2R_2SiCl_2 + H_2O \rightarrow R_2SiOSiR_2 + 2HCl$$

silicone *hydrogen chloride*

(In these equations, R denotes an alkyl group, i.e., a chain of carbon atoms and associated hydrogen atoms.) The hydrogen chloride generated in these reactions has a variety of applications including cleaning, as a chlorine substitute in some newer organic oxychlorination processes, and for the generation of chlorine by catalytic oxidation or electrolysis. In the electrolytic decomposition of aqueous hydrochloric acid, a process which is finding increasing favor, the reactions are:

ANODIC REACTION: $2Cl^- - 2e^- \rightarrow Cl_2(g)$

CATHODIC REACTION: $2H^+ + 2e^- \rightarrow H_2(g)$

The oxides of nitrogen are atmospheric pollutants which play an important role in the formation of photochemical smog. Molecular nitrogen and oxygen from the air react in the high temperature zones of furnaces to form the thermodynamically favored nitrogen oxide (nitric oxide), NO. As the combustion gases cool, part of the nitric oxide is oxidized to nitrogen dioxide, NO_2. If fuels contain organic nitrogen, as in coal and oil, part of this is also converted to oxides of nitrogen. Thermal fixation (oxidation) of atmospheric nitrogen depends both on the local gas temperature and on the local gas composition, so temperature reduction techniques (such as flue gas recirculation and water injection) and oxygen availability reduction provide some control over nitrogen oxides production (Perishing and Berkau, 1973). Nevertheless, research into scrubbing techniques is necessary (Chappell, 1973) and is complicated by the interaction and different requirements of NO, NO_2, and SO_2. In principle, catalytic conversion of nitric oxide to nitrogen dioxide and subsequent absorption yields nitric acid, in the same way that tail gases from nitric acid plants are recovered, but the relatively low levels in stack gases usually make this procedure less attractive economically, and discourage recovery.

Activated carbon is a particularly valuable absorbent for low concentration levels of various gases. Carbon disulfide can be removed by activated carbon for recovery; carbon catalyzes air oxidation of hydrogen sulfide to elemental sulfur, and vapors such as mercury that are not removed by adsorption on carbon can be removed in trace quantities by means of specially impregnated carbons.

WATER REUSE

Natural water is never pure in the sense of being the pure chemical compound H_2O. It always contains a range of solids, liquids, and gases of both inorganic and organic origin, but we say that it is ''polluted'' if it contains one or more substances in amounts that restrict the intended use of the water (ACS, 1969).

There are occasions where the total absence of some solutes is a disadvantage. Very ''soft'' water is inclined to be both corrosive and ''tasteless''; dissolved oxygen is essential to aquatic life; and it is believed that a certain concentration of calcium sulfate helps bakers to achieve a golden crust on bread. Here the amount rather than the presence is the vital factor, but there are many other cases (heavy metal ions and pesticide residues, for example) where complete absence (or at least nondetectability by our most sensitive analytical techniques) is desirable.

From one point of view, it could be said that we have always recycled water, or at least taken advantage of and assisted the natural recycling processes. On the other hand, there is considerable ignorance of the fundamental physical, chemical, and biological processes in the various water cycles, and our approach to water treatment is largely empirical.

Water is such a good solvent for such a range of chemicals, as well as being the basic biological medium, that it is involved somewhere in all agricultural and industrial processes. If all other materials were carefully recycled, then we would have automatically fairly pure natural bodies of water. In other words, care in the recovery of chemicals and organic matter in any industrial, agricultural, or domestic system is accompanied by the bonus of improved water quality. Water itself becomes a useful and valuable by-product from a well-designed process. Such an argument should not be taken to extremes; in order to recover or concentrate any one material as completely as possible, it is necessary to dissipate others, and a suitable balance must be achieved. In the past, industrial water supplies frequently were considered inexhaustible, and little attempt was made to limit use. Even today there are factories which carry out cooling processes by circulating mains water to waste (Malina, 1970; WWT, 1975b).

Some iron and steel works and papermills used almost limitless supplies of water, with entire rivers passing through them, so that dilution was sufficient to meet any effluent disposal standards. Modification of existing plants to a recirculation system probably would be prohibitively expensive, but for *new* plants total recirculation might well be no more expensive than no recirculation (IUPAC, 1963; WWT, 1976f). From the point of view of the total river or lake it is far more efficient to treat wastes at the point of origin than to permit dilution, necessitating subsequent purification at the intake points for other users. An alternative procedure is the construction of special sewage canals to run parallel with rivers, collecting industrial waste waters to be conducted not to the sea but to central treatment plants. This method has the advantage of economy of scale in waste treatment, but the disadvantage of mixing the effluents. In either case, the value of recovered materials compensates for at least a portion of the treatment cost.

Amounts and Nature of Contaminants

Within a factory or industrial complex practicing water recycling, a detailed inventory is available of the nature and quantities of all contaminants entering the waste stream, and of the efficiency of each of the purification stages of the cycle, so that control over the whole system can be maintained. More commonly, however, one is concerned not with a closed system but with a large natural body of water with complex transport, chemical, and biological processes which is receiving wastes from a wide range of sources: sewage, industrial wastes, mine drainage, solid waste landfill leachate, agricultural run-off, and watercraft effluent. Only a few natural water systems have even a partial pollution inventory (ACS, 1969).

Inorganic wastes include cations such as sodium, potassium, ammonium, calcium and magnesium ions, and anions such as chloride, nitrate, bicarbonate, sulfate, and phosphate. More toxic materials such as cyanide and mercury are more likely to have been removed within the plants before water is discharged, but this should never be taken for granted. Important organic compounds appearing in wastewaters include pesticides, detergents, phenols and carboxylic acids, as well as carbohydrates, amino acids, esters, amino sugars, and amides from sewage. Other classes of pollutants are suspended solids and microbiological organisms.

Most inorganic ions may be determined relatively easily, but the difficulty in identifying and measuring the various organic substances results in reliance often being placed on collective parameters. The 5-day *biological oxygen demand,* BOD_5, measures the mass of dissolved oxygen utilized by microorganisms as they degrade or transform the compounds of nitrogen and carbon during 5 days of incubation, which is normally 70-80% of the total biological oxygen demand, BOD_{total}. The *chemical oxygen demand,* COD, measures the oxygen demand of all organic compounds that can be oxidized to carbon dioxide and water by strong oxidizing agents, and the *total organic carbon,* TOC, is a closely related parameter. Although they are important measures of contamination, these parameters provide no information on the specific natures of the organic compounds present. As an example, the only reliable measure of the concentration of cotton-mill wastes is their total organics as measured by reaction with hot permanganate solution. Most of the organic compounds in this case undergo biochemical decomposition very slowly, whereas permanganate oxidation proceeds readily.

Biodegradation

Because of its potential application cheaply and on a large scale, the degradation of chemical wastes, both organic and inorganic, by microorganisms has received considerable attention (Chapter 5). One significant advance in many countries has been the conversion of detergents from branched-chain alkyl benzene sulfonates to the much more readily biodegradable linear alkyl sulfonates. Although some aspects of biodegradation can be studied in the laboratory, in natural microbial communities there are complicated interactions so the total effect is not necessarily the sum of the individual effects. Biological and microbiological techniques for water purification are still receiving attention, although there is a trend in advanced water treatment towards physicochemical methods.

Advanced Wastewater Treatment

The term ''advanced wastewater treatment'' (AWT) is applied to any treatment that extends the normal waste purification beyond that provided by primary treatment (screening, flocculation, sedimentation) (Lash and Kominek, 1975) and secondary treatment (biological oxidation), by chemical or physical treatment which either follows or replaces the biological processes. These physicochemical processes include many of those mentioned in Chapters 3 and 4: improved chemical flocculation and filtration of suspended matter, adsorption on activated carbon, chemical oxidation (with reagents such as ozone) of dissolved organics, and the removal of inorganic ions by electrodialysis, ion exchange and reverse osmosis.* It is expected that reverse osmosis will become especially important, particularly as pesticide rejection is high in this process (Chian et al., 1975). Recently there have been suggestions for removing asbestos fibers and

*For example: ACS, 1969; Arden, 1976; Bailey et al., 1974; Bennett and Lash, 1974; Cruver, 1975; Cruver and Nusbaum, 1974; *Ecos,* 1975a; Hills 1976; Illner-Paine, 1970; Kremen, 1975; Kugelman, 1972; Lawrence, 1971; Leitz, 1976; McBain, 1976; Malina, 1970; Masters, 1974; Moore and Moore, 1976; Peeler, 1975; Porges, 1960; Sammon and Stringer, 1975; Sherwood and Sebastian, 1970; Stephens, 1976; Truby and Sleigh, 1970.

viruses (adsorbed on magnetite particles) by high-gradient magnetic separation (CEN, 1974c), and for eutectic freezing (Emmerman et al., 1973).

After advanced treatment, ex-sewage water is used for many purposes (*Ecos,* 1975a; Schmidt et al., 1975): irrigation and hydroponic gardening (Sias and Nevin, 1973), industry (Gregg et al., 1975; Horstkotte et al., 1974; Trussel, 1975), recreation (Stevens, 1967), or for the ultimate in total recovery, tap water supply (Greiser, 1971; Sebastian, 1970; Wechsler, 1975).

There is an initial instinctive dislike by most of us to the suggestion of wastewater being used for the domestic supply (''direct'' water recycle), but a little thought reminds us that we fairly readily accept purified wastewater that comes to our taps via a natural body of water—a lake, river, or underground aquifer (''indirect'' water recycle). It is estimated that 50 million people in the United States are drinking water of a quality lower than that which is produced from an advanced wastewater treatment plant (Sebastian, 1974). The incremental cost of advanced over secondary treatment of wastewater is usually less than the market value of tap water, so advanced treatment of wastewater up to a reusable quality may be preferable economically as well as environmentally. A dual water supply, one supply potable and the other disinfected but not up to potable standard might be more publicly acceptable (EST, 1975i; Mayer, 1976). The lower quality water could be used for garden watering and the flushing of wastes, uses which typically account for nearly one-half of domestic use. However, there are risks associated wih possible accidental cross-linkage of supply lines, and infections have been spread by the watering of gardens and lawns with wastewater (Pringle, 1974). ''Conjunctive'' use of high-quality, directly recycled water with ''natural'' water supply reservoirs could improve the economy of water storage. Reservoirs often are built at high capital cost to meet conditions of prolonged drought which occur only occasionally; it may be possible to provide short-term supplementary supplies from wastewater with higher operating costs per unit volume but with lower capital costs and consequent overall economies (Bailey et al., 1974). Already, municipal wastewater is reused at over 350 locations in the United States, most being in the arid southwest (Schmidt et al., 1975).

Most public authorities, however, still consider that there is insufficient scientific information about acute and long-term effects on human health resulting from the direct intimate use of reclaimed wastewaters, and that fail-safe technology for this purpose is unavailable. Some of the factors that must be considered have been summarized by Pringle (1974):

- The infectious dose of viruses is very low;
- Clinical illness is observed in only a small fraction of those who become infected, but subclinical infections have a significant effect on health;
- The killing rate of chlorine for different viruses is variable and much slower than for bacteria, particularly for viruses absorbed on particles likely to be present in wastewater;

- Even some types of bacteria appear resistant to chlorination;
- Bacterial growth can occur on the dialysis and osmosis membranes, ion exchange columns, and carbon adsorption columns used in AWT;
- Chlorine-resistant protozoa in municipal water supplies can act as carriers for microorganisms;
- Low-level chemical contaminants can have chronic effects on human health;
- Accidental or malicious spilling of toxic chemicals into the sewer system could have acute toxic effects;
- Toxic products can result from the chlorination of otherwise relatively innocuous contaminants when wastewater is treated with chlorine;
- Potentially harmful nonbiodegradable organics are not removed by advanced water treatment processes, including carbon adsorption;
- Experience with the poor quality of some present water supplies derived from "natural" sources provides little confidence in the reliability of water treatment plants.

The conclusion that must be reached is that while agricultural and industrial in-plant waster reuse and the careful reuse of municipal wastewater for industrial, agricultural, and recreational purposes should be developed, the time has not yet come for direct wastewater reuse as domestic potable water supplies to the general public.

Plant Nutrients

In many areas particular efforts are being made now to remove compounds of nitrogen and phosphorus from wastewater because of their role in eutrophication (Kugelman, 1972); enrichment with nutrients supports the growth of algae and other plant life. In some circumstances phosphorus is more critical than nitrogen compounds, because some blue-green algae can utilize atmospheric nitrogen and their growth is limited by the availability of phosphorus only. Aluminum sulfate (alum) or calcium hydroxide (lime) added to the effluent of secondary waste treatment processes can be used to precipitate phosphates as aluminum phosphate or hydroxyphosphates in neutral to slightly acidic solutions, or as calcium phosphates or hydroxyapatite in slightly alkaline conditions:

$$Al^{3+}(aq) \quad + \quad PO_4^{3-}(aq) \quad \rightarrow \quad AlPO_4(c)$$

aluminum ion in solution | *phosphate ion in solution* | *aluminum phosphate precipitate*

$$3HPO_4^{2-}(aq) \; + \; 5Ca^{2+}(aq) \; + \; 4OH^-(aq) \; \rightarrow \; Ca_5(OH)(PO_4)_3(c) + 3H_2O$$

hydrogen phosphate ion in solution | *calcium ion in solution* | *hydroxide ion in solution* | *calcium hydroxyapatite precipitate*

Alternatively, both nitrate and phosphate concentrations can be reduced by irrigating croplands or by growing algae in ponds before the water is discharged into waterways where algae are not wanted; some examples are included in Chapter 5.

Reuse of Chemicals in Water Treatment

If water is to be recycled on a large scale, it is also necessary to consider the recovery or regeneration of water treatment chemicals. This includes the recovery of alum by acidic or alkaline treatment of the sludge (Chen et al., 1976; Wang and Yang, 1975; Westerhoff, 1973) and the recovery of lime by re-calcination and dry separation from the ash by air classification (Sherwood and Sebastian, 1970). Carbon granules can be reactivated by volatilizing adsorbed impurities in a furnace (Sherwood and Sebastian, 1970; Sebastian and Isheim, 1970).

Alternatively, chemicals which are themselves wastes from other processes can be used for water treatment. For example, calcium hydroxide leached from pulverized fuel ash during hydraulic transportation has been used to reduce the natural acidity due to bicarbonate in cooling water (Van Eeden, 1975).

In-Plant Water Recycle

Many industries now are finding it rewarding economically to recycle water extensively within their plants (Appleyard and Shaw, 1974; Brandon, 1976; IUPAC, 1963; Lacy, 1974; Nemerow, 1963; Porter and Brandon, 1976; Sherlock, 1976). Pulp, paper, and fiberboard mills, which use large quantities of water and can produce a considerable amount of chemical and organic waste, are making good progress in water recycling (Cornell, 1975; Haynes, 1974; Rapson, 1976; WWT, 1976f) and there is interest in water recycling in aggregate and concrete production (Monroe, 1973). Reverse osmosis has proved valuable in pulp and paper mills, but it is not the complete solution that was anticipated at one stage; it has been observed that an increase in ionic concentration resulting from water recycle (particularly chloride and sulfate ion) inhibits slime growth but accelerates corrosion. The ultimate aim is complete water recycle, which would remove water supply and waste disposal as two of the major factors determining mill location.

Potato processing wastes and similar materials have high BOD and suspended solids, but suitable treatment permits water recycle, the costs being offset by animal feed by-product recovery (Lash et al., 1973). Löf and Kneese (1968) have considered in detail water recycling in the beet sugar industry; and Appleyard and Shaw (1974) have collected examples of water recycling from the food, electronics, textile, steel, and metal-finishing industries.

Desalination

Allied to the problem of potable water recovery from wastewater is the use as a fresh water source of brackish water (500-10,000 parts per million of dissolved solids) or seawater (3% or 30,000 ppm dissolved solids). At present, distillation is the usual method of desalinating seawater, but it has corrosion and scaling problems. Combined steam power generation and multistage flash distillation plants are more efficient than separate units (Jacques, 1975), but electric power utilities usually aim to maximize electricity generation rather than optimize the total system.

Reverse osmosis has an energy requirement 30% of that for distillation, largely avoids scaling problems by not requiring liquid-vapor phase changes, and has fewer corrosion problems because of the lower temperatures (Kremen, 1975). Reverse osmosis is particularly suitable for brackish water, and although at high separation rates the product does contain 300 ppm of residual salts, this level is within most health standards. Ion exchange is economic for brackish water (Blesing, 1971), and electrodialysis is excellent for removal of small amounts of salts but electrical energy consumption is directly dependent on the ionic content of the water.

Water Recycle Literature

More has been written about the "treatment" of water than about the recycling of any other material, and a comprehensive guide to the literature is not possible here. Useful introductions are provided by the section on "The Water Environment" in an American Chemical Society publication "Cleaning Our Environment: The Chemical Basis for Action" (ACS, 1969), and by the *Chemistry in Britain* (1976) issue on "Water." "Environmental Chemistry" by Manahan (1972) includes the chemistry of water pollution and treatment at an intermediate level. Weber (1972) has written a comprehensive account of "Physicochemical Processes for Water Quality Control." General titles include "Industrial Waste Water" (Proceedings of an international congress on industrial waste water) (Göransson, 1972); "Progress in Water Technology" (1972) (a series of volumes of conference proceedings of the International Association on Water Pollution Research); "Water Science and Technology" (Tebbutt, 1973); "Water Quality and Treatment" (AWWA, 1971); and "Water Treatment" (James, 1971). Bibliographies have been published recently on "Water Pollution and its Control" (Fish, 1972) and "Ground Water Pollution" (Summers and Spiegel, 1974).

There are also numerous journals of water treatment, and one of the best up-to-date indications of the range of research effort in this area is the annual Literature Review Issue of the "Water Pollution Control Federation Journal." Other journals are "Water and Wastes Engineering," "Water Pollution

Control," "Water and Waste Treatment," "Water Research," "Industrial Water Engineering," "Water," and "Effluent and Water Treatment Journal."

ORGANIC LIQUIDS

Water is our most important solvent, but within the liquid state there is a great variety of solvents with a wide range of properties (Barton, 1974). Although the quantities of solvents and other organic liquids to be disposed of or reused are insignificant compared with the amounts of solid waste such as metals, slags, ash, and paper, they present more significant environmental hazards (Treacher, 1973). Solvent vapors from the manufacture and use of coatings and adhesives are generated widely in industry, and solvents are present frequently in liquid wastes resulting from cleaning, degreasing, and solvent extraction processes. The chlorinated solvents are nonflammable but toxic both to man and to the bacteria in sewage works; ketones, alcohols, esters, and petroleum solvents are all flammable and their vapors in air must be maintained at levels well below the explosive limits. Some solvents contribute to photochemical smog formation.

Organic liquids need not create pollution problems, because their recovery is relatively simple (Howard, 1975, 1976; Sittig, 1975). Solvent vapor levels in air can be reduced by 98-99% by adsorption in beds of activated carbon granules in-line with the equipment producing the vapor (CE, 1976; Ross, 1968). When the carbon bed is saturated, superheated steam can be used to release the solvent for condensation and recovery, either by direct phase separation in the case of water-insoluble solvents, or by fractionation. Liquid mixtures of solvents with other materials may be distilled to recover one or more components. For example, ethylene glycol is now being recovered from "used" antifreeze solutions (WWT, 1975f). High boiling point liquids, such as silicone fluids from diffusion pumps, and other thermally sensitive materials can be recovered by distillation using a vacuum centrifugal method (Rees, 1976). The residues from organic distillations are available for incineration or pyrolysis; coatings wastes form a particularly useful energy source because synthetic polymers have high calorific values (Table 6.6) (Alter and Ingle, 1974). Consumption may be reduced to such an extent by adsorption or distillation recovery systems that capital costs are recovered within a year, particularly with the rising costs of petroleum-based materials, but it is estimated (Coleman, 1975) that in the United Kingdom the total solvent recovery rate is less than 0.1% of production.

The "Solvents" section of "Reports on the Progress of Applied Chemistry" each year usually contains up-to-date information on methods of solvent recovery.

METALS

This section is a brief survey of the general recovery of materials in the metallic state. Recvovery of metal compounds from solutions, sludges, and dusts has been discussed previously, and details of the recovery of specific metals may be found in the references listed in Appendix 1.

It was pointed out in Chapter 1 (Table 1.3) that some metals are recycled to a considerable extent already, the exceptions including the metals used primarily as coatings (tin, chromium, zinc). The scrap metal trade is well established, and although its technology necessarily lags behind primary metal production when new alloys are introduced, it is relatively efficient in redistributing scrap between metal-using industries. The simplest division of scrap metals is into ferrous and nonferrous categories, stainless steel being in the nonferrous class for some purposes because its nickel and/or chromium are more valuable than its iron content. The Standard Classification of the National Association of Recycling Industries defines 120 varieties of nonferrous scrap metals, including 39 for copper-based alloys, 12 for lead and lead-tin, 9 for zinc, and 21 for aluminum, these being the main scrap metals (Lyons, 1972). Copper is frequently alloyed with zinc (in brass) and with tin (bronze). Other metals recovered are antimony, nickel, tin, tungsten, and of course the precious and semiprecious metals. The scrap metal trade has been described in detail by many authors (Cooper, 1975; Cutler and Goldman, 1973; Jackson, 1975; Kakela, 1975; Mantell, 1975; Mantle, 1976; Neal, 1971; Nussbaum, 1976; Sittig, 1975; Thomas, 1974), and there are many journals, for example *Phoenix Quarterly* (Institute of Scrap Iron and Steel, Inc.), *Recycling Today—Secondary Raw Materials, Metal Bulletin, American Metal Market,* and *Industrial Recovery*.

Nevertheless, it is estimated that for most metals much less than half of the obsolete scrap supplies available for recycling are recovered, the figures for the United States in 1969 being: aluminum, 13% (NASMI, 1972); copper, 40%; lead, 54%; and zinc, 4% (USEPA, 1972a, 1974a). The figure for lead is 38% if the tetraethyl lead used in gasoline and other chemical compounds of lead are included. Considering the significant proportion of the world's metal output going into automobiles, which have only a short useful life, the development of increased recovery rates from this source is of major importance (Chapter 6). Metal catalysts (copper, nickel, cobalt, molybdenum) tend to be abandoned because of recovery difficulties, but this problem is under investigation (Pearson, 1973); as with the new complex alloys, sophisticated hydrometallurgical separation is required in many cases. The situation with regard to urban waste is also unsatisfactory, with only the heavier ferrous waste being utilized to any extent. An estimate (Cannon and Smith, 1974) of the metal content of United States urban waste is: ferrous metal, 7%; aluminum, 0.9%; copper, 0.5%; lead, 0.04%; and tin, 0.02%. Although these quantities correspond to only small fractions of the total production of these metals, they are still significant. The high energy content of aluminum (Table 2.1) and the limited world reserves of copper, lead, and tin justify continued efforts to improve recovery methods (Miles and Douglas, 1972). A cryogenic fragmentation method can be employed to assist in the recovery of tin and aluminum from cans (*New Scientist,* 1975c), but it appears that recovery of ferrous metal for its own sake from raw or pulverized urban waste is still uneconomic (Dudley et al., 1975).

Technical problems associated with metal recycling should not be underestimated. For example, tin and copper contaminants cause steel to be brittle, and lead (for example, from solder) moves to the bottom of steel furnaces and can infiltrate the refractory linings. There are also health hazards associated with the use of recycled metals if impurity levels are not controlled. Cadmium appears to be particularly dangerous in this respect (Schroeder, 1971) as it is widely used as a plating metal, and on recovery from mixed metals, cadmium is separated with the zinc and copper.

SYNTHETIC POLYMERS

The synthetic polymer or plastics industry has grown rapidly and is still expanding (CEN, 1975d), particularly in packaging (Berry and Makino, 1974; Bruins, 1974; Makino and Berry, 1973; Wharton and Craver, 1975) and in automobile construction (CEN, 1974e; Dean et al., 1974b). The nature, sources, and quantities of plastic wastes in the United States, and the United Kingdom, Australia, and other industrially developed countries have been reviewed in some detail.* In general, the packaging industry consumes 20-30% of the plastics produced, and most of this appears in the waste stream within a year. The building industry uses a similar proportion, notably in plumbing and in surface treatments, and this has a considerably longer life (at least 10 years). The useful life of consumer "durables" and family goods lies somewhere between (say, 5 years), and again accounts for 20-30% of production. Other major applications are agriculture, with estimates of use ranging from 2 to 7% of total plastics production, and the automobile industry.

There are two problems that are common to all consumer products, but which are more apparent in the polymer industry than elsewhere. Although the proportion of the petroleum production consumed by the plastics industry is low

**For example, in the United States:* Baum and Parker, 1974b; Jackson, 1975; Jensen et al., 1974; Lederman, 1974; Milgrom, 1972, 1975; USEPA, 1974a; *in the United Kingdom:* Bessant and Staudinger, 1973; Fergusson, 1975, 1976; Flintoff, 1974b; McRoberts, 1973; Staudinger, 1970, Thomas, 1974; *in Australia:* Pausacker, 1975.

(between 2% and 3% in the United Kingdom (Bessant and Staudinger, 1973)), rapid cost increases and shortages are eventually inevitable. The second problem, waste disposal (Baum and Parker, 1974a; Lederman, 1974; McRoberts, 1973; Showyin, 1972), involves the space occupied by plastic containers in landfills and the drainage barriers formed by plastic sheeting, environmental presures (CEN, 1971b; Mark, 1974; Staudinger, 1970, 1974), and the inadequacy of many incinerators in coping with a high plastics content (due to high heat release, melting and clogging, and emissions). Although photodegradation (Chapter 4) and biodegradation (Chapter 5) have been proposed to assist waste disposal, the ideal solution (in principle, at least) is recycling: remolding; depolymerization, upgrading, and repolymerization; conversion by solvolysis or pyrolysis; or energy-recovery incineration. This reuse of polymer waste has been reviewed recently by Sperber and Rosen (1974) with reference to over 200 publications, and by many other authors.* U.S. patents on the reclamation of rubber and other polymers have been collected by Szilard (1973). (The waste rubber field is dominated by scrap tires, discussed separately in Chapter 6.)

The chemical bases of polymer recycling were described in Chapter 4; in this section some examples are collected to illustrate the various possibilities.

Wastes in the Synthetic Polymer Industry

Plastics production and consumption is comprised of six distinct stages, and recycling can take place most readily in the early stages where it can occur internally within one manufacturing organization (Bessant and Staudinger, 1973; Fergusson, 1976; Flintoff, 1974b; Jackson, 1975; McRoberts, 1973; Milgrom, 1972), and with very much more difficulty in the final stage, consumer waste.

Raw materials manufacturers produce wastes in the form of subspecification (''off-grade'') materials, which can be reused by blending, or can be sold as lower grade material. *Compounders and converters* prepare specific grades in terms of size, filler, and color, and can reuse their wastes internally. *Manufacturers of semifinished goods* (sheet, film, moldings, foams) produce wastes in the form of trimmings and offcuts which can be recycled within the factory or sold to a reclamation merchant. *Manufacturers of end-use goods* (blow moldings, extrusions, etc.) also produce trimmings and offcuts which, if kept separate, can be returned to the process or sold back to a manufacturer of a product at an earlier stage in the prcess. The automobile and furniture *assembly industries* use large quantities of plastics and create waste, which can be utilized if kept separate or sorted. Finally, the *consumer* usually discards plastic into the urban waste stream, from which separation is difficult and usually uneconomic.

*For example, Baum and Parker, 1974b; Cheater, 1973; Flintoff, 1974b; Jensen et al., 1974; McRoberts, 1973; NCRR, 1974c; Staudinger, 1970; Zerlaut and Stake, 1974.

Thus home and prompt plastic scrap recycling is well established and moderately efficient, but there is essentially no recovery of obsolete scrap. Most of the remainder of this section is devoted to methods of utilizing obsolete plastic scrap.

Recycling Poly(tetrafluoroethylene)

Poly(tetrafluoroethylene), PTFE, is unique among polymers at present in its relatively low and constant production rate, its high value, and the fact that a significant proportion of it is recycled (Arkles, 1973). The particular chemical inertness of PTFE was mentioned in Chapter 4.

PTFE is sorted manually or mechanically and returned in granulated form to fabricators, where it is processed by the simultaneous application of pressure and heat. Some physical properties of reprocessed (''repro'') PTFE are degraded, but its electrical insulating properties are not affected significantly. As PTFE components are not used often in load-bearing applications, the recycled material is used widely.

Various physical and chemical methods are available for removing contaminants from PTFE. As this polymer is chemically inert, severe chemical purification processes may be used, for example fuming nitric acid and perchloric acid, but these are hazardous, and physical techniques such as zig-zag air classification or underwater milling are preferable. Finely divided PTFE floats on water because it is resistant to wetting, although its density is over 2 grams per cubic centimeter. Consequently, all materials with densities greater than 1 gram per cubic centimeter can be removed by underwater milling. Superheated steam at 330°C (below the PTFE melting point) passed through a fluidized bed of mixed polymers volatilizes and decomposes most polymers other than PTFE, permitting its recovery.

Reuse of Polymer Containers

In a relatively few cases, plastic containers are reused in applications similar to the original purpose: some industrial drum-liners and containers are reused; bulk domestic detergent bottles are refilled; and many containers for ice-cream, margarine, and yoghurt, for example, find uses in home kitchens and workshops. However, in most cases the commercial reuse of containers for the original purpose, as is done commonly with glass bottles, is usually uneconomic: staining, tainting, and scuffing occur more readily in the case of plastics, and washing is more difficult (Schickler et al., 1973). Some plastic beverage containers are reused; eight million refillable polyethylene containers were used for about 288 million fillings in the United States in 1970 (NCRR, 1973d). The use of plastic containers to store petroleum products and similar chemicals for even a short time makes them unusable for beverages; but a very sensitive gas-sampling device has been developed to disclose the presence of hydrocarbons in bottles as they move along a filling line (Pavoni et al., 1975).

Separation

The main area where the need arises for the separation of synthetic polymers from other materials is in urban waste and in automobile shredder waste, both of which were discussed in some detail in Chapter 6. The subject of polymer separation has been reviewed on numerous occasions*; the major difficulty in many polymer-recycling schemes is the low concentration of polymers in the mixed waste. It should be noted that some institutions generate wastes that are richer in plastics than those from the householder. Airline waste, for example, contains a high proportion of polyethylene and polystyrene (Milgrom, 1972).

Almost all the physical separation techniques described in Chapter 3 have been used to some extent. Solvent extraction can be used to obtain various components of mixed polymers in approximately 98% purity (Sperber and Rosen, 1975) but this technique still has to be proved to be commercially viable. Because polymers and paper are less dense than most of the other solid waste components, density-difference methods are particularly valuable: the U.S. Bureau of Mines has been active in this area (Holman et al., 1972, 1974; Jensen et al., 1974). The separation of paper from plastic can be accomplished then by electrostatic or thermal techniques. Different plastics can be separated from each other only with difficulty, but methods based on density differences (Jensen et al., 1974; Kenahan et al., 1973; Nagaya and Adichi, 1972; Saito, 1975; Valdez, 1976) and surface-wetting differences (Saito and Izumi, 1976; Saitoh et al., 1976; Trevitt, 1976) hold some promise. When metals and plastics occur in combination, prior cryogenic fragmentation may be useful. The recent development of laminated structures, involving either two types of plastic, or plastic plus metal or paper welded together in a single extrusion or molding process, is an added complication. Thick, heavy containers can be recovered more readily than thin packaging films.

Effective separation of plastics from urban waste is difficult to achieve, and it has been said that plastic is what is left after everything else has been removed (NCRR, 1974c). Segregation of plastics in the household, with separate municipal collections, could be established and although this has been done in Japan, it would meet opposition or at least apathy in many communities. Hand-

*Baum and Parker, 1974b; Fergusson, 1975, 1976; Flintoff, 1974b; Holman et al., 1974; Jackson, 1975; Jensen et al., 1974; McRoberts, 1973; NCRR, 1974c; Sperber and Rosen, 1974; Staudinger, 1970.

sorting of special items, such as polystyrene beverage cups in large institutions, is feasible in some circumstances. Once separated, plastics present a further problem: their low densities mean that freight rates per unit mass are high.

Remolding of Thermoplastics

Hot pressing, which is a simple method of fabricating objects from thermoplastics, is well suited to waste plastics because the dies are usually of simple design, and the mold is confined so that high pressure can be sustained to allow voids to be filled by plastic flow. *Injection molding* is less flexible and accommodating of variations in the plastic, and the absence of a confining die in *extrusion* requires even greater uniformity in the plastic (Jensen et al., 1974). Plastic fabrication methods and their relationships to remolding have been reviewed by Milgrom (1972).

Thermoplastic wastes of the same chemical composition can be reextruded or remolded readily after partial melting (Fergusson, 1975; Milgrom, 1975; Owen and Bevis, 1975; Schickler et al., 1973; Sperber and Rosen, 1974). The wastes must satisfy four requirements; they must be in a satisfactory form (granules, pellets, powder), they must be homogeneous, they must be uncontaminated, and they must be protected from oxidation during reprocessing (Flintoff, 1974b; Scott, 1976). Plastic milk bottles collected by the existing milk distribution network, and film spools collected by processing laboratories, for example, could meet these standards, but secondary products may still meet with discriminatory legislation (NCRR, 1974c). If a polymer is contaminated, filtration may be used to remove solid impurities after size reduction and melting. A limiting factor which must be considered when plastics are recycled directly is the degree of degradation which takes place every time a polymer is extruded or raised in temperature.

A process has been reported for the recycling of polyethylene film scrap which heats the film to near its melting point and then quenches it with water. The product from the resulting explosion is a popcorn-like, dense, free-flowing granular material that can be reextruded readily (Milgrom, 1975).

Other technical problems arise when *mixed* plastics are remolded, but if the proportions are controlled to some extent, products such as industrial pallets, pipes, poles, shoe soles, bicycle saddles, garden and household utensils, and toys can be made (Baum and Parker, 1974b; Fergusson, 1976; ICI, 1975; Milgrom, 1975; Rath, 1976; Trevitt, 1976). There is usually a minimum wall thickness (about 3 millimeters) that is possible in the product. The Japanese company Mitsubishi has developed the ''Reverzer'' process, in which waste plastics are forced by a transport screw through a plasticizing cone before they enter the injection cylinder. It appears that 20 or 30 ''Reverzers'' are operating in Japan, and one is being installed in Britain by Laporte Industries (*Industrial Recovery*, 1976e; MRW, 1976a, 1976b).

Two-shot injection molding is possible, with exterior virgin plastic and interior scrap (Donovan et al., 1975). Similarly, reclaimed polymers such as poly(ethylene terephthalate) can be co-extruded with the virgin polymer to give a laminated film with properties comparable to those of the original polymer film.

Waste plastic particle board (for example, Reclamat International's "Tufbord": Rath, 1976) has features making it suitable for many applications, particularly on farms: it does not absorb water, it is easy to clean, and it is maintenancefree. Scrap plastics can be reinforced with other materials which may themselves be wastes. Wood chips or fibers (up to at least 50%) can be incorporated for building board, and "contaminants" such as paper and container residues are acceptable in many waste plastic based low-grade building products, chemical bonds being formed between the cellulose and the polymer molecules (Gaylord, 1976). Bricks, concrete, and road foundations can be made with waste plastic and glass fillers (Kinnersly, 1973; Milgrom, 1972, 1975).

The mechanical milling which is necessary before polymers are reprocessed can be assisted by prior treatment with stress-crack inducing liquids. Also, the addition of typically 25% of a chlorinated polyethylene "compatibilizer" tends to improve the poor mechanical properties that may arise from the mutual insolubility of the components (Milgrom, 1975; Sperber and Rosen, 1974). In this way, thermoplastic polymer melts can be recovered which are capable of being spun into fibers (for pressing into sheets) (Szilard, 1973); alternatively, small amounts of solvent can act as plasticizers, aiding extrusion or facilitating spinning. It has been found that treatment for several hours under ambient temperature and pressure conditions with ferrous chloride/phenylhydrazine systems considerably softens natural and synthetic rubbers.

Solvents can be used more extensively, in complete solvent extraction. Thus, polyethylene can be dissolved in xylene or other low molecular mass alkane (Szilard, 1973), and this is followed by filtration of the solution and the stripping with steam of the solvent from the polymer for subsequent reuse. Solvent extraction is being investigated widely (Sittig, 1975; Szilard, 1973); nylon can be dissolved in alcohol-water mixtures, polystyrene is soluble in aromatic liquids like ethylbenzene or toluene, and polyurethane dissolves in carboxylic acids and dimethylformamide. Solubility parameters are valuable aids in the prediction of solubilities (Barton, 1975; Seymour and Stahl, 1976). Polymer mixtures may be separated by mixed solvents (Sperber and Rosen, 1975), and plasticized poly(vinyl chloride) can be extracted from a fiber substrate by solvent extraction in a closed reclamation unit without any effluent (Milgrom, 1975). Porous polymer powders can be recovered subsequently from some polymer solutions by flash evaporation.

Milgrom (1975) has devised a plastic "pyramid," the products at the top havng very rigid specifications so that the substitution of recycled plastic for virgin plastic would be difficult. (The pyramid is reprinted here, with acknowledgment to The Plastics and Rubber Institute, London, and to *Solid Waste,* the journal of the Institute of Solid Waste Management, London.) There is considerable scope in applications at the base of the pyramid where virgin plastics are too expensive (although full utilization of all waste plastics probably would flood the market).

wire
cable, film
bottles, first line pipe
sheet, tubes, coated fabrics
coated paper and paperboard

molded goods, utility pipe, low-quality
profiles and molded goods, flooring, foam,

wood and paperboard substitutes, cement and stone substitutes

Depolymerization and Repolymerization

The hydrolyses of condensation polymers such as polyurethane, polyamide, polycarbonate, and poly(ethylene terephthalate) have been investigated in some detail, Taylor (1973) having reviewed the patent literature. Poly(ethylene terephthalate) and other polymers have been depolymerized to forms suitable for repolymerization, by dissolution in the corresponding monomer, the addition of an alcohol such as ethylene glycol, and refluxing (Szilard, 1973). High monomer yields from some addition polymers are available by thermal methods, and polystyrene may be a good candidate for this process as the cost of styrene monomer is increasing rapidly (McRoberts, 1973). Substances may be added that yield "stable" free radicals capable of reacting with polystyrene free radicals to prevent chain cross-linking and so increase the monomer yield (Szilard, 1973). The monomers subsequently are repolymerized.

Although these processes may be sound in principle, in practice they suffer from the same disadvantage as remolding, that is, the difficulty of obtaining sufficient pure material in one place for an economic operation, and so they are suitable only for specialized applications. One of these is polyurethane from discarded automobiles, which is available in large amounts in auto wreckers yards (Mahoney et al., 1974).

Rather than attempt to recover the original monomers for repolymerization, it may be preferable to aim for graft copolymerization of mixed materials, by utilizing the remaining chemically reactive groups on polymer molecules. It may even prove possible to use the chemical groups present in charred residue, which may act in the same way as the carbon used now in reinforcing plastics (Zerlaut and Stake, 1974). Appropriate components may be incorporated during the initial manufacturing process to facilitate subsequent graft copolymerization, for example, chlorinated polyethylene as a polyethylene additive.

Problems of separation and compatibility are avoided if the plastics are downgraded to products of a less well-specified nature, as in pyrolysis.

Pyrolysis and Other Chemical Conversion

Most pyrolytic and related oxidation/reduction processes (Chapters 4 and 6), although not designed specifically for polymers, are well suited to recovering fuels and chemicals from them (Sperber and Rosen, 1974). Pyrolysis of polymers between 700 and 800°C in a molten salt or fluidized bed under an inert atmosphere yields olefins such as ethylene as well as aromatic compounds, for example (CEN, 1975e; Menzel et al., 1973). The problems of sorting plastics from combustible nonplastics and of segregating the various types of plastic do not arise. As the proportion of plastics in urban wastes increases, the nature of the pyrolysis products will be determined to an increasing extent by the plastics content.

Equipment specially designed by Union Carbide for plastics pyrolysis is also available. A screen extruder fitted with a pyrolysis chamber at the discharge end and automatic temperature control can yield a range of products from hard wax to liquid with fairly uniform molecular masses and viscosities (Baum and Parker, 1974b; EST, 1968c). Inert gases such as nitrogen can be used as heat carriers in the thermal decomposition of polymers (Jackson, 1975). Recovered products are potentially useful as fuels, solvents, lubricants, heavy-duty asphalt, or as raw materials for chemical and microbiological industries.

Waste poly(vinyl chloride) (PVC) on pyrolysis between 300 and 400°C yields about one-half its mass of anhydrous hydrogen chloride gas, as well as a residual char suitable as a fuel (Holman et al., 1974; Jackson, 1975; Jensen et al., 1974). (The pure PVC resin contains approximately 60% hydrogen chloride, but the yield is only 50% because PVC products usually contain additives.) In fact, on the basis of one set of reported freight rates, it costs twice as much to ship concentrated hydrochloric acid in tank cars as it does to transport an equivalent amount of hydrogen chloride in box cars in the form of PVC. Microwave heating has been used in the decomposition of PVC (as well as of other plastics) to provide a more uniform temperature (Tsutsumi, 1974), but considerable development of such pyrolysis processes is necessary before they become more generally used.

Applications such as hydrogen chloride production from PVC require a very low-cost starting material, and low-cost separation techniques for use with urban waste are still under development. The technology of mixed plastic pyrolysis for fuel is more advanced.

Combustion for Energy Recovery

If poly(vinyl chloride) could be removed easily from urban waste to prevent hydrogen chloride formation in incinerators (for example, by utilizing its high density in a float-sink procedure: see Table 3.2), direct use of the combustion energy of other plastics together with any other accompanying combustible waste would be a reasonably efficient procedure in our present economy based on fossil fuels. Given this basis, it is justifiable and profitable as an interim measure to "use petroleum twice" by first making plastics, then burning them. In the long term, however, when the true value of fossil fuels becomes more apparent, this procedure will be seen to be unsatisfactory. Plastics have heat contents comparable to those of the best fuels and far higher than that of the average urban waste (Table 6.6). The problems of incineration of plastics, particularly PVC (Baum and Parker, 1974a; EST, 1971c; Gutfreund, 1971) have been discussed in Chapter 6. In mixed wastes, hydrogen chloride tends to be neutralized by the alkaline components (ammonia from proteins, oxides of the alkali metals). Incinerators can be designed to cope with the high heat content and tendency of thermoplastics to melt and flow, and wet scrubbers are available to cope with gaseous and particulate emissions.

TEXTILES

The recovery of rags is one of the oldest recycling operations, and although the old rag-man is almost a thing of the past, waste textiles are still widely utilized. In the home, clothes and linen are still converted into cleaning rags, one of the few examples of significant domestic recycling in affluent societies.

A study carried out by the U.S. Environmental Protection Agency, the National Association of Secondary Material Industries, and the Battelle Memorial Institute to identify opportunities for increased waste utilization (NASMI, 1972; USEPA, 1972a, 1974a) included a consideration of textiles. It reported an annual 3×10^9 pounds (1.3 megatonnes) of natural and synthetic textiles recycled in 1969, approximately 20% of the annual textile production. About one-half of this was received by textile dealers. Stock (1975) has discussed textile recycling in Britain.

New textile wastes from factories (knitting, yarn spinning, carpet, and apparel manufacturers) are collected usually by merchants, often without payment being made, and are sorted into a number of categories. Mixtures of different materials, including synthetics, are usually used as flock (filling for chairs, mattresses, and so on). Cotton rags and cotton-rich blends are widely used as wipers, but the industrial wiping rag market is declining as more disposable paper wipers are used, and the growing proportion of synthetic and permanent press fabrics results in fewer suitable absorbent cotton textiles (Golueke, 1975). Rags of natural fibers are salvaged (usually directly from textile or garment factories) also for use in the manufacture of fine writing paper. Rags from urban waste are not generally used for this purpose, as they contain synthetic fibers, and separation would be uneconomic. (If cotton and wool could be extracted selectively by a suitable solvent, this source might be utilized.) Lower grades of rags are used in the manufacture of roofing felts. Woollens (Pederson, 1975) may be re-pulled and respun for further use, for example, in carpets.

A significant role is played in textile recycling by social service organizations which collect or receive garments and rags. Wearable clothes are sorted out for cleaning and resale or for distribution by charity groups, and the remainder provide wiping rags or are sold to waste textile dealers. Beause of the declining traditional markets for rags, alternative uses are being explored (Neal, 1971). The presence of a carbon skeleton is common to the molecules in all fibers, so the production of active carbon products is a possibility, particularly as the demand for them is likely to expand as air and water pollution are controlled to an increasing extent.

CELLULOSE, CARBOHYDRATE, AND RELATED MATERIALS

Sources

Most organic wastes are predominantly cellulose and carbohydrate, and this section deals with those which are recoverable, the breadth of the term "recoverable" depending on the method of recovery. The paper industry, for example, has fairly stringent requirements for its cellulose fiber, while at the other extreme, microbiological processes can cope with a wide range of agricultural and industrial organc wastes.

Agriculture gives rise to a huge amount of cellulose waste: grain stubble, tree prunings, pea vines, sugarcane bagasse, vegetable tops, and so on. Agricultural crop waste in the United States was estimated at about 500 megatonnes annually, around 1970 (Anderson, 1972; Inglett, 1973; Leatherwood, 1973; Mantell, 1975; OST, 1969; Poole, 1975). Most of this is burned at present to prevent the spread of diseases, but a considerable proportion is returned to the farmland. Animal wastes form an even larger source of cellulose and carbohydrate (Miller, 1973b), estimated in the United States at 2000 megatonnes in 1970, and food processing industries also produce considerable quantities.* As mechanical harvesting and processing methods have developed, greater percentages of crops are transported to central processing plants rather than being plowed in or burned in the fields. Because of the disease risk, most farmers are reluctant to receive

*For example, Buckle and Edwards, 1972; Garner and Ritter, 1974; Grossman and Thygeson, 1974b; Inglett, 1973; Koziorowski and Kucharski, 1972; Loehr, 1974; Loehr et al., 1973; Mantell, 1975; Nemerow, 1963.

back materials not originating from their own farms, and pesticides in the skins of fruit and vegetables reduce their desirability as cattle feed.

Considering the tree as a whole, less than one-half of the potential forest material has been utilized in the past, the waste forming another very large potential source of cellulose (APS, 1976; Dewhurst, 1976; Goldstein, 1975, 1976). Much of this waste is still left in the forests and is ultimately recycled by natural means, but this is a slow and inefficient process. The wide distribution of wood arising from building demolition means that this source of cellulose fiber is not used to any great extent. Occasionally, items such as doors and window frames are reused, but the proportion is insignificant, and frequently the waste timber is not used even as fuel. Forest wastes are being investigated and uses are being developed in a number of areas (APS, 1976; Bootle, 1972; Currier and Laver, 1973): various construction boards and moldings, compost, tannin and adhesives from bark, weed-suppressing mulches in parks and gardens, and dry bark and sawdust for cleaning up oil slicks and removing heavy metal ions from water. The adhesives are made by mixing tannin with formaldehyde, and an interesting extension of this is the possibility of a building board made by hot-pressing a mixture of bark, sawdust, and paraformaldehyde: using the tannin in the bark without prior extraction (*Ecos,* 1975b).

Cellulose in the form of paper, wood, and cotton constitutes about one-half of all urban waste (Chapter 6), and most of this is paper. In the United States, about 10% of the newsprint fiber discarded in urban waste is recycled at present. Virtually all the in-house cellulose fiber scrap is recycled, and in general the paper industry can use all the suitable paper scrap it can get. It follows that only low-grade consumer waste cellulose is available for other uses, such as hydrolysis to glucose, unless they have an economic advantage over the paper industry use.

As in other areas of recycling, it is necessary to be aware of possible hazards. Thus, selenium, which occurs naturally, is concentrated in plants that are the source of cellulose for paper production, and its levels in various materials are being monitored (Breidenbach and Floyd. 1970).

Simple Extraction and Processing Recovery Methods

This section illustrates the application of some of the techniques of Chapter 3 to the recovery of carbohydrate, cellulose, and protein, a topic reviewed recently in Birch et al. (1976), and by Mantell (1975) and Sittig (1975). The use of wastes for animal feed is considered separately, below.

Reverse osmosis and ultrafiltration are proving to be valuable in the recovery of protein and carbohydrates from various food-processing wastes (Griffith et al., 1975; Oosten, 1976; Sittig, 1975). After protein has been removed from cheese whey by ultrafiltration, lactose can be concentrated by reverse osmosis, followed by spray drying or crystallization for recovery, or by fermentation (Bennett and Lash, 1974; Coton, 1976; Horton et al., 1972). Reverse osmosis also recovers chocolate and sucrose from confectionery and preserved fruit industries; and protein has been recovered from meat, dairy, and food processing industry wastes with ion exchange resins (Grant, 1974, 1976; Jones, 1976; Sittig, 1975; WWT,1975c). Sedimentation or centrifugation after heating recovers protein from potato chip processing waste water (Meister and Thompson, 1976).

Simple grinding of fruit stones, seeds, and the woody rings of corncobs produces materials with a variety of abrasive, polishing and absorbing applications (Arnold, 1975b; Bennett and Lash, 1974). The stems of tobacco leaves, unsuitable for normal use, can be pulped and made into sheet for use in cigar manufacture. Agriculture and forestry wastes such as sugar cane bagasse may be resin-bonded for the manufacture of various building and sound-insulating boards (Michell et al., 1975), or pulped for paper and paperboard; and sawdust may be incorporated in clay bricks. Combustion of rice hulls has been shown to provide a reactive silica ash suitable for special cements (Mehta and Pitt, 1976). Wood waste can be used in controlled-release pesticides (Allan et al., 1973).

Waxes (higher aliphatic acids) can be extracted from wastes such as used tea leaves (Mathew et al., 1974); pectin can be removed from citrus peelings (Miyazaki, 1974); and nut shells can yield chemicals which are the bases of resins and coatings (Potnis and Aggarwal, 1974). Chocolate fat is pressed out of mixed chocolate wastes, and nut wastes and seed residues (for example, from mustard seed) can be pressed to yield vegetable oils. The residual meal is dried and used in animal feed. Fats can be removed from aqueous wastes by relatively simple air flotation, and processed to yield tallow for soap-making and as a railway and ship-launching lubricant. Finely divided protein can be precipitated with purified sodium lignosulfonate (a product recovered from wood-pulping) in acid conditions before air flotation: the Alwatech process (Hopwood and Rosen, 1972), which is suitable for slaughterhouse, poultry-packing, and oil and soap refinery wastes. Suint grease can be recovered from wool-scouring wastes by centrifugation; by coagulation with sulfuric acid, calcium salts, or aluminum salts; or by flotation; and processed into lanolin (Koziorowski and Kucharski, 1972; Nemerow, 1963). If tannery sludge can be kept fairly free of chromium (for example, by recycling the chrome tanning mixture) the protein in the waste can be recovered as stock feed or fertilizer (*Ecos,* 1974c). Wastes from meat-packing, poultry-packing, and related industries are processed into various forms of pet food and nutritional meal (Grant, 1974, 1976; Hamm et al., 1973). Crop wastes are potential sources of leaf protein, which can be extracted from plant wastes by processes which shear or rub the leaves (Pirie, 1976).

As well as yielding valuable by-products, the recovery processes have the added benefit of minimizing the severe effluent problems of some of these industries.

Paper Fiber Recycling

Secondary paper fibers have some advantages over virgin fibers. They have been pulped and purified, the energy required to convert them into useful products is considerably less than that needed to process wood into products, and the environmental impact is usually lower unless significant amounts of additives are present (Berry and Makino, 1974; Hunt and Franklin 1973; Makino and Berry, 1973). The secondary fiber industry is organized relatively well, and processes have been developed that can accommodate unsorted paper waste including many foreign items (Dow, 1972; EST, 1970g; Jackson, 1975). However, the problems that often outweigh these benefits are those that also confront other recycling endeavors: dispersed and fluctuating supplies of the waste material; high collection and transport costs; composition less predictable than ''new'' fibers; and public and official prejudice.

Effective recycling of paper, more than that of any other material, depends on the establishment of markets before collection takes place: because of its present place in the economic system, the axiom ''scrap materials are purchased, not sold'' particularly applies to paper (Clark, 1971).

Waste paper and paperboard recovery rates range from close to 50% in Japan and West Germany, to less than 10%. The United States, the largest single consumer of paper products, has a rate which is falling and is now less than 20% (NCRR, 1972; USEPA, 1974b). There is a general appreciation that these figures must be improved if severe shortages are not to occur (EST, 1970g; NCRR, 1974c; Pausacker, 1975; Thomas, 1974). Organizations such as the Forest Products Laboratory of the U.S. Department of Agriculture (Auchter, 1971, 1973; Carr, 1971) are carrying out research on the nature of fiber in urban waste (Myers, 1971b), the removal of additives (Klungness, 1974; Mohaupt and Koning, 1972), separation techniques (Laundrie and Klungness, 1973a, 1973b), and secondary fiber processing methods. Recent innovations in fiber cleaning, bleaching and deinking include the use of sodium silicate (''water glass'') and soluble silicate mixtures (Falcone and Spencer, 1975). Paper was included in the Battelle Memorial Institute survey of opportunities for increased waste utilization in the United States (Jackson, 1975; NASMI, 1972; Ness, 1972; USEPA, 1972a, 1974a); the economics of paper residuals management have been reviewed and analyzed by Spofford (1971); and the U.S. Environmental Protection Agency has published an analysis of supply and demand of paper fiber (Little, 1976).

Paper fibers can be classified broadly into those derived from mechanical treatment of wood (groundwood pulp) which are short or low grade, and the superior fibers obtained from chemical pulping. Groundwood fiber is used for newsprint and much of this finds its way rapidly into household waste. Strong brown papers (including bags and container board) are made by chemical pulping of softwood fibers, in particular from the kraft or sulfate process, and enter the urban waste from both commercial and domestic sources. The specifications for secondary paper grades have been published by Bond and Straub (1973).

The cost of grading mixed waste paper is usually prohibitively high, so the potential uses of the recycled fiber corresponds to that of the lowest grade of its constituents. (Multiple stock paper-making does permit inferior, dark fibers to be used as the filler in combination paper board with white, strong fibers on the surfaces.) Considerable quantities (in proportions up to 100%) of waste paper fiber are used in the manufacture of roofing felt, flooring paper, and paper insulation. Research on methods of upgrading secondary fibers is continuing. Another aspect of cellulose fiber recycling is the use of agricultural and forestry waste products such as cornstalks and sugarcane bagasse and even garden waste or feedlot waste fiber for lower grades of paperboard and construction fiberboard (Arnold, 1975a; Barbour et al., 1974; Mantell, 1975; Miller, 1971b, 1973a; Miller and Wolff, 1975; Sittig, 1975; Sloneker et al., 1973). A new system of sawing wood along the fiber direction produces (instead of sawdust) ribbons of wood 2mm thick which can be pulped to make paper with a strength only 10-30% less than that of paper made from standard pulp (Ivanova et al., 1975). An alternative approach also receiving attention is the development of new markets for low-grade fiber: single-cell protein, carbohydrate, animal feed, paint and resin extenders, and pyrolysis products.

Some degradation of fibers always occurs during repulping, but at the present recycling rates this does not cause concern. The tensile strength of recycled newsprint fiber, although less than that of the original product, is adequate for use in newspaper presses, and the "printability" and tear strength are frequently superior. A hint of a new recycling method that would minimize this degradation is the development of a solvent for cellulose, a combination of dimethyl sulfoxide and paraformaldehyde which dissolves without degradation molecular chains containing up to 8000 glucose units (CEN, 1975f). (Dimethyl sulfoxide is a by-product of paper pulping; see "Chemical Products," below.) Mixtures of dinitrogen tetroxide with ethyl acetate or with dimethylformamide also dissolve cellulose (Grinshpan et al., 1975).

Paper Additives

Many of the additives incorporated in paper for water resistance, strength improvement, gloss, and printing become contaminants which cause problems for the secondary fiber industry and which have high water pollution potential (Hunt and Franklin, 1973). They include wax, polymers, inks (particularly color-printing), adhesives, asphalt, clay, and metal foils; and the levels of residual antimony and polychlorobiphenyls, which are incorporated in some inks, should be monitored also. Wax has been removed by shredding the paper, washing with hot water, and screening (Mohaupt and Koning, 1972); alternatively, organic solvents have been used to remove plastic and wax coatings and inks. Waste paper containing a protein glue has been disintegrated successfully by treatment with the enzyme protease.

It appears that degradation in the properties of the recycled fibers results not from the presence of the contaminants, but from the processes used in their removal (Klungness, 1974), which include pulping the printed paper stock with deinking chemicals, squeezing it through rollers, and rinsing. Hydrogen peroxide has been used recently in waste paper deinking and bleaching, and in the ''Biocel'' process, deinking chemicals (dolomitic lime, CaOMgO; and ammonia) are not discarded, but finish up in fertilizer (Jackson, 1975).

There are few technological limitations to the reuse of newspaper and paperboard, and the technological and economic problems in the reuse of magazines and mixed papers can be overcome. However, because many of the recovery systems are old, relatively small, and operate with low profit margins, the modernization of plants to allow them to cope with a growing number of contaminants is difficult.

Cellulose Fiber Products from Urban Waste

Paper constitutes the largest single category of material in urban waste, and it is estimated that two-thirds of the paper and paperboard going into urban waste is potentially recoverable. (Miami, 1971; Myers, 1971a; Nuss et al., 1975; Sittig, 1975). Nevertheless, the total United States recovery rate is 20%, and, like that of many other countries, it is falling.

It was pointed out in Chapter 6 that reclamation costs could be lowered and product quality could be improved by separation of paper ''at source.'' One test of volunteer paper separation in the United States (Myers, 1971a) provided estimates of the rate of generation of various paper product: 0.24 kilogram (0.53 pound) per person per day, comprising 47% newspaper, 13% magazines, 12% strong paper, and 28% mixed paper. It also indicated that householders have difficulty in distinguishing between high- and low-grade wood fiber products. Separate collection of newspapers from households (which already takes place in a reasonable number of municipalities) does facilitate the subsequent separation of this fiber from the higher-quality chemical pulp fiber.

At present the two alternative methods of cellulose fiber removal from mixed urban waste are wet shredding, screening, and hydropulping (the Black-Clawson process); or dry shredding followed by air classification (for example, the Bureau of Mines method and the Trezek or ''Cal'' Recovery System). Removal of the nonpaper fraction prior to the addition of water minimizes the pollution of the water, and the Trezek process recovers about 85% of the paper fiber, considerably more than the wet separation process (Golueke, 1975).

In dry separation methods, the need arises to separate wastepaper from plastic film. Most plastics are thermoplastic, softening in the temperature range 65-120°C which cellulose can withstand readily for short periods, and two methods have been developed to take advantage of these properties. In one, the thermoplastic film adheres to a heated surface; and in the other, hot gases cause the thermoplastic films to contract upon themselves (Laundrie and Klungness, 1973a, 1973b).

An alternative reuse of paper in urban waste which has minimal separation problems is its contribution to the fuel value, as discussed in Chapters 4 and 6.

Hydrolysis

The acid hydrolysis of cellulose to carbohydrate, described in Chapter 4, has been applied to agricultural waste and wood. It has been suggested that cellulose separated from urban waste may be treated by an improved technique (sulfuric acid in a reactor at about 200°C for a few minutes), possibly followed by fermentation of the resulting carbohydrate to ethanol (Fagan et al., 1971a, 1971b; Grethlein, 1975; Millett, 1975; Porteous, 1969, 1975a, 1975b). A high BOD waste stream would have to be treated after the hydrolysis and fermentation.

An alternative approach to the production of sugars and ethanol from cellulose is enzymatic hydrolysis, using the cellulase enzyme produced by bacteria or fungi (Chapter 5). This has advantages over acid hydrolysis in that acid-proof reactors are not required, and the enzymes are specific for cellulose whereas acids react with impurities. The economic, microbiological, and kinetic factors have been studied in detail,* and some of the data have been collected by Bond and Straub (1973) and Mandels and Sternberg (1976).

Sterile conditions are not essential, but the pH must be kept within certain limits for optimum yields. It appears that cellulolytic organisms capable of attacking native cellulose, such as cotton, produce at least two enzymes. The first, described as C_1, breaks up the aggregates in the fibers to produce linear chains which are hydrolyzed by the second enzyme, called C_x. C_x enzymes are fairly common, but C_1 enzymes are quite rare, the best source known being *Trichoderma viride*. Enzymatic hydrolysis of cellulose often shows a rapid first stage followed by a relatively slow second stage, and this is explained by the two-phase nature of cellulose: hydrolysis of the amorphous (noncrystalline) phase proceeds more rapidly than that of the crystalline regions. It is observed that the hydrolysis rate decreases with an increasing proportion of lignin (as

*Andren et al., 1975; Brandt, 1974; Brandt et al., 1973; Brown and Fitzpatrick, 1976; Dunlap, 1973; Dunlap and Callihan, 1974; Griffin et al., 1974; Hughes, 1975b; Spano et al., 1976; Stuck, 1973; Su and Paulavicus, 1975.

found in an older plant); the lignin appears to protect the cellulose from enzymatic attack. Thus, to optimize the cellulose biodegradation rate it is necessary to pretreat the cellulosic materials to reduce the particle size, to decrease the relative crystallinity of the cellulose, and to remove lignin or disrupt its physical structure (Brown and Fitzpatrick, 1976; Klopfenstein and Koers, 1973; Rogers and Spino, 1973). Chemical treatment with a mild alkali (for example, a 10% sodium hydroxide solution) and an oxidation catalyst has the effect of increasing the subsequent hydrolysis rate (Dunlap and Callihan, 1974). Photochemical pretreatment (exposure to ultraviolet radiation in the presence of certain salts such as sodium nitrite) (Franklin et al., 1973b; Rogers, 1976) or microwave irradiation also assist the degradation process. Newsprint, made by mechanical pulping of softwoods, contains a considerable amount of ligno-cellulose complex which is attacked only with great difficulty by cellulolytic enzymes. Chemical pulping, on the other hand, partially delignifies the wood pulp, so the resulting paper is more readily attacked. Native cellulosic material, such as sugarcane bagasse, contains small amounts of hemicellulose or carbo-hydrate, so the microbial population can increase rapidly. If soluble carbohydrate is not added, pure cellulose exhibits an initial lag in enzymatic hydrolysis.

Conversion of more than 50% of the cellulose content of newspaper and other wastes to glucose solutions (typically 4-7% after 24 hours at 50°C) has been achieved on a laboratory scale by the cellulase enzyme from mutant strains of the fungus *Trichoderma viride* in the U.S. Army's Natick laboratory (Brandt et al., 1973; CEN, 1974h; EST, 1975b; Spano et al., 1976). Various other cellulose wastes and enzymes have been investigated in this laboratory, and pilot operations are being developed.

Trichoderma viride also has been shown to utilize two-thirds of the cellulose and hemicellulose in feedlot waste, producing protein and cellulase (Griffin et al., 1974; Sloneker et al., 1973). In fermentation of feedlot waste with *T. viride* there is a 3-day period during which the cellulase activity develops. By the fourth day, the manure odor is replaced by an earthy odor, and after the sixth or seventh day the rapid growth phase is completed and stabilization occurs.

More significant than the technical problems in glucose production from waste cellulose are the problems of economics. One application of enzymatic or acid hydrolysis which is less difficult in these respects is the conversion of lactose from whey byproducts into a glucose/galactose sweetening agent (Horton et al., 1972).

Protein Production

Proteins are made up of different combinations of several amino acids as the basic units. The human body can synthesize some amino acids, but others required for normal functioning should be present in food. Any one cereal food, for example, is deficient in one or more amino acids, and quality of diet in terms of amino acid content is one aspect of the world's starvation problem. Adequate meat protein cannot be made available in sufficient quantity for everyone, but single-cell protein (SCP) may fill the need (Ghose, 1969; Hughes, 1975b; Humphrey, 1966; Kihlberg, 1972; Parker, 1973; Rolfe, 1976; Tannenbaum and Matelas, 1968; Tannenbaum and Pace, 1976).

The most ancient method of using carbohydrate and cellulose to provide the energy and some of the chemicals for high-quality protein production is as *animal feed* (Rook, 1976). The use of the organic fraction (mainly cellulose) of urban waste for this purpose was described in Chapter 6. Other wastes that have been fed successfully include straw and other agricultural residue (Klopfenstein and Koers, 1973), wood wastes, paper (Coombe and Briggs, 1974; *Ecos,* 1974b), poultry manure and other manure (Grossman and Thygeson, 1974b; Smith, 1973), wastes from the food processing industries: chocolate and confectionery, potato chip (Grames and Kueneman, 1968) and other vegetable and fruit waste (White, 1973), mustard seed residue, spent tea leaves from instant tea production (Croyle et al., 1974), corncob pith, and brewery grain wastes (Bennett and Lash, 1974). Sulfite wood pulp liquor containing sugars is used as a binder for animal feed pellets. Single-cell protein, derived either directly from the waste of fermentation industries such as brewing, or by fermentation from the wastes of other industries, is also valuable stock feed.

After animal feed, the next most firmly established methods of utilizing cellulose wastes for (indirect) protein production are *composting* and *anaerobic digestion* (Chapter 5). They have been applied to sugars, starch, and cellulose from a wide range of urban, forestry, and agricultural wastes, and from some food-processing industry wastes. Production of animal feed ingredients and fuel gas by thermophilic anaerobic fermentation appears to be an effective way of utilizing animal feedlot wastes (Coe and Turk, 1973), and composting can be applied to both agricultural and industrial organic wastes (Troth, 1973).

Selective fermentation using bacteria, algae, yeasts and fungi can produce *single-cell protein* for human food and animal feed.* Protein content of the product is variable, depending both on species variations and on cultural conditions, but it is generally 50-55% in yeasts and 50-80% in bacteria. In fungi (15-45% protein) and algae (20-60% protein) the cell wall constitutes a greater proportion of the cell. Bacteria and yeasts are usually more suitable as foods, although fungi and algae usually can be recovered more readily. Microorganisms can be found with amino acid ratios close to the ideal for human food, or

*For example: Aguirre et al., 1976; Birch et al., 1976; Brown and Fitzpatrick, 1976; Bunker, 1963, 1964; Callihan and Dunlap, 1973; Callihan et al., 1975; Daly and Ruiz, 1975; Dunlap, 1973; Dunlap and Callihan 1974; Gounelle de Pontanel, 1972; Imrie, 1976; Imrie and Rightelato, 1976; Kosaric, 1973; MacLaren, 1975; MacLennan, 1974, 1975; *New Scientist,* 1975; Rightelato et al., 1976; Rogers, 1976; Rogers and Spino, 1973; Sittig, 1975; Yen, 1974b.

organisms can be selected which have complementary amino acid patterns or which supplement stable foods that are deficient in particular amino acids.

One problem is that rapidly growing cells characteristically have a high nucleic acid content, and this is undesirable in SCP for food: humans, unlike other mammals, cannot produce the enzyme uricase which oxidizes uric acid. Uric acid is a decomposition product of nucleic acid which appears to precipitate in tissues, joints, kidney stones, and bladder stones (Imrie, 1976; Kihlberg, 1972). The nucleic acid content of SCP can be reduced by hydrolysis with ribonuclease; but full testing for other possible toxic effects must be carried out also (MacLaren, 1975).

The relative amounts synthesized of lipids (including fats) and protein depend on the carbon-to-nitrogen ratio of the medium in which the microorganisms are growing. When low nitrogen levels limit growth, energy reserves are stored and the lipid content may be high, notably in some algae (for example, *Chlorella*) and the yeasts, especially *Rhodotorula* (Bunker, 1964) which also contains valuable vitamins; the feasibility of production of single-cell fats and oils is equivalent to that of single-cell protein (Ratledge, 1976).

For nutritional purposes it is possible to vary the protein:lipid ratio as required. Although it is easier to produce animal feed than human food, it is apparent that the developing, underfed countries require SCP for direct use as food, and that raising animal protein is generally an extravagant and wasteful luxury (Tudge, 1975). Edible SCP has been produced from alkali-treated sugarcane bagasse and from whey, but producing acceptable SCP from most other forms of waste cellulose is obviously difficult (Skinner, 1975). The consistency of the product is important; food must be palatable as well as edible (Imrie, 1976; Yudkin, 1972).

Some examples of currently used or potentially useful processes are now introduced briefly, for various types of microorganisms.

Actinomyces. Thermophilic actinomyces have proved suitable for the production of animal feed from feedlot wastes (Bellamy, 1973a).

Yeasts. Yeasts such as *Candida* (formerly *Torulopsis) utilis* (torula yeast) and *Saccharomyces fragilis* grown on carbohydrate wastes such as whey, molasses, and sulfite liquor are used as poultry and stock feeds (Bennett and Lash, 1974; Koziorowski and Kucharski, 1972; Muller, 1969; Pearl, 1969; Skogman, 1976; Sobkowicz, 1976). Fermenting yeasts use both aerobic and anaerobic mechanisms, so adequate oxygen must be provided to encourage aerobic processes. *Candida* can utilize pentose sugars, which occur in pulp waste liquor, as well as the hexose sugars, while *Saccharomyces* cannot make use of pentoses. The yeast *Rhodotorula gracilis* can be acclimatized to utilize pentose sugars.

Bacteria. Bacteria of the genus *Cellulomonas* have been studied in detail for the fermentation of sugarcane bagasse and agricultural and forestry wastes: The protein was found to have a good amino acid profile (Dunlap and Callihan, 1974). *Cellulomonas flavigena* does not have a C_1 enzyme to bring about a preliminary disruption of cellulose fibers, so pretreatment with alkali is necessary (Callihan et al., 1975).

Fungi. High-quality fungal protein (*Aspergillus fumigatus*) can be produced from waste cellulose following a photochemical pretreatment (Rogers et al., 1972). The fungus *Trichoderma viride* has been referred to previously in connection with enzymatic hydrolysis (Griffin et al., 1974), and *Myrothecium verrucaria* was found to provide the maximum rate of protein biosynthesis from ball-milled newpaper in extensive screening tests (Updegraff, 1971). Runners-up were *Trichoderma lignorum* and *Aspergillus fumigatus,* and although all these fungi were superior to bacteria such as *Cellulomonas* on the newspaper substrate, the yields and rates were considerably lower than those referred to previously on more readily hydrolyzed sources of cellulose. A strain of *Aspergillus niger* and a *Fusarium* species have been found to have an unusually high protein content (Imrie and Rightelato, 1976; Rightelato et al., 1976), and protein production by *Fusarium semitectum, Aspergillus oryzae,* and *Trichoderma viride* from various wastes has been reported (Worgan, 1976). The filamentous nature of fungi resembles the structure of conventional foods more closely than do the consistencies of bacteria and yeasts, so fungi are more acceptable from this point of view.

Higher fungi such as mushrooms are, in fact, desirable foods now, and they grow in composted waste (Biddlestone and Gray, 1973; Kosaric, 1973) but they tend to be expensive protein sources when produced indoors. In China, the Padi Straw mushroom has remained an outdoor garden crop, and it offers many advantages: It grows rapidly (10 to 12 days) on a wide range of agricultural and chemical waste materials, and has about 40% of very high-quality protein (Roper, 1972).

Algae. Algae grown on simple shallow lagoons, coupled with fish culture, form a suitable low-technology system for developing countries. The inhabitants of Chad and Niger in Africa and, allegedly, also the Aztecs in Mexico, have eaten the blue-green alga *Spirulina maxima* over a long period of time. These algae cling together because of their shape, and can be harvested easily and made into cakes or soup (Kosaric, 1973; Tannenbaum and Mateles, 1968). Green algae of the genus *Chlorella* are also being studied (Bennett and Lash, 1974; Ghose, 1969; Priestley, 1976; Vincent, 1969). Algae grown on sewage lagoons can be used to feed chickens and pigs, but if alum is used in the water treatment process it must be recovered from the algae slurry (by acidification with sulfuric acid, for reuse as a flocculant) as it has a toxic effect (WWE, 1973).

Chemical Products

Bleached pulp, paper mill sludge, and cotton waste can be converted into chemicals such as tartaric acid and glycine by oxidation and hydrolysis (Chapter 4). Lignin derivatives, produced as described in the section "Pulp and Paper Mill Liquors" in this chapter, or by acid hydrolysis, can be processed chemically and (after photochemical treatment) microbiologically to provide a wide range of products*: synthetic tanning agents, ion exchangers, activated carbon, oxalic acid, the flavoring vanillan, the preservative ethyl vanillate, phenols, insecticides, drugs, polyesters, and a slow release fertilizer.

The future demands for phenol from lignin and for furfural from the hemicellulose in wood appear to be particularly promising (CEN, 1975g; Goldstein, 1975a). Glacial acetic acid, formic acid, turpentine, beta pinene and other tall oil products, dimethyl sulfide and its derivative, the increasingly important solvent dimethyl sulfoxide, are among the products obtained from sulfite waste liquor (Pearl, 1968; Ward, 1975). Activated carbon can be made by heating various cellulosic wastes such as peanut hulls in a controlled, oxidizing atmosphere at temperatures of 300-1000°C, and quenching (Bennett and Lash, 1974; EST, 1972g). Tannin from barks of many trees, including *Pinus radiata* and *Acacia mollisima* can be converted into adhesives suitable for plywood and particle board, by combining it with formaldehyde, for example (*Ecos,* 1975b). Tannin-rich waste products such as bark and husks can be used directly to scavenge oil spills (*Ecos,* 1975b) and heavy-metal ions (Randall et al., 1974).

There is a great variety of chemical products, particularly solvents, that can be formed by fermentation of carbohydrates, and some of these were mentioned in Chapters 5 and 6. Anaerobic digestion is carried out on a wide range of carbohydrate and cellulosic materials to yield methane: animal manure, sewage, algae from sewage sludge ponds, the organic component of urban waste, and brewery and distillery wastes. Ethanol is a particularly valuable fermentation product, from sugars in wood pulp liquors by *Saccharomyces cerevisiae,* for example, for use as an industrial solvent, a fuel (Trevelyan, 1975), or for conversion into acetaldehyde and acetic acid. Other possible fermentation products are butanol; acetone; propionic, butyric, formic and lactic acids; and amines (Beesch and Tanner, 1974; Castor, 1974; Michaelis, 1975; Miller, 1961; Scott, 1976). Several of these fermentation products can be converted into vinyl compounds for subsequent polymerization (Table 7.4). Amino acids, also obtained by fermentation, may be polymerized to a product with excellent film- and fiber-forming properties, but which is currently too costly for general use (Elias, 1976).

*For example, Allan et al., 1973; Browning, 1975; EST, 1975b; Falkehag, 1975; Flaig, 1973; Goldstein, 1975a, 1975b, 1975c, 1976; Koziorowski and Kucharski, 1972; Lindberg et al., 1975; Nakano, 1975; Nemerow, 1963; Pearl, 1968; Sittig, 1975; Ward, 1975.

Table 7.4. Synthetic Polymers from Carbohydrate and Cellulose

Carbohydrate	Fermentation Products	Vinyl Monomers	Polymers
$C_6H_{12}O_6$	CH_3CH_2OH ethanol →	$CH_2{=}CH_2$ ethylene →	polyethylene; poly(vinyl chloride); poly(vinyl acetate)
	CH_3COCH_3 acetone →	$CH_2{=}CHCH_3$ propylene →	polypropylene; poly(acrylonitrile)
$(C_6H_{10}O_5)_n$	$CH_3CH_2CH_2CH_2OH$ butanol →	$CH_2{=}CHCH_2CH_3$ butylene →	polybutylene
		↓ $CH_2{=}CHCH{=}CH_2$ butadiene →	polybutadiene

Source. After Scott, 1976.

The possibility of obtaining all these chemical products implies that wood and other sources of cellulose could replace petroleum as a major plastics feedstock (Deanin, 1975; Elias, 1975). The technology is available now, and as petroleum prices increase, the economics are becoming more favorable. Low-grade cellulose material unsuitable for solid wood products or for pulping is suitable for chemical processing, so it is considered that there is adequate wood available (CEN, 1975g; Goldstein, 1975c, 1976). In some countries that rely on wood as a fuel, however, reserves are being depleted rapidly, and a balance between chemical feedstock and fuel must be reached for cellulose, in the same way as for petroleum and coal. Eventually, it may be possible to make extensive use of an expanded carbon cycle (Scott, 1976):

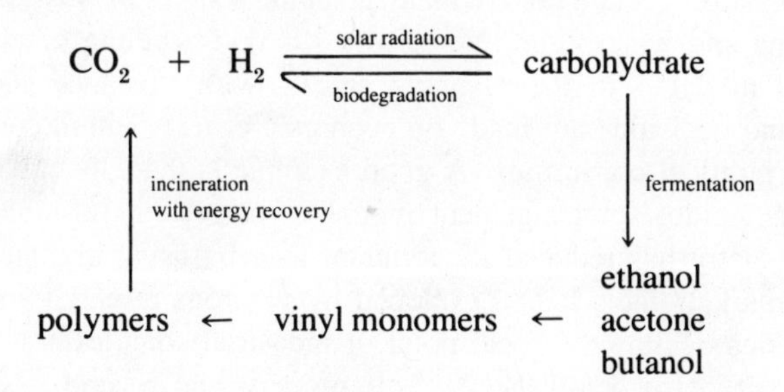

Energy Sources

Cellulose and related materials as sources of fuel and energy have been considered in previous chapters: incineration, pyrolysis, hydrogasification, wet combustion and similar processes in Chapter 4; photosynthetic methods of energy production in Chapter 5; and application to urban wastes in Chapter 6. An interesting idea not included in previous discussion is the possibility of a biological fuel cell (Videla and Arvia, 1975), in which the electrochemical reactants in a cell would be generated by microbiological activity.

The subjects of the energy potential of cellulosic organic wastes (Anderson, 1972; Bailie, 1971; IGT, 1976; NSF, 1975; Wender et al., 1974) including forestry wastes (Finney and Sotter, 1975), agricultural wastes (Appell and Miller, 1973; Schlesinger et al., 1973), cellulose photosynthetic energy storage (Chedd, 1975; EST, 1976a; Kemp and Szego, 1974; Kemp et al., 1975; Povich, 1976; Szego and Kemp, 1973), and cellulose-to-oil conversion (Barbour et al., 1974) are receiving attention in many places.

HYDROCARBONS

Hydrocarbons, their extraction and transport as crude oil and coal, their refining for fuels, and conversion to chemical products make up a significant proportion of the chemical industry, as well as being major potential contributors to environmental problems.

Crude *oils* are mixtures of alkane (paraffin) hydrocarbons of different molecular sizes and arrangements, together with various "impurities" such as sulfur and metals. They have high viscosities and high flash points, and must be separated according to boiling point to provide more useful products. The natural proportion of the fraction suitable for internal combustion engines is small, so the long hydrocarbon molecules must be "cracked." Impurities must be removed, with sulfur being of particular significance. These processes require numerous complex operations, and there are many potential sources of waste and pollution, both airborne and waterborne (Nemerow, 1963). Recycling of all residuals is investigated in detail in the planning stages, with "balance sheets" for all materials, and decisions arc made on economic criteria within certain environmental and political constraints. A good example is the application by Russell (1973) of a residuals management model to petroleum refining. The petrochemical industry has reduced its pollutant load by using organic materials as fuels until markets have been developed by vigorous research and marketing. Most refineries occupy the focal point of industrial complexes which utilize a wide range of petrochemicals. The oil refinery and related systems are not perfect, but they have been developed systematically, and are in a more favorable position with respect to recycling and material utilization than most other industries (Mantell, 1975).

The waste oil situation is less well organized. Of the 9.5×10^9 liters (2.5×10^9 U.S. gallons) of lubricating oil used for industrial and automotive purposes in the United States, it is estimated that about 6×10^9 liters is consumed, and of the remaining 4×10^9 liters only 10% is recycled (APR, undated; Sittig, 1975), that is, 4% of total consumption. The corresponding figure for the United Kingdom is 6.7% (Tinker, 1976). Although "additives" cause some problems, waste oil can be re-refined satisfactorily (APR, 1975; Crocker, 1976; EST, 1974b; Maizus, 1975; Martin, 1972; Mitchell, 1976; Sittig, 1975; Teknekron, 1975; Weinstein, 1974; Whisman et al., 1974) and developments, such as clarification with propane (Quang et al., 1975) in place of acid and clay treatment, are still being made. However, tax disincentives and collection costs often make the economics of re-refining only marginally attractive. The problem of dispersion of waste oil has increased markedly in the past 15 years or so, with the tendency of consumers to purchase replacement oil from retail stores at discount prices rather than have crankcase oil changed at service stations (Cukor et al., 1973b). The energetic and economic implications of recycling oil have been reviewed recently (Cukor, 1975b; Cukor and Hall, 1975; Cukor et al., 1973c; Sorrentino and Whinston, 1975), and a conference on waste oil recovery and reuse (APR, 1975) considered various economic and technical aspects. Ultrafiltration and reverse osmosis are being applied to concentrate the oil in waste streams containing small proportions of water-emulsified or water-soluble oils such as cutting oils (Markind et al., 1974; Nordstrom, 1974).

An alternative application of waste oil is pelletization with coal dust for use as a fuel (Haynes et al., 1974). Oil recovered from ocean spills also tends to be used as fuel in industrial or naval power stations (Wayne and Perna, 1971).

Fortunately, there are bacteria which degrade alkane hydrocarbons, even methane (Harrison, 1976), and microorganisms are the agents by which petroleum wastes in water and soil are degraded. These bacteria are also potential food sources (Gounelle de Pontanel, 1972), but the economics of petroleum as a food-base chemical are less attractive than they were a few years ago.

Coal also contains a high proportion of hydrocarbons, but of much more complex structure than those in oil, and combined with various organic nitrogen and sulfur compounds. Coal-mining wastes, which form the largest store of untapped waste material in many countries, were discussed earlier in this chapter in the section on building and construction materials.

The coal gas (town gas) industry always found markets for its waste products: tar for roadmaking; dyestuffs, and explosives; ammonia for fertilizers; sulfur for sulfuric acid; coke as fuel; and clinker and ash as aggregate. Natural gas has replaced coal gas in many places, but because of the recent increased pressure on oil reserves it is probable that a large, new, improved coal gasification industry will develop (Bodle and Vyas, 1974; CEN, 1975h; Perry, 1974), including the manufacture of methanol from coal gas (Harney, 1975; Mills and Harney, 1974). The design of these plants with respect to residuals management will have much in common with oil refineries. In particular, the market for various forms of filtration and adsorption carbon products derived from coal (Bowling et al., 1972) will expand as more gaseous and aqueous effluents are purified before discharge.

The roles to be played by our remaining reserves of oil, natural gas, and coal as sources of chemicals and energy will need to be considered carefully (Elias, 1975), with close attention paid to residuals management and resource conservation.

RADIATION SOURCES

Fissionable materials such as uranium and plutonium are recovered from depleted nuclear power station fuels, because only a fraction is consumed in the nuclear reactions. The fuel is dissolved, in nitric acid for example, and the various components are recovered by solvent extraction and chemical processes (Kehler and Miles, 1972; Nemerow, 1963; Ross, 1968). Other isotopes used as auxiliary power sources for remote locations may also be extracted. A possible application of nuclear wastes is the use of gamma radition to carry out polymerization, either of pure monomers, or with the incorporation of waste materials for the preparation of new products. Sterilization is another possible use of ionizing radiation from ''waste'' radioactive sources.

8

THERMODYNAMICS OF RECYCLING

It is probably apparent from everyday observation that the general trend or tendency in nature is away from order, and toward disorder. Many ordered arrangements such as packs of cards, gardens, and children's toys, spontaneously tend to become disordered, and effort (energy) is required to reorder them. On the other hand, there are certain exceptions. Life itself involves a highly ordered arrangement of molecles, and energy must be supplied to preserve this order. Eventually, however, death is accompanied by decay of the ordered structure.

When a living system is formed, the extent of increase in disorder in the surroundings is, in fact, greater than the increase in order in the living system, so even life processes result in a *net increase* in disorder. Thus, any spontaneous change is accompanied by increasing overall disorder in a system plus its surroundings. (When expressed in scientific terms, this observation is known as the Second Law of Thermodynamics.)

ENTROPY

Just as it is necessary to use a temperature scale when describing precisely how hot something is, the concept of *entropy* is introduced to describe the extent of disorder. It is possible, therefore, to state the "increasing entropy" principle:

> For processes that take place spontaneously, the total entropy (system plus surroundings) increases.

The environmental and resource "crises" may be interpreted as entropy crises. We are attempting to maintain or increase order in our "system" (body, house, city, planet), but this requires an even greater increase in entropy or disorder in the environment (that is, in everything outside the system). The only processes in the universe that are not accompanied by entropy increases are those that are carried out reversibly, that is, processes in which the driving forces for the change are offset only infinitesimally from a state of balance. An example of a reversible process is a solid-liquid system precisely at the melting point (for example, an ice-water mixture at exactly 0°C). An infinitesimal increase in temperature causes melting; a corresponding decrease causes freezing. Such reversible processes can never be achieved *exactly,* and most processes in our normal environment are not even approximately reversible. The melting of an ice-block in a 20°C glass of water is not a reversible process, and is accompanied by an entropy increase.

In order to discuss this subject further, it is necessary to introduce clearly defined quantities. Whenever a physical or chemical process occurs, it is possible to define as follows the total (universal) entropy change, ΔS_{total}, which accompanies it:

$$\Delta S_{total} = \Delta S_{system} + \Delta S_{environment}$$

(The Greek letter Δ, "delta," denotes the extent of change in a quantity: in this case, the change in entropy.) The conclusion stated previously, that the entropy of the universe must increase or remain the same in any process, but never decrease, may be expressed mathematically as

$$\Delta S_{total} \geq 0$$

where the symbol "≥" means "is greater than or equal to". It follows that

$$\Delta S_{system} + \Delta S_{environment} \geq 0$$

If one wishes to focus attention on the *system,* which is usually the case, the second term, which deals with the change in the environment, can be rearranged as follows (for a process which occurs at constant temperature and constant pressure) (Moore and Moore, 1976). The relationship between entropy change (ΔS) and heat absorbed (q) for such a process is

$$\Delta S_{environment} = \frac{q}{T}$$

where q is the heat absorbed by the environment and T is the absolute temperature. The heat absorbed by the system is called the *enthalpy* change, ΔH_{system}, and this is equal and opposite to q, so

$$\Delta S_{environment} = -\frac{\Delta H_{system}}{T}$$

The subscript "system" usually is omitted, so ΔH is the heat absorbed by the system at constant pressure and

$$\Delta S_{\text{total}} = \Delta S - \frac{\Delta H}{T} \geq 0$$

total entropy change — *entropy change of the system* — *entropy change of the environment*

The entropy change accompanying conversion of resources into products and wastes is represented diagrammatically in Figure 8.1. Entropy is increased in three ways during processing of raw materials (Stumm and Davis, 1974):

1. Heat production and dissipation (the ΔH terms in the preceding equations);
2. Dilution and mixing of materials;
3. Destruction of information (loss of genetic information; simplification of ecosystems).

From the point of view of entropy, we can say that industrial processes transform

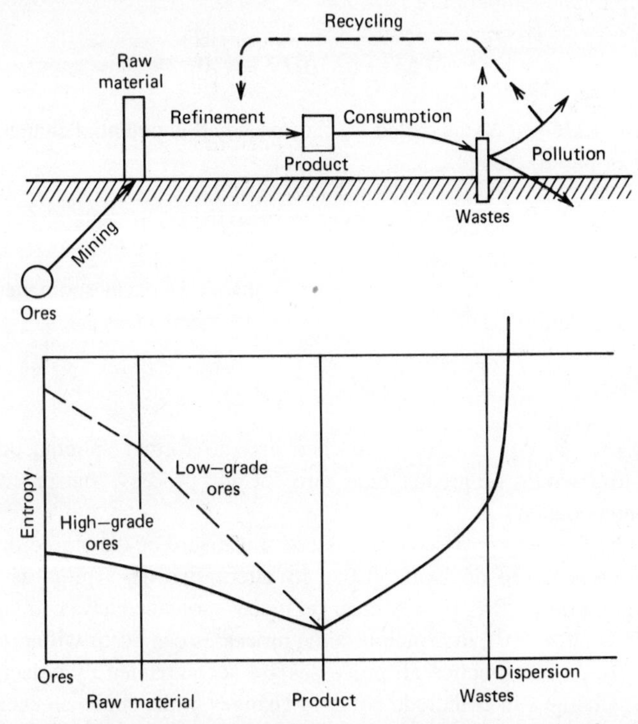

Figure 8.1. Diagrammatic illustration of the entropy change accompanying the conversion of resources into products and wastes. (From Stumm and Davis, 1974, with permission.)

resources (ores) into useful products of lower entropy (higher order and higher concentration), at the same time causing a greater entropy increase in the surroundings as a result of burning fossil fuels. Ultimately, dispersion and pollution convert the useful products into dissipated waste products with an associated entropy state that is even greater (more disordered) than that of the original resources

ENERGY

So far, emphasis has been placed on the entropy changes associated with various processes, based on the relationship

$$\Delta S_{total} = \Delta S - \frac{\Delta H}{T} \geq 0$$

It is frequently more convenient to express this relation in energy units by multiplying by the temperature T:

$$T\Delta S - \Delta H \geq 0$$

The quantity ($\Delta H - T\Delta S$) is called the thermodynamic potential change or Gibbs free energy change, ΔG:

$$-\Delta G = T\Delta S - \Delta H \geq 0$$

This means that a necessary condition for a change to occur spontaneously in a system (at constant pressure) is

$$\Delta G < 0$$

(When $\Delta G = 0$, equilibrium is reached and no further spontaneous change occurs; if ΔG would be greater than zero for any process, that process cannot occur spontaneously.)

The Gibbs free energy G can be used as a measure of the stored or potential ability of a system to do work: hence its alternative description as "thermodynamic potential." No processes take place spontaneously with an overall increase in G ($\Delta G > 0$); in principle some processes can occur with no change in G ($\Delta G = 0$); and in practice all processes are accompanied by a decrease in G ($\Delta G < 0$). Changes in G include not only changes in the chemical energy stored in materials such as fuels and foods (associated with the enthalpy change, ΔH), but also changes in the capacity of a system to be useful because of its order and organization (the $T\Delta S$ term):

$$\Delta G \quad = \quad \Delta H \quad - \quad T\Delta S$$

Gibbs free energy change (thermodynamic potential change)	*enthalpy change (change in chemical energy)*	*energy term resulting from entropy change*

A loss of G (negative ΔG) results from a negative enthalpy change (utilization of chemically stored energy) or a positive entropy change (an increase in disorder).

Berry (1971) has defined the annual consumption of thermodynamic potential the *survival function*. From the preceding discussion, it should be apparent that the survival function is concerned not only with the rate of consumption of the chemical energy stored as fuel, but also with the decrease in useful capacity of any system because of reduction in its degree of order and organization. When a vein of copper ore is mined, and the copper is dispersed in electric fittings, for example, there has been a reduction in the thermodynamic potential. Mixed urban waste of food, paper, glass, and metal has a lower thermodynamic potential than the original food containers; and an automobile in working order loses thermodynamic potential when it becomes rusty and inoperative. The stability of our way of life may be described by the manner in which our survival function develops with time. The total available thermodynamic potential is finite: a ''bank balance'' stored on earth in a chemical form, plus an ''annual salary'' from the sun. One way to conserve the thermodynamic potential is to recycle materials while their thermodynamic potential values are still relatively high, rather than discarding them in such a way that they are transformed to an energy state far lower than that of the original ore. If we do not conserve them in this way the materials may become thermodynamically ''lost'' just as efficiently as helium is lost from our atmosphere: ''entropy pollution'' (Bent, 1971; Moore and Moore, 1976).

A higher state of technological development usually implies a higher rate of consumption of thermodynamic potential, but any society can be thrifty or prodigal in its use of thermodynamic potential (Berry, 1971). If we draw on the reserves for long enough, as we have been doing, they must disappear eventually, and unless in the meantime we have been able to develop a society which can survive on the annual solar ''income,'' catastrophic changes must occur.

ENERGY EXPENDITURES FOR PRODUCTION AND RECYCLING

The recent awareness of an imminent energy ''crisis'' has focussed attention on the energetic implications of recycling (Beal and Haselar, 1972; Berry, 1971, 1972; Chapman, 1975; Hannon, 1972; 1973; Khazzoom et al., 1976; Large, 1973; Purcell and Smith, 1976) and of environmental clean-up operations (Hirst,

1973a, 1973b, 1975). Thermodynamic considerations indicate that we should consider both energy and entropy factors.

One method of portraying energy and entropy costs of various processes is shown in Figure 8.2. The horizontal axis corresponds to increasing material concentration, from the fully dispersed state averaged over the earth's crust (the "Clarke level"), through the more concentrated ore bodies currently available, and on to pure materials and components. Intermediate on this scale lie our artificial systems (assemblies of pure materials and components), and wastes. Alternatively, this scale can be described in terms of changes in entropy of the system due to increase or decrease in concentration. The vertical axis is an arbitrary energy scale, the height of each "loop" depending on the energy that must be expended for a particular process (or the equivalent amount of entropy that must be generated in the environment). Minerals and natural fuels are found concentrated in veins or domes in states of thermodynamic potential much higher than if they were dispersed completely. Industrial processes expend energy (generate entropy in the environment) raising these materials to high thermodynamic potentials by extracting, purifying and ordering them. During use by consumers, the manufactured objects decrease in thermodynamic potential to a small extent, but the biggest loss occurs when items are discarded as waste; the thermodynamic potential of the material falls below that which existed originally in the mineral bodies. The vertical and horizontal extents of each loop provide measures of the total entropy production, or thermodynamic potential expenditure, for each process. It should be noted that the vertical scale, chemical energy expended, is always positive regardless of the direction of movement on the concentration scale. Movement in either direction is accompanied by an overall production of energy and expenditure of thermodynamic potential. Once we have built a "system" (house, car, etc.) we have the alternatives of: repair; waste disposal followed by recycling; or discard and pollution. The last choice involves a subsequent large entropy production or energy expenditure if we later wish to recover the materials for reuse.

From Figure 8.2 it is clear that our aim should be to minimize the size of each loop, and that this can be done firstly by avoiding the dissipation of resources, and secondly by choosing the energetically most economical processing methods. There are moves towards the use of thermodynamic analysis when alternative options in manufacture, disposal, and recycle processes are considered (IFIAS, 1974).

Berry (1972) and Berry and Fels (1972, 1973) have carried out such thermodynamic calculations for the various processes associated with the manufacture, recycle, and degradation of automobiles, and their results have been summarized by Moore and Moore (1976) and Wade (1976). The most costly step (in both economic and energetic terms) in the manufacture of a car is the production of about 1½ tonnes of steel and its fabrication into the various components. The total thermodynamic potential consumed is of the order of 37,000 kilowatt-hours

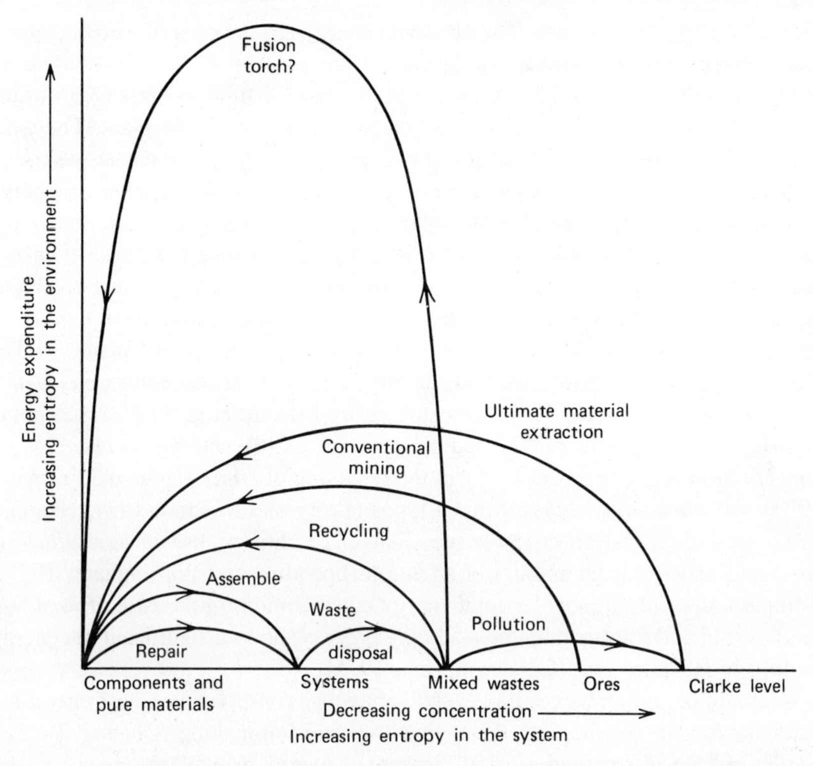

Figure 8.2. Diagrammatic illustration of dilution entropy and energy expenditure entropy changes for various processing paths. (After Rose et al., 1972.)

or 133 × 10^9 joules per car (which is comparable to the annual energy consumption in operating it). Maximum recycling would save about 30% of the energy cost, and tripling the useful lifetime would save about 60%. However, there is a large discrepancy between the actual and theoretical energy costs of a new car; that is, the thermodynamic potential changes of the materials during manufacture are small compared with the actual energy expended, so the greatest energy savings could be achieved by improvement in basic fabrication techniques.

In the case of steel production, the vertical component in Figure 8.2 is the dominant one because iron is not at a low concentration on a global scale, and available iron ores are still of high grade. The same is true for glass bottles (considered in more detail in Chapter 6), but in the case of some elements which are rare and more uniformly dispersed, it is more important to prevent dissipation of resources than to develop more efficient production methods; the entropy term dominates the thermodynamics.

Many other recycling operations are now receiving attention from the point of view of energy expenditure. The electrical energy requirement of *water recovery* from sewage depends strongly on the size of the plant, with major economies of scale possible (Hirst, 1975), as illustrated by the 28 million liters (7.5 million U.S. gallons) per day Lake Tahoe Advanced Waste Treatment plant. There are energy costs associated with *solid waste* disposal in landfills, and energy savings to be made in recovering metals, glass, and paper, or in incineration for energy recovery, so there are considerable energetic advantages in recycling (Broussaud, 1976; Franklin, 1975; Hirst, 1975). Various methods of energy recovery from solid wastes have been reviewed already, by chemical processes (Chapter 4) and biological processes (Chapter 5), with particular reference to urban solid waste in Chapter 6 and to other organic sources in Chapter 7. The energy expenditures associated with the production of various *nonferrous metals* from ore and from scrap have been studied by Bravard et al. (1972), Chapman (1973a, 1973b, 1974, 1975), and Khazzoom (1976), among others. Energy conservation is a significant factor in the recycling of *paper* (Hunt and Franklin, 1973) and *packaging material* of all types (Berry and Makino, 1974; Hannon, 1972; Makino and Berry, 1973: see Chapter 6). Energy use in *agriculture* is receiving an increasing amount of attention (Spedding and Walsingham, 1975). Comparison of the figures for total energy consumption for the case of *waste oil* used as fuel and rerefined by the acid/clay process shows a significant net energy savings in the latter case (Cukor and Hall, 1975).

It should be remembered, however, that energy conservation is just one of the many factors to be considered in recycling decision-making (Chapter 2). The results of careful calculation of all aspects of energy expenditure must be considered and weighed against resource levels, environmental factors, social considerations, and ultimately economic criteria. Considerable popular recycling effort has been directed towards recovering bottles and paper, but the net *energy* benefits often have been only marginal.

EFFICIENT THERMAL ENERGY UTILIZATION

When fuel is used to generate electricity at a power station, the achievement of maximum efficiency relies on the reuse of heat at successively lower temperatures, but even so, only about one-third of the thermodynamic potential is utilized. The remainder is transferred to rivers, perhaps causing thermal pollution, or to the atmosphere as steam from cooling towers or as hot air. Rivers can be several degrees warmer after passing through a power station, in some cases becoming too warm for the survival of some fish. Within limits, the temperature increase may have local beneficial effects. The example is quoted by Beal and Haselar (1972) of Southampton Water in the United Kingdom, where a clam export industry developed, based on clams accidentally introduced and now thriving in heat rejected from a power station. This is exceptional, and only serves to emphasize the potential of the waste energy for other purposes.

Water, after driving steam turbines, is still hot enough to heat a town's buildings, and it is possible to pipe hot water over long distances. Other uses are for melting snow, desalinating seawater, increasing the growth rates of plants, fish-farming, and airport defogging: Coutant (1976), Engdahl (1969) and Mandell (1974) have collected and reviewed a variety of such applications. If heat is reclaimed from a power station in this way, the overall thermal efficiency increases to about 90% (although the efficiency of electricity generation decreases to some extent). The total heat wasted is already enormous, and is increasing rapidly as nuclear power stations are commissioned. A big proportion of the domestic electrical load is for water and space heating, contributing markedly to the peak load in cold weather, so the electrical generation capacity could be used more efficiently if hot water was distributed by pipeline. The capital cost, of course, is high, but the possibility of a "total energy" complex producing direct central heating and hot water as well as electrical power should certainly be considered seriously when any new installation is planned. One barrier at present is that electricity generation is the responsibility of a commercial organization or an authority which aims to maximize electricity sales rather than to provide a total energy service.

If an incineration unit forms part of a community's solid waste treatment plant, this should contribute to the generation of electricity and piped hot water. In such a scheme in the Swedish town of Malmo, hot air generated by the incineration of solid waste is first used to heat water for circulating through the homes, and then to treat the sewage from an adjacent plant (Beal and Haselar, 1972). Waste incineration and seawater desalting can be integrated also (Cernia, 1975).

Turning from heating to cooling, we should note that vaporizing natural gas after transport in the liquid state can provide a significant heat sink which can be used for air-conditioning and for industrial purposes (CEN, 1974b; Witwer et al., 1976).

In making decisions on suitable waste heat applications, many factors must be considered (Mandell, 1974): Can the thermal energy be used year-round? Are back-up systems required to ensure a continuous supply of heat? How do the economics of utilization compare with the cost of rejection of waste heat by means such as cooling towers? It is our responsibility to optimize the utilization of both material and fuel sources.

The International Energy Agency set up by the Organization for Economic Coopertion and Development has made a start in coordinating energy research and development, allocating topics to various member countries: coal technology (British National Coal Board and Department of Energy); solar energy (Japanese Ministry of International Trade and Industry); nuclear safety and energy conservation (U.S.); waste heat utilization (Germany); and municipal and industrial waste utilization for energy conservation (Netherlands) (Kenward, 1975c).

APPENDIX 1

RECYCLING BY ELEMENTS

Usually it is preferable to recycle matter at as high a level as possible in the systems hierarchy (Chapter 1, Table 1.1) in the form of a devise, component, or compound, but if this is not possible it is necessary to recycle it in its elemental form (level 1 of the systems hierarchy). In principle, transmutation of elements by nuclear reaction is possible, but this can be carried out on only a small scale and at very great expense, so it is essential to conserve each elementary material in high concentration and minimize its dissemination.

Some widely used elements such as aluminum are abundant on Earth; other elements such as helium are scarce and must not be wasted. In this appendix, information sources are listed for some of the more important elements. Brief discussions are included of the recycling of carbon (emphasizing that we should not take for granted any element in the continuous cycle of chemical reactions that is our life) and of helium (a special case, in that apart from transmutation it is a truly nonrenewable resource). Full bibliographic details of all references are included in Appendix 5, in alphabetical order by author.

CARBON

Chemical Cycles and the Global Environment, Garrels et al., 1975.

Clean Liquid and Gaseous Fuels from Organic Solid Wastes, Wender et al., 1974.

Electrochemical Carbon Regeneration, Owen and Barry, 1972.

Organic Wastes, Hughes, 1975b.

Perpetual Methane Economy. Is it Possible? Klass, 1974.

Preparation and Evaluation of Activated Carbon Produced from Municipal Refuse, Stevenson et al., 1973.

Recycling CO_2 is Goal of Research at IGT, CEN, 1974i.

Solar Energy by Photosynthesis, Calvin, 1974.

It has been suggested that the ultimate fuel problem is not the supply of coal, or oil, or gas, but the supply of carbon. When carbon compounds are burned in an oxygen-rich atmosphere there is produced carbon dioxide—carbon in its highest oxidation state—which can make no further contribution to reserves of energy. Given sufficient energy, carbon dioxide can be converted by one of several techniques into forms of carbon that will react with oxygen exothermically (to evolve heat). Biological conversion in the expanded carbon cycle using solar energy is an obvious possibility (Chapter 7). Another is the conversion of carbon dioxide and water to carbon monoxide and methane with a catalyst such as ruthenium, but there are probably superior ways of storing energy. It is even possible biologically to synthcsize hydrogen rather than carbon compounds.

Although the "greenhouse effect," associated with the increasing carbon dioixde concentration in the atmosphere, may eventually require that carbon dioxide be removed from fuel combustion gases, it is unlikely to be recycled as a fuel to any extent. Carbon dioxide could be removed chemically:

$$\underset{\textit{hydroxide ion in solution}}{2OH^-(aq)} + \underset{\textit{carbon dioxide dissolved in water}}{CO_2(aq)} \rightarrow \underset{\textit{carbonate ion in solution}}{CO_3^{2-}(aq)} + \underset{\textit{water}}{H_2O}$$

$$\underset{\textit{carbonate ion}}{CO_3^{2-}(aq)} + \underset{\textit{carbon dioxide}}{CO_2(aq)} + \underset{\textit{water}}{H_2O} \rightarrow \underset{\textit{bicarbonate ion}}{2HCO_3^-(aq)}$$

Alternatively, it could be removed biologically from combustion gases. Plant growth is stimulated in high carbon dioxide atmospheres; and methane-producing microorganisms consume carbon dioxide, a typical overall reaction being:

$$\underset{\textit{ethanol}}{2CH_3CH_2OH} + \underset{\textit{water}}{H_2O} + \underset{\textit{carbon dioxide}}{CO_2(aq)} \rightarrow \underset{\textit{acetic acid}}{2CH_3COOH(aq)} + \underset{\textit{methane}}{CH_4(g)} + \underset{\textit{water}}{2H_2O}$$

All such processes would require energy, and one criticism of the development of biodegradable plastics is that valuable solid resources would be converted into a gas from which carbon can be "fixed" only with difficulty.

If carbon wastes are pyrolyzed in an oxygen-deficient atmosphere rather than burned in an oxygen-rich atmosphere, a proportion of the stored energy is conserved, and "degradation" to the lowest level of carbon dioxide is delayed. These processes are described in Chapters 4 and 6.

Recycling of carbon as the element in its graphite form is worthwhile, because it is useful and moderately valuable. Graphite turnings or offcuts from the machining of electrodes are used as fillers in plastics, for lubrication, for the manufacture of other graphite products, and as recarburizers in the iron and steel industry. Carbon black is recovered from waste tires, and used in the manufacture of new tires. Activated carbon, used widely in the purification of gases, water, and other liquids can be regenerated thermally in furnaces with limited oxygen, or electrochemically using the carbon as an inert electrode for electrochemical processes involving the adsorbed species.

HELIUM

First Annual Report of the Secretary of the Interior Under the Mining and Minerals Policy Act of 1970, USDI, 1972.

New Helium Policy, Chem Tech, 1975.

Resources and Man, CRM, 1969.

Who Pays Is Key to Helium's Future, Anderson, 1976.

World helium reserves are restricted essentially to that which is trapped in geological structures with natural gas deposits, and much of the natural gas production is not accompanied by helium recovery. When helium enters the Earth's atmosphere from these sources, it diffuses upwards and is eventually lost into space.

Helium has a unique combination of properties including low density, noninflammability, and low freezing point, and its use in cryogenics, energy transfer, and underwater breathing mixtures will become increasingly important just as the reserves are becoming exhausted.

STRUCTURAL AND PLATING METALS

Development and Commercialization of the Waste-Plus-Waste Process for Recovering Metals from Cyanides and Other Wastes, George et al., 1976.

The Economics of Recovery of Materials from Industrial Waste—A Case Study, Bridgwater, 1975.

The Economics of Recycling Waste Materials, U.S. Congress, 1972.

Energy Expenditures Associated with the Production and Recycle of Metals, Bravard et al., 1972.

First Annual Report of the Secretary of the Interior Under the Mining and Minerals Policy Act of 1970, USDI, 1972.

Generation, Collection and Merchanting of Scrap, Cooper, 1975.

Industrial Waste Disposal, Koziorowski and Kucharski, 1972.

Recovery of Metals from a Variety of Industrial Wastes, Cochran and George, 1976a.

Recovery of Metals from Electroplating Wastes by the Waste-Plus-Waste Method, George and Cochran, 1970.

Resource Recovery and Recycling Handbook of Industrial Wastes. Sittig, 1975.

Resource Recovery and Source Reduction, USEPA, 1974a.

A Review of the Problems Affecting the Recycling of Selected Secondary Materials, NASMI, 1972

Solid Waste, Bond and Straub, 1973.

Solid Waste Processing, Engdahl, 1969.

Solid Wastes: Origin, Collection, Processing, and Disposal, Mantell, 1975.

Study to Identify Opportunities for Increased Solid Waste Utilization, USEPA, 1972a.

Theories and Practices of Industrial Waste Treatment, Nemerow, 1963.

The Treatment of Industrial Wastes, Besselievre and Schwartz, 1976.

The Waste-Plus-Waste Process for Recovering Metals from Electroplating and Other Wastes, Cochran and George, 1976b.

Aluminum

Air Pollution Control of Aluminum and Copper Recycling Processes, Lauber et al., 1973.

Aluminum Extraction from Impure Sources by Vapor Transport with Magnesium Fluoride, Layne et al., 1974.

Base Line Forecasts of Resource Recovery, 1972 to 1990; Final Report, Nuss et al., 1975.

The Beverage Container Problem: Analysis and Recommendations, Bingham and Mulligan, 1972.

The Chemistry of Recycling Aluminum, Blayden, 1974b.

Economic Realities of Reclaiming Natural Resources in Solid Waste, Clark, 1971.

Electromagnetic Separation of Aluminum and Non-Ferrous Metals, Campbell, 1974.

An Electromagnetic System for Dry Recovery of Non-Ferrous Metals from Shredded Municipal Solid Waste, Sommer and Kenny, 1974.

Energy Conservation and Recycling of Copper and Aluminium, Chapman, 1974.

The Energy Costs of Producing Copper and Aluminium from Primary Sources, Chapman, 1973a.

The Energy Costs of Producing Copper and Aluminium from Secondary Sources, Chapman, 1973b.

Extraction of Aluminum from Mixed Solid Waste, Dale, 1974.

Glass and Aluminum Recovery System, Franklin, Ohio, Cummings, 1974.

Incentives for Recycling and Reuse of Plastic, Milgrom, 1972.

Potential Energy Conservation from Recycling Metals in Urban Solid Waste, Franklin et al., 1975.

Processing Slag from an Aluminum Dross Furnace to Recover Fluxing Salt, Aluminum Metal, and Aluminum Oxide, Caldwell et al., 1974.

The Production and Consumption of Automobiles, Berry and Fels, 1972.

Reclamation of Non-Ferrous Metals from Waste, Jackson, 1973.

Recovery of Aluminum from Shredded Municipal and Automotive Wastes, Dean et al., 1975.

Recovery of Non-Ferrous Metals by Means of Permanent Magnets, Spencer and Schlömann, 1975.

Recycling Aluminum from Consumer Waste, Blayden, 1974a.

Recycling and Reclaiming of Municipal Solid Wastes, Jackson, 1975.

Recycling of Glass and Metals: Container Materials, Cannon and Smith, 1974.

Recycling of Used Aluminum Products, Testin, 1971.

The Resource Potential of Demolition Debris in the United States, Wilson, 1975.

Resource Recovery from Municipal Solid Waste, NCRR, 1974c.

Separation of Nonmagnetic Metals from Solid Waste by Permanent Magnets, Schlömann, 1975.

Specifications for Materials Recovered from Municipal Refuse, Alter and Reeves, 1975.

Upgrading Junk Auto Shredder Rejects, Froisland et al., 1972.

Cadmium

Solvent Extraction in Processes for Metal Recovery from Scrap and Waste, Fletcher, 1973.

Chromium

Chemical Reclaiming of Superalloy Scrap, Brooks et al., 1969.

Chromatic [Chromic] Acid Recovery, WWT, 1975g.

Chromium and Nickel Wastes—A Survey and Appraisal of Recycling Technology, Dressel et al., 1976.

Investigation of Commercial Chrome Tanning Systems, Davis and Scroggie, 1973.

An Ion-Exchange Process for Recovery of Chromate from Pigment Manufacturing, Robinson et al., 1974.

Metallic Recovery from Waste Waters Utilizing Cementation, Case, 1974.

Recovery of Chromium Ions from Electroplating Waste Solutions, Nishimura and Inoue, 1974.

Recovery of Metals and Other Materials From Chromium Etching and Electrochemical Machining Wastes, George and Cochran, 1974.

Recovery of Metals from Electroplating Wastes by the Waste-Plus-Waste Method, George and Cochran, 1970.

The Treatment of Nickel and Chromium Plating Liquors by Reverse Osmosis, Clark and Fowler, 1974.

Cobalt

Chemical Reclaiming of Superalloy Scrap, Brooks et al., 1969.

Extraction of Cobalt by Kelex 100 and Kelex 100/Versatic 911 Mixtures, Lakshmanan and Lawson, 1973.

Molten Salts: New Route to High Purity Metals, CE, 1968.

Solvent Extraction for Recovery of Metal Waste, Reinhardt, 1975.

Solvent Extraction in Processes for Metal Recovery from Scrap and Waste, Fletcher, 1973.

Copper

Air Pollution Control of Aluminum and Copper Recycling Processes, Lauber et al., 1973.

Bureau of Mines Research Programs on Recycling and Disposal of Mineral-, Metal-, and Energy-Based Wastes, Kenahan et al., 1973.

Copper Cementation in a Revolving-Drum Reactor, A Kinetic Study, Fisher and Groves, 1976.

Copper Industry Quarterly, Bureau of Domestic Commerce, U.S. Department of Commerce.

Copper-Lead Scrap Separation, Bigg, 1970.

The Electrochemical Treatment of Industrial Effluents, Surfleet, 1970.

The Electrowinning of Copper from Dilute Copper Sulphate Solutions with a Fluidised Bed Cathode, Flett, 1971.

Energy Conservation and Recycling of Copper and Aluminium, Chapman, 1974.

The Energy Cost of Automobiles, Berry and Fels, 1973.

The Energy Costs of Producing Copper and Aluminium from Primary Sources, Chapman 1973a.

The Energy Costs of Producing Copper and Aluminium from Secondary Sources, Chapman 1973b.

The Extraction of Copper and Accompanying Metals from Liquid Slags with Fused Melts and the Electrodeposition of Metals from Salt Melts, Bogacz et al., 1975.

The Extraction of Copper(II) and Iron(II) from Chloride and Sulphate Solutions with LIX 64N in Kerosene, Eccles et al., 1976.

The Extraction of Copper from Aqueous Chloride Solutions with LIX-70 in Kerosene, Lakshmanan and Lawson, 1975.

Hydrometallurgical Process for Recovering Copper from Bearing Scrap and Similar Copper-Clad Materials, Pearson, 1972.

Industrial Waste Disposal. Excess Sulfuric Acid Neutralization with Copper Smelter Slag, Twidwell et al., 1976.

Metal Extraction from Waste Materials, Fletcher, 1971.

Metal Recovery from Effluents and Sludges, Jackson, 1972.

Metal Recovery from Waste Waters Utilizing Cementation, Case. 1974.

Molten Salts: New Route to High Purity Metals, CE, 1968.

Packed Bed Electrodes, Chu et al., 1974.

Potential Energy Conservation from Recycling Metals in Urban Solid Wastes, Franklin et al., 1975.

The Production and Consumption of Automobiles, Berry and Fels, 1972.

A Quantitative Assessment of Old and New Copper Scrap Collected in the U.K., Tron, 1976.

Reclamation of Non-Ferrous Metals from Waste, Jackson, 1973.

Recovery and Utilization of Copper Scrap in the Manufacture of High Quality Cathode, Burson and Morgan, 1974.

Recovery of Metals and Other Materials from Chromium Etching and Electrochemical Machining Wastes, George and Cochran, 1974.

Recovery of Metals from Electroplating Wastes by the Waste-Plus-Waste Method, George and Cochran, 1970.

Recovery of Non-Ferrous Metals by Means of Permanent Magnets, Spencer and Schlömann, 1975.

Recovery of Non-Ferrous Metals from Domestic Refuse, Miles and Douglas, 1972.

Recovery of Secondary Copper and Zinc in the United States, Carillo et al., 1974.

The Recovery of Soluble Copper from an Industrial Chemical Waste, Keating and Williams, 1976.

Recovery of the Metal Values from Effluents, Flett and Pearson, 1975.

Recovery of Zinc, Copper, and Lead-Tin Mixtures from Brass Smelter Flue Dusts, Powell et al., 1972a.

Recycling and Reclaiming of Municipal Solid Wastes, Jackson, 1975.

Recycling as an Industry, Ness, 1972.

Recycling: Fundamentals and Concepts, Bundi and Wasmer, 1976.

Recycling Trends in the United States: A Review, Spendlove, 1976.

The Resource Potential of Demolition Debris in the United States, Wilson, 1975 (copper electrical and plumbing components).

A Review of Methods of Controlling Impurities in Electrolytes of Electroytic Copper Refineries, Dolan, 1975.

A Review of the Electrolytic Recovery of Copper from Dilute Solutions, Kuhn, 1971a.

Separating Copper from Scrap by Preferential Melting. Laboratory and Economic Evaluation, Leak et al., 1974.

Separation of Nonmagnetic Materials from Solid Waste by Permanent Magnets, Schlömann, 1975.

Solvent Extraction in Processes for Metal Recovery from Scrap and Waste, Fletcher, 1973.

Specifications for Materials Recovered from Municipal Refuse, Alter and Reeves, 1975.

Upgrading Junk Auto Shredder Rejects, Froisland et al., 1972.

Upgrading of Waste of the Metallurgical Industry, De Cuyper, 1973.

Use of Cryogenics in Scrap Processing, Bilbrey, 1974.

Iron

An Approach to Ferrous Solid Waste, Regan, 1972.

Automotive Scrap Recycling, Sawyer, 1974.

Base Line Forecasts of Resource Recovery, 1972 to 1990; Final Report, Nuss et al., 1975.

The Beverage Container Problem: Analysis and Recommendations, Bingham and Mulligan, 1972.

Bright Outlook for Recycling Ferrous Scrap from Solid Waste, Ostrowski, 1974, 1975.

Bureau of Mines Research on the Recovery and Reuse of Post-Consumer Ferrous Scrap, Kaplan, 1974.

Bureau of Mines Research Programs on Recycling and Disposal of Mineral-, Metal-, and Energy-Based Wastes, Kenahan et al., 1973.

Continuous Regeneration of Hydrochloric Acid Pickle Liquors, Taylor, 1975c.

Converting Stainless Steel Furnace Flue Dusts and Wastes to a Recyclable Alloy, Powell et al., 1975.

Copper Industry Uses Much Scrap Iron, EST, 1973b.

Cryogenic Scrap Processing: Production of High Purity Steel Scrap, Iron and Steel, 1971.

The Energy Cost of Automobiles, Berry and Fels, 1973.

Environmental Chemistry, Manahan, 1972, Chapter 5, "Microorganisms—the Catalysts of Aquatic Chemical Reactions."

Evaluation of Steel Made with Ferrous Fractions from Urban Waste, Makar et al., 1975.

The Extraction of Copper(II) and Iron(II) from Chloride and Sulphate Solutions with LIX 64N in Kerosene, Eccles et al., 1976.

Identifiction of Opportunities for Increased Recycling of Ferrous Solid Waste, Regan et al., 1972.

Is Recycling the Solution? Pausacker, 1975, Chapters 6 and 7.

A Linear Programming Model of Residuals Managment for Integrated Iron and Steel Production, Russell and Vaughan, 1974.

Magnetic Separation: A Basic Process, NCRR, 1974a.

Magnets as Applied to Ferrous Scrap Recovery, Sealy, 1975.

Metal Recovery from Effluents and Sludges, Jackson, 1972.

Metal Recovery Starts Tin Can Separation at Benwell, MRW, 1976d.

The Metallurgical Upgrading of Automotive Scrap Steel, Carlson and Schmidt, 1973.

Phoenix Quarterly (journal of the Institute of Scrap Iron and Steel, Inc.).

Pigment Grade Iron Oxides: Recovery from Iron-containing Waste Liquors, DeWitt et al., 1952.

Potential Energy Conservation from Recycling Metals in Urban Solid Wastes, Franklin et al., 1975.

The Production and Consumption of Automobiles, Berry and Fels, 1972.

Railroad Freight Rates: Ferrous Scrap: An Equitable Solution, ISIS, 1974.

Railroading Scrap, Kakela, 1975.

Raw Material Outlook Today and Tomorrow, Landau, 1974.

Recent Example of By-Product Utilization by B.H.P., Tegart and Mowat, 1975.

Recovery of Phosphates and Metals from Phosphates Sludge by Solvent Extraction, Powell et al., 1972b.

Recycling and Reclaiming of Municipal Solid Wastes, Jackson, 1975.

Recycling of Ferrous Scrap From Incinerator Residue, Ostrowski, 1973b.

Recycling of Ferrous Scrap from Solid Wastes, Ostrowski, 1973a.

Recycling of Glass and Metals: Container Materials, Cannon and Smith, 1974.

Recycling of Tin Free Steel Can, Tin Cans, and Scrap from Municipal Incinerator Residue, Ostrowski, 1971b.

Recycling of Tinplate, Dudley et al., 1975.

Recycling Refuse—New Life for Ironworks, Davis, 1974.

Recycling, Thermodynamics, and Environmental Thrift, Berry, 1972.

The Resource Potential of Demolition Debris in the United States, Wilson, 1975.

Resource Recovery from Municipal Solid Waste, NCRR, 1974c.

Scrap—Demand versus Newly Available Supply: 1975-1985, Nussbaum, 1976.

Single Unit Conversion of Unprocessed Auto and Appliance Scrap into Refined Steel, Grott, 1974.

Some Problems in the Recycling of Waste Ferruginous Dusts from Iron and Steel Production, Meikle and Nicol, 1975.

Specifications for Iron and Steel Scrap, ISIS, 1973.

Specifications for Materials Recovered from Municipal Refuse, Alter and Reeves, 1975.

Sponge Iron from Dry and Oilly Mill Scale, Steelmaking Furnace Dust, and Tires, Dean and Valdez, 1972.

Steel from Ferrous Can Scrap, Cramer and Makar, 1976.

Steel: The Recyclable Material, Cannon, 1973,

Studies on Upgrading of Automotive Scrap by Vacuum Melting and Electroslag Remelting, Carlson et al., 1974.

Transportation: Bugaboo of Scrap Iron Recycling, Cutler and Goldman, 1973.

Urban Refuse: New Source for Energy and Steel, Willson, 1974.

Use of Ferrous Scrap from Municipal Solid Waste Incinerators in Steelmaking, Ostrowski, 1971a.

The Use of Ferrous Scrap in Steelmaking, Leek, 1975.

Utilization of Automobile and Ferrous Refuse Scrap in Cupola Iron Production, Daellenbach et al., 1974.

Utilization of Refuse Scrap in Cupola Gray Iron Production, Daellenbach et al., 1976.

Waste Recovery and Pollution Abatement, Rickles, 1965.

Lead

Chemical Cycles and the Global Environment, Garrels et al., 1975.

Copper-Lead Scrap Separation, Bigg, 1970.

Experiments in Treating Zinc-Lead Dusts from Iron Foundries, Valdez and Dean, 1975.

Method for Recovery of Zinc and Lead from Electric Furnace Steelmaking Dusts, Higley and Fukubayashi, 1974.

A New Sulfur Dioxide-Free Process for Recovering Lead from Battery Scrap, Wilson, 1976.

Processing Lead Recovery Furnace Slag, Spedding, 1974.

Reclamation of Non-Ferrous Metals from Waste, Jackson, 1973.

Recovery of Non-Ferrous Metals by Means of Permanent Magnets, Spencer and Schlömann, 1975.

Recovery of Non-Ferrous Metals from Domestic Refuse, Miles and Douglas, 1972.

Recovery of Zinc, Copper, and Lead-Tin Mixtures from Brass Smelter Flue Dusts, Powell et al., 1972a.

Specifications for Materials Recovered from Municipal Refuse, Alter and Reeves, 1975.

Recycling and Reclaiming of Municipal Solid Wastes, Jackson, 1975.

Upgrading of Wastes of the Metallurgical Industry, De Cuyper, 1973.

Waste Management in the Lead-Zinc Industry, Ratter, 1974.

Manganese

Chemical Cycles and the Global Environment, Garrels et al., 1975.

The Utilization of Manganiferous Industrial Wastes and By-Products in the Manufacture of Coloured Bricks, Beretka and Brown, 1974.

Molybdenum

Chemical Reclaiming of Superalloy Scrap, Brooks, et al., 1969.

Recovery of Metals and Other Materials from Chromium Etching and Electrochemical Machining Wastes, George and Cochran, 1974.

Wastes Reclamation 2: Economic Aspects, Pearson, 1973.

Nickel

Chemical Reclaiming of Superalloy Scrap, Brooks et al., 1969.

Chromium and Nickel Wastes—A Survey and Appraisal of Recycling Technology, Dressel et al., 1976.

Metal Recovery from Effluents and Sludges, Jackson, 1972.

Molten Salts: New Route to High Purity Metals, CE, 1968.

Recovery of Metals and Other Materials from Chromium Etching and Electrochemical Machining Wastes, George and Cochran, 1974.

Recovery of Metals from Electroplating Wastes by the Waste-Plus-Waste Method, George and Cochran, 1970.

Recovery of the Metal Values from Effluents, Flett and Pearson, 1975.

Solvent Extraction for Recovery of Metal Waste, Reinhardt, 1975.

Solvent Extraction in Processes for Metal Recovery from Scrap and Wastes, Fletcher, 1973.

The Treatment of Nickel and Chromium Plating Liquors by Reverse Osmosis, Clark and Fowler, 1974.

Tin

Metal Recovery Starts Tin Can Separation at Benwell, MRW, 1976d.
Reclamation of Non-Ferrous Metals from Waste, Jackson 1973.
Recovery of Zinc, Copper and Lead-Tin Mixtures from Brass Smelter Flue Dusts, Powell et al., 1972a.
Specifications for Materials Recovered from Municipal Refuse, Alter and Reeves, 1975.
Recycling of Tinplate, Dudley et al., 1975.
Upgrading of Wastes of the Metallurgical Industry, De Cuyper, 1973.
Tin Can Recycling Plant Starts Up, Solid Wastes, 1976b.

Titanium

Recovery of Metals and Other Materials from Chromium Etching and Electrochemical Machining Wastes, George and Cochran, 1974.
Use of Waste Products Formed During the Manufacture of Titanium and Titanium Dioxide, Czerwonko, 1974.

Tungsten

Recovery of Sodium Tungstate from Scrap Tungsten Carbide, Powers, 1974.

Zinc

Elimination of Zn from Wastewater of Rayon Plants by Precipitation, De Jong, 1976.
The Energy Cost of Automobiles, Berry and Fels, 1973.
Experiments in Treating Zinc-Lead Dusts from Iron Foundries, Valdez and Dean, 1975.
The Extraction of Zinc(II) from Sulphate and Chloride Solutions with Kelex 100 and Versatic 911 in Kerosene, Harrison et al., 1976.
Metal Recovery from Effluents and Sludges, Jackson, 1972.
Method for Recovery of Zinc and Lead from Electric Furnace Steelmaking Dusts, Higley and Fukubayashi, 1974.
The Production and Consumption of Automobiles, Berry and Fels, 1972.
Recovery of Corrosion Resisting Zinc from Junked Automobiles, Dean et al., 1974a.
Recovery of Non-Ferrous Metals by Means of Permanent Magnets, Spencer and Schlömann, 1975.
Recovery of Non-Ferrous Metals from Domestic Refuse, Miles and Douglas, 1972.
Recovery of Phosphates and Metals from Phosphate Sludge by Solvent Extraction, Powell et al., 1972b.

Recovery of Secondary Copper and Zinc in the United States, Carillo et al., 1974.

Recovery of Zinc, Copper, and Lead-Tin Mixtures from Brass Smelter Flue Dusts, Powell et al., 1972a.

Recycling in the Zinc Industry, Barter, 1975.

Separation of Nonmagnetic Metals from Solid Waste by Permanent Magnets, Schlömann, 1975.

Solvent Extraction for Recovery of Metal Waste, Reinhardt, 1975.

Solvent Extraction in Processes for Metal Recovery from Scrap and Waste, Fletcher, 1973.

Specifications for Materials Recovered from Municipal Refuse, Alter and Reeves, 1975.

Upgrading Junk Auto Shredder Rejects, Froisland et al., 1972.

Upgrading of Wastes of the Metallurgical Industry, De Cuyper, 1973.

Waste Management in the Lead-Zinc Industry, Ratter, 1974.

Zinc Precipitation and Recovery from Vicose Rayon Waste Water, USEPA, 1971.

PRECIOUS AND SEMIPRECIOUS METALS

Current Silver Recovery Practices in the Photographic Industry, Cooley and Dugan, 1976.

First Annual Report of the Secretary of the Interior under the Mining and Minerals Policy Act of 1970, USDI, 1972.

Gold Recovery from Scrap Electronic Solders by Fused Salt Electrolysis, Keespies et al., 1970.

Recovering Metals from Waste Photographic Processing Solutions, Bober and Leon, 1975.

Recoverying Silver from Photographic Materials, Eastman Kodak Co., publication J10.

Recovery of Metals from Electroplating Wastes by the Waste-Plus-Waste Method, George and Cochran, 1970.

Recovery of Silver from Aqueous Solutions, Machida and Jujita, 1975.

Recycling and Reclaiming of Municipal Solid Wastes, Jackson, 1975.

Resource Recovery and Recycling Handbook of Industrial Wastes, Sittig, 1975.

Silver Recovery Process, Morrison and Lu,1975.

Solid Waste Processing, Engdahl, 1969.

Solid Wastes: Origin, Collection, Processing and Disposal, Mantell, 1975.

A Successful Chemical Regeneration Process, WWT, 1975h.

LESS COMMON METALS

Chemical Cycles and the Global Environment, Garrels et al., 1975, Chapters 10, 12 (mercury).

Electrolytic Recovery of Metallic Gallium, Sleppy and Gohen, 1975.

First Annual Report of the Secretary of the Interior under the Mining and Minerals Policy

Act of 1970, USDI. 1972.

Physics Looks at Waste Management, Rose et al., 1972, (flowchart of mercury in U.S. society).

Resource Recovery and Recycling Handbook of Industrial Wastes, Sittig, 1975.

Solid Waste Processing, Engdahl, 1969.

Solid Wastes: Origin, Collection, Processing, and Disposal, Mantell, 1975.

Survey of Indigenous Coals, Fly Ashes, and Flue Dusts as a Potential Source of Ge, Singh and Mathur, 1973.

GROUP FIVE ELEMENTS (including nitrogen and phosphorus)

Chemical Cycles and the Global Environment, Garrels et al., 1975, Chapters 8, 9, 12.

Converting Sewage into Savings, Chemical Week, 1976.

The Earth Phosphate Cycle, Loran, 1973.

The Ecology of Resource Degradation and Renewal, Chadwick and Goodman, 1975.

Environmental Chemistry, Manahan, 1972, Chapter 5.

First Annual Report of the Secretary of the Interior under the Mining and Minerals Policy Act of 1970, USDI, 1972.

Resource Recovery and Recycling Handbook of Industrial Wastes, Sittig, 1975.

Solid Waste Processing, Engdahl, 1969.

Solid Wastes: Origin, Collection Processing, and Disposal, Mantell, 1975 (particularly phosphorus).

SULFUR (Calcium sulfate is discussed in detail in Chapter 7)

The Bacterial Sulphur Cycle as a Source of Sulphur, Madigan, 1971b.

Bureau of Mines Research Programs on Recycling and Disposal of Mineral-, Metal-, and Energy-Based Wastes, Kenahan et al., 1973.

Chemical Cycles and the Global Environment, Garrels et al., 1975, Chapter 7.

Coal Can Be a Clean Fuel, EST, 1975g.

Environmental Chemistry, Manahan, 1972, Chapter 5.

First Annual Report of the Secretary of the Interior under the Mining and Minerals Policy Act of 1970, USDI, 1972.

Industrial Waste Disposal. Excess Sulfuric Acid Neutralization with Copper Smelter Slag, Twidwell et al., 1976.

Recycling of Lime-Sulfide Unhairing Liquors, Money and Adminis, 1974.

Recycling Processes for Sulfur Compounds, Heitman and Reher, 1974.

Resource Recovery and Recycling Handbook of Industrial Wastes, Sittig, 1975.

Sulfur is Tight Despite Excess Production, CEN, 1975c.

Sulfur Removal and Recovery, ACS, 1976.

Sulfur Utilization in Pollution Abatement, Sullivan and McBee, 1974.

Titanium Dioxide—Recycling of Sulfuric Acid, Kern et al., 1974.

Utilization of Secondary Sulfur in Construction Materials, McBee and Sullivan, 1976.

HALOGENS

Continuous Regeneration of Hydrochloric Acid Pickle Liquors, Taylor, 1975c.

First Annual Report of the Secretary of the Interior under the Mining and Minerals Policy Act of 1970, USDI, 1972.

Hydrochloric Acid Recovery from Chlorinated Organic Waste, Russell and Mraz, 1974.

Recovery of Methyl Bromide from Wheat Fumigation by Adsorption, Martin et al., 1975.

Recycling and Disposal of Waste Plastics, Jensen et al., 1974 (recovery of hydrogen chloride from PVC).

Recycling of Plastics from Urban and Indsutrial Refuse, Holman et al., 1974 (recovery of hydrogen chloride from PVC).

Recycling Poly(tetrafluoroethylene), Arkles, 1973.

Resource Recovery and Recycling Handbook of Industrial Wastes, Sittig, 1975.

Salt Solution, Jacques, 1975 (bromine recovery).

Waste Recovery and Pollution Abatement, Rickles, 1965, (fluorides from phosphate fertilizer manufacture).

APPENDIX 2

RECYCLING REVIEW PUBLICATIONS

The titles of books, journals, films, conference proceedings, and reviews which survey recycling generally are collected here for convenience. Full bibliographical details in most cases are included in Appendix 5, in alphabetical order by author.

Advances in Solid Waste Treatment Technology, Hershaft, 1972.

An Assessment of Energy Recovery Methods Applicable to Domestic Refuse Disposal, Porteous, 1975b.

Automotive Scrap Recycling, Sawyer, 1974.

Base Line Forecasts of Resource Recovery, 1972 to 1990; Final Report, Nuss et al., 1975.

Bibliography of WSL Publications on Waste Recovery, WSL, 1975g.

Bureau of Mines Research Activities in Secondary Resource Recovery, Falkie, 1975.

Bureau of Mines Research and Accomplishments in Utilization of Solid Waste, Cservenyak and Kenahan, 1970.

Bureau of Mines Research Programs on Recycling and Disposal of Mineral-, Metal-, and Energy-Based Wastes, Kenahan et al., 1973.

Chemical Abstracts, (American Chemical Society, Columbus, Ohio 43202). Section 60, "Sewage and Wastes."

The Chemical Conversion of Solid Wastes to Useful Products, Barbour et al., 1974.

Chemical Cycles and the Global Environment, Garrels et al., 1975.

ChemTech (Chemical Technology), American Chemical Society 1155 16th St. N.W., Washington, D.C. 20036 (a journal which includes articles on recycling).

Compost Science—Journal of Waste Recycling, Rodale Press, Inc., 33 E Minor St., Emmaus, Pennsylvania 18049.

Conservation and Recycling (International Journal for the Rapid Publication of Critical Reviews and Original Research Communications in the Science and Technology of Materials Conservation and Recycling), Ed. M. B. Bever (M.I.T., Cambridge, Massachusetts) and Michael E. Henstock (University of Nottingham, U.K.), Pergamon Press, Headington Hill Hall, Oxford OX3 OBW, U.K.

Contemporary Chemistry: Science, Energy and Environmental Change, C. G. Wade, Macmillan, New York, 1976, Chapter 7, "Solid Waste and Recycling."

Conversion of Refuse to Energy, Kirov, 1975b.

Disposal of Solid Wastes, CE, 1973.

Ecologist (journal), Wadebridge, Cornwall, U.K.

Ecology Lady, 16 mm color film, 11 minutes, Stuart Finley, Inc., 3428 Mansfield Road, Falls Church, Virginia 22041.

The Ecology of Resource Degradation and Renewal, Chadwick and Goodman, 1975.

Economic Aspects of the Conservation of Resources in the Chemical and Allied Industries, SCI, 1976a.

Economics of Recycling Waste Materials, U.S. Congress, 1972.

The Energy Conservation Papers, Williams, 1975.

Energy Expenditures Associated with the Production and Recycle of Metals, Bravard et al., 1972.

Energy from Solid Waste, Jackson, 1974.

Energy Potential from Organic Wastes: A Review of the Quantities and Sources, Anderson, 1972.

Energy Recovery from Solid Waste, Hiraoka, 1975.

Energy Recovery from Solid Waste, RRC, 1976.

Environmental Chemistry, Moore and Moore, 1976, Chapter 13, "Solid Wastes."

Environmental Pollution Management (journal), McDonald, 268 High St., Uxbridge, Middlessex, UB8 1UA, U.K.

Environmental Science and Technology, American Chemical Society, 1155 16th St. N. W., Washington, D.C. 20036 (a journal which frequently includes articles on recycling).

Fertility from Town Waste, Wylie, 1955.

First Annual Report of the Secretary of the Interior under the Mining and Minerals Policy Act of 1970, USDI, 1972.

Fluid Bed Reactors in Solid Waste Treatment, Bailie and Burton 1972.

Food from Waste, Birch et al., 1976.

Garbage as You Like It, Goldstein, 1969 (composting of urban waste).

Garbage in America, Neil N. Seldman, Institute for Local Self-Reliance, Washington, D.C.

Guidelines for Procurement of Products that Contain Recycled Material, USEPA, 1976a.

Handbook of Solid Waste Disposal: Materials and Energy Recovery, Pavoni et al., 1975.

How to Make Waste Pay, Higginson, 1976.

IGT Weighs Potential of Fuels from Biomass, CEN, 1976a.

Industrial Recovery, monthly journal of the National Industrial Materials Recovery Association, U.K.

Industrial Waste Disposal Made Profitable, Bennett and Lash, 1974.

Industrial Waste and By-Products Generated in Australia: The Results of a Survey, Beretka, 1973.

International Biomass Energy Conference Proceedings (Biomass Energy Institute, Winnipeg, Canada).

Is Recycling the Solution? Pausacker, 1975.

Journal of Environmental Planning and Pollution Control, Mercury House Business Publications, London.

Kann Recycling die Untweltsbeeinträchtigung vermindern? Stumm and Davis, 1974.

Living Today, 16 mm film produced by the Central Office of Information, U.K.

Material Gains—Reclamation, Recycling, and Reuse, Thomas, 1974.

Materials Reclamation Weekly, Distinctive Publications, Croydon, U.K.

Mineral Waste Utilization Symposium, Fourth, Aleshin, 1974.

Mineral Waste Utilization Symposium, Fifth, Aleshin, 1976.

National Center for Resource Recovery Bulletin, National Center for Resource Recovery, Inc., Washington, D.C.

National Center for Resource Recovery Fact Sheets.

National Industrial Materials Recovery Association Reprint Pamphlets.

A National Policy Toward Recycling, Boyd, 1976.

A Nationwide Survey of Resource Recovery Activities, Hopper, 1975; McEwen, 1976.

New Alchemy Institute Newsletter.

The New Prospectors, director Brian Earley, 16 mm film, U.K. Central Office of Information, for the Warren Spring Laboratory of the Department of Industry, U.K. Central Film Library Catalogue No. U.K. 3290.

Novel Feeds and Foods, SCI, 1976b.

Phoenix Quarterly, Journal of the Institute of Scrap Iron and Steel, Inc., Washington D.C.

Physics Looks at Waste Management, Rose et al., 1972.

Plastics and the Environment, Staudinger, 1974.

Pollution Engineering (journal), Technical Publishing Company, 35 Mason Street, Greenwich, Connecticut 06830.

Processing Agricultural and Municipal Wastes, Inglett, 1973.

Quantity and Composition of Post-Consumer Solid Waste: Material Flow Estimates for 1973 and Baseline Future Projections, Smith, 1976.

The Realities of Recycling, 16 mm color film, 38 minutes, Stuart Finley, Inc., 3428 Mansfield Road, Falls Church, Virginia 22041.

Recent Advances in the Recovery of Useful Materials from Industrial Waste, SCI, 1976c.

Reclaiming Rubber and Other Polymers, Szilard, 1973 (review of patent and report literature).

Recycle; in Search of New Policies for Resource Recovery, League of Women Voters, 1972.

Recycling, 16 mm color film, 21 minutes, Stuart Finley, Inc., 3428 Mansfield Road, Falls Church, Virginia 22041.

Recycling: A Guide to Effective Solid Waste Utilization, National Association of Recycling Industries, 1973, 24pp.

Recycling and Disposal of Solid Waste, Henstock, 1975a.

Recycling and Disposal of Solid Wastes—Industrial, Agricultural, Domestic, Yen, 1974a,

Recycling and Reclaiming of Municipal Solid Wastes, Jackson, 1975.

Recycling and the Consumer; Solid Waste Management, USEPA, 1974d.

Recycling and the Consumer, Solid Waste Management, U.S. Environmental Protection Agency, SW-117.1, (1974) (display material).

Recycling and Waste Disposal, a journal previously titled *Waste Disposal,* ed. E. Rhodes, Marylebone Press, Manchester, U.K.

Recycling as an Industry, Ness, 1972.

Recycling: Assessment and Prospects for Success, Darnay, 1972.

Recycling—Backwards to an Optimum, Kinnersly, 1973.

Recycling: Fundamentals and Concepts, Bundi and Wasmer, 1976.

Recycling—Möglichkeiten zur Ruckgewinning von Lösungsmittel und Stoffen (Recycling Possibilities for the Recovery of Solvents and Other Materials), Hartinger, 1975.

Recycling Our Resources, McGough, 1972.

Recycling: Practical Answer to the Problems of Air Pollution, Water Pollution, Solid Waste, American Metal Market, 1970.

Recycling, Recovery, Reuse, special features in *Water and Waste Treatment,* **18**(6), (1975); **19**(6), (1976).

Recycling: the Modern Day Phoenix, Keller, 1973.

Recycling Today (Secondary Raw Materials) (journal), Market News Publishing Corporation, New York.

Report of the World Trends Committee, HMSO, London, 1976.

Resource Recovery, ASME, 1975.

Resource Recovery and Conservation (journal), Elsevier, Amsterdam, ed. J. G. Albert and H. Alter (National Center for Resource Recovery, Inc., Washington, D.C).

Resource Recovery and Recycling Handbook of Industrial Wastes, Sittig, 1975 (review of patent and report literature).

Resource Recovery and Source Reduction, USEPA, 1974a, 1974b.

Resource Recovery and Utilization, Alter and Horowitz, 1975.

Resource Recovery and Waste Reduction, USEPA, 1975.

Resource Recovery. Catalog of Processes, Franklin et al., 1973b.

Resource Recovery from Municipal Solid Waste, NCRR, 1974c.

Resource Recovery from Solid Waste, Rosich, 1975a.

The Resource Recovery Industry, Ganotis and Hopper, 1976 (survey of contractors, consultants, equipment manufacturers)

Resource Recovery in Solid Waste Management, Rosich, 1975b.

Resource Recovery. State of Technology, Franklin et al., 1973b.

Resource Recovery Technology Update from U.S.E.P.A.: Status Report on Resource Recovery Technology: Demonstrating Resource Recovery, Lingle, 1976.

Resources Policy (journal), J.P.C. Science and Technology Press Ltd., 32 High Street, Guildford, Surrey GU1 3EW, U.K.

Reuse and Recycle of Wastes, ANERAC, 1971.

Re-Use of Polymer Waste, Sperber and Rosen, 1974.

Review of Advanced Solid Waste Processing Technology, Wilson, 1974.

A Review of the Problems Affecting the Recycling of Selected Secondary Materials, NASMI, 1972.

The Salvage and Recycling of Useful Materials, Pearson and Webb, 1973.

The Salvage Industry; What It Is—How It Works, Kiefer, 1973b.

Solid Waste, Bond and Straub, 1973.

Solid Waste Demonstration Projects, Stump, 1972.

Solid Waste Disposal: Incineration and Landfill, Baum and Parker, 1974a.

Solid Waste Disposal: Recycling and Pyrolysis, Baum and Parker, 1974b.

Solid Waste Management, OST, 1969.

Solid Waste Management: Available Information Materials, USEPA, 1974c, 1976c (bibliographies).

Solid Waste Processing, Engdahl, 1969 (includes extensive bibliography).

Solid Waste Recycling Projects; a National Directory, Hansen, 1973.

Solid Waste: Resources Out of Place, Kenahan, 1971.

Solid Waste Treatment and Disposal, Kirov, 1972.

Solid Waste, monthly journal of the Institute of Solid Wastes Management, London.

Solid Wastes, ACS, 1971 (reprint volume).

Solid Waste II, ACS, 1973 (reprint volume).

Solid Wastes Management, (Refuse Removal Journal and Liquid Wastes Management), Communication Channels Inc., 461 Eighth Ave., New York, 10001.

Source Separation for Materials Recvoery; Guidelines, USEPA, 1976b.

Study to Identify Opportunities for Increased Solid Waste Utilization, USEPA, 1972a.

Thermal Power Plant Waste Heat Utilization, Mandell, 1974.

Thermal Processing of Municipal Solid Waste for Resource and Energy Recovery, Weinstein and Toro, 1976.

Treatment and Disposal of Solid Waste—Resource Recovery Aspects, Patrick, 1973.

Verwertung des Wertlosen (Utilization of the Worthless), Ungewitter, 1938.

The Village Green, 16 mm film, sound, color, 15 minutes, U.S. Environmental Protection Agency Publication SW-8tg, 1974. (Documents a successful and self-sustaining recycling center in New York City Sponsored by the Environmental Action Coalition).

Waste Age (journal).

The Wastebin (newsletter on low-technology recycling), P.O. Box 14012, Portland, Oregon 97214.

Waste Disposal and Recycling Bulletin, Science and Technology Department, Birmingham Central Libraries, Birmingham B3 3HQ, U.K. (intended for local government personnel responsible for the management of waste disposal).

Waste Management, Control, Recovery, and Reuse, Kirov, 1975a.

Waste Management Research, Occasional Papers Series, Department of the Environment,

Headquarters Library, London, 1975 (bibliography of research, particularly reclamation projects).

Waste Materials, Golueke and McGauhey, 1976.

Waste Materials, audio tape from *An Introduction to Materials,* Open University, U.K., TS 251/07.

Waste Reclamation 1: A Source of Raw Materials, Douglas and Jackson, 1973.

Waste Reclamation 2: Economic Aspects, Pearson, 1973.

Waste Reduction and Resource Recovery—There's Room for Both, Humber, 1975.

Waste Recovery and Pollution Abatement, Rickles, 1965.

Water and Waste Treatment (journal), D. R. Publications Ltd., 111 St. James's Road, Croydon, CR9 2TH, Surrey, U.K.

What about Wastes and How to Reuse Them, Rasmuson, 1973.

What You Can Do to Recycle More Paper, U.S. Environmental Protection Agency Publication SW-143, 1975 (suitable for elementary school).

APPENDIX 3

ADDRESSES OF ORGANIZATIONS

Listed here are the addresses noted on the most recent publications cited in Appendix 5 or in personal communications. The accuracies of the addresses or even the currrent names of the organizations are not guaranteed, nor is this an exhaustive list of organizations engaged in recycling activities. In some cases an indication of the type of recycling is provided in italics, but this does not necessarily describe the full range of activities.

Abcor, Inc.,
341 Vassar St.,
Cambridge,
Massachusetts 02139.
(membrane separation processes)

A.C.I. Technical Center Pty, Ltd.,
Australian Consolidated Industries,
Waterloo,
New South Wales 2017,
Australia.
(solid waste survey)

Agricultural Research Service,
U.S. Department of Agriculture,
Agricultural Environmental Quality Institute,
Agricultural Research Center—East,
Beltsville,
Maryland 20705.

Eastern Regional Research Laboratory,
Philadelphia,
Pennsylvania 19104.

Northern Regional Research Laboratory,
1815 North University Street,
Peoria,
Illinois 61604.

Richard B. Russell Agricultural Research Center,
Athens,
Georgia 30603.

Western Regional Research Laboratory,
Berkeley,
California 94704.
(utilization of various agricultural wastes)

Aluminum Association,
750 3rd Avenue,
New York,
New York 10017.
(recycling of aluminum from consumer waste)

Aluminum Co. of America (Alcoa),
Alcoa Center,
Pittsburgh,
Pennsylvania 15219.

Anaconda American Brass Co.,
Engineering Environments,
P.O. Box 747,
Waterbury,
Connecticut 06720.
(metal recovery)

Andco, Inc.,
Buffalo,
New York 14225.
(pyrolysis of urban waste)

Associated Portland Cement
Manufacturers Ltd.,
Engineering R and D,
Barnstone,
Nottingham NG13 9JT,
U.K.
(urban waste as an auxiliary fuel in cement manufacture)

Association of Petroleum Re-Refiners,
1730 Pennsylvania Avenue N.W.,
Washington, D.C. 20006.

Association of Reclaimed Textile
Processors,
Midland Bank Chambers,
Market Place,
Dewsbury,
Yorkshire,
U.K.

Atomic Energy Research
Establishment,
Harwell,
Berkshire,
U.K.

Atomics International,
A Division of Rockwell International
Corp.,
P.O. Box 309,
Canoga Park,
California 91304.

Australian Mineral Development
Laboratories (AMDEL),
Flemington St.,
Frewville,
South Australia 5063.
(reclamation in the mineral industry)

Battelle Memorial Institute,
Columbus Laboratories,
505 King Ave.,
Columbus,
Ohio 43201.
(surveys of recycling opportunities)

Battelle Pacific—Northwest
Laboratories,
Richland,
Washington 99352.

B.C.Research,
British Columbia Research Council,
3650 Westbrook Crescent,
Vancouver,
Canada V6S 2L2.
(pyrolysis of urban waste)

Berkeley Recycling Center,
3029 Bienvenue Ave.,
Berkeley,
California 94705.

Biomass Energy Institute,
Winnipeg,
Canada.

Biomechanics Ltd.,
Smarden,
Ashford,
Kent,
U.K.
(anaerobic treatment of industrial wastes: the Bioenergy® Process)

Black Clawson Co.,
Middletown,
Ohio 45042.
(urban waste utilization)

Boise Cascade Corp.,
Chemical Research Laboratory,
Portland,
Oregon 97201.
(chemicals from waste cellulose)

British Secondary Metals Association,
12 Henrietta St.,
London WC2,
U.K.

British Steel Corporation,
Forward Technology Unit,
London,
U.K.

British Waste Paper Association,
21 Devonshire Street,
London Wl,
U.K.

Building Research Establishment,
Department of the Environment,
Watford WD2 7JR,
U.K.
(recycled materials in construction)

Bureau of Mines,
U.S. Department of the Interior,
Albany Metallurgy Research Center,
P.O. Box 70,
Albany,
Oregon 97321.

Boulder City Metallurgy Research Laboratory,
500 Date St.,
Boulder City,
Nevada 89005.

College Park Metallurgy Research Center,
College Park,
Maryland 20740.

Division of Nonmetallic Minerals,
Ballston Towers No. 3,
4015 Wilson Boulevard,
Arlington,
Virginia 22203.

Division of Solid Wastes,
Washington, D.C. 20241

Morgantown Energy Research Center,
P.O. Box 880,
Morgantown,
West Virginia 26505.

Pittsburgh Energy Research Center,
4800 Forbes Ave.,
Pittsburgh,
Pennsylvania 15213.

Rolla Metallurgy Research Center,
P.O. Box 280,
Rolla,
Missouri 65401.

Salt Lake City Metallurgy Research Center,
1600 East 1st South,
Salt Lake City,
Utah 84112.

Tuscaloosa Metallurgy Research Laboratory,
Tuscaloosa,
Alabama 35486.

Twin Cities Metallurgy Research Center,
P.O. Box 1660,
Twin Cities,
Minnesota 55111.

Bureau of Solid Waste Management,
Department of Health, Education and Welfare.
(now the Office of Solid Waste Management Programs,
Environmental Protection Agency)
Washington, D.C. 20460.

Calspan Corporation,
Environmental Systems Department,
Buffalo,
New York 14221.

Canada Centre for Inland Waters,
Wastewater Technology Centre,
Environmental Protection Service,
Department of the Environment,
P.O. Box 5050,
Burlington,
Ontario,
Canada.

Canadian Forestry Service,
Deparment of the Environment,
Western Forest Products Laboratory,
Vancouver,
British Columbia V6T 1X2,
Canada.

Canberra College of Advanced Education,
School of Applied Science,
P.O. Box 381,
Canberra City,
A.C.T. 2601,
Australia.
(survey of solid wastes)

Carnegie-Mellon University,
Department of Chemical Engineering,
Pittsburgh, Pennsylvania 15213.
(polymer waste utilization)

Center for Advanced Computation,
Energy Library,
University of Illinois,
Urbana,
Illinois 61801.

Chemfix, Inc.,
505 McNeilly Rd.,
Pittsburgh,
Pennsylvania 15225.

PD Chemfix,
33 Kingston Crescent,
North End,
Portsmouth P02 8AD,
U.K.
(fixation of wastes by soluble silicates)

Chemical Recovery Association,
6 Kingscote Road,
Dorridge,
Solihull,
West Midlands B93 8RA,
U.K.

Colorado School of Mines,
Research Institute,
P.O. Box 112,
Golden,
Colorado 80401.

Combustion Power Co., Inc.,
1346 Willow Road,
Menlo Park,
California 94025.

Commonwealth Scientific and
Industrial Research Organisation
(CSIRO),Australia.
(work in several Divisions on various aspects of recycling)

Complete Tree Institute,
University of Maine,
Orono,
Maine 04473.

The Conservation Foundation,
1717 Massachusetts Ave. N.W.,
Washington, D.C. 20036.

Continental Can Company Inc.,
1200 West 76th Street,
Chicago,
Illinois 60631.

Cornell University,
(including College of Agriculture and Life Sciences, and Departments of Food Science, Agricultural Engineering, Water Resources Engineering),
Ithaca,
New York 14850.
(food and agricultural waste recovery)

Council on Environmental Quality,
722 Jackson Place N.W.,
Washington, D.C. 20006.

Dartmouth College,
Thayer School of Engineering,
Hanover,
New Hampshire 03755.

Division of Building Research,
Commonwealth Scientific and
Industrial Research Organization,
P.O. Box 56,
Highett,
Victoria 3190,
Australia.
(recycled materials for building)

Dow Chemical,
Environmental Research Laboratory,
Midland,
Michigan 48640,

Drexel University,
Department of Chemical
Engineering,
Philadelphia,
Pennsylvania 19104.

E.I. du Pont de Nemours and Co., Inc.,
Engineering Research and Development Division,
Wilmington,
Delaware 19898.

Savannah River Laboratory,
Aiken,
South Carolina 29801.

Dynatech Research and Development Co.,
Dynatech Corporation,
Cambridge,
Massachusetts 02139.
(urban waste utilization)

Eastman Kodak Co.,
Rochester,
New York 14650.
(recycling of plastics, silver, and chemicals)

Ebara Manufacturing Co.,
Resource Recovery Project,
Central Research Institute,
Fujisawa 251,
Japan.
(urban waste recycling)

Ecologist,
Wadebridge,
Cornwall,
U.K.

Ecology, Inc.,
Brooklyn,
New York 11201.
(urban waste utilization)

Economic Research Service,
U.S. Department of Agriculture,
Washington, D.C.

Ecotech Systems (UK) Ltd.,
Balcna Closc,
Creekmoor,
Poole,
Dorset,
U.K.

Effluent Disposal Ltd.,
Walsall Wood,
West Midlands,
U.K.

Eidal International Corp.,
Box 2087,
Albuquerque,
New Mexico 87101.

Energy Research and Development Institute,
New Mexico State University,
Las Cruces,
New Mexico 88003.
(solar energy research)

Environics,
Huntington Beach,
California 92646.

Environmental Protection Agency (U.S.),see: Industrial Environmental Research Laboratory; Office of Research and Monitoring; Office of Solid Waste Managements Programs; Resource Recovery Division; Solid and Hazardous Waste Research Laboratory.

Environmetrics Systems Pty, Ltd.,
Perth,
Western Australia.
(urban waste utilization)

Envirotech Corporation,
Eimco Division,
537 West Sixth South,
P.O. Box 300,
Salt Lake City,
Utah 84110.

Wemco Division,
721 North B. Street,
Sacramento,
California 95814.

Esso Research and Engineering Co.,
Government Research Laboratory,
Linden,
New Jersey 07036.

Exxon Corporation,
1251 Avenue of the Americas,
New York,
New York 10020.

Federal Energy Administration,
Washington, D.C. 20461.

Ford Motor Co.,
Dearborn,
Michigan 48121.

Forest Products Laboratory,
U.S. Department of Agriculture
Forest Service,
P.O. Box 5130,
Madison,
Wisconsin 53705.
(waste paper utilization)

Franklin City Manager,
Franklin,
Ohio 45005.

Friends of the Earth Ltd.,
9 Poland Street,
London WlV 3DG,
U.K.

Garrett Research and Development Co., Inc. (now Occidental Research Corporation),
1855 Carrion Road,
La Verne,
California 91750.
(urban waste recycling)

General Electric Co.,
Research and Development Center,
P.O. Box 8,
Schenectady,
New York 12301.
(cellulose utilization)

General Motors Research
Laboratories,
Warren,
Michigan 48090.

Glass Container Manufacturers'
Institute, Inc.,
1800 K St., N.W.,
Washington, D.C. 20006.
(waste glass utilizaton)

The Goodyear Tire and Rubber Co.,
Akron,
Ohio 44316.
(waste tire utilization)

Groveton Papers Division,
Diamond International Corp.,
Groveton,
Northumberland,
New Hampshire 03582.
(incineration of urban waste with paper mill waste)

Grumman Aerospace Corporation,
Bethpage,
New York 11714.

Gulf Environmental Systems,
see Universal Oil Products.

Hercules Research Center,
Hercules, Inc.,
910 Market Street,
Wilmington,
Delaware 19801.
(urban waste composting and reclamation)

Highway Materials Laboratory,
Department of Civil Engineering and Engineering Mechanics,
McMaster University,
Hamilton,
Ontario L8S 4L7,
Canada.
(recycled materials for road construction)

Horner and Shifrin, Inc.,
5200 Oakland Ave.,
St. Louis,
Missouri 63110.
(urban waste recycling)

IIT Research Institute,
P.O. Box 4963,
Chicago,
Illinois 60680.
(includes polymer waste utilization)

Imperial College School of Mines,
Department of Metallurgy and Materials Science,
Nuffield Research Group,
London SW7 2AZ,
U.K.
(molten salt solvent extraction of wastes)

Imperial Metal Industries Ltd.,
Witton,
Birmingham,
U.K.
(industrial boiler utilizing domestic refuse)

Industrial Environmental Research Laboratory,
U.S. Environmental Protection Agency,
Cincinnati,
Ohio 45268.
Reasearch Triangle Park
North Carolina 27711.

Institute for Local Self-Reliance,
1717 18th St. N.W.,
Washington, D.C. 20009.
(waste utilization)

Institute for Mineral Research,
Michigan Technological University,
Houghton,
Michigan 49931.

Institute for Solid Wastes,
1313 E 10th Street,
Chicago,
Illinois 60637.

Institute of Gas Technology,
3424 S State Street,
Chicago,
Illinois 60616.
(includes fuel gas from anaerobic digestion)

Institute of Scrap Iron and Steel, Inc.,
1729 H Street N.W.,
Washington, D.C. 20006.

Institute of Solid Wastes Management,
28 Portland Place,
London W1N 4DE,
U.K.

Institute of Storage and Food Technology,
Agricultural Academy,
ul. C. Norwida 25,
50-375 Wroclaw,
Poland.

Institute of Wood Chemistry and Cellulose Technology,
Abo Adademi SF-20500,
ABO, 50, Finland.

Inter-Technology Corporation,
Box 340,
Warrenton,
Virginia 22186.
(energy from organic materials)

ISWA Research Secretariat,
(International Solid Wastes and Public Cleansing Association),
Überlandstrasse 133,
8600 Dübendorf,
Switzerland.

Kansas State University,
Department of Chemical Engineering,
Manhattan,
Kansas 66506.
(pyrolysis of feedlot wastes)

A.M. Kinney, Inc.,
2912 Vernon Place,
Cincinnati,
Ohio 45219.

K.T. Lear Associates, Inc.,
P.O. Box 288,
Manchester,
Connecticut 06040.
(urban waste recycling)

Louisiana State University,
Chemical Engineering Department,
Baton Rouge,
Louisiana 70803.
(protein from cellulose)

Massachusetts Institute of Technology,
Cambridge,
Massachusetts 02139.
(urban and organic waste recycling)

Materials Associates,
Washington, D.C. 20037.

Midwest Research Institute,
425 Volker Boulevard,
Kansas City,
Missouri 64110.

1522 K Street N.W.,
Washington, D.C. 20005.
(economic and environmental effects of recycling)

Mitsui Mining and Smelting Co. Ltd.,
8-7-1, Shimo-renjaku,
Mitaka,
Tokyo,
Japan.

Monsanto Enviro-Chem Systems, Inc.,
800 N Lindbergh Boulevard,
St. Louis,
Missouri 63166.
(utilization of urban waste)

Monsanto Research Corporation,
Dayton,
Ohio 45407.

Montana College of Mineral Science and Technology,
Metallurgy-Mineral Processing Department,
Butte,
Montana 59701.

National Analysts, Inc.,
1015 Chestnut St.,
Philadelphia,
Pennsylvania 19107.
(recycling surveys)

National Ash Association, Inc.,
Washington, D.C. 20013.

National Association of Recycling Industries, Inc. (NARI),
330 Madison Ave.,
New York,
New York, 10017.

National Association of Secondary Material Industries, Inc. (NASMI),
now National Association of Recycling Industries, Inc.

National Center for Resource Recovery, Inc. (NCRR),
1211 Connecticut Ave. N.W.,
Washington, D.C. 20036.

National College of Food Technology,
University of Reading,
St. George's Avenue,
Weybridge,
Surrey KT13 ODE,
U.K.

National Industries Materials Recovery Association (NIMRA),
York House,
Westminster Bridge Rd.,
London SE1 7UT,
U.K.

National Solid Wastes Management Association,
1145 19th St. N.W.,
Washington, D.C. 20036.

National Steel Corporation,
Process Metallurgy Research Center,
Weirton,
West Virginia 26062.
(ferrous scrap recycling)

New Alchemy Institute
Box 432,
Woods Hole,
Massachusetts 02543.

New Alchemy Institute West,
P.O. Box 2206,
Santa Cruz,
California 95063.

North Carolina State University,
School of Forest Resources,
P.O. Box 5488,
Raleigh,
North Carolina 27607.

Nottingham University,
Department of Metallurgy and
Materials Science,
University Park,
Nottingham NG7 2RD,
U.K.

Oak Ridge National Laboratory,
Oak Ridge,
Tennessee 37830.
(ORNL-NSF Environmental Program: energy analysis)

Occidental Research Corporation,
(a subsidiary of Occidental
Petroleum Corp.),
1855 Carrion Road,
La Verne,
California 91750.
(urban waste recycling)

Office of Research and Monitoring,
National Environmental Research
Center,
U.S. Enviromental Protection
Agency,
Research Triangle Park,
North Carolina 27711.

Office of Solid Waste Management
Programs,
Environmental Protection Agency,
Washington, D.C. 20460.

Open University,
Energy Research Group,
Walton Hall,
Milton Keynes MK7 6AT,
U.K.

Oregon State University,
Department of Agricultural
Chemistry,
Corvallis,
Oregon 97331.

Owens-Illinois, Inc.,
North Technical Center,
P.O. Box 1035, Toledo,
Ohio 43666.
(urban waste and plastic recycling)

Pennsylvania State University,
Department of Animal Science,
University Park,
Pennsylvania 16802.

School of Chemical Engineering,
Philadelphia,
Pennsylvania 19104.

Philadelphia Quartz Co.,
Research and Development Center,
P.O. Box 258,
Lafayette Hill,
Pennsylvania 19444.
(soluble silicates used in materials recycling)

Plastic Recycling Ltd.,
Kentford,
Newmarket,
Suffolk, CB8 7QB,
U.K.

The Plastics and Rubber Institute,
11 Hobart Place,
London SWlW OHS.
U.K.

QMC Wolfson Recycle Unit,
QMC Industrial Research Ltd.,
Queen Mary College,
University of London,
229 Mile End Rd.,
London El 4AA,
U.K.

Raytheon Research Division,
Waltham,
Massachusetts 02154.

Raytheon Service Company
12 Second Avenue,
Burlington,
Massachusetts 01830.

Re-Chem International Ltd.,
80 Shirley Rd.,
Southampton SO1 3EY,
U.K.

Reclamation Industries Council,
16 High Street,
Brampton,
Huntingdon,
Cambridgeshire,
U.K.

Rensselaer Polytechnic Institute,
Department of Chemical and
Environmental Engineering,
Troy,
New York 12181.

Research Triangle Institute,
Research Triangle Park,
North Carolina 27711.

Resource Recovery Division,
Office of Solid Waste Management
Programs,
U.S. Environmental Protection
Agency,
Washington, D.C. 20460.

Resource Recovery Services, Inc.,
Woodbridge,
New Jersey 07095.

Resource Recovery Systems,
Barber-Coleman Co.,
Irvine,
California 92713.

Resources for the Future, Inc.,
1755 Massachusetts Ave., N.W.,
Washington D.C. 20036.

Reynolds Metals Co.,
Reclamation Products,
6601 West Board Street,
Richmond,
Virginia 23218.

Rubber and Plastics Research
Association, (RAPRA),
Shawbury,
Shrewsbury,
Shropshire SY4 4NR,
U.K.

City of St. Louis,
4100 S First St.,
St. Louis,
Missouri 63118.

Sanitary Engineering Research
Laboratory, (SERL),
Department of Engineering,
University of California,
Richmond,
California 94804.
(organic waste utilization)

Scrap Metal Research and Education Foundation,
1729 H Street N.W.,
Washington D.C. 20006.

Shell Research Ltd.,
Woodstock Laboratory,
Sittingbourne Research Centre,
Sittingbourne,
Kent ME9 8AG,
U.K.
(SCP from methane)

Simsmetal Pty. Ltd.,
Australia.
(metal recycling)

Solid and Hazardous Waste Research Laboratory,
National Environmental Research
Center,
U.S. Environmental Protection
Agency,
Cincinnati 45268,
Ohio.
(utilization of cellulosic wastes)

Southampton University,
Department of Chemistry,
Southampton SO9 5NH,
U.K.
(electrochemical recycling techniques)

Southern Research Institute,
2000 Ninth Ave, South,
Birmingham,
Alabama 35205.

Stanford Research Institute,
Menlo Park,
California 94025.

Steel Can Group,
470 Bourke St.,
Melbourne,
Australia.
(ferrous metal recycling from urban waste)

Stone and Webster Engineering Corp.,
P.O. Box 2325,
Boston,
Massachusetts 02107.
(cellulose utilization)

Tate and Lyle Ltd.,
Philip Lyle Memorial Research
Laboratory,
P.O. Box 68,
Reading, RG6 2BX,
U.K.
(utilization of carbohydrate wastes)

Technical Association of the Pulp
and Paper Industry (TAPPI),
Dunwoody Park,
Atlanta,
Georgia 30341.
(reclamation in the pulp and paper industry)

Teknekron, Inc.,
2118 Milvia Street,
Berkeley,
California 94704.

Torrax Systems Inc.,
64 Erie Ave.,
North Tonawanda,
New York 14120.
(urban waste utilization)

U.K. Waste Materials Exchange,
Department of Industry,
P.O. Box 51,
Stevenage,
Hertfordshire, SG1 2DT,
U.K.

Union Carbide,
Linde Division,
270 Parke Ave.,
New York, 10017.

Old Saw Mill River Rd.,
Tarrytown,
New York 10591.

Union Electric Co.,
1901 Gratiot Street,
P.O. Box 149,
St. Louis,
Missouri 63166,
(urban waste utilization)

United Reclaim Ltd.,
Speke Hall Road,
Liverpool L24 1UZ,
U.K.
(tire recycling)

Universal Oil Products Co. (UOP),
Fluid Systems Division
(incorporating Roga Division),
San Diego,
California 92101.
(water reclamation)

University of Alabama,
Department of Civil and Mineral
Engineering,
P.O. Drawer 1468,
University,
Alabama 35486.

University of Aston in Birmingham
Biodeterioration Information
Centre,
Department of Biological
Sciences,
80 Coleshill Street,
Birmingham B4 7PF,
U.K.

Department of Chemical
Engineering,
Birmingham B4 7ET,
U.K.

University of Birmingham,
Department of Minerals Engineering
(Wolfson Secondary Metals
Research Group),
P.O. Box 363,
Birmingham B15 2TT,
U.K.

University of Calgary,
Department of Civil Engineering,
Calgary,
Alberta, T2N 1N4,
Canada.
(recycled construction materials)

University of California,
Departments of Chemical Biodynamics, Chemical Engineering, Chemistry, Civil Engineering, Mechanical Engineering, and Lawrence Berkeley Laboratory,
Berkeley,
California 94720.
(recycling of urban and cellulose wastes)

University of California at Los Angeles,
School of Engineering and Applied Sciences,
Los Angeles,
California 90024.

University of Chicago,
Department of Chemistry,
Chicago,
Illinois 60637,
(energy analyses)

University of Illinois,
Department of Civil and Ceramic Engineering; and Center for Advanced Computation,
Urbana,
Illinois 61801.
(anaerobic conversion of urban waste; energy analyses)

University of Manchester Institute of Science and Technology (UMIST),
Department of Chemical Engineering,
P.O. Box 88,
Sackville Street,
Manchester M60 1QD,
U.K.

University of Missouri
Department of Chemical Engineering,
Columbia,
Missouri 65201.
(singe-cell protein)

University of Missouri,
Departments of Civil Engineering, Ceramics Engineering, and Environmental Health,
Rolla,
Missouri 65401.
(waste glass utilization)

University of Montana,
Wood Chemistry Laboratory, Department of Chemistry, and School of Forestry,
Missoula,
Montana 59801.

University of New South Wales,
School of Engineering,
Kensington,
New South Wales 2033,
Australia,

University of Sherbrooke,
Sherbrooke,
Quebec,
Canada.
(Project PUDDING: Potential Uses for Discarded or Detrimental Industrial and Natural Garbage)

University of Southern California,
Departments of Chemical Engineering and Environmental Engineering Sciences,
Los Angeles,
California 90007.

University of Western Ontario,
Department of Chemical and Biochemical Engineering,
London 72,
Ontario,
Canada.

University of Wisconsin, Departments of Sanitary Engineering, Mechanical Engineering, and Chemical Engineering,
Madison,
Wisconsin 53709.

U.S. Army Natick Laboratories,
Natick,
Massachusetts 01760.
(microbiological cellulose conversion)

U.S. Bureau of Mines, *see* Bureau of Mines.

U.S. Bureau of Reclamation,
Denver,
Colorado 80225.

U.S. Department of Agriculture, *see* Agricultural Research Service.

U.S. Environmental Protection Agency, *see* Environmental Protection Agency.

Virginia Polytechnic Instiute and State University, Center for Environmental Studies, and Psychology Department,
Blacksburg,
Virginia 24061.
(psychology in recycling)

Warren Spring Laboratory,
Department of Industry,
P.O. Box 20,
Gunnels Wood Rd.,
Stevenage,
Hertfordshire SG1 2BX,
U.K.
(recovery from industrial and urban waste)

Washington State University,
Environmental Science Research Center,
Pullman,
Washington 99163.

Water Research Center,
Medmenham Laboratory,
P.O. Box 16,
Marlow,
Buckinghamshire SL7 2HD,
U.K.

Stevenage Laboratory,
Elder Way,
Stevenage,
Hertfordshire SG1 1TH,
U.K.

WCS International,
Santa Rosa,
California 95404.
(utilization of urban waste)

Western America Ore Co.,
P.O. Box 1237,
Coolidge,
Arizona 85228.
(ferrous wate utilization)

Westvaco Corp. Research Center,
Box 5207,
N. Charleston,
South Carolina 29406.

West Virginia University,
Coal Research Bureau,
Morgantown,
West Virginia 26505.
(utilization of fly ash)

College of Mineral and Energy Resources,
Morgantown,
West Virginia 26506.

Department of Chemical Engineering,
Morgantown,
West Virginia 26505.
(gasification of solid wastes)

Wheelabrator-Frye Inc.,
Mishawaka,
Indiana 46544.
New York,
New York 10017.
(urban waste utilization)

Worcester Polytechnic Institute,
Department of Chemical Engineering,
Worcester,
Massachusetts 01609.
(cellulose liquefaction)

APPENDIX 4

GLOSSARY OF SCIENTIFIC TERMS

Some terms that do not appear here may be included in the index and explained in the text.

ABS (acrylonitrile butadiene styrene) plastic or resin. A polymer blend of acrylonitrile (vinyl cyanide), butadiene, and styrene which combines the hardness and strength of a vinyl resin with the toughness and impact resistance of a rubber.

Accelerometer. A device that measures rate of acceleration or deceleration.

Actinomyces. Members of the bacterial family *Actinomycetaceae*, which are characterized by the formation of a true mycelium (a mass of filaments).

Activated carbon (activated charcoal). A powdered, granular, or pelletized form of amorphous carbon which has been treated to increase its adsoprtion ability, and which is characterized by a very large surface area per unit volume because of a large number of fine pores.

Activated sludge. Organic sludge (particularly sewage sludge) which has undergone an aerobic fermentation process; it is semiliquid and flocculent, with an earthy odor.

Addition polymer. A polymer such as polyethylene, polypropylene, or polystyrene formed by the chain addition of unsaturated monomer molecules without the elimination of other molecules being necessary.

Adsorbate. The gaseous or dissolved or suspended substance which is condensed in the form of a film of molecules on the surface of a solid or liquid such as charcoal, silica, water or mercury.

Adsorbent. A solid or liquid that adsorbs other substances.

Adsorption. A condensation in the form of a surface film of molecules on the surface of a solid or liquid (as opposed to *absorption,* which is the penetration of substances into the interior of a solid).

Aerobic bacteria. Bacteria requiring free oxygen for the metabolic breakdown of materials.

Aerosol. A suspension of very fine particles of a liquid or solid within a gas.

Aggregate (in the sense used here). The sands, gravels, and crushed stone used for mixing with cementing materials in the preparation of mortars and concretes.

Air classification (air elutriation, winnowing). Separation of materials according to mass and size by means of a flow of air.

Algae. General name for the chlorophyll-bearing organisms in a plant subkingdom characterized by the absence of specialized organs.

Aliphatic. Describing any organic compound of carbon and hydrogen with a chain of carbon atoms

Alum. A double sulfate of a trivalent metal such as aluminum, chromium, or iron, and a univalent ion such as potassium, sodium, or ammonium.

Alumino-silicate. A crystalline combination of silicate and aluminate.

Amino acid. A class of organic compounds that contain at least one carboxylic acid group and one basic amino group and which are polymerized to form peptides and proteins.

Anaerobic. Able to live without free (gaseous) oxygen; used to describe the process of fermentation in a closed, oxygenfree vessel, or bacteria that exist in the partial or complete absence of oxygen.

Anaerobic digester. A closed vessel with a controlled environment in which anaerobic fermentation takes place.

Anion. Negatively charged ion.

Annealing. Treating (a metal, alloy, glass, etc.) to remove internal stresses and so make less brittle, usually by heating and slow cooling.

(aq) (in chemical equations). Denoting an aqueous solution.

Aquaculture (aquiculture). Cultivation of fauna in water.

Aquifer. A subsurface zone that yields well water.

Autoclave. A strong, gas-tight vessel in which can be carried out heating under pressure.

Autotrophic. (Microorganism) capable of synthesizing organic nutrients directly from simple inorganic substances such as carbon dioxide and inorganic nitrogen.

Back-end recovery. Recovery of materials from urban waste after shredding, incineration, or other treatment; for example recovery of organic materials by composting or anaerobic digestion, inorganic materials such as iron and glass after incineration, or energy.

Bacteria. Extremely small, relatively simple microorganisms with a primitive nucleus in which the deoxyribonucleic acid-containing region lacks a limiting membrane.

Bagasse. Remains of sugarcane after the juice has been extracted by pressure between the rollers of a mill; frequently used as a fuel in the sugar mill.

Ballistics. Concerned with the motion and behavior of projectiles or thrown bodies.

Buffered (in the sense used here). Used to describe a solution containing chemical materials chosen to minimize changes in acidity or alkalinity which would otherwise occur.

(c) (in chemical equations). Denoting the solid, crystalline state.

Calcination. The process of heating to a high temperature; particularly the conversion of metals to their oxides by heating in air or the decomposition of ores by the volatilization of some components.

Carbohydrate. An organic compound of carbon, hydrogen, and oxygen with the ratio of hydrogen and oxygen atoms usually 2:1, and including sugars, starches, and cellulose.

Carbothermic reduction. Heating an oxide in an electric arc furnace in the presence of coke.

Casting. Forming a liquid or plastic substance into a fixed shape by allowing it to cool in a mold.

Catalyst. A substance that alters (usually increases) the rate at which a chemical reaction occurs, but which itself is not consumed in the reaction (or is regenerated) and which may be recovered essentially unaltered in nature and quantity at the end of the reaction.

Cation. Positively charged ion.

Cellulase. An extracellular enzyme produced during the growth of fungi, bacteria, insects, and other lower animals, that hydrolyzes cellulose.

Cellulolytic. Describing microorganisms that have the ability to hydrolyze cellulose.

Cellulose. Structural tissue of long-chain carbohydrates (polysaccharides of approximately 1000-10,000 sugar units) which forms the skeletal structure of plant cell walls.

Cement (in the sense used here). A dry powder made from silica, alumina, lime, iron oxide, and magnesia which, after being mixed with water, sets to a hard mass.

Cementation (in the sense used here). The precipitation of a metal from solution by the sacrificial dissolution of a less noble metal.

Centrifuge. A rotating device for separating liquids of different densities or for separating suspended colloidal particles according to size fraction by centrifugal force.

Ceramic. A product made by baking or firing a nonmetallic mineral.

Chemoautotrophs. Bacteria and protozoa that do not carry out photosynthesis and which can obtain energy by the oxidation of inorganic compounds.

Chemoheterotrophs. Organisms that require preformed organic compounds as sources of energy an carbon.

Classification (in the sense used here). Separation of materials by stratification in a flow of gas or liquid.

Clinker. Burnt or vitrified stony material formed, for example, by partial melting of ash in a furnace or incinerator.

Coal shale. The sedimentary rock material with which coal seams are associated together with some unseparated coal.

Colloidal state. An intimate mixture of a dispersed phase (colloid) uniformly distributed in a finely divided state through a dispersion medium (gas, liquid, or solid).

Complex (in the sense used here). A chemical compound in which some of the molecular bonds are formed by pairs of electrons supplied predominantly by one of the bound atoms.

Condensation polymer. A polymer formed by the linking of monomer molecules accompanied by the elimination of simple molecules.

Cryogenics. The production and maintenance of very low temperatures and the study of phenomena at these temperatures.

Cullet. Crushed glass used in the manufacture of new glass to assist the melting process.

Culm. Finely divided waste coal, separated by screening from larger pieces.

Dialysis. Separation of solutes on the basis of their unequal diffusion rates through a membrane, separating low molecular mass solutes from colloidal material and high molecular mass solutes.

Diatomaceous earth (kieselguhr, tripolite). A siliceous mineral composed of the fossilized skeletons of microscopic water creatures.

Disaccharide. A compound sugar which yields two monosaccharide units on hydrolysis.

Distillation. Heating a mixture so that one or more components vaporize and subsequently condense on a cooled surface.

Dross. An impurity (usually an oxide) floating on the surface of molten metals.

Ecology (environmental biology). The study of organisms and the relationships between them, considered in relation to their nonliving environment.

Eddy current. An electrical current which is induced in a conductor when the electromagnetic field through it is changing.

Electrode. Electrical conductor by which electrons enter or leave a medium (liquid, solid, gas, or vacuum).

Electrodialysis. Separation of cations and anions on the basis of a membrane able to pass selectively either cations or anions.

Electrolyte. A compound which conducts an electric current when molten or dissolved in a suitable solvent.

Electromagnetic radiation. A disturbance with electrical and magnetic components which propagates outward from any electric charge which experiences a change of velocity, and is associated with electromagnetic energy with a wide range of possible frequencies.

Element. A substance consisting entirely of atoms with the same chemical properties (the same atomic number).

Elutriation (elution). Purification by washing; separation by means of a water stream.

Endothermic reaction. A chemical process that absorbs heat.

Entropy. A thermodynamic function reflecting the extent of disorder in a system.

Enzyme. A catalytic protein produced by living cells, which increases the rate at which a chemical reaction occurs without itself being destroyed.

Ester. A compound formed by the bonding reaction of an alcohol and an organic acid with the elimination of water.

Eutectic mixture. A solid solution of two or more substances, having the lowest freezing point of all possible mixtures of those components.

Exothermic reaction. A chemical process that envolves heat.

Exponential curve. A graph of the function $y = a^x$, where a is a positive constant.

Facultative anaerobe. An anaerobic microorganism that can grow under aerobic conditions; it is characterized by two alternative mechanisms for obtaining energy, aerobic respiration and anaerobic fermentation.

Fat. An ester of glycerol with a fatty acid.

Fatty acid (aliphatic acid). An organic compound derived from a saturated aliphatic hydrocarbon and containing an acid group.

Feedlot wastes. Animal manure, hair, uneaten food, and washdown water collected from intensively penned stock feedlots.

Fermentation. A microbiological transformation of organic substances, especially carbohydrates, generally with the evolution of a gas.

Ferromagnetism. The strongly magnetic properties exhibited by certain metals, alloys, and compounds of the transition elements, rare earth elements, and actinide elements resulting from the spontaneous alignment of individual magnetic moments of large groups of atoms.

Ferrous. Pertaining to iron.

Floc. Small masses formed in a fluid by coagulation or biochemical reaction of fine suspended particles.

Flocculation. The aggregation of fine particles into larger, cloudlike, non-crystalline masses.

Fluidized bed. A device in which a gas, often air, is blown through the porous base of a container to float a bed of particles, causing them to act as a fluid.

Flux (in the sense used here). A substance used to promote the fusing (melting) of solids.

Fly ash. The very fine flue ash from pulverized coal furnaces which is collected in cyclones or electrostatic precipitators; also know as pulverized fuel ash, pfa.

Free radical. An atomic or molecular species that possesses at least one unpaired electron, and is therefore chemically reactive.

Front-end recovery. Recovery from urban waste of inorganic materials such as iron, aluminum, and glass and perhaps of some paper, before conventional shredding and incineration or other disposal process.

(Froth) flotation. Separation of a mixture by the action of bubbles rising through a liquid and buoying up solid particles, which can be removed as part of the froth.

Fuel cell. A device to convert chemical energy directly into electrical energy, in which the chemical reactants are stored outside the device.

Fumes. Aerosols resulting from pyrometallurgical operations and made up of solid particles generated by condensation from the gaseous state, generally after volatilization from molten metals and subsequent oxidation.

Fungi. Nucleated, spore-bearing organisms devoid of chlorophyll, and often filamentous.

(g) (in chemical equations). Denoting the gaseous state

Galactose. A monosaccharide (simple sugar) occurring as a constituent of plant and animal disaccharides and polysaccharides.

Garbage. Domestic food waste.

Glasphalt. Glass-containing asphalt.

Glass-ceramic. A glass which has been devitrified (crystallized) in a controlled manner so that it consists of a very large number of very small crystals.

Glucose. A monosaccharide which is the most commonly occurring sugar.

Hemicellulose (hexosan). A polysaccharide made up of less sugar units than cellulose (typically 50-150 units rather than the 1000-10,000 of cellulose) which is extractable by dilute alkaline solutions and is more susceptible than cellulose to attack by microorganisms.

Hexose. A monosaccharide (simple sugar) containing six carbon atoms in the molecule.

Home scrap (in-house scrap). The nonproduct output of an industry manufacturing a material such as steel or plastic.

Homogeneous. Uniform in composition or structure.

Hydrapulping (hydropulping). Shredding of organic materials and slurrying in water to assist the recovery of cellulose fibers and separation from inorganics.

Hydrogasification. Reaction of carbonaceous material with hydrogen gas at high temperatures and high pressures to form gaseous compounds; a technique to manufacture synthetic pipeline gas from coal.

Hydrolysis. Chemical reaction of a substance with water or with an aqueous solution.

Hydrometallurgy. Treatment of metals and metal-containing materials by solvent processes.

Hydrophilic. Describing a material that tends to interact with water.

Hydrophobic. Describing a material that tends not to interact with water.

Hydroponics. The cultivation of plants in nutrient solutions with the mechanical support of an inert medium such as a bed of gravel or sand which at intervals is flooded with the solution.

Incineration. Burning or oxidation as a method of waste treatment so that only ashes remain; the resulting heat may or may not be utilized.

In-house scrap (home scrap). The nonproduct output of an industry manufacturing a material such as steel or plastic.

Inorganic. Of mineral origin; not belonging to the large class of carbon compounds which are termed *organic*.

Ion. An atom or group of atoms with either fewer or more electrons than necessary for the atom or group to be electrically neutral.

Ion exchange. The reversible exchange of mobile ions of like charge between a solid (usually in the form of beads) and a liquid, or between two immiscible liquids.

Ionization. A process in which a neutral atom or molecule loses or gains electrons to acquire a net charge and became an ion.

Isomerization. The interconversion between two chemical compounds that have the same molecular formula but which differ in the sequence or relative arrangement of their atoms.

Kraft pulp (sulfate pulp). A wood pulp produced by a process in which sodium sulfate is used in the caustic soda digestion liquor.

(l) (in chemical equations). Denoting the liquid state.

Leaching. Dissolving into a suitable solvent a soluble constituent of a mixture.

Lignin. A three-dimensional polymer of aromatic phenylpropane units that acts as a cement between cellulose fibers in plant cell walls.

Ligninolytic. Describing microorganisms that can degrade lignin.

Lignosulfonic acid. A compound of lignin and sulfate formed during the pulping of woody tissue.

Lipid. An organic compound containing a long aliphatic hydrocarbon chain, for example fatty acids and alcohols, waxes, and fats.

Liquefaction (in the sense used here). Reaction of carbonaceous material with a reducing gas such as carbon monoxide at high temperatures and high pressures to form a heavy oil.

Magnetic susceptibility. Ratio of the intensity of magnetization produced in a substance to the intensity of the magnetic field to which it is subjected.

Mega (prefix). Million, 10^6.

Mesophilic. Describing microorganisms that thrive in the intermediate temperature range (generally 25-45°C).

Metabolite. Chemical product of a biological process.

Mold (in the microbiological sense). A woolly fungus growth.

Monomer. Chemical compound consisting of single molecules, which is capable of being built up by repeated union of monomer molecules to form a polymer.

Monosaccharide. A simple carbohydrate (a sugar) which cannot be hydrolyzed to a more simple carbohydrate.

Municipal waste. Urban waste (which see); sometimes used in a more restricted sense referring to office, shop, and street waste, but excluding domestic waste.

Nonferrous. Not including iron or its alloys or compounds.

Nuclear fission. The division of an atomic nucleus into parts of similar mass, with liberation of energy.

Nuclear fusion. The combination of atomic nuclei, with the liberation of energy.

Nucleic acid. A large, chainlike molecule containing phosphoric acid, sugar, and basic organic nitrogen groups,

Obligate anaerobe. A microorganism restricted to an anaerobic mechanism for obtaining energy.

Obsolete (applied to scrap or waste). Generated after production and use; postconsumer.

Open hearth furnace. A melting furnace with a shallow hearth and low roof in which the heating is by direct flame and by radiation from the roof and walls.

Organic. Comprised of chemical compounds containing carbon chains or rings combined with hydrogen, and often also with oxygen, nitrogen, and other elements.

Oxidation. The loss of electrons by an element or radical, for example by the addition of oxygen.

Paramagnetism. The property of substances which when placed in a magnetic field are magnetized parallel to the field to an extent proportional to the field; associated with small, positive magnetic susceptibilities.

Parasite. An organism that makes, at some stage in its life, some connection with the tissues of an individual of a different species (the host) from which it derives food.

Pathogen. A disease-producing organism.

Pectin. A carbohydrate obtained from the inner portion of the rinds of citrus fruits.

Pentose. A monosaccharide (simple sugar) containing five carbon atoms in the molecule.

pH. A measure of the hydrogen ion concentration in a solution, that is the acidity or alkalinity. A value of 1 indicates high acidity, and a solution with a pH of 14 is highly alkaline.

Phosphogypsum. Gypsum (calcium sulfate dihydrate) which is a by-product in the treatment of rock phosphate with sulfuric acid to manufacture fertilizer.

Photolysis. The initiation of chemical reaction by light.

Photosynthesis. The production by plants of carbohydrates such as glucose from carbon dioxide and water, liberating oxygen, using energy obtained from light with the aid of the green pigment chlorophyll.

Pickle liquor. Spent pickling solution, usually containing sulfuric acid, which has been used to remove oxide scale from metals.

Pigment (in the sense used here). A colored substance, which when applied to a surface, colors it (but is distinct from a dye which penetrates the fibers or tissues of the surface).

Plasma (in physics). A highly ionized gas which contains equal numbers of ions and electrons.

Plastic (in the sense used here). A polymeric material, usually organic, of large molecular mass, which can be shaped by causing it to flow under pressure.

Plasticizer (flexibilizer). Substance added to an otherwise rigid material to improve its plastic properties (softness and flexibility).

Polyelectrolyte. A natural or syntheitc electrolyte of high molecular mass (for example, proteins and polysaccharides), which on dissociating in solution has ions of one sign bound to the polymer and ions of the other sign diffusing through the solution.

Polymer. A material consisting of molecules made up of a number of individual (monomer) units.

Polysaccharide. A carbohydrate with molecules composed of many monosaccharide units.

Polyvalent. An atomic or molecular species with a valence other than one.

Pozzolan. A material that by itself is not a cement, but which when it reacts with calcium hydroxide or similar metal hydroxides in the presence of water, forms a cement.

Precipitate. An insoluble substance formed in a solution as the result of a chemical reaction or physical change.

Producer gas. A gas high in carbon monoxide and hydrogen formed by burning solid fuel in limited oxygen or by passing a mixture of air and steam through a bed of incandescent carbon.

Prompt scrap. The nonproduct output of a fabrication operation in which intermediate products are converted into consumer products.

Protein. High molecular mass condensation polymers of amino acids which form a large part of all living matter.

Pulverized fuel ash (pfa, fly ash). The very fine flue ash from pulverized coal furnaces which is collected in cyclones or electrostatic precipitators.

Pyrolysis (destructive distillation). Chemical decomposition of a carbonaceous material by heat in the absence of oxygen or in a controlled oxygen atmosphere.

Pyrometallurgy. High-temperature metal extraction processes.

Radical (in chemistry). A group of atoms that remain bonded together while undergoing reactions as a group.

Reactor. A vessel with a controlled environment in which a chemical or biochemical reaction takes place.

Recycling. The process of utilizing one or more of the components (including chemical energy) of discarded or waste material.

Reduction (in chemistry). A reaction in which atoms of an element gain electrons.

Refractory. Describing a material which is not damaged by heating to high temperatures.

Refuse (in the sense used here). Solid waste.

Residual (in the sense used here). A nonproduct output of an operation with zero price in the current market.

Residuals management. Waste treatment, pollution control.

Reverse osmosis. The use of a pressure gradient on a solution to cause only the solvent to pass through a semipermeable membrane, thus concentrating the solute on the high-pressure side.

Rubbish (in the sense used here). Domestic nonfood waste.

Ruminant animal. An animal possessing a rumen, an anaerobic digester in which the cellulose is broken down with the assistance of enzymes produced by microorganisms.

(s) (in chemical equations). Denoting the sorbed state.

Salvage. The extraction of homogeneous waste material from a mixture of wastes.

Saprophyte. A microorganism capable of feeding only on soluble matter from dead plants and animals.

Semipermeable membrane. A membrane allowing the passage of some substances (generally solvents) but not of others (generally solutes).

Single-cell protein (SCP). Animal feed and human food protein derived from single-cell microorganisms grown on various resources and wastes (a term coined to avoid words such as "bacterial" or "microbial" which have unpleasant connotations).

Sintered. Describing solid particles that have been heated so that they have partially fused together into a coherent, bonded mass without complete melting.

Sitall. A Russian term for glass-ceramic, that is a glass that has been vitrified (crystallized) in a controlled manner so that it consists of a very large number of very small crystals.

Slag. Liquid or vitrified (glassy) nonmetallic component of metal smelting operation; a complex mixture of molten or vitreous salts.

Slagging incineration (high-temperature incineration, HTI). Incineration carried out at a temperature sufficiently high to melt the noncombustible components.

Solvolysis. The chemical reaction of a substance with the solvent in which it is dissolved.

Specular reflection. Perfect or regular reflection of electromagnetic radiation which occurs when the reflecting surface is flat to within approximately one-eighth of a wave length of the radiation, as in the reflection of visible radiation by a mirror.

Starch. A carbohydrate (polysaccharide) of the general composition $(C_6H_{10}O_5)_n$, occurring as organized granules in many plant cells.

Superalloy. A high tensile strength alloy for use at elevated temperatures, usually containing a high proportion of chromium and nickel.

Surface-active agent (surfactant). A soluble compound that reduces the interfacial tension between liquid and gas, liquid and liquid, or liquid and solid.

Synthesis gas (in the sense used here). A gas consisting of a mixture of carbon monoxide and hydrogen formed by the high-temperature reaction of steam with carbon.

System. The assemblage of components being considered, as distinct from the surrounding components.

Tailings. Wastes generated in the extraction of ores during mining and similar operations.

Terrazzo. A mosaic surface made by embedding chips of marble, granite, or similar material in mortar, and polishing the surface.

Terminal waste treatment. Waste treatment after collection by local bodies (municipalities) or their agents.

Thermodynamics. The study of the general laws governing the properties of matter and the processes which involve energy change.

Thermophilic. Describing microorganisms that thrive at high temperatures (generally 45-75°C).

Thermoplastic resin. A material that repeatedly softens on heating and hardens on cooling.

Thermosetting resin. A material which hardens irreversibly on heating and which cannot be melted or remolded without destruction of its characteristics.

Ton (long ton, as used in U.K.). 2240 pounds, 1.01605 (metric) tonnes.

Ton (short ton, as used in U.S.). 2000 pounds, 0.90719 (metric) tonne.

Tonne (metric). A unit of mass equal to 1000 kilograms or 2204.62 pounds.

Ultrafiltration. Separation of large dissolved molecules or suspended particles from the solvent on the basis of their inability to pass through membrane pores while the solvent passes under a pressure gradient.

Unsaturated compound. A chemical compound with more than one bond between adjacent atoms (usually carbon atoms) and therefore reactive to the addition of other atoms.

Urban waste. All solid waste collected from private dwellings, commercial premises, trade, and some industrial premises by local body or municipal collection services or by private contractors acting on their behalf.

Valence. In chemistry, a number that characterizes the combining power of an atom or group of atoms.

Vitreous. Glassy.

Waste. Solid, liquid, or gaseous material discarded during community activities.

Water gas. A fuel gas consisting of carbon monoxide (together with some hydrogen and carbon dioxide) formed by the action of steam on hot carbonaceous material.

Wax. Any substance resembling beeswax and distinguished by a composition of esters and higher alcohols and the absence of fatty acids.

Winnowing (air elutriation, air classification). Separation of objects of lower density and/or higher air resistance which are entrained in a rising air stream, from those with higher density and/or lower air resistance.

Yeast. A collective name for those fungi that usually contain a vegetative body incorporating simple cells.

APPENDIX 5

BIBLIOGRAPHY

The institutions with which the authors were affiliated at the time the reports were written are included here in parentheses in the citations to facilitate exchange of further information; full addresses of some institutions are collected in Appendix 3. When a reference includes an experession such as "in ANERAC, 1971," the full bibliographic details will be found in this bibliography in the appropriate alphabetical position (under "ANERAC" in this example).

Abert, James G., Harvey Alter, and J. Frank Bernheisel (1974), "The Economics of Resource Recovery from Municipal Solid Waste," *Science,* **183,** 1052-1058. (National Center for Resource Recovery, Washington, D.C.)

Abrahams, John H., Jr. (1971), "Utilization of Packaging Wastes," in ANERAC, 1971, pp. 150-161. (Glass Container Manufacturers' Institute, Inc., Washington, D.C.)

Abrahams, John H., Jr. (1973), "Road Surfacing with Waste Glass," in Albuquerque, 1973, pp. 14-34. (Glass Container Manufacturers' Institute, Inc., Washington, D.C.)

ACI (1973), *The First National Survey of Community Solid Waste Practices, Australia, 1972-1973,* Australian Consolidated Industries Ltd; *see also* summary by Varjavandi and Fischof, 1975.

ACS (1969), *Cleaning Our Environment: The Chemical Basis for Action,* American Chemical Society, Washington, D.C.

ACS (1971), *Solid Wastes* (reprints from *Environmental Science and Technology),* American Chemical Society, Washington, D.C.

ACS (1973), *Solid Wastes II* (reprints from *Environmental Science and Technology*), ed. S. S. Miller, American Chemical Society, Washington, D.C.

ACS (1976), "Sulfur Removal and Recovery," *Advances in Chemistry* (139), American Chemical Society, Washington, D.C.

Adami, A., and M. J. Ridge (1968a), "Observations on Calcium Sulphate Dihydrate Formed in Media Rich in Phosphoric Acid. I. Precipitation of Calcium Sulphate Dihydrate," *Journal of Applied Chemistry,* **18,** 361-363. (Division of Building Research, CSIRO, Australia)

Adami, A., and M. J. Ridge (1968b), "Observations on Calcium Sulphate Dihydrate Formed in Media Rich in Phosphoric Acid. II. Derived Calcium Sulphate Hemihydrate," *Journal of Applied Chemistry,* **18,** 363-365. (Division of Building Research, CSIRO, Australia)

AERE (1976), *Membrane Separation Processes,* Atomic Energy Research Establishment, Harwell, U.K. (Details of reverse osmosis and ultrafiltration techniques developed by AERE and Paterson Candy International Ltd; available from K.W. Carley-Macaulay, Process Technology Division, B 353, AERE)

Aguirre, Francisco, Oscar Maldonado, Carlos Rolz, Juan F. Menchú, Rodolfo Espinos, and S. de Cabrera (1976), "Protein from Waste: Growing Fungi on Coffee Waste," *ChemTech,* **6**(10), 636-642. (Applied Research Division, Central America Research Institute for Industry, P.O. Box 1552, Guatemala City, Guatemala)

AIChE (1973), 73rd National Meeting, American Institute of Chemical Engineers, Minneapolis, Minnesota, 27-30 August 1972; *AIChE Symposium Series,* **69** (133).

Aïtcin, Pierre Claude, and Claude Poulin (1974), "Bubble Aggregates," in Aleshin, 1974, pp. 265-268. (Department of Civil Engineering, University of Sherbrooke, Quebec)

Albrecht, Oscar W. (1976), "Shipping Wastes to Useful Places," *Environmental Science and Technology,* **10** (5), 440-442. (U.S. EPA, Cincinnati, Ohio)

Albuquerque (1973), *Proceedings of Symposium on Utilization of Waste Glass in Secondary Products,* Albuquerque, New Mexico, 24-25 January, Report NASA-CR-135792.

Aleshin, Eugene (1974), Ed., *Proceedings of the Fourth Mineral Waste Utilization Symposium,* Chicago, 7-8 May, U.S. Bureau of Mines and IIT Research Institute, Chicago, Illinois.

Aleshin, Eugene (1976), Ed., *Proceedings of the Fifth Mineral Waste Utilization Symposium,* Chicago, 13-14 April, U.S. Bureau of Mines and IIT Research Institute, Chicago, Illinois.

Alexander, W. O. (1976), "New Materials Created from Waste," in Henstock, 1975a. (University of Aston, U.K.)

Allaby, Michael (1976), "Home-Made Paper," *New Scientist,* **71**(1019), 652.

Allan, G. G., et al. (1973), *Wood Waste Reuse in Controlled Release Pesticides,* U.S. Environmental Protection Agency; U.S. NTIS PB-222,051.

Allison, Ian T. (1975), "Total Resource Recovery and Fuel Production from Municipal Solid Waste," in Johnston, 1975, pp. T92-T94. (Waste Control Science, Santa Rosa, California)

Alpert, S. B., F. A. Ferguson, et al. (1972), *Pyrolysis of Solid Waste: A Technical and Economic Assessment,* Stanford Research Institute Report WVU-ENG-CHE-73-01 for West Virginia University; U.S. NTIS PB-218,231.

Alter, H., and E. Horowitz (1975), Ed., *Resource Recovery and Utilization,* American Society for Testing and Materials, Proceedings of National Materials Conservation Symposium and Workshop on Resource Recovery and Utilization, April-May, 1974, National Bureau of Standards, Gaithersburg, Maryland.

Alter, Harvey, and George W. Ingle (1974), "Coatings Wastes as Energy Sources," *American Paint and Coatings Journal,* **58**(54), 58-63. (National Center for Resources Recovery, Washington, D.C.; and Monsanto Co., U.S.A.)

Alter, H., and W. R. Reeves (1975), *Specifications for Materials Recovered from*

Municipal Refuse, U.S. Environmental Protection Agency; U.S. NTIS PB-242,540. (National Center for Resources Recovery, Inc.)

Alter, Harvey, Stuart L. Natof, Kenneth L. Woodruff, Wilfred L. Freyberger, and Ellery L. Michaels (1974), "Classification and Concentration of Municipal Solid Waste," in Aleshin, 1974, pp. 70-76. (National Center for Resources Recovery, Inc., Washington, D.C.; and Institute for Mineral Research, Michigan Technological University, Houghton)

Alter, Harvey, Stuart Natof, and Lee C. Blayden (1976a), "Pilot Studies Processing MSW and Recovery of Aluminum Using an Eddy Current Separator," in Aleshin, 1976, pp. 161-168. (National Center for Resources Recovery, Inc., Washington D.C.; and Aluminum Company of America Laboratories, Pennsylvania)

Alter, Harvey, Kenneth L. Woodruff, Abraham Fookson, and Brian Rogers (1976b), "Analysis of Newsprint Recovered from Mixed Municipal Waste," *Resource Recovery and Conservation,* **2**(1), 79-84. (NCRR, Washington, D.C.; Resource Recovery Services, Woodbridge, New Jersey; and Gilette Research Institute, Rockville, Maryland 20850)

American Metal Market (1970), "Recycling: Practical Answer to the Problems of Air Pollution, Water Pollution, Solid Waste," special issue, March 16.

Anbar, M., D. F. McMillen, and R. D. Weaver (1975), "Electrochemical Power Generation Using a Liquid Lead Electrode as a Catalyst for the Oxidation of Carbonaceous Fuels," *Record Intersociety Energy Conversion Engineering Conference,* 10th, 48-55. (Stanford Research Institute, Menlo Park, California)

Anderson, Larry L. (1972), *Energy Potential from Organic Wastes: A Review of the Quantities and Sources,* Bureau of Mines Information Circular 8549, U.S. Department of the Interior, Washington, D.C.

Anderson, Earl V. (1976), "Who Pays Is Key to Helium's Future," *Chemical and Engineering News,* **54**(50), 11-13.

Andren, Robert K., Mary H. Mandels, and John E. Medeiros (1975), "Production of Sugars from Waste Cellulose by Enzymatic Hydrolysis. I. Primary Evaluation of Substrates," in APS, 1975, pp. 205-219. (U.S. Army Natick Development Center, Massachusetts)

ANERAC (1971), *Reuse and Recycle of Wastes,* Proceedings, 3rd Annual North Eastern Regional Antipollution Conference, Rhode Is., 21-23 July, 1970, Technomic Publishing Co., Westport, Connecticut.

ANERAC (1976), *Energy from Solid Waste Utilization,* Proceedings, 6th Annual North Eastern Regional Antipollution Conference, University of Rhode Is., Kingston, Rhode Is., 8-9 July, 1975, ed. Stanley M. Barnett, Donald Sussman, and C. J. Wilson, Technomic Publishing Co., Wesport, Connecticut.

Anyos, Tom, Dean B. Parkinson, Irvin A. Illing, and Joseph G. Berke (1973), "A Novel Thermoplastic Paving Material," in Albuquerque, 1973, pp. 188-197. (Stanford Research Institute, California).

Appell, Herbert R., and Ronald D. Miller (1973), "Fuel from Agricultural Wastes," in Inglett, 1973, Chapter 8, pp. 84-92. (Pittsburgh Energy Research Center, U.S. Bureau of Mines)

Appell, Herbert R., Irving Wender, and Ronald D. Miller (1971), "Conversion of Municipal Refuse to Oil," in ANERAC, 1971, pp. 225-231. (Pittsburgh Energy Research Center, U.S. Bureau of Mines)

Appell, H. R., Y. C. Fu, E. G. Illig, F. W. Steffgen, and R. D. Miller (1975), *Conversion of Cellulosic Wastes to Oil,* Bureau of Mines Report of Investigations 8013, U.S. Department of the Interior, Washington, D. C. (Pittsburgh Energy Research Center)

Appleyard, Colin J., and Martin G. Shaw (1974), "Reuse and Recycle of Water in Industry," *Chemistry and Industry* (6), 240-246. (Bostock Hill and Rigby Ltd, Birmingham, U.K.)

APR (1975), *Waste Oil. Headache or Resource?* Proceedings of the Second International Conference on Waste Oil Recovery and Reuse, Cleveland, Ohio, February 24-26, Association of Petroleum Re-refiners, Washington, D. C.

APR (undated), *Lubricating Oil Never Wears Out: It Just Gets Dirty,* Association of Petroleum Re-refiners, Washington, D.C.

APS (1975), "Proceedings of the Eighth Cellulose Conference," Syracuse, New York, 19-23 May, *Applied Polymer Symposia,* **28,** ed. T. E. Timell, Part I, "Wood Chemicals—A Future Challenge."

APS (1976), *ibid.*, Part II, "Complete-Tree Utilization," and "Biosynthesis and Supermolecular Structure of Cellulose."

Arden, T. V. (1976), "Ion Exchange," *Chemistry in Britain,* **12**(9), 285-288. (Portals Water Treatment Ltd, Maidenhead, Berkshire, U.K.)

Arella, D. G. (1974), *Recovering Resources from Solid Waste Using Wet-Processing: EPA's Franklin, Ohio, Demonstration Project,* Publication SW-47d, U.S. Environmental Protection Agency.

Arkles, Barry (1973), "Recycling Poly(tetrafluoroethylene)," *Polymer Science and Technology,* **3,** 121-137. (Liquid Nitrogen Processing Corporation, 412 King St., Malvern, Pennsylvania 19355)

Arnold, Lionel K. (1975a), "The Commercial Utilization of Cornstalks," in Mantell, 1975, pp. 377-391. (Department of Chemical Engineering, Iowa State University, Ames)

Arnold, Lionel K. (1975b), "The Commercial Utilization of Corncobs," *ibid.*, pp. 393-401.

Ashworth, Gavin A., Peter J. Ayre, and Robert E. W. Jansson (1975), "Production of Metal Powders in a Pump Cell," *Chemistry and Industry* (London), 3 May, 382-383. (Department of Chemistry, University of Southampton, U.K.)

Asimov, Isaac (1956), "The Last Question," *Science Fiction Quarterly,* November, Columbia Publications; republished in *Nine Tomorrows,* Dennis Dobson, London, 1963.

ASME (1975), *Resource Recovery,* Proceedings 15th Annual Symposium, American Society of Mechanical Engineers, 6-7 March, Albuquerque, New Mexico.

Auchter, Richard J. (1971), "Future Wood Needs for Papermaking Fibers Should Not Be a Problem," *Paper Trade Journal,* 23 August, 58-61; presented in Miami, 1971. (Forest Products Laboratory, USDA, Madison, Wisconsin)

Auchter, Richard J. (1973), "Recycling Forest Products Retrieved from Urban Waste," *Forest Products Journal,* **23**(2), 12-16. (Forest Products Laboratory, U.S. Department of Agriculture, Madison, Wisconsin)

Augenstein, D. C., D. L. Wise, R. L. Wentworth, and C. L. Cooney (1976), "Fuel Gas Recovery from Controlled Landfilling of Municipal Wastes," *Resource Recovery and Conservation,* **2,** 103-117. (Dynatech Research and Development Co., Cambridge, Massachusetts; and M. I. T., Cambridge, Massachusetts)

Avila, A. J., H. A. Sauer, T. J. Miller, and R. E. Jaeger (1973), "Freeze Drying of Spent Plating and Etching Baths to Recover Metals," *Plating,* **60**(3), 239-241; presented at 59th Annual Technical Conference, American Electroplaters' Society, 21 June 1972, Cleveland, Ohio. (Western Electric Co., Chicago, Illinois 60623; and Bell Telephone Laboratories, Inc., Murray Hill, New Jersey 07974)

AWWA (1971), *Water Quality and Treatment* (A Handbook of Public Water Supplies), The American Water Works Association, Inc., 3rd ed., McGraw Hill, New York.

Bailey, D. A., K. Jones, and Charmaine Mitchell (1974), "The Reclamation of Water from Sewage Effluents by Reverse Osmosis," *Water Pollution Control,* **73**(4), 353-366. (Water Pollution Research Laboratory, Department of the Environment, U.K.)

Bailie, Richard C. (1971), "Wasted Solids as an Energy Resource," in American Association for the Advancement of Science Symposium, *The Energy Crisis: Some Implications and Alternatives*. (Department of Chemical Engineering, West Virginia University)

Bailie, Richard C., and Seymour Alpert (1973), "Conversion of Municipal Waste to a Substitute Fuel," *Public Works,* August, 76-79, 98. (Department of Chemical Engineering, West Virginia University; and Stanford Research Institute, Menlo Park, California)

Bailie, Richard C., and Robert S. Burton, III (1972), "Fluid Bed Reactors in Solid Waste Treatment," *American Institute of Chemical Engineers Symposium Series,* **68**(122), 140-151. (Department of Chemical Engineering, West Virginia University)

Bailie, R. C., and D. M. Doner (1975), "Evaluation of the Efficiency of Energy Resource Recovery Systems," *Resource Recovery and Conservation,* **1,** 177-187. (Department of Chemical Engineering, West Virginia University)

Bailie, Richard C., and Masaru Ishida (1972), "Gasification of Solid Waste Materials in Fluidized Beds," *American Institute of Chemical Engineers Symposium Series,* **68**(122), 73-80. (Department of Chemical Engineering, West Virginia University)

Balas, A. (1974), "Use and Recovery of Chromium in the Tannery," *Technicuir,* **8**(10), 152-155 (French). (Tanneries Grosjean, France)

Ball, F. J., G. N. Brown, J. E. Davis, A. J. Repik, and S. L. Torrence (1972), "Recovery of Sulfur Dioxide from Stack Gases as Elemental Sulfur by a Dry, Fluidized Activated Carbon Process," in Jimeson and Spindt, 1973, Chapter 16, pp. 183-194. (Westvaco Corp., North Charleston, South Carolina)

Baloh, Anton (1976), "Reverse Osmosis in the Sugar Industry," *Sugar Journal,* **38**(12), 19-25.

Barbour, James F., Robert R. Groner, and Virgil H. Freed (1974), *The Chemical Con-*

version of Solid Wastes to Useful Products, Report EPA-670-2-74-027, U.S. Environmental Protection Agency, Washington, D.C.; U.S. NTIS PB-233,178; summary in Jackson, 1975. (Department of Agricultural Chemistry, Oregon State University, Corvallis)

Bard, C. C., H. C. Baden, and S. E. Vincent (1975), "Method for Recovery and Reuse of Iron Cyanides Used in Photographic Processing," *Research Disclosure,* (134), 46-47, No. 13454.

Barrett, D., "Pyrolysis of Organic Wastes," in Kirov, 1975a, pp. 203-206. (Department of Fuel Technology, University of New South Wales, Australia)

Barter, M. A. (1975), "Recycling in the Zinc Industry," in Johnston, 1975, pp. T19-T22. (Electrolytic Zinc Co., Melbourne Australia)

Barton, William R. (1967), "Raw Materials for Manufacture of Cement," in Faber et al., 1967, pp. 46-51. (U.S. Bureau of Mines, Washington, D.C.)

Barton, A.F.M. (1974), *The Dynamic Liquid State,* Longman, London.

Barton, Allan F. M. (1975), "Solubility Parameters," *Chemical Reviews,* **75**(6), 731-753. (Victoria University of Wellington, New Zealand; currently Murdoch University, Western Australia)

Baum, Bernard, and Charles H. Parker (1974a), *Solid Waste Disposal,* Vol. 1, *Incineration and Landfill,* Ann Arbor Science, Ann Arbor, Michigan.

Baum, Bernard, and Charles H. Parker (1974b), *Solid Waste Disposal,* Vol. 2, *Recycling and Pyrolysis,* Ann Arbor Science, Ann Arbor, Michigan.

Beal, Charles Francis, and A. E. Haselar (1972), *Waste into Profit,* Faculty of Building, Boreham Wood, England, presented to the Royal Society, London.

Beesch, Samuel C., and Fred W. Tanner, Jr. (1974), "Industrial Fermentation Processes," in Kent, 1974, Chapter 7, pp. 156-192.

Bell, P. R., and J. J. Varjavandi (1975), "Pyrolysis—Resource Recovery from Solid Waste," in Kirov, 1975a, pp. 207-210. (A.C.I. Technical Centre Pty Ltd., Australia)

Bellamy, W. D. (1973a), "Use of Thermophilic Microorganisms for the Recycling of Cellulosic Wastes," in AIChE, 1973, pp. 138-140; General Electric Co. Reprint 7470. (General Electric Co., Schenectady, New York)

Bellamy, W. Dexter (1973b), "Conversion of Insoluble Agricultural Wastes to SCP by Thermophilic Microorganisms," International Conference on SCP at MIT, Cambridge, Massachusetts, 30 May. (General Electric Co., Schenectady, New York)

Beller, Morris, and Meyer Steinberg (1973), "Glass-Polymer Composites," in Albuquerque, 1973, pp. 250-276. (Department of Applied Science, Brookhaven National Laboratory, Upton, New York (1973).

Bennett, G. F., and L. Lash (1974), "Industrial Waste Disposal Made Profitable," *Chemical Engineering Progress,* **70**(2), 75-85. (University of Toledo, Ohio; and Envirotech Corp., Salt Lake City, Utah)

Bent, Henry A. (1971), "Haste Makes Waste. Pollution and Entropy," *Chemistry,* **44**(9), 6-15. (North Carolina State University, Raleigh)

Beretka, J. (1973), "Industrial Wastes and By-Products Generated in Australia: The

Results of a Survey,'' *Proceedings of the Royal Australian Chemical Institute,* **40**, 357-362. (Division of Building Research, CSIRO, Australia)

Beretka, J., and T. Brown (1974), ''The Utilization of Manganiferous Industrial Wastes and By-Products in the Manufacture of Coloured Bricks,'' *Journal of the Australian Ceramic Society,* **10**(1), 4-7. (Division of Building Research, CSIRO, Australia)

Berry, R. Stephen (1971), ''The Option for Survival,'' *Science and Public Affairs (Bulletin of the Atomic Scientists)*, **27**(5), 22-27. (Department of Chemistry, University of Chicago)

Berry, R. Stephen (1972), ''Recycling, Thermodynamics, and Environmental Thrift,'' *Science and Public Affairs* (*Bulletin of the Atomic Scientists*), **28**(5), 8-15. (Department of Chemistry, University of Chicago)

Berry, R. Stephen, and Margaret Fulton Fels (1972), *The Production and Consumption of Automobiles. An Energy Analysis of the Manufacture, Discard, and Reuse of the Automobile and Its Component Materials,* Report to Illinois Institute for Environmental Quality. (Department of Chemistry, University of Chicago)

Berry, R. Stephen, and Margaret F. Fels (1973), ''The Energy Cost of Automobiles,'' *Science and Public Affairs (Bulletin of the Atomic Scientists),* December, 11-17, 58-60. (Department of Chemistry, University of Chicago)

Berry, R. Stephen, and Hiro Makino (1974), ''Energy Thrift in Packaging and Marketing,'' *Technology Review,* **76**(4). (Department of Chemistry, University of Chicago)

Bessant, Kenneth H. C., and J. J. Peter Staudinger (1973), ''The Role of Plastics in the Conservation of Resources,'' *Chemistry and Industry* (12), 548-551. (Bessant: BP Chemicals International Ltd, Devonshire House, Mayfair Place, Piccadilly, London W1X 6AY, U.K.)

Besselievre, Edmund, and Max Schwartz (1976), *The Treatment of Industrial Wastes,* McGraw-Hill, New York, 2nd ed.

Biddlestone, A. J., and K. R. Gray (1973), ''Composting—Application to Municipal and Farm Wastes,'' in CE, 1973, pp. 76-80. (Department of Chemical Engineering, University of Birmingham, U.K.)

Brigg, A.C.T. (1970), ''Copper-Lead Scrap Separation,'' *Industrial Recovery,* **16,** 29-31. (Warren Spring Laboratory, Stevenage, U.K.)

Bilbrey, J. H., Jr. (1974), ''Use of Cryogenics in Scrap Processing,'' in Aleshin, 1974, pp. 424-430. (Salt Lake City Metallurgy Research Center, U.S. Bureau of Mines)

Bingham, Tayler H., and Paul F. Mulligan (1972), *The Beverage Container Problem: Analysis and Recommendations,* Report for Environmental Protection Agency, EPA-R2-72-059; U.S. NTIS PB-213,341. (Research Triangle Institute, North Carolina)

Birch, G. G., K. J. Parker, and J. T. Worgan (1976), Eds, *Food from Waste,* proceedings of an industry-university cooperation symposium, National College of Food Technology, University of Reading, U.K., 7-9 April, 1975, Applied Science Publishers Ltd, Barking, U.K.

Blatt, W. F. (1976), ''Principles and Practice of Ultrafiltration,'' in Meares, 1976, Chapter 3, pp. 81-120.

Blaustein, Bernard D., and John S. Tosh (1975), "Energy from Wastes: Solid, Liquid; and Gaseous Fuels," *Annual Symposium on Coal Gasification, Liquefaction, and Utilization, Proceedings,* **2,** XIII. (Pittsburgh Energy Research Center, U.S. Energy Research and Development Adm., Pittsburgh)

Blayden, L. C. (1974a), "Recycling Aluminum from Consumer Waste," Erie County and City of Buffalo Resources Recovery Seminar, 10 June. (Alcoa, Pittsburgh, Pennsylvania)

Blayden, Lee C. (1974b), "The Chemistry of Recycling Aluminum," 78th National Meeting, American Institute of Chemical Engineers, 20 August, Salt Lake City, Utah. (Alcoa, Pittsburgh, Pennsylvania)

Blesing, N. V. (1971), "Ion Exchange Applications in Desalination and Hydrometallurgy," *Bulletin of the Australian Mineral Development Laboratories,* (12), October, 23-44. (AMDEL, South Australia)

Blunden, J. R., C. G. Down, and J. Stocks (1974), "The Economic Utilization of Quarry and Mines Wastes for Amenity Purposes in Britain," in Aleshin, 1974, pp. 255-264. (Royal School of Mines, Imperial College of Science and Technology, London)

Bober, T. W., and R. B. Leon (1975), "Recovering Metals from Waste Photographic Processing Solutions," *Research Disclosures,* **137,** 5-6.

Boden, P. J. (1975), "Plating and ECM Sludges as Raw Materials," in Henstock, 1975a, (University of Nottingham, U.K.)

Bodle, William W., and Kirit C. Vyas (1974), "Clean Fuels From Coal," *Oil and Gas Journal,* 26 August, 73-88. (Institute of Gas Technology, Chicago, Illinois)

Boettcher, R. A. (1972), "Air Classification of Solid Wastes; Performance of Experimental Units and Potential Applications for Solid Waste Reclamation," U.S. Environmental Protection Agency Solid Waste Management Report SW-30c; U.S. NTIS PB-214,133; summary in Jackson, 1975. (Stanford Research Institute, California)

Bogacz, Aleksander, Jerzy Gospos, Wojciech Szklarski, and Bozena Ziolek (1975), "The Extraction of Copper and Accompanying Metals from Liquid Slags with Fused Melts and the Electrodeposition of Metals from Salt Melts" (Polish), *Prace Naukowe Instytutu Chemii Nieorganicznej i Metalurgii Pierwiastkow Rzadkich Politechniki Wroclawskiej,* **27,** 57-75; *Chemical Abstracts,* **83,** 119036. (Breslau Polytechnic, Poland)

Bohn, Hinrich (1971), "Methane From Waste, *Environmental Science and Technology,* **5**(7), 573. (Department of Agricultural Chemistry and Soils, University of Arizona, Tucson)

Bolto, B. A. (1976), "Desalination by Thermally Regenerable Ion Exchange. Part I," *Proccedings of the Royal Australian Chemical Institute,* **43**(11), 345-349. (Division of Chemical Technology, CSIRO, South Melbourne, Australia)

Bolto, B.A., K. W. V. Cross, R. J. Eldridge, E. A. Swinton, and D. E. Weiss (1974), "Magnetic Filter Aids," *Filtration and Separation,* **11,** 461-464. (Division of Chemical Technology, CSIRO, South Melbourne, Australia)

Bond, Richard G., and Conrad P. Straub (1973), *Handbook of Environmental Control,* Vol. 2, *Solid Waste,* Chemical Rubber Co., Cleveland, Ohio.

Bootle, K. R. (1972), "Can We Make Better Use of Sawmill Waste?"in Kirov, 1972, pp. 191-196. (Division of Wood Technology, Forestry Commission of New South Wales, Australia)

Boulding, Kenneth E. (1972), "The Economics of the Coming Spaceship Earth," in Toffler, 1972, pp. 235-243.

Bourne, Arthur (1972), *Pollute and Be Damned,* Dent, London.

Bowen J. R. (1975), "Pyrolysis of Garbage—the Landgard Solid Waste Disposal System," in Johnston, 1975, pp. T95-T97. (Monsanto Australia Ltd.)

Bowling, K. McG., D. H. Philipp, and H. Rottendorf (1972), "Coal Derivatives for Effluent Treatment," in Kirov, 1972, pp. 143-149. (Division of Mineral Chemistry, CSIRO, North Ryde, N.S.W., Australia)

Boyd, James (1976), "A National Policy Toward Recycling," (discussion of final report of the National Commission on Materials Policy, "Material Needs and the Environment: Today and Tomorrow," June 1973), *Environmental Science and Technology,* **10**(5), 422-424. (Materials Associates, Washington D.C.,)

Brachi, Philip (1974), "Sun on the Roof," *New Scientist,* **63** (915), 712-714.

Brackett, C. E. (1967), "Availability, Quality, and Present Utilization of Fuel Ash," in Faber et al., 1967, pp. 16-36. (Southern Electric Generating Co., Birmingham, Alabama)

Brackett, C. E. (1970), "Production and Utilization of Ash in the United States," in Bureau of Mines Information Circular 8488, U.S. Department of the Interior, Washington, D.C., pp. 11-16.

Brand, B. G. (1974), *Scrap Rubber Tire Utilization in Road Dressings,* U.S. Environmental Protection Agency; U.S. NTIS PB-232, 559.

Brandon, Craig A. (1976), "Reuse of Total Composite Wastewater Renovated by Hyperfiltration in Textile Dyeing Operation," *Industrial Water Engineering,* **12**(6), 14-18. (Clemson University, Clemson, South Carolina)

Brandt, Dixon (1974), "Remarks on the Process Economics of Enzymatic Conversion of Cellulose to Glucose," in NSF, 1975. (Stone and Webster, Boston, Massachusetts)

Brandt, Dixon, Lloyd Hontz, and Mary Mandels (1973), "Engineering Aspects of the Enzymatic Conversion of Waste Cellulose to Glucose," in AIChE, 1973, pp. 127-133. (Stone and Webster, Boston, Massachusetts; and U.S. Army Natick Laboratories, Massachusetts)

Braton, Norman R., and James A. Koutsky (1974), "Cryogenic Recycling," in Aleshin, 1974, pp. 33-39. (University of Wisconsin, Madison)

Bravard, J. C., H. B. Flora, II, and Charles Portal (1972), *Energy Expenditures Associated with the Production and Recycle of Metals,* Report ORNL-NSF-EP-24, Oak Ridge National Laboratory, Tennessee.

Breeling, James L., and Donald G. Moore (1972), "Pollutants as Resources," in P. L. White, Ed., *Proceedings, 3rd Western Hemisphere Nutrition Congress, 1971,* Futura Publishing Co., Mount Kisco, New York, pp. 300-304. (Department of Foods and Nutrition, Americal Medical Association; and Chemetron Corporation)

Breidenbach, A. W., and E. P. Floyd (1970), *Needs for Chemical Research in Solid Waste Management,* U.S. Government Printing Office, Washington, D.C.

Bridgwater, A. V. (1975), "The Economics of Recovery of Materials from Industrial Waste—A Case Study," *Resource Recovery and Conservation,* **1,** 115-127. (Department of Chemical Engineering, University of Aston in Birmingham, U.K.)

Bridgwater, A. V., S. A. Gregory, C. J. Mumford, and E. L. Smith (1975), "A Systems Approach to the Economics of Waste Handling," *Resource Recovery and Conservation,* **1,** 3-23. (Department of Chemical Engineering, University of Aston in Birmingham, U.K.)

Brink, Russell H. (1974), "Use of Waste Sulfate on Transpo '72 Parking Lot," in Bureau of Mines Information Circular 8640, U.S. Department of the Interior, Washington, D.C., pp. 197-207.

Brooks, P. T., G. M. Potter, and D. A. Martin 1969), *Chemical Reclaiming of Superalloy Scrap,* Bureau of Mines Report of Investigations 7316, U.S. Department of the Interior, Washington, D.C.

Brothers, John E. (1973), "Reclaimed Rubber," in Maurice Morton, Ed., *Rubber Technology,* Van Nostrand-Reinhold, New York, Chap. 9, pp. 496-514. (Midwest Rubber Reclaiming Co., East St. Louis, Illinois)

Broussaud, A. (1976), "Indirect Energy Savings Generated by Urban Refuse Recovery," in Aleshin, 1976, pp. 153-160. (Département Mineralurgie, Service Géologique National, Bureau de Recherches Géologiques et Minières, B. P. 6009-45018 Orleans Cédex, France)

Brown, Harry D. (1973), "Pyrolysis System for Recycling of Refuse," *U.S. Patent* 3,770,419; *Chemical Abstracts,* **80,**137054; summary in Sittig, 1975.

Brown, D. E., and S. W. Fitzpatrick (1976), "Food from Waste Paper," in Birch et al., 1976, pp. 139-155. (Department of Chemical Engineering, University of Manchester Institute of Science and Technology, U.K.)

Brown, Bernard S., John Mills, and John M. Hulse (1974), "Chemical and Biological Degradation of Waste Plastics," *Nature,* **250** (5462), 161-163. (Department of Medical Biochemistry, Oxford Rd, Manchester M13 9PT, U.K.)

Brown, B., B. Crouse, D. Etter, and W. Schattner (1976a), "Paper Chemical Reclamation and Reuse via Reverse Osmosis," *Research Disclosure* (142) 46; *Chemical Abstracts,* **84,** 152493.

Brown, Paul Wencil, James R. Clifton, Geoffrey Frohnsdorff, and Richard L. Berger (1976b), "The Utilization of Industrial By-Products in Blended Cements," in Aleshin, 1976, pp. 278-284. (Center for Building Technology, B-348 Building Research, National Bureau of Standards, Washington, D.C. 20234; and Department of Civil and Ceramic Engineering, University of Illinois, Urbana)

Browning, William C. (1975), "The Lignosulfonate Challenge," in APS, 1975, pp. 109-124. (SCT Associates, Houston, Texas)

Bruins, Paul F. (1974), Ed., *Packaging with Plastics,* Gordon and Breach, New York, from seminar at Polytechnic Institute of Brooklyn, 1971.

Buckle, K. A., and Edwards, R. A. (1972), "Disposal and Utilization of Food Processing Wastes," in Kirov, 1972, pp. 121-127. (Department of Food Technology, University of New South Wales, Australia)

Bundi, U., and H. R. Wasmer (1976), "Recycling: Fundamentals and Concepts," *ISWA Information Bulletin* (International Solid Wastes and Public Cleansing Association, Zurich)(18), 2-8. (Eidgenössische Technishe Hochschule, Eidgenössische Anstalt für Wasserversorgung, Abwasserreinigung and Gewässerschutz (EAWAG), 8600 Dübendorf, Switzerland)

Bunker, H. J. (1963), "Microbial Food," in Rainbow and Rose, 1963, Chapter 3, pp. 34-67. (Consultant in microbiology, Twickenham, Middlesex, U.K.)

Bunker, H. J. (1964), "Microbial Food," in *Global Impacts of Applied Microbiology,* Stockholm, Sweden, 29 July-3 August, 1963, John Wiley, New York, pp. 234-240. (Industrial microbiology consultant, Twickenham, U.K.)

Burson, W. R., and F. B. Morgan, III (1974), "Recovery and Utilization of Copper Scrap in the Manufacture of High Quality Cathode," in Aleshin, 1974, pp. 392-396. (Southwire Co., Carrollton, Georgia)

Burton, Robert S., III, and Richard C. Bailie (1974), "Fluid Bed Pyrolysis of Solid Waste Materials," *Combustion,* **45**(8), 13-19. (Department of Chemical Engineering, West Virginia University)

Bynum, D., J. F. Evertson, and H. O. Fleisher (1972), "Scrap Tire-Beer Bottle Concrete," *Matériaux et Constructions,* **5**(27), 151-157. (College of Engineering, Texas A & M Universtiy)

Caldwell, H. S. Jr., J. F. Hogan, and T. H. Elkins (1974), "Processing Slag from an Aluminum Dross Furnace to Recover Fluxing Salt, Aluminum Metal, and Aluminum Oxide," in Aleshin, 1974, pp. 309-315. (U.S. Bureau of Mines, College Park, Maryland)

Callihan, C. D., and C. E. Dunlap (1973), *Single-Cell Proteins from Cellulosic Wastes,* U.S. Environmental Protection Agency; U.S. NTIS PB-223,873.

Callihan, Clayton D., George H. Irwin, James E. Clemmer, and Oliver W. Hargrove (1975), "Proteins from Waste Cellulose," in APS, 1975, pp. 189-196. (Chemical Engineering Department, Louisiana State University, Baton Rouge)

Calmon, Calvin, and Harris Gold (1976), "New Directions in Ion Exchange," *Environmental Science and Technology,* **10**(10), 980-984. (Water Purification Associates, Cambridge, Massachusetts 02142)

Calvin, Melvin (1974), "Solar Energy by Photosynthesis," *Science,* **184,** 375-381. (Laboratory of Chemical Biodynamics, University of California, Berkeley)

Campbell, Jay A (1974), "Electromagnetic Separation of Aluminum and Nonferrous Metals," in Aleshin, 1974, pp. 95-102; and in Proceedings, 103rd Annual Meeting, American Institute of Mechanical Engineers, Dallas, Texas, 24-28 February. (Combustion Power Company, Menlo Park, California)

Campbell, William J. (1976), "Metals in the Wastes We Burn?" *Environmental Science and Technology,* **10**(5), 436-439. (U.S. Bureau of Mines, College Park, Maryland)

Campbell, Gregory A., and William C. Meluch (1976), "Polyurethane Foam Recycling: Superheated Steam Hydrolysis," *Environmental Science and Technology,* **10**(2), 182-185. (General Motors Research Laboratories, Warren, Michigan)

Campbell, Howard, and T.C. Shutt (1973), "Vibrocasting Glass-containing Construction Panels," in Albuquerque, 1973, pp. 217-227.

Cannon, James (1973), "Steel: the Recyclable Material," *Environment,* **15**(9), 11-20; reprinted with minor changes from *Economic Priorities Report* (Council on Economic Priorities), **4**(3). (Fellow, Council on Economic Priorities)

Cannon, H.S., and M. L. Smith (1974), "Recycling of Glass and Metals: Container Materials," in Yen, 1974a, Chapter 10, pp. 301-334. (Continental Can. Co., U.S.A.)

Cardenas, Raul R., Jr., and Stephen Varro, Jr. (1973), "Disposal of Urban Solid Waste by Composting," in Inglett, 1973, Chapter 16, pp. 183-204. (Department of Civil Engineering, University of New York, Bronx; and Ecology, Inc., Brooklyn, New York)

Carlson, O. N., and F. A. Schmidt (1973), *The Metallurgical Upgrading of Automotive Scrap Steel,* U.S. Environmental Protection Agency; U.S. NTIS PB-223,740.

Carlson, O. N., F. A. Schmidt, J. K. McClusky, C. V. Owen, R. E. Shaw, and R. R. Lichtenberg (1974), "Studies on Upgrading of Automotive Scrap by Vacuum Melting and Electroslag Refining," in Aleshin, 1974, pp. 397-408. (Ames Laboratory, USAEC, Iowa State University, Ames)

Carpenter, Lewis V. (1971), *Annotated Bibliography on Incineration, Carbonization, and Reduction of Garbage, Rubbish, and Sewage Sludge,* originally published 1939, reprinted, Burt Franklin, Lennox Hill, New York.

Carr, Wayne F. (1970), "Value Recovery from Wood Fiber Refuse," in *Proceedings, 2nd Mineral Waste Utilization Symposium, Chicago*. (Forest Products Laboratory, USDA, Madison, Wisconsin)

Carr, Wayne F. (1971), "Many Problems Involved in Increasing Utilization of Waste Paper," *Paper Trade Journal,* 17 May, 48-52. (Forest Products Laboratory, USDA, Madison, Wisconsin)

Carrillo, Fred. V., Mark H. Hibpshman, and Rodney D. Rosenkranz (1974), *Recovery of Secondary Copper and Zinc in the United States,* Bureau of Mines Information Circular 8622, U.S. Department of Interior, Washington, D.C.; U.S. NTIS PB-233,215; summary in Jackson, 1975.

Carter, W. C. (1975), "The Bangkok Composting Plant," in Kirov, 1975a, pp. 137-139. (Clarke Chapman, John Thompson Ltd., Wolverhampton, England)

Case, Oliver P. (1974), *Metallic Recovery from Waste Waters Utilizing Cementation,* U.S. Environmental Protection Agency Report EPA-670/2-74-008. (Anaconda American Brass Co., Waterbury, Connecticut)

Casida, L.E., Jr. (1968), *Industrial Microbiology,* Wiley, New York.

Castor, Gaylord B. (1974), "Gasoline Substitutes," *Science,* **183,** 698.

Cavanna, M., E. Riaño, and J. Sanchez Almarez (1974), "Installation and Results of the First Spanish Pilot Plant for the Treatment of Raw Refuse from Madrid (Spain) with U.S.B.M. Technology," in Aleshin, 1974, pp. 142-149. (J. Sanchez Almarez Empresa Nacional ADARO de Investigaciones Mineras, S. A. Serrano, 116, Madrid-6, Spain)

Cavanna, M. M., E. Riaño, J. Sanchez Almarez, and H. García Ramírez (1976), ''Latest Developments in Processing Spanish Urban Raw Refuse,'' in Aleshin, 1976, pp. 141-145. (Empresa Nacional ADARO de Investigaciones Mineras, S. A., Serrano, 116, Madrid-6, Spain)

CCEQ (1973), *Proceedings of the Joint Conference on Recycling Municipal Sludges and Effluent on Land,* Champaign, Illinois, 9-13 July, Co-ordinating Committee on Environmental Quality; U.S. NTIS PB-227,106.

CE (1968), ''Molten Salts: New Route to High Purity Metals,'' *Chemical Engineering* (26 August) 36, 38.

CE (1969), ''Nuclear Fusion Dividend: Solid Waste Treatment,'' *ibid.*, **76** (15 December) 56-57.

CE (1971), ''Carbon Monoxide Wrings Oil from Lignite or Garbage,'' *ibid.*, **78**(24), 86, 88.

CE (1973), ''Disposal of Solid Wastes,'' Institution of Chemical Engineers Midlands Branch Symposium, Birmingham, 27 September 1972, *Chemical Engineer,* (270), 55.

CE (1976), ''Processes Remove Effluent Solvent-Vapors,'' *Chemical Engineering,* **83**(6), 67-68.

CEN (1971a), ''Process Converts Animal Wastes to Oil,'' *Chemical and Engineering News,* **49** (16 August) 43; reprinted in Giddings and Monroe, 1972, pp. 238-239.

CEN (1971b), ''Plastics Face Growing Pressure from Ecologists,'' *ibid.*, **49** (29 March) 12-15; reprinted in Giddings and Monroe, 1972, pp. 240-241.

CEN (1974a), ''Projects Seek Energy from Wastes,'' *ibid.*, **52**(7), 21.

CEN (1974b), ''Japan Aims at Making Waste Heat Useful,'' *ibid.*, **52**(7), 21.

CEN (1974c), ''Magnetic Separations Near Market Breakthrough,'' *ibid.*, **52**(4), 21-22.

CEN (1974d), ''Fly Ash Solves Two Pollution Problems,'' *ibid.*, **52**(14), 33-34.

CEN (1974e), ''Surge in Detroit's Use of Plastics,'' *ibid.*, **52**(20), 10-11.

CEN (1974f), ''Goodyear, Tosco Team to Recycle Scrap Tires,'' *ibid.*, **52**(23), 5.

CEN (1974g), ''Cryogenic Process Recycles Used Auto Tires,'' *ibid.*, **52**(24), 21-22.

CEN (1974h), ''Waste Cellulose Possible Glucose Source,'' *ibid.*, **52**(21), 20.

CEN (1974i), ''Recycling CO_2 is Goal of Research at IGT,'' *ibid.*, **52**(15), 24.

CEN (1974j), ''Recycled Wastes Minor Source for Minerals,'' *ibid.*, **52**(19), 17.

CEN (1975a), ''Coke, Pepsi Choose Plastic Bottles,'' *ibid.*, **53**(23), 6.

CEN (1975b), ''Pollutant Recovery, Water Re-use Mulled,'' *ibid.*, **53**(22), 19-21.

CEN (1975c), ''Sulfur is Tight Despite Excess Production,'' *ibid.*, **53**(23), 9-10.

CEN (1975d), ''Plastics to Maintain Their Competitive Edge,'' *ibid.*, **53**(45), 12-13.

CEN (1975e), ''Plastic Wastes Yield Valuable Chemicals,'' *ibid.*, **53**(30), 17.

CEN (1975f), ''New Solvent for Cellulose Developed,'' *ibid.*, **53**(21), 4.

CEN (1975g), ''Wood Draws Attention as Plastics Feedstock,'' *ibid.*, **53**(16), 13-14.

CEN (1975h), ''Coal Conversion Activities Picking Up,'' *ibid.*, **53**(48), 24-25.

CEN (1976a), ''IGT Weighs Potential of Fuels from Biomass,'' *ibid.*, **54**(8), 24-26.

(Review of Institute of Gas Technology Symposium on fuels from biomass, sewage, urban refuse, and agricultural wastes, Orlando, Florida, January.)

CEN (1976b), "Scrap Tires Pyrolyzed in Molten Salts," *ibid.*, **54**(21), 26-28.

CEN (1976c), "New Cement Uses Flyash, Costs Less to Make," *ibid.*, **54**(14), 16.

Cernia, Enrico (1975), "Integration of Municipal Wastes Incineration—Sea Water Desalting Processes," *Resource Recovery and Conservation,* **1**,192-194. (Snam Progetti, Milan)

Chadwick, M. J., and G. T. Goodman (1975), *The Ecology of Resource Degradation and Renewal* (15th Symposium of the British Ecological Society, University of Leeds, 10-12 July, 1973), Blackwell, Oxford, U.K.

Chamberlain, C. T. (1973), "Incineration—Some Theoretical Aspects," in CE, 1973, pp. 60-64. (Universal Machinery and Services Ltd., Millshaw, Leeds, LS11 8EQ, U.K.)

Chamberlain, Clive T., and Rolph Müller (1976), "Integrated Heat and Materials Recovery Using Incineration," *Environmental Pollution Management,* **6**(2), 55-56. (Universal Machinery and Services Ltd., Millshaw, Leeds LS11 8EQ, U.K.)

Chandler, Peter (1976), "The Importance of Heat Recovery in Thermal Incineration," *Environmental Pollution Management,* **6**(2), 41-43, 45. (Hygrotherm Engineering Ltd., Botanical House, Botanical Avenue, Talbot Rd., Manchester M16 OHL, U.K.)

Chapman, Peter F. (1973a), *The Energy Costs of Producing Copper and Aluminium From Primary Sources,* Open University Energy Research Group Report, ERG 001, August. (Open University, Milton Keynes, U.K.)

Chapman, Peter F. (1973b), *The Energy Costs of Producing Copper and Aluminium From Secondary Sources,* Open University Energy Research Group Report, ERG 002, September. (Open University, Milton Keynes, U.K.)

Chapman, Peter F. (1974), "Energy Conservation and Recycling of Copper and Aluminium, *Metals and Materials,* June, 311-319. (Open University, Milton Keynes, U.K.)

Chapman, Peter F. (1975), "Energy Budgets. 4. The Energy Costs of Materials," *Energy Policy,* **3**(2), 47-57. (Open University, Milton Keynes, U.K.)

Chapman, Richard A., and F. Richard Wocasek (1974), "CPU-400 Solid-Waste-Fired Gas Turbine Development," *Proceedings, National Incinerator Conference,* **6**, 347-357. (National Environmental Research Center, U.S. EPA, Cincinnati, Ohio)

Chappell, G. A. (1973), "Aqueous Scrubbing of Nitrogen Oxides from Stack Gases," in Jimeson and Spindt, 1973, Chapter 18, pp. 206-217. (Esso, Linden, New Jersey)

Cheater, G. (1973), "Disposal of Plastics Waste," in CE, 1973, pp. 85-89. (RAPRA, Shrewsbury, U.K.)

Chedd, Graham (1975), "Cellulose from Sunlight," *New Scientist,* **65**(939), 572-575.

ChemTech (1975), "New Helium Policy," **5,** 391.

Chemical Week (1976), "Converting Sewage into Savings," **118**(2), 47.

Chemistry in Britain (1975), "Recycling and Reclamation," **11**(9), 310.

Chemistry in Britain (1976), "Water," **12**(9), 278-295.

Chen, Ben H. H., Paul H. King, and Clifford W. Randall (1976), "Alum Recovery from

Representative Water-Treatment-Plant Sludges,'' *Journal American Water Works Association,* **68**(4), 204-207. (R. M. Towill Corp., Honolulu, Hawaii; and Virginia Polytechnic Institute and State University, Blacksburg)

Cheremisinoff, P. E. (1976), ''Resource Recovery Status Report,'' *Pollution Engineering,* **8**(5), 40-41.

Chian, Edward S. K., Willis N. Bruce, and Herbert H. P. Fang (1975), ''Removal of Pesticides by Reverse Osmosis,'' *Environmental Science and Technology,* **9**(1), 52-59. (Department of Civil Engineering, University of Illinois, Urbana; and Illinois State Natural History Survey, Urbana, Illinois 61801)

Chu, A. K. P., M. Fleischmann, and G. J. Hills (1974); A. K. P. Chu and G. J. Hills (1974), ''Packed Bed Electrodes. I. The Electrochemical Extraction of Copper Ions from Dilute Aqueous Solutions. II. Anodic Stripping and the Recovery of Copper from Dilute Solutions,'' *Journal of Applied Electrochemistry,* **4,** 323-330, 331-336. (Department of Chemistry, The University, Southampton, U.K.)

CIG (1975a), ''Old Tyres—A Cheap New Raw Material,'' *Gases in Research and Industry* (Commonwealth Industrial Gases Ltd., Australia) (12), 4-5.

CIG (1975b), ''Cryogenic Breaking: New Way to Recover Scrap Metals,'' *ibid.,* (13), 6-7.

Clark, Thomas D. (1971), *Economic Realities of Reclaiming Natural Resources in Solid Waste,* Publication SW-93ts.j, U. S. Environmental Protection Agency.

Clark, B. L., and R. T. Fowler (1976), ''The Treatment of Nickel and Chromium Plating Liquors by Reverse Osmosis,'' *Process and Chemical Engineering* (Australia), **27**(12), 15-18. (School of Engineering, University of New South Wales, Australia)

Clarke, A. B. (1974), ''Recovery of Trace Metals from Aqueous Solution—Review Article,'' *Birmingham University Chemical Engineer,* **25**(1), 19-23. (Department of Chemical Engineering, University of Birmingham, U.K.)

Cochran, C. Norman (1974), ''Recovery of Hydrogen Fluoride Fumes in Alumina in Aluminum Smelting,'' *Environmental Science and Technology,* **8**(1), 63-66. (Alcoa Laboratories, Pittsburgh, Pennsylvania)

Cochran, Andrew A., and Lawrence C. George (1976a), ''Recovery of Metals from a Variety of Industrial Wastes,'' *Resource Recovery and Conservation,* **2**(1), 57-65. (Rolla Metallurgy Research Center, U.S. Bureau of Mines)

Cochran, A.A., and L. C. George (1976b), ''The Waste-Plus-Waste Process for Recovering Metals from Electroplating and Other Wastes,'' *Plating and Surface Finishing,* July, 38-43. (Rolla Metallurgy Research Center, U.S. Bureau of Mines)

Coe, Warren B., and Michael Turk (1973), ''Processing Animal Waste by Anaerobic Fermentation,'' in Inglett, 1973, Chapter 4, pp. 29-37. (Hamilton Standard, Windsor Locks, Connecticut)

Coldrick, John (1975), ''Sewage as a Resource,'' *New Scientist,* **68**(973), 276-278. (Wessex Water Authority, Bristol, U.K.)

Coleman, Arthur K. (1975), ''New Concepts in the Handling of Industrial Wastes,'' *Chemistry and Industry* (London), (13), 534-544. (Re-Chem International Ltd., Southampton, U.K.)

Collings, R. K., A. A. Winer, D. G. Feasby, and N. G. Zoldners (1974), "Mineral Waste Utilization Studies," in Aleshin, 1974, pp. 2-12. (Mines Branch, Department of Energy, Mines, and Research, Ottawa, Canada)

Collins, L. W., W. R. Downs, E. K. Gibson, and G. W. Moore (1974), "An Evaluation of Discarded Tires as a Potential Source of Fuel," *Thermochimica Acta,* **10**(2), 153-159. (Lyndon B. Johnson Space Center, NASA, Houston, Texas 77058; and Lockheed Electronics, Houston, Texas 77058)

Collins, R. J. (1976), "Use of Dredged Silt in the Construction Industry," *Water and Waste Treatment,* **19**(6), 30. (Building Research Establishment, Department of Environment, U.K.)

Complete WateReuse, Industry's Opportunity (1973), ed. Lawrence K. Cecil, American Institute of Chemical Engineers National Conference on Complete WateReuse, April, Washington, D.C.

Complete WateReuse, Waters Interface with Energy, Air, and Solids (1975), American Institute of Chemical Engineers National Conference on Complete WateReuse, May, Chicago, Illinois; review in CEN, 1975b.

Complete WateReuse, Symbiosis as a Means of Abatement for Multi-Media Pollution (1976), American Institute of Chemical Engineers National Conference on Complete WateReuse, June, Cincinnati, Ohio.

Compost Science (1976), "Recycling Human Waste," Sixth Annual Composting and Waste Recycling Conference, 11-14 May, Portland, Oregon, **17**(3), 2-32.

Conner, Jesse R. (1974), "Ultimate Disposal of Liquid Wastes by Chemical Fixation," *Engineering Bulletin of Purdue University, Engineering Extension Series,* 145, Part 2, 906-922; *Chemical Abstracts,* **83,** 151763. (Chemfix Inc., a Division of Environmental Science, Inc., Pittsburgh, Pennsylvania)

Connor, Jesse R., and Lawrence P. Gowman (undated), "Chemical Fixation of Activated Sludge and End Product Applications," Chemfix, Inc., Pittsburgh, Pennsylvania.

Consumer (New Zealand) (1976), "Bottles and Cans: Right Now, Your Best Buy is a Returnable Bottle," (129), 145-147.

Converse, A. O., H. E. Grethlein, S. Karandiker, and S. Kuhrtz (1973), *Acid Hydrolysis of Cellulose in Refuse to Sugar and Its Fermentation to Alcohol,* U.S. Environmental Protection Agency; U.S. NTIS PB-221,239.

Cooley, Austin C., and Thomas J. Dagon (1976), "Current Silver Recovery Practices in the Photographic Processing Industry," *Journal of Applied Photographic Engineering,* **2**(1), 36-41. (Eastman Kodak Co., Rochester, New York)

Coombe, J. B., and Anne L. Briggs (1974), "Use of Waste Paper as a Feedstuff for Ruminants," *Australian Journal of Experimental Agriculture and Animal Husbandry,* **14** (68), 292-301. (Division of Plant Industry, CSIRO, Canberra, Australia)

Cooney, Charles L., and Donald L. Wise (1975), "Thermophilic Anaerobic Digestion of Solid Waste for Fuel Gas Production," *Biotechnology and Bioengineering,* **17**(8), 1119-1135. (Department of Nutrition and Food Science, Massachusetts Institute of Technology)

Cooper, J. B. (1975), "Non-Ferrous Metals Recycling," in Henstock, 1975a. (Coopers (Metals), Swindon, U.K.)

Corey, Richard C. (1969), Ed., *Principles and Practices of Incineration,* Wiley-Interscience, New York.

Cornell, Conrad F. (1975), "Salt-Recovery Process Allows Reuse of Pulp-Bleaching Effluent," *Chemical Engineering,* **82**(24), 136-137. (Erco Envirotech Ltd., Salt Lake City, Utah)

Coton, S. G. (1976), "Recovery of Dairy Waste," in Birch et al., 1976, pp. 221-231. (Milk Marketing Board, Thames Ditton, Surrey, U.K.)

Coulson, Guy S. (1976), "An Economic Alternative for Municipal Incineration," *Environmental Pollution Management,* **6**(2), 47-49. (Robert Jenkins Systems Ltd, Wortley Rd., Rotherham, South Yorkshire S61 1LT, U.K.)

Coutant, Charles C. (1976), "How to Put Waste Heat to Work," *Environmental Science and Technology,* **10**(9), 868-871. (Environmental Sciences Division, Oak Ridge National Laboratory, Tennessee)

Cover, A. E., and W. C. Schreiner (1975), "The Kellogg Molten Salt Process," *Energy Communications,* **1**(2), 135-156. (M. W. Kellogg Co., Houston, Texas 77046)

Coyle, Bernard H., Jr., Judith Koperski, and Robert N. Anderson (1976), "A Technical and Economic Analysis of Processes for the Recovery of Metals in the Non-ferrous Portion of Automobile Shredder Refuse," in Aleshin, 1976, pp. 350-362. (NL Industries, Inc., Hightstown, New Jersey; Department of Applied Earth Sciences, Stanford University, California; and Department of Materials Science, San Jose State University, California)

Cramer, S. D., and H. V. Makar (1976), "Steel from Ferrous Can Scrap," in Aleshin, 1976, p. 398-406. (College Park Metallurgy Research Center, U.S. Bureau of Mines)

Crane, T. H., H. N. Ringer, and D. W. Bridges (1975), *Production of Gaseous Fuel by Pyrolysis of Municipal Solid Waste,* Report NASA-CR-141791 to National Aeronautics and Space Administration, Lyndon B. Johnson Space Center, Houston, Texas. (Reource Recovery Systems, Barber-Colman Co., Irvine, California)

Crites, Ronald W., and Charles E. Pound (1976), "Land Treatment of Municipal Wastewater," *Enivronmental Science and Technology,* **10**(6), 548-551. (Metcalf and Eddy, Inc., Palo Alto, California 94303)

CRM (1969), *Resources and Man,* Committee on Resources and Man, W. H. Freeman, San Francisco.

Crocker, Frank H. (1976), "Putting the Record Straight" [on used oil recyling], *Industrial Recovery,* **22**(3), 10, 12. (Tenneco Organics Ltd.)

Cross, Frank L., Jr. (1972), *Handbook on Incineration,* Technomic Publishing Co., Westport, Connecticut.

Croyle, R. D., L. L. Wilson, and T. A. Long (1974), "Potential of Spent Tea Leaves for Animal Feeds and Composting," *Compost Science,* **15**(2), 28-30. (Pennsylvania State University)

Cruver, J. E. (1975), "Waste-Treatment Applications of Reverse Osmosis," *Journal of Engineering for Industry* (Transactions of the American Society of Mechanical

Engineers), **97**(1), 246-251; presented at Intersociety Conference on Environmental Systems, July 29-August 1, 1974, Seattle, Washington. (Roga Systems Division, General Atomic Co., San Diego, California)

Cruver, J. E., and I. Nusbaum (1974), "Application of Reverse Osmosis to Wastewater Treatment," *Journal of the Water Pollution Control Federation,* **46**(2), 301-311; Water Pollution Control Federation Meeting, Atlanta, Georgia, 8-13 October, 1972. (Gulf Environmental Systems, California)

Cruz, Ibarra E. (1975), "Studies on the Production and Utilisation of Gas from Coconut Wastes in the Philippines," in Kirov, 1975a, pp. 211-217. (College of Engineering, University of the Philippines)

Cservenyak, F. J., and C. B. Kenahan (1970), *Bureau of Mines Research and Accomplishments in Utilization of Solid Waste,* Bureau of Mines Information Circular 8460, U.S. Department of the Interior, Washington, D.C.

Cukor, Peter M. (1975a), "A Field Test of the Quality of Re-refined Lube Oils," in Teknekron, 1975, Part V.

Cukor, Peter M. (1975b), "A Review of Re-refining Economics," in Teknekron, 1975, Part VI.

Cukor, Peter M., and Timothy Hall (1975), "Energy Consumption in Waste Oil Recovery," in Teknekron, 1975, Part IV.

Cukor, Peter M., Michael John Keaton, and Gregory Wilcox (1973a), "Federal Research on Oil from Automobiles," in Teknekron, 1973, Part I; U.S. NTIS PB-237,618.

Cukor, Peter M., Michael John Keaton, and Gregory Wilcox (1973b), "An Investigation of Dispersed Sources of Used Crankcase Oils" in Teknekron, 1973, Part II; U.S. NTIS PB-237,619.

Cukor, Peter M., Michael John Keaton, and Gregory Wilcox (1973c), "Economic, Technical, and Institutional Barriers to Waste Oil Recovery" in Teknekron, 1973, Part III; U.S. NTIS PB-237,620.

Cummings, John P. (1973), "Waste Glass in Road Construction," in Albuquerque, 1973, pp. 73-95. (Owens Illinois, Inc., Toledo, Ohio)

Cummings, John P. (1974), "Glass and Aluminum Recovery System, Franklin, Ohio," in Aleshin, 1974, pp. 106-115. (Owens Illinois, Inc., Toledo, Ohio)

Cummings, John P. (1976), "Glass and Non-Ferrous Metal Recovery Subsystem at Franklin, Ohio—Final Report," in Aleshin, 1976, pp. 175-183. (Owens Illinois, Inc., Toledo, Ohio)

Currier, R. A., and M. L. Laver (1973), *Utilization of Bark Waste,* Report EPA-670-2-73-005, U.S. Environmental Protection Agency, Washington, D.C.; U.S. NTIS PB-221,876. (Department of Forest Products, Oregon State University, Corvallis)

Cutler, Herschel, and Gerald S. Goldman (1973), "Transporation: Bugaboo of Scrap Iron Recycling," *Environmental Science and Technology,* **7,** 408-411. (Institute of Scrap Iron and Steel, Washington, D.C.

Czerwonko, Anna (1974), "Use of Waste Products Formed During the Manufacture of Titanium and Titanium Dioxide," *Prace Naukowe Instytutu Chemii Nieorganicznej i Metalurgii Pierwiastkow Rzadkich Politechniki Wroclawskiej,* **22,** 351-402 (Polish);

Chemical Abstracts, **83,** 32590. (Wroclaw Technical University, Poland)

Daellenbach, Charles B., Warren M. Mahan, and James J. Drost (1974), ''Utilization of Automobile and Ferrous Refuse Scrap in Cupola Iron Production,'' in Aleshin, 1974, pp. 417-423. (Twin Cities Metallurgy Research Center, U.S. Bureau of Mines)

Daellenbach, Charles B., Richard R. Lindeke, and Warren M. Mahan (1976), ''Utilization of Refuse Scrap in Cupola Gray Iron Production,'' in Aleshin, 1976, pp. 234-240. (Albany Metallurgy Research Center and Twin Cities Metallurgy Research Center, U.S. Bureau of Mines)

Dale, J. C. (1974), ''Extraction of Aluminum from Mixed Solid Waste,'' in Aleshin, 1974, pp. 103-105. (The Aluminum Association, New York)

Daly, W. H., and L. P. Ruiz (1975), *Fabrication of Single Cell Protein from Cellulosic Wastes,* U.S. Environmental Protection Agency; U.S. NTIS PB-239,502.

Dalyell, Tam (1973), ''Recycling the Car,'' in ''Forum,'' *New Scientist,* **58**(850), 698.

Darnay, Arsen J., Jr., (1969), ''Throwaway Packages—A Mixed Blessing,'' *Environmental Science and Technology,* **3**(4); reprinted in ACS, 1971, pp. 28-33. (Midwest Research Instiutte, Washington, D.C.)

Darnay, Arsen (1972), *Recycling: Assessment and Prospects for Success,* presented to seminar, Denver, Colorado; Report SW-8i, U.S. Environmental Protection Agency, Washington, D.C.; U.S. NTIS PB-213,961; summary in Jackson, 1975. (U.S. EPA, Cincinnati, Ohio)

Darnay A. J., and W. E. Franklin (1969), *The Role of Packaging in Solid Waste Management, 1966-1976,* Public Health Service Publication No. 1855, Washington, D.C.

Davis, Norman (1974), ''Recycling Refuse—New Life for Ironworks?'' *New Scientist,* **62**(897), 319-320. (British Steel Corporation, Forward Technology Unit, London)

Davis, Thomas A., and Don B. Hooks (1974), ''Utilization of Waste Kiln Dust from the Cement Industry,'' in Aleshin, 1974, pp. 354-363. (Southern Research Institute, Birmingham, Alabama)

Davis, M. H. and J. G. Scroggie (1973), ''Investigation of Commercial Chrome-Tanning Systems. III. Re-cycling of Used Chrome Liquors. IV. Re-cycling of Chrome Liquors and Their Use as a Basis for Pickling. V. Re-cycling of Chrome Liquors in Commercial Practice,'' *Journal of the Society of Leather Technologists and Chemists,* **57,** 53-58, 81-83, 173-176. (Division of Protein Chemistry, CSIRO, Parkville, Victoria 3052, Australia)

Dean, K. C., and E. G. Valdez (1972), ''Sponge Iron from Dry and Oily Mill Scale, Steelmaking Furnace Dust, and Tires,'' American Institute of Mechanical Engineers, Annual Meeting, Chemical and Process Metallurgy II, San Francisco, 21-24 February. (Salt Lake City Metallurgy Research Center, U.S. Bureau of Mines)

Dean, K. C., C. J. Chindgren, and LeRoy Peterson (1971), *Preliminary Separation of Metals and Nonmetals from Urban Refuse,* Bureau of Mines Solid Waste Research Program, Technical Progress Report 34, U.S. Department of Washington, D.C.

Dean, K. C., C. J. Chindgren, and E. G. Valdez (1972), ''Innovations in Recycling of Automotive Scrap,'' Institute of Scrap Iron and Steel Annual Meeting, Washington, D.C., 15-18 January. (Salt Lake City Metallurgy Research Center, U.S. Bureau of Mines)

Dean, K. C., E. G. Valdez, and J. W. Sterner (1974a), "Recovery of Corrosion-resisting Zinc from Junked Automobiles," Society of Automotive Engineers, Automotive Engineering Congress, Detroit, Michigan, 25 February-1 March, Paper 740032. (U.S. Bureau of Mines)

Dean, K. C., J. W. Sterner, and E. G. Valdez (1974b), *Effect of Increasing Plastics Content on Recycling of Automobiles,* Bureau of Mines Solid Waste Research Program, Technical Progress Report 79, U.S. Department of the Interior, Washington, D.C.

Dean, K. C., E. G. Valdez, and J. H. Bilbrey, Jr. (1975), "Recovery of Aluminum from Shredded Municipal and Automotive Wastes," *Resource Recovery and Conservation,* **1**, 55-66; presented at American Institute of Mechanical Engineers Annual Meeting, Dallas, Texas, 23-28 February, 1974. (Salt Lake City Metallurgy Research Center, U.S. Bureau of Mines)

Deanin, Rudolph D. (1975), "Plastics from Wood," in APS, 1975, pp. 71-76. (Plastics Department, Lowell Institute of Technology, Lowell, Massachusetts, 01854)

De Cuyper, J. (1973), "Valorisation des Résidues de l'Industrie Métallurgique," ("Upgrading of Wastes of the Metallurgical Industry"), *Industrie Minerale-Mineralurgie* (2), 133-141, (Catholic University, Louvain, France)

Degner, Vernon R., and Bob McChesney (1974), "Performance of Advanced R C Separator on Municipal Solid Waste," in Aleshin, 1974, pp. 63-69. (Wemco Division, Envirotech Corp., Sacramento, California)

De Jong, H. C. (1976), "Elimination of Zn from Wastewater of Rayon Plants by Precipitation," *Resource Recovery and Conservation,* **1,** 369-371. (Akzo Research Laboratories, Arnhem, Netherlands)

Denver Regional Council of Governments and Urban Drainage and Flood Control District (1972), "Recycling Activity Description," U.S. NTIS PB-218,990; summary in Jackson, 1975.

Dewhirst, L. (1976), "Raw Material Conservation in the Pulp and Paper Industry," in SCI, 1976c, pp. 721-725. (W. S. Atkins and Partners, Woodcote Grove, Ashley Road, Epsom, Surrey KT18 5BW, U.K.)

DeWitt, C. C., M. D. Livingood, and K. G. Miller (1952), "Pigment Grade Iron Oxides: Recovery from Iron-containing Waste Liquors," *Industrial and Engineering Chemistry,* **44,** 673-678. (Michigan State College, East Lansing, Michigan)

Diaz, Luis F. (1975a), "Three Key Factors in Refuse Size Reduction," *Resource Recovery and Conservation,* **1,** 111-113. (Department of Mechanical Engineering, University of California, Berkeley)

Diaz, Luis F. (1975b), "Discussion of 'Domestic Cellulose Waste' (Golueke, 1975)," *Biotechnology and Bioengineering Symposia,* (5), 23-26. (Department of Mechanical Engineering, University of California, Berkeley)

Diaz, L. F., F. Kurz, and G. J. Trezek (1974), "Methane Gas Production as Part of a Refuse Recycling System," *Compost Science,* **15**(3). (Department of Mechanical Engineering, University of California, Berkeley)

Dickson, Philip F. (1973), "Cold Weather Paving with Glasphalt," in Albuquerque, 1973, pp. 97-115. (Department of Chemical and Petroleum Refining Engineering, Colorado School of Mines, Golden, Colorado)

Dolan, David S. (1975), "A Review of Methods of Controlling Impurities in Electrolytes of Electrolytic Copper Refineries," in Johnston, 1975, pp. T16-T18. (Bechtel Pacific Corporation, Melbourne, Australia)

Donovan, R. C., K. S. Rabe, W. K. Mammel, and H. A. Lord (1975), "Recycling Plastics by Two-Shot Molding," *Polymer Engineering and Science,* **15**(11), 774-780. (Western Electricity Co., Princeton, New Jersey)

Douglas, E., and Peter R. Birch (1974), "Warren Spring to Set Up Re-Sorting Pilot Plant," *Municipal Engineering,* 13 December. (Warren Spring Laboratory, Stevenage, U.K.)

Douglas, E., and P. R. Birch (1976), "Recovery of Potentially Re-usable Materials from Domestic Refuse by Physical Sorting," *Resource Recovery and Conservation,* **1**, 319-344. (Warren Spring Laboratory, Stevenage, U.K.)

Douglas, E., and D. V. Jackson (1973), "Waste Reclamation 1: A Source of Raw Materials," *Journal of Environmental Planning and Pollution Control,* **1**(2), 12-18. (Warren Spring Laboratory, Stevenage, U.K.)

Douglas, E., and C. P. Sayles (1970), "Dry Sorting Using Pneumatically Fluidised Powders," American Intitute of Chemical Engineers, Annual Meeting, Chicago, December, *AIChE Symposium Series,* **67**(116), 202 (1971). (Warren Spring Laboratory, Stevenage, U.K.)

Douglas, E., M. Webb, and G. R. Daborn (1974), "The Pyrolysis of Waste and Product Assessment," in Institute of Solid Wastes Management Symposium, *Treatment and Recycling of Solid Waste,* Manchester, U.K., January. (Warren Spring Laboratory, Stevenage, U.K.)

Douglas, E., M. Webb, and C. Power (1976), "Developments Leading to the Design of an Urban Refuse Pyrolysis Unit for Gas Production," in Aleshin, 1976, pp. 241-250. (Warren Spring Laboratory, Stevenage, U.S.)

Dow, A. B. (1972), "Salvaging and Reutilization of Waste Paper," in Kirov, 1972, pp. 175-177. (Australian Paper Manufacturers Ltd., Sydney, Australia)

Drake, H. J., and J. E. Shelton (1974), "Disposal of Iron and Steel Slag," in Aleshin, 1974, pp. 303-308. (Division of Nonmetallic Minerals, U.S. Bureau of Mines, Arlington, Virginia)

Dreissen, H. H., and A. T. Basten (1976), "Reclaiming Products from Shredded Junked Cars by the Water-Only and Heavy-Medium Cyclone Processes," in Aleshin, 1976, pp.377-385. (Stamicarbon b.v., Geleen, The Netherlands)

Dressel, W. M., L. C. George, and M. M. Fine (1976), "Chromium and Nickel Wastes—A Survey and Appraisal of Recycling Technology," in Aleshin, 1976, pp.262-270. (Rolla Mettalurgy Research Center, U.S. Bureau of Mines)

Dudley, K., B. D. Linley, and W. R. Laws (1975), "Recycling of Tinplate," *Solid Wastes,* **65**(7), 297-306. (The Metal Box Co.; Batchelor Robinson and Co. Ltd.; and British Steel Corporation Corporate Laboratories, U.K.)

Dunlap, Charles E. (1973), "Single-Cell Protein Production from Waste Cellulose," *Proceedings, International Biomass Energy Conference,* 14, Biomass Energy Institute, Winnipeg, Canada, 13-15 May. (Department of Chemical Engineering, University of Missouri, Columbia)

Dunlap, Charles E., and Clayton D. Callihan (1974), "Single Cell Protein Production from Cellulosic Wastes," in Yen, 1974a, Chapter 11, pp. 335-347. (Department of Chemical Engineering, University of Missouri, Columbia; and Department of Chemical Energy, Louisiana State University, Baton Rouge)

Dureau, Michael B. (1975), "Inplant Waste Abatement or Process Modification for Zero Discharge," in Johnston, 1975, pp. T37-T38. (Envirotech Australia Pty Ltd., Artarmon, Australia)

Eastlund, Bernard J., and William C. Gough (1970), "The Fusion Torch: A New Approach to Pollution and Energy Usage," *Chemical Engineering Progress Symposium Series* (American Institute of Chemical Engineers), **66**(104), 175-186. (U.S. Atomic Energy Commission, Washington, D.C.)

Easton, Eric B. (1976), "Resource Recovery from Municipal Solid Wastes: A Governmental Perspective?" *Resource Recovery and Conservation,* **2**(1), 85-88. (Solid Waste Report, Silver Spring, Maryland 20910)

Eccles, H., G. J. Lawson, and D. J. Rawlence (1976), "The Extraction of Copper(II) and Iron(II) from Chloride and Sulphate Solutions with LIX 64N in Kerosene," *Hydrometallurgy,* **1,** 349-359. (Department of Minerals Engineering, University of Birmingham, U.K.)

Ecologist (1972), *A Blueprint for Survival,* "Ecologist" staff, Tom Stacey Ltd., London.

Ecos (Journal of CSIRO Environmental Research, Australia) (1974a), "Polymer Beads versus Dead Diatoms," (2), 23.

Ecos (1974b), "Paper as Sheep Feed," (2), 21-22.

Ecos (1974c), "Recycling Comes to Tanning," (2), 10-11.

Ecos (1975a), "Re-using Sewage Water," (3), 14-18.

Ecos (1975b), "Sawmill Wastes: Cleaner Disposal, New Uses," (4), 13.

Ecos (1976), "Wastes: Other People's Raw Materials," (7), 19-23.

Eggen, Alfred C. W., and Ronald Kraatz (1974a), "Gasification of Solid Wastes in Fixed Beds," American Society of Mechanical Engineers, Winter Annual Meeting, New York, 17-22 November, ASME Publication 74-WA/Pwr-10. (K.T. Lear Associates, Manchester, Connecticut)

Eggen, Alfred C. W., and Ronald Kraatz (1974b), "Relative Value of Fuels Derived from Solid Wastes," *Resource Recovery thru Incineration,* Proceedings, National Incineration Conference, American Society of Mechanical Engineers, **6,** 19-32. (K.T. Lear Associates, Manchester, Connecticut)

Eichholz, Bernard F. (1972), "Fiber Recovery through Hydropulping," in Stump, 1972, pp. 25-35. (Franklin City, Ohio)

Ekiner, Okan M. (1975), "Chlorinated Butadiene/Alkyl Acrylate Grafted Co- Polymers," *Deferred Publications, U.S. Patent Office,* T937,006; *Chemical Abstracts,* **84,** 136527.

Elias, Hans-Georg (1975), "Raw Materials for the Polymer Industry. Part I. The Conventional Situation," *ChemTech,* **5**(12), 748-751. (Midland Macromolecular Institute, 1910 West St. Andrews Drive, Midland, Michigan 48640)

Elias, Hans-Georg (1976), "Raw Materials for the Polymer Industry. Part II. The

Unconventional Situation,'' *ibid.*, **6**(4), 244-249.

Emery, John J. (1975), ''Waste and Byproduct Utilization in Highway Construction,'' *Resource Recovery and Conservation,* **1,** 25-43. (Highway Materials Laboratory, McMaster University, Hamilton, Ontario)

Emery, John J. (1976a), ''Trends in Waste Utilization in Construction,'' in Aleshin, 1976, pp. 26-32. (Department of Civil Engineering and Engineering Mechanics, McMaster University, Hamilton, Ontario)

Emery, John J. (1976b), ''Slags,'' in Aleshin, 1976, pp. 291-300. (Department of Civil Engineering and Engineering Mechanics, McMaster University, Hamilton, Ontario)

Emery, John J. and Chang S. Kim (1974), ''Trends in the Utilization of Wastes for Highway Construction,'' in Aleshin, 1974, pp. 22-32. (Highway Materials Laboratory, McMaster University, Hamilton, Ontario)

Emmerman, Dieter K., Marvin B. Ziering, and Harold E. Davis (1973), ''New Freezing Process for Industrial WateReuse,'' in Complete WateReuse, 1973, pp. 125-129. (Avco Systems Division, Wilmington, Massachusetts)

Energy Digest (1976), ''The Edmonton Refuse Incineration Plant,'' **5**(2), 10-12.

Energy Primer: Solar, Water, Wind, and BioFuels (1976), Portola Institute, in cooperation with New Alchemy West, Whole Earth Truck Store, Ecology Action (Palo Alto) and Alternative Sources of Energy Newsletter, Fifth printing. (Portola Institute, 558 Santa Cruz Avenue, Menlo Park, California 94025)

Engdahl, R.B. (1969), *Solid Waste Processing* (A State-of-the-Art Report on Unit Operations and Processes), Bureau of Solid Waste Management Report SW-4c, Public Health Service Publication No. 1856, Battelle Memorial Institute Columbus Laboratories; U.S. NTIS PB-216,653; summary in Jackson, 1975.

Engler, Cady R., Walter P. Walawender, and Liang-tseng Fan (1975), ''Synthesis Gas from Feedlot Manure. Conceptual Design Study and Economic Analysis,'' *Environmental Science and Technology,* **9**(13), 1152-1157. (Department of Chemical Engineering, Kansas State University, Manhattan)

EPM (1975a), ''Scrap Tyre Recycling Firm Faces a Waste Problem,'' *Environmental Pollution Management,* **5**(3), 74-76.

EPM (1975b), ''Biodegradable Plastics Arrives,'' *ibid.*, **5**(5), 135.

EPM (1975c), ''Fluidised Bed Incinerators—Developments from Two Firms,'' *ibid.*, **5**(5), 145, 147.

EPM (1976a), ''Cement Firm Will Cut Fuel Costs by Burning Refuse,'' *ibid.*, **6**(3), 83.

EPM (1976b), ''Foresight Pays Off In Power—from Refuse Plant,'' *ibid.*, **6**(3), 84-85.

Essenhigh, R. H. (1970), ''Incineration—A Practical and Scientific Approach,'' *Environmental Science and Technology,* **4**(6); reprinted in ACS, 1971, pp. 59-69. (Pennsylvania State University, University Park)

EST (1967), ''Ingenuity and Incinerators,'' *Environmental Science and Technology,* **1**(8); reprinted in ACS, 1971, pp. 70-71.

EST (1968a), ''Sewage Sludge and Refuse Composting,'' *ibid.*, **2**(8); reprinted in ACS, 1971, pp. 42-44.

EST (1968b), ''Fluid Bed Incinerators Studied for Solid Waste Disposal,'' *ibid.*, **2**(7); reprinted in ACS, 1971, pp. 75-77.

EST (1968c), "Plastic Wastes Yield to Pyrolysis," *ibid.*, **2**(7); reprinted in ACS, 1971, p. 58.

EST (1969a), "Bureau Attacks Nation's Solid Waste," *ibid.*, **3**(8); reprinted in ACS, 1971, pp. 13-15.

EST (1969b), "Aluminum Scraps Find Second Life," *ibid.*, **3**(11); reprinted in ACS, 1971, pp. 26-27.

EST (1969c), "Scrap Tires: Materials and Energy Source," *ibid.*, **3**(2); reprinted in ACS, 1971, p. 22.

EST (1970a), "Solid Wastes," *ibid.*, **4**(5), 5-12; reprinted in ACS, 1971, pp. 5-12.

EST (1970b), "Auto Hulk Disposal—A Growing Business," *ibid.*, **4**(1); reprinted in ACS, 1971, pp. 23-25.

EST (1970c,d), "Converting Solid Wastes to Electricity," *ibid.*, **4**(8), 631; reprinted in ACS, 1971, pp. 85-87 and in ACS, 1973, pp. 92-94.

EST (1970e), "Reclaiming Solid Wastes for Profit," *ibid.*, **4**(9); reprinted in ACS, 1971, pp. 81-82.

EST (1970f), "Fly Ash Utilization Climbing Steadily," *ibid.*, **4**(3); reprinted in ACS, 1971, pp. 34-36.

EST (1970g), "Waste Recycling Really Works," *ibid.*, **4**(10); reprinted in ACS, 1971, pp. 83-84 and ACS, 1973, pp. 56-57.

EST (1971a), "Composting Municipal Solid Waste," *ibid.*, **5,** 1088; reprinted in ACS, 1973, pp. 22-24.

EST (1971b), "Plants Burn Garbage, Produce Steam," *ibid.*, **5,** 207; reprinted in ACS, 1973, pp. 103-105.

EST (1971c), "Can Plastics be Incinerated Safely?" *ibid.*, **5,** 667; reprinted in ACS, 1973, pp. 61-63.

EST (1971d), "Pyrolysis of Refuse Gains Ground," *ibid.*, **5,** 310; reprinted in ACS, 1973, pp. 112-114.

EST (1971e), "Federal Demonstration Projects: What Has Been Achieved?" *ibid.*, **5,** 498; reprinted in ACS, 1973, pp. 14-17.

EST (1972a), "Recycling Sludge and Sewage Effluent by Land Disposal," *ibid.*, **6,** 871; reprinted in ACS, 1973, pp. 72-74.

EST (1972b), "Hard Road Ahead for City Incinerators," *ibid.*, **6,** 992; reprinted in ACS, 1973, pp. 110-111.

EST (1972c), "Waste Heat Recovery Makes Sense," *ibid.*, **6,** 219; reprinted in ACS, 1973, p. 109.

EST (1972d), "Local Trash Cuts Downtown Fuel Bills," *ibid.*, **6,** 780; reprinted in ACS, 1973, p. 21.

EST (1972e), "Putting Industrial Sludges in Place," *ibid.*, **6,** 874; reprinted in ACS, 1973, pp. 70-71.

EST (1972f), "Building Bricks From the Waste Pile," *ibid.*, **6,** 502; reprinted in ACS, 1973, pp. 52-53.

EST (1972g), "Latest Solid Waste Challenge: Reclaiming Peanut Hulls," *ibid.*, **6,** 592; reprinted in ACS, 1973, pp. 54-55.

EST (1972h), "Federal Redirections in Solid Wastes," *ibid.*, **6,** 318; reprinted in ACS, 1973, pp. 18-20.

EST (1973a), "Recycled building products. . .," *ibid.*, **7**(1), 12.

EST (1973b), "Copper Industry Uses Much Scrap Iron," *ibid.*, **7,** 100-102; reprinted in ACS, 1973, pp. 64-66.

EST (1973c), "First Lesson in Resource Recovery," *ibid.*, **7,** 300; reprinted in ACS, 1973, pp. 106-108.

EST (1973d), "Scrap Tires Can Yield Marketable Products," *ibid.*, **7,** 188; reprinted in ACS, 1973, pp. 67-69.

EST (1973e), "Repairing Streets in Cold Weather. . .," *ibid.*, **7**(1), 13.

EST (1974a), "The Electroplaters are Polishing Up," *ibid.*, **8**(5), 406-407.

EST (1974b), "Oil: It Never Wears Out, It Just Gets Dirty," *ibid.*, **8**(4), 310-311.

EST (1975a), "The Hang-Ups on Recycling," *ibid.*, **9**(12), 1015.

EST (1975b), "Some Trash Can Really be Sweet," *ibid.*, **9**(12), 1011.

EST (1975c), "Green Systems for Wastewater Treatment," *ibid.*, **9**(5), 408-409.

EST (1975d), "How is the Bottle Battle Going?" *ibid.*, **9**(10), 906.

EST (1975e), "Tennessee's War on Solid Waste," *ibid.*, **9**(5), 406.

EST (1975f), "Disposing of Solid Wastes by Pyrolysis," *ibid.*, **9**(2), 98-99.

EST (1975g), "Coal Can Be a Clean Fuel," *ibid.*, **9**(1), 18-19.

EST (1975h), "Recovering Chlorine from Waste HCl," *ibid.*, **9**(1), 16-17.

EST (1975i), "Water Reuse Considered," *ibid.*, **9**(8), 708-709.

EST (1976a), "Green Plants as Solar Energy Converters," *ibid.*, **10**(6), 526, 528.

EST (1976b), "How Trash is Being Turned into Useful Heat," *ibid.*, **10**(9), 860-861.

EST (1976c), "Energy Recovery from Pulping Wastes," *ibid.*, **10**(8), 735.

EST (1976d), "Ion Exchange: New Design Offers Coninuous Process," *ibid.*,**10**(10), 976-977.

Evans, E. M. (1974), "Degradability of Plastics," in Staudinger, 1974, pp. 157-182. (B.P. Chemicals, U.K.)

Faber, John H., John P. Capp, and John D. Spencer (1967), Eds., *Symposium on Fly Ash Utilization,* Proceedings, Edison Electric Institute, National Coal Association, Bureau of Mines Symposium, 14-16 March, Pittsburgh, Pennsylvania, Bureau of Mines Information Circular 8348, U.S. Department of the Interior, Washington, D.C.

Fagan, R. D., A. O. Converse, and H. E. Grethlein (1971a), "The Economic Analysis of the Acid Hydrolysis of Refuse," in ANERAC, 1971, pp. 124-143. (Thayer School of Engineering, Dartmouth College, Hanover, New Hampshire)

Fagan, Robert D., Hans E. Grethlein, Alvin O. Converse, and Andrew Porteous (1971b), "Kinetics of the Acid Hydrolysis of Cellulose Found in Paper Refuse," *Environmental Science and Technology,* **5**(6), 545-547. (Thayer School of Engineering, Dartmouth College, Hanover, New Hampshire)

Falcone, James S., and Robert W. Spencer (1975), "Silicates Expand Role in Waste Treatment, Bleaching, Deinking," *Pulp and Paper,* **49**(14), 114-117. (Philadelphia

Quartz Co., Lafayette Hill, Pennsylvania)

Falkehag, S. Ingemar, (1975), "Lignin in Materials," in APS, 1975, pp. 247-257. (Westvaco Research Center, North Charleston, South Carolina)

Falkie, Thomas V. (1975), "Bureau of Mines Research Activities in Secondary Resource Recovery," *Phoenix Quarterly,* **7**(3), 2-9. (U.S. Bureau of Mines)

Fan, Dah-Nien (1975), "On the Air Classified Light Fraction of Shredded Municipal Solid Waste. I. Composition and Physical Characteristics," *Resource Recovery and Conservation,* **1**, 141-150. (Department of Mechanical Engineering, Howard University, Washington, D.C. 20059)

Fenton, Richard (1975), "Current Trends in Municipal Solid Waste Disposal in New York City," *Resource Recovery and Conservation,* **1**, 167-176. (Environmental Protection Administration, City of New York, 10007)

Fergusson, W. C. (1974), "Plastics, Their Contribution to Society and Considerations of Their Disposal," in Staudinger, 1974, pp. 1-52. (ICI Ltd., U.K.)

Fergusson, W. C. (1975), "Plastics Waste Recovery and Recycling," in Henstock, 1975a. (ICI Plastics Division)

Fergusson W. C. (1976), "Recovery of Resources from Plastics Industrial Waste," in SCI, 1976c, pp. 725-729. (The Plastics and Rubber Institute, London, U.K.)

Fife, James A. (1973), "Solid Waste Disposal: Incineration or Pyrolysis," *Environmental Science and Technology,* **7,** 308; reprinted in ACS, 1973, pp. 98-102. (Metcalf and Eddy, Inc., Boston, Massachusetts 02116)

Finney, C. S., and D. E. Garrett (1974), "Flash Pyrolysis of Solid Wastes," *Energy Sources,* **1**(3), 295-314. (Garrett Research and Development Co., La Verne, California)

Finney, C. S., and D. E. Garrett (1975), "Flash Pyrolysis of Solid Wastes," *Industrial Process Design for Pollution Control,* **6,** 108-117, Proceedings, American Institute of Chemical Engineers Workshop. (Garrett Research and Development Co., La Verne, California)

Finney, C. S., and J. G. Sotter (1975), "Pyrolytic Oil from Tree Bark: Its Production and Combustion Properties," *American Institute of Chemical Engineers Symposium Series,* **71** (146, Forest Product Residuals), 51-60. (Garrett Research and Development Co., La Verne, California)

Fish, H. (1972), *Water Pollution and its Control: A Bibliography,* Gothard House, Henley-on-Thames, Oxford, U.K.

Fisher, M. J. (1976), "Fluidized Bed Refuse Incineration Systems," *Solid Wastes,* **66**(7), 312-318. (Heenan Environmental Systems Ltd., P.O. Box 14, Worcester WR4 9HA, U.K.)

Fisher, Walter W., and Rees D. Groves (1976), *Copper Cementation in a Revolving-Drum Reactor, A Kinetic Study,* Bureau of Mines Report of Investigations 8098, U.S. Department of the Interior, Washington, D.C. (Salt Lake City Metallurgy Research Center)

Fisher, T. F., M. L. Kasbohm, and J. R. Rivero (1976), "The PUROX System," in IGT, pp. 447-459. (Linde Division, Union Carbide Corporation, Tonawanda, New York)

Flaig, Wolfgang (1973), "Slow Releasing Nitrogen Fertiliser from the Waste Product

Lignin Sulphonates,'' *Chemistry and Industry* (London), (12), 553. (Institut für Biochemie des Bodens, Bundessallee 50, 3301 Braunschweig, Germany)

Fleischmann, Martin, and Derek Pletcher (1975), ''Industrial Electrosynthesis, '' *Chemistry in Britain,* **11**(2), 50-54. (Chemistry Department, Southampton University, U.K.)

Fletcher, A. W. (1971), ''Metal Extraction from Waste Materials,'' *Chemistry and Industry* (London), (28), 776-780. (Warren Spring Laboratory, Stevenage, U.K.)

Fletcher, A. W. (1973), ''Solvent Extraction in Processes for Metal Recovery from Scrap and Waste,'' *Chemistry and Industry* (London), (9), 414-419. (Warren Spring Laboratory, Stevenage, U.K.)

Fletcher, A. W. (1976), ''Hydrometallurgy,'' *Industrial Recovery,* **22**(2), 10, 12, 24; **22**(3), 14, 16, 18. (Warren Spring Laboratory, Stevenage, U.K.)

Flett, Douglas S. (1971), ''The Electrowinning of Copper from Dilute Copper Sulphate Solutions with a Fluidised-Bed Cathode,'' *Chemistry and Industry* (London), (11), 300-302. (Warren Spring Laboratory, Stevenage, U.K.)

Flett, Douglas S., and Douglas Pearson (1975), ''Recovery of the Metal Values from Effluents,'' *Chemistry and Industry* (London), 2 August, 639-645. (Warren Spring Laboratory, Stevenage, U.K.)

Flintoff, F. L. D. (1974a), ''The Disposal of Solid Wastes,'' in Staudinger, 1974, pp. 85-131.

Flintoff, F. L. D. (1974b), ''Recycling and Recovery of Plastics,'' in Staudinger, 1974, pp. 133-155.

Flintoff, Frank, and Ronald Millard (1969), *Public Cleansing,* Maclaren, London.

Folk, Hugh (1973), *Two Papers on the Effects of Mandatory Deposits on Beverage Containers,* U.S. NTIS PB-227,884.

Forster, C. F., and J. C. Jones (1976), ''The Bioplex Concept.'' in Birch et al., 1976, pp. 278-291. (Wessex Water Authority, Redcliffe Way, Bristol BS1 6NY, U.K.)

Fox, Robert D. (1973), ''Pollution Control at the Source,'' *Chemical Engineering,* **80**(18), 72-82. (Dow Chemical, U.S.A)

Fox R. D., R. T. Keller, and C. J. Pinamont (1973) *Recondition and Reuse of Organically Contaminated Waste Sodium Chloride Brines* U.S. Environmental Protection Agency Report EPA-R2-73-200. (Environmental Research Laboratory, Dow Chemical, Midland, Michigan)

France, H. G. (1975), ''Recycle of Tan Liquor from Organic Acid Pickle/Tan Process,'' *Journal of the American Leather Chemists Association,* **70**(5), 206-219. (Union Carbide Corporation, South Charleston, West Virginia)

Franklin, W. E., D. Bendersky, L. J. Shannon, and W. R. Park (1973a), *Resource Recovery. State of Technology,* Midwest Research Institute Report to Council on Environmental Quality, Washington, D.C., U.S. NTIS PB-214,149; summary in Jackson, 1975.

Franklin, W. E., D. Bendersky, L. J. Shannon and W. R. Park (1973b), *Resource Recovery. Catalog of Processes,* Midwest Research Institute Report to Council on Environmental Quality, Washington, D.C., U.S. NTIS PB-214,148; summary in

Jackson, 1975. (Midwest Research Institute, Kansas City, Missouri)

Franklin, William E., David Bendersky, William R. Park, and Robert G. Hunt (1975), ''Potential Energy Conservation from Recycling Metals in Urban Solid Wastes,'' in Williams, 1975, Chapter 5, pp. 171-218.

Froisland, L. J., K. C. Dean, and C. J. Chindgren (1972), *Upgrading Junk Auto Shredder Rejects,* Bureau of Mines Solid Waste Research Program, Technical Progress Report 53, U.S. Department of the Interior, Washington, D.C.

Fry, L. John, and Richard Merrill (1973), ''Methane Digesters for Fuel Gas and Fertilizer,'' *New Alchemy Institute Newsletter,* (3). (Newsletter #3 of the New Alchemy Institute is available for $4.00 from New Alchemy Institute West, P.O. Box 2206, Santa Cruz, California 95063)

Fukui, Saburo, Yoshihiko Nishimoto, Yoshiaki Mizumoto, and Shigeo Hosegawa (1974), ''Recovery of Polyolefins from Wastes,'' *Japanese Kokai,* 74, 107,070; *Chemical Abstracts,* **82,** 141230. (Mitsubishi Heavy Industries, Japan)

Fulweiler, W. H. (1930), ''Use of Sewage Gas as City Gas,'' *Sewage Works Journal,* **2**(3), 424-434.

Funk, Harvey D., and Stuart H. Russell (1976), ''Energy and Materials Recovery System, Ames, Iowa,'' in Aleshin, 1976, pp. 133-140. (Henningson, Durham and Richardson, Inc., 8404 Indian Hills Drive, Omaha, Nebraska 68114)

Futurist (1974), ''The Ark: A Solar-Heated, Wind-Powered Greenhouse and Fish Pond Complex,'' **8**(6), 296-298.

Ganotis, Chris G., and Richard E. Hopper (1976), ''The Resource Recovery Industry,'' *Environmental Science and Technology,* **10**(5), 425-429 . (MITRE Corp., Bedford, Massachusetts 01730; and Office of Solid Waste Management, EPA, Washington, D.C.)

Garner, William, and Ronald R. Ritter (1974), ''Animal Waste Management,'' in Yen, 1974a, Chapter 3, pp. 101-146. (U.S. Environmental Protection Agency)

Garrels, Robert M., Fred T. Mackenzie, and Cynthia Hunt (1975), *Chemical Cycles and the Global Environment,* William Kaufmann, Los Altos, California.

Gaylord, Norman G. (1976), ''Role of Compatibilization in Polymer Utilization,'' *Chem-Tech,* **6**(6), 392-395. (Gaylord Research Institute, 273 Ferry St., Newark, New Jersey 07105)

Geller, E. Scott (1975), ''Increasing Desired Waste Disposals with Instructions,'' *Man-Environment Systems,* **5**(2), 125-128. (Psychology Department, Virginia Polytechnic Institute and State University, Blacksburg)

Geller, E. Scott, John C. Farris, and David S. Post (1973), ''Prompting a Consumer Behavior for Pollution Control,'' *Journal of Applied Behavior Analysis,* **6**(3), 367-376. (Virginia Polytechnic Institute and State University, Blacksburg)

Geller, E. Scott, Jeanne L. Chaffee, and Richard E. Ingram (1975), ''Promoting Paper Recycling on a University Campus,'' *Journal of Environmental Systems,* **5**(1), 39-57. (Virginia Polytechnic Institute and State University, Blacksburg)

George, L. C., and Andrew A. Cochran (1970), *Recovery of Metals from Electroplating Wastes by the Waste-Plus-Waste Method,* Bureau of Mines Solid Waste Research

Program, Technical Progress Report 27, U.S. Department of the Interior, Washington, D.C.

George, L. C., and A. A. Cochran (1974), "Recovery of Metals and Other Materials from Chromium Etching and Electrochemical Machining Wastes," in Aleshin, 1974, pp. 346-353. (Rolla Metallurgy Research Center, U.S. Bureau of Mines)

George, L. C., L. N. Hjersted, A. A. Cochran, and D. R. Allan (1976), "Development and Commercialization of the Waste-Plus-Waste Process for Recovering Metals from Cyanide and Other Wastes," presented at 82nd Meeting, American Institute of Chemical Engineers, 29 August-1 September, Atlantic City, New Jersey. (Rolla Metallurgy Research Center, U.S. Bureau of Mines; and Conservation Chemical Co., Kansas City, Missouri)

Ghose, T. K. (1969), "Foods of the Future," *Process Biochemistry,* **4** (December), 43-46. (Department of Biochemical Engineering, Jodarpur University, India; and U.S. Army Natick Laboratories, Massachusetts)

Ghosh, Sambhunath, and Donald L. Klass (1974), "Conversion of Urban Refuse to Substitute Natural Gas by the Biogas™ Process," in Aleshin, 1974, pp. 196-211; summary in *Public Works,* 1976. (Institute of Gas Technology, Chicago, Illinois)

Giddings, J. Calvin, and Manus B. Monroe (1972), *Our Chemical Environment,* Canfield Press, San Francisco.

Gilbert, R. J., and D. W. Lovelock (1975), Eds., *Microbial Aspects of the Deterioration of Materials,* Society for Applied Bacteriology Technical Series No. 9, Academic Press, London.

Godfrey, D. E., and L. C. Tupper (1974), "An Approach to the Simulation of a Solid Waste Separation Plant," in Aleshin, 1974, pp. 40-46. (General Electric Co., Schenectady, New York)

Göransson, B. (1972), Ed., *Industrial Waste Water* (Plenary and Main Technical Lectures, IUPAC International Congress on Industrial Waste Water, November, 1970, Stockholm, Sweden), Butterworths, London; *Pure and Applied Chemistry* (1972), **29**(1-3).

Gogineni, M. R., W. C. Taylor, A. L. Plumley, and James Jonakin (1973), "Wet Scrubbing of Sulphur Oxides from Flue Gases," in Jimeson and Spindt, Chapter 12, pp. 135-151. (Combustion Engineering, Inc., Windsor, Connecticut 06095)

Goldstein, Jerome (1969), *Garbage as You Like It,* Rodale Books, Emmaus, Pennsylvania.

Goldstein, Irving S. (1975a), "Perspectives on Production of Phenols and Phenolic Acids from Lignin and Bark," in APS, 1975, pp. 259-267. (Department of Wood and Paper Science, North Carolina State University, Raleigh)

Goldstein, Irving S. (1975b), "Chemicals from Lignocellulose," in Symposium on Enzymatic Conversion of Cellulosic Materials: Technology and Applications, Boston, Massachusetts, 8-10 Septemter. (Department of Wood and Paper Science, North Carolina State University, Raleigh)

Goldstein, Irving S. (1975c), "Potential for Converting Wood into Plastics," *Science,* **189**, 847-852; presented at 169th American Chemical Society Meeting, 7 April, Philadelphia, Pennsylvania. (School of Forest Resources, North Carolina State

University, Raleigh)

Goldstein, Irving S. (1976), "Wood as a Source of Chemical Feedstocks," 69th Annual Meeting, American Institute of Chemical Engineers, December, Chicago, Illinois. (Department of Wood and Paper Science, North Carolina State University, Raleigh)

Golueke, Clarence G. (1975), "Domestic Cellulose Waste," *Compost Science,* **16**(1), 16-19; in NSF, 1975.

Golueke, Clarence G., and P. H. McGauhey (1976), "Waste Materials," in Hollander, 1976, pp. 257-277. (SERL, University of California, Richmond)

Golueke, Clarence G., and William J. Oswald (1959), "Biological Conversion of Light Energy to the Chemical Energy of Methane," *Applied Microbiology,* **7**, 219-227. (SERL, University of California, Richmond)

Golueke, Clarence G., and William J. Oswald (1963), "Power from Solar Energy via Algae Produced Methane," *Solar Energy,* **7**(3), 86-92. (SERL, University of California,Berkeley)

Golueke, Clarence G., William J. Oswald, Gordon L. Dugan, Charles E. Rixford, and Stanley Scher (1973), *Photosynthetic Reclamation of Agricultural Solid and Liquid Wastes,* Sanitary Engineering Research Laboratory, University of California, Report to Environmental Protection Agency, EPA-R3-73-031, U.S. NTIS PB-222,454.

Goode, Alan H., M. E. Tyrell, and I. L. Field (1973), "Glass Wool and Other Ceramic Products from Waste Glass," in Albuquerque, 1973, pp. 235-249. (Tuscaloosa Metallurgy Research Laboratory, U.S. Bureau of Mines)

Gough, William C., and Bernard J. Eastlund (1971), "The Prospects of Fusion Power," *Scientific American,* **224**(2), 50-64.

Gounelle de Pontanel, P. (1972), Ed., *Proteins from Hydrocarbons,* Proceedings of the 1972 Symposium, Aix-en-Provence, and Relevant Guidelines of the U.N. Advisory Group, distributed in English throughout the world by Academic Press, London.

Grames Lloyd M., and Ray W. Kueneman (1968), "Primary Treatment of Potato Processing Wastes with By-Product Feed Recovery," 41st Annual Conference, Water Pollution Control Federation, 22-27 September, Chicago, Illinois. (Eimco Corporation, Salt Lake City, Utah; and J. R. Simplot Co., Caldwell, Idaho)

Grant, R. A. (1974), "Protein Recovery from Process Effluents Using Ion-Exchange Resins," *Process Biochemistry,* **9**(2), 11-14. (Tasman Vaccine (UK) Ltd, Creekmoor, Poole, Dorset, U.K.)

Grant, R. A. (1976), "Protein Recovery from Meat, Poultry, and Fish Processing Plants," Birch et al., 1976, pp. 205-220. (Ecotech Systems (UK) Ltd., Creekmoor, Poole, Dorset, U.K.)

Gray, K. R., and A. J. Biddlestone (1973), "Composting—Process Parameters," *Chemical Engineer,* (270), 71-76. (Department of Chemical Engineering, University of Birmingham, U.K.)

Greenberg, Jacob (1972), "Method of Catalytically Inducing Oxidation of Carbonaceous Materials by the Use of Molten Salts," *U.S. Patent,* 3,647,358.

Greenberg, Jacob, and Douglas C. Whitaker (1972), "Treatment of Sewage and Other Contaminated Liquids with Recovery of Water by Distillation and Oxidation," *U.S. Patent,* 3,642,583.

Gregg, John S., R. Rhodes Trussel, and James M. Montgomery (1975), ''Utility/Industry Joint Effort to Use Reclaimed Water for Cooling Water,'' *Proceedings AWWA Annual Conference, 95th,* 15-2, 22pp. (Water Supply Division, Contra Costa County Water District, California)

Greiser, C. L. (1971), ''Total Recovery Possible?'' *Water and Wastes Engineering,* **8**(7), 34-35. (Research Division, Lockheed Missile and Space Co., Palo Alto, California)

Grethlein, Hans E. (1975), ''Acid Hydrolysis of Refuse,'' *Biotechnology and Bioengineering Symposium,* **5**,(Cellulose Chemical and Energy Resources), 303-318. (Thayer School of Engineering, Dartmouth College, Hanover, New Hampshire)

Griffin, H. L., J. H. Sloneker, and G. E. Inglett (1974), ''Cellulose Production by *Trichoderma viride* on Feedlot Waste,'' *Applied Microbiology,* **27**(6), 1061-1066. (Northern Regional Research Laboratory, USDA, Peoria, Illinois)

Griffith, D. C., A. G. Fane, C. J. D. Fell, and R. T. Fowler (1975), ''Use of Membrane Processes in the Treatment of Aqueous Wastes,'' in Kirov, 1975a, pp. 105-112. (School of Chemical Engineering, University of New South Wales, Australia)

Grindrod, Barry (1971), ''The Marvellous Chicken-powered Motorcar,'' *Mother Earth News,* (10), July, 14-19.

Grinshpan, D. D., F. N. Koputskii, and I. N. Ermolenko (1975), ''Dissolution of Cellulose in Organic Media Containing Nitrogen Tetroxide,'' *Tezisy Dokladov Vses. Konf. Khim. Fiz. Tsellyul.,* 1st, **2,** 180-182 (Russian), Ed. V. S. Gromov, ''Zinatne,'' Riga, USSR; *Chemical Abstracts,* **85**(22), 162166. (Beloruss. Gos. Univ. im Lenina, Minsk, USSR)

Grossman, E. D., and J. R. Thygeson (1974a), ''Closed-Cycle Drying of Manure,'' American Institute of Chemical Engineers, 67th Annual Meeting, Washington, December. (Department of Chemical Engineering, Drexel Universtiy, Philadelphia, Pennsylvania)

Grossman, E. D., and J. R. Thygeson (1974b), ''Farm and Field Wastes,'' in Yen 1974a, Chapter 4, pp. 147-173. (Department of Chemical Engineering, Drexel University, Philadelphia, Pennsylvania)

Grott, G. J. (1974), ''Single Unit Conversion of Unprocessed Auto and Appliance Scrap into Refined Steel,'' in Aleshin, 1974, pp. 431-437. (Western America Ore Company, Coolidge, Arizona)

Grubbs, M. R., M. Paterson, and B. M. Fabuss (1976), ''Air Classification of Municipal Refuse,'' in Aleshin, 1976, pp. 169-174. (Raytheon Service Co., Burlington, Massachusetts)

Grzymek, Jerzy (1974), ''Alumina-from-Clay Plant Gives Cement Bonus,'' *Process Engineering,* February, 43, 45; *Chemical Abstracts,* **82**(8), 47169. (Acad. Min. Metall., Krakow, Poland)

GSA (1976), *Policies and Programs Being Developed to Expand Procurement of Products Containing Recycled Materials,* Report to U.S. Congress by the Comptroller General of the United States, General Services Administration and Department of Defense, PSAD-76-139, May 18.

Guillet, James (1973), *Bio-cyclic Plastics,* University of Toronto Institute of Environmental Studies, Publication EF-17.

Gutfreund, Kurt (1971), *Feasibility Study of the Disposal of Polyethylene Plastic Waste,* U.S. Environmental Protection Agency Solid Waste Management Program Report SW-14c, Washington, D.C.; also numbered as Public Health Service Publication No. 2010; condensed version by Kiefer, 1973a. (IIT Research Institute, Chicago, Illinois)

Gutt, W. (1972), "Aggregates from Waste Materials," *Chemistry and Industry* (London), 3 June, 439-447; reprinted as current Paper 14/72, Building Research Establishment, Department of Environment, U.K.

Gutt, W. (1974), "The Use of By-Products in Concrete," *Resources Policy,* **1**(1), 29-45; published as Current Paper 53/74, Building Research Establishment, Department of Environment, U.K.; presented at *The Conservation of Materials* Conference, Institution of Chemical Engineers, 26-27 March, Harwell, U.K.

Gutt, W., and M. A. Smith (1973), "Utilization of By-Product Calcium Sulphate," *Chemistry and Industry* (London), 7 July, 610-619; included in Gutt et al., 1974, Chapter 5. (Building Research Establishment, Watford, U.K.)

Gutt, W. H., and M. A. Smith (1976), "Aspects of Waste Materials and Their Potential for Use in Concrete," *Resource Recovery and Conservation,* **1,** 345-367. (Building Research Establishment, Watford, England)

Gutt, W., P. J. Nixon, M. A. Smith, W. H. Harrison, and A. D. Russell (1974), *A Survey of the Locations, Disposal, and Prospective Uses of the Major Industrial By-Products and Waste Materials,* Current Paper 19/74, Building Research Establishment, Department of Environment, U.K.; portion published in Gutt and Smith, 1973.

Gutt, W. H., M. A. Smith, and P. J. Nixon (1976), "Utilization of Industrial By-Products in Building and Construction in Great Britain," in Aleshin, 1976, pp. 271-277. (Building Research Establishment, Watford, England)

Ham, Robert K., Warren K. Porter and John J. Reinhardt (1972), "Refuse Milling for Landfill Disposal," in Stump, 1972, pp. 37-72. (Department of Sanitary Engineering, University of Wisconsin; and city of Madison, Wisconsin)

Hamm, D., R. E. Childs, and A. J. Mercuri (1973), "Management and Utilization of Poultry Processing Wastes," in Inglett, 1973, Chapter 10, pp. 101-118. (Richard B. Russell Agricultural Research Center, USDA, Athens, Georgia)

Hammond, V. L., and L. K. Mudge (1975), *Feasibility Study of Use of Molten Salt Technology for Pyrolysis of Solid Waste,* Battelle Pacific-Northwest Laboratories, Report to Environmental Protection Agency, EPA-68-03-0145, U.S. NTIS PB-238,674.

Hammond, V. L., L. K. Mudge, C. H. Allen, and G. F. Schiefelbein (1972), *Energy from Solid Waste by Pyrolysis Incineration,* Battelle Pacific-Northwest Laboratories Report BNWL-SW-4471.

Hammond, V. L., L. K. Mudge, C. H. Allen, and G. F. Schiefelbein (1974), "Energy from Forest Residuals by Gasification of Wood Wastes," *Pulp and Paper,* **48**(2), 54-57. (Chemical Technology Department, Battelle Pacific-Northwest Laboratories Richland, Washington)

Haneman, D. (1977), "Chemical Methods of Solar Energy Conversion," *Proceedings of the Royal Australian Chemical Institute,* **44**(2), 37-41; text of a plenary lecture to RACI National Meeting, May 1976. (School of Physics, University of New South

Wales, Australia)

Hannon, Bruce M. (1972), "Bottles, Cans, Energy," *Environment*, **14**(2), 11-21; summaries in Moore and Moore, 1976; and in Wade, 1976. (Center for Advanced Computation, University of Illinois, Urbana)

Hannon, Bruce (1973), *System Energy and Recycling: A Study of the Beverage Industry*, U.S. NTIS PB-233,183.

Hansen, P. (1973), (compiler), *Solid Waste Recycling Projects; a National Survey*, Publication SW-45, U.S. Environmental Protection Agency.

Hansen, P. (1975), *Residential Paper Recovery: a Municipal Implementation Guide*, Publication SW-155, U.S. Environmental Protection Agency.

Harney, Brian M. (1975), "Methanol From Coal—A Step Towards Energy Self-Sufficiency," *Energy Sources*, **2**(3), 233-249. (U.S. Bureau of Mines, Washington, D.C.)

Harris, F. L., G. B. Humphreys, and K. S. Spiegler (1976), "Reverse Osmosis (Hyperfiltration) in Water Desalination," in Meares, 1976, Chapter 4, pp. 121-186.

Harrison, J. S. (1963), "Bakers' Yeast," in Rainbow and Rose, 1963, Chaper 2, pp. 9-33. (Research and Development Department, The Distillers Co. Ltd., Epsom, Surrey, U.K.)

Harrison, David E. F. (1976), "Making Protein from Methane,"*ChemTech*, **6**(9), 570-574. (Shell Research Ltd., Sittingbourne, Kent, U.K.)

Harrison, G., V. I. Lakshmanan, and G. J. Lawson (1976), "The Extraction of Zinc(II) from Sulphate and Chloride Solutions with Kelex 100 and Versatic 911 in Kerosene,"*Hydrometallurgy*, **1**, 339-347. (Wolfson Secondary Metals Research Group, University of Birmingham, U.K.)

Hartinger, Ludwig (1975), "Recycling-Möglichkeiten zur Rückgewinnung von Lösungsmitteln und Stoffen" (Recycling Possibilities for the Recovery of Solvents and Other Materials), *Metalloberfläche*, **29**(5), 221-231.

Hartley, R. G. (1975), "Domestic Waste as an Energy Source," in *Seminar on Energy Resources for Domestic Use* (W. A. Division of the Institution of Engineers; W. A. Chapter of the Royal Australian Institute of Architects; Western Australian Institute of Technology), 21 October, Perth, Western Australia. (Maunsell & Partners Pty Ltd., Perth, Western Australia)

Harvey, K. (1973), "Incineration of Municipal Refuse—Practical Aspects," in CE, 1973, pp. 65-70. (City of Birmingham Salvage Department, 124 Edmund St., Birmingham B3 2EY, U.K.)

Haynes, Dewey C. (1974), "Water Recycling in the Pulp and Paper Industry," *Tappi* (Journal of the Technical Association of the Pulp and Paper Industry), **57**(4), 45-52. (Buckeye Cellulose Corporation, Memphis, Tennessee)

Haynes, C. D., D. C. Hagood, and G. S. Walker (1974), "Pelletized Waste Oil-Coal Dust Mixtures as a Fuel Source," in Aleshin, 1974, pp. 47-49. (Department of Civil and Mineral Engineering, University of Alabama)

Hays, John T. (1973), "Composting of Municipal Refuse," in Inglett, 1973, Chapter 17, pp. 205-215. (Hercules, Inc., Wilmington, Delaware)

Hearon, W. M., Cheng Fan Lo., and John F. Witte (1975), "Chemicals from Cellulose," in APS, 1975, pp. 77-84. (Boise Cascade Corp., Portland, Oregon)

Heitmann, Aloys, and Peter Reher (1974), "Recycling—Prozesse für Schwefel-Verbindungen" (Recycling Processes for Sulfur Componds), *Chemie Ingenieur Technique,* **46**(14), 589-594. (Bayer AG, Leyerkusen, Germany)

Helliwell, P. R. (1972), *Incineration of Refuse and Sludge,* Symposium Proceedings, Southampton University Press, U.K.

Henn, John J. (1975), *Updated Cost Evaluation of a Metal and Mineral Recovery Process for Treating Municipal Incinerator Residues,* Bureau of Mines Information Circular 8691, U.S. Department of the Interior, Washington, D.C.

Henstock, Michael E. (1975a), Ed., *Recycling and Disposal of Solid Waste,* Proceedings of a Course Organized by the Department of Metallurgy and Materials Science, University of Nottingham, 1-5 April, 1974, Pergamon, Oxford.

Henstock, M. E. (1975b), "The Economic Potential of Recycling," in Henstock, 1975a. (University of Nottingham, U.K.)

Henstock, M. E. (1975c), "The Separation of Solid Waste into Useful Fractions," in Henstock, 1975a. (University of Nottingham, U.K.)

Henstock, Michael E. (1976), "Realities of Recycling," in SCI, 1976c, pp. 709-713. (Department of Metallurgy and Materials Science, Nottingham University, U.K.)

Hershaft, Alex (1972), *Advances in Solid Waste Treatment Technology,* Grumman Research Report RE-437 J, U.S. NTIS AD-749,409; in abridged form as "Solid Waste Treatment Technology," *Environmental Science and Technology,* **6,** 412; reprinted in ACS, 1973, pp. 82-91; summary in Jackson, 1975. (Grumman Aerospace Corp., Bethpage, New York)

Heylin, Michael (1971), "Consumer Packaging," *Chemical and Engineering News,* **49**(16), 20-23. (Staff, CEN)

Higginson, A. E. (1974), "The Storage and Collection of Refuse," in Staudinger, 1974, pp. 53-83. (Design and Development, Refuse Disposal, Greater London Council, U.K.)

Higginson, A. E. (1976), "How to Make Waste Pay," Report on Conference, 11-12 November, 1975, London, *Solid Wastes,* **66**(1), 22-25.

Higley, L. W., and H. H. Fukubayashi (1974), "Method for Recovery of Zinc and Lead from Electric Furnace Steelmaking Dusts," in Aleshin, 1974, pp. 295-302. (Rolla Metallurgy Research Center, U.S. Bureau of Mines)

Hills, Graham (1976), "Electrochemistry," *Chemistry in Britain,* **12**(9), 291-293. (Chemistry Department, Southampton University, U.K.)

Hiraoka, Masakatsu (1975), "Energy Recovery from Solid Waste," *Kagaku To Kogyo* (Tokyo), **28**(6), 417-420 (Japanese); *Chemical Abstracts,* **84**, 124360. (Faculty of Engineering, Kyoto University, Japan)

Hirst, Eric (1973a), "The Energy Cost of Pollution Control," *Environment,* **15**, 37-44; summary of Hirst, 1973c. (Oak Ridge National Laboratory, Tennessee)

Hirst, Eric (1973b), "Pollution Control Energy Costs," Paper 73-WA/Ener-7 in American Society of Mechanical Engineers, Winter Annual Meeting, Detroit, Michigan, 11-15

November. (Oak Ridge National Laboratory, Tennessee)

Hirst, Eric (1973c), *Energy Implications of Several Environmental Quality Strategies,* ORNL-NSF-EP-53, Oak Ridge National Laboratory, Tennessee.

Hirst, Eric (1975), "Energy Implications of Cleanup Operations," *Environmental Science and Technology,* **9**(1), 25-28. (Federal Energy Administration, Washington, D.C.)

Hitchell, G. (1975), "Pulverisation of Municipal Refuse," in Kirov, 1975a, pp. 79-82. (Municipality of Canterbury, N.S.W., Australia)

Hitte, Stephen (1976), "Anaerobic Digestion of Solid Waste and Sewage Sludge into Methane," *Compost Science,* **17**(1), 26-30 (1976), (Office of Solid Waste Management Programs, EPA, Washington, D.C.)

Hoad, Brian (1974a), "Recycling: Just a Waste of Time," *Bulletin* (Australia), **96** (4887), 28-31.

Hoad, Brian (1974b), "Why Spend $700 on Garbage?" *Bulletin* (Australia), **96** (4930), 23.

Hoge, William H. (1976), *Mineral Papers, Porous Films and Synpulps* (A Report on New Concepts Including Low-Energy, Low-Capital, Pollution-Free Papermaking by Extrusion), Paperfilm Associates, Box 10, R.D. 2, Flemington, New Jersey 08822.

Hollander, Jack M. (1976), Ed., *Annual Review of Energy,* **1,** Annual Reviews, Inc., Palo Alto, California.

Holloway, Clifford C., and Lawrence J. Scanlan (1975), *Municipal Solid Waste Recycling Treatment Plant,* Report by Environmetric Systems, Perth, Western Australia.

Holman, James L., James B. Stephenson, and James W. Jensen (1972), *Processing the Plastics from Urban Refuse,* U.S. Bureau of Mines; U.S. NTIS PB-208,014; summary in Jackson, 1975.

Holman, James L., J. B. Stephenson, and M. J. Adam (1974), *Recycling of Plastics from Urban and Industrial Refuse,* Bureau of Mines Report of Investigations 7955, U.S. Department of the Interior, Washington, D.C.

Honeysett, J. D. (1975), "The Packaging Industry and Solid Waste Management," in Kirov, 1975a, pp. 163-166. (Packaging Industry Environmental Council, Australia)

Hopper, Richard E. (1975), *A Nationwide Survey of Resource Recovery Activities,"* Report SW-142, U.S. Environmental Protection Agency; see also McEwen, 1976.

Hopwood, A. P., and G. D. Rosen (1972), "Protein and Fat Recovery from Effluents, " *Process Biochemistry,* **7**(March), 15-17. (Alwatech a.s., Oslo)

Horner and Shifrin, Inc. (1973), *Solid Waste as Fuel for Power Plants,* Report SW-36d, U.S. Environmental Protection Agency.

Horstkotte, G. A., D. G. Niles, D. S. Parker, and D. H. Caldwell (1974), "Full-Scale Testing of a Water Reclamation System," *Journal of Water Pollution Control Federation,* **46**(1), 181-197; presented at 45th Annual Conference, 12 October 1972, Atlanta, Georgia. (Central Contra Costa Sanitary District, Walnut Creek, California; and Brown and Caldwell, San Francisco)

Horton, Bernard S., Robert L. Goldsmith, and Robert R. Zall (1972), "Membrane Processing of Cheese Whey Reaches Commercial Scale," *Food Technology,* **26**,

30-35. (Abcor, Inc., Cambridge, Massachusetts; and Department of Food Science, Cornell University, New York)

Howard, K. R. (1972), "Composting of City Refuse — Auckland's Experiences," in Kirov, 1972, pp. 51-60. (Auckland City Council, New Zealand)

Howard, Frank A. (1975), "Advances in Chemical and Solvent Recovery," *Water and Waste Treatment,* **18**(6), 18. (Crewe Chemicals Ltd., Hall Lane, Sandbach, Cheshire, U.K.)

Howard, Frank A. (1976), "Chemical Recovery," in SCI, 1976a, pp. 239-242. (Crewe Chemicals Ltd., Hall Lane, Sandbach, Cheshire, U.K.)

Howden, Patrick (1974), "Ecological Housing—A Detailed Look at Closed-Circuit Living," *Shelter* (Australian Department of Housing and Construction) (13), October. (Eco-Tech Workshop, University of Sydney, Australia)

Howden, Patrick (1975), "The Role of Electronics in 'Closed Cycle' Living: Sydney University Builds an 'Ecology House,' " *Electronics Australia,* June, 26-29. (Eco-Tech Workshop, University of Sydney, Australia)

Hughes, Michael A. (1975a) "The Recovery of Metals from Secondary Sources: Solvent Extraction Routes," *Chemistry and Industry* (London), (24), 1042-1046. (School of Chemical Engineering, University of Bradford, Bradford BD7 1DP, U.K.)

Hughes, D. (1975b), "Organic Wastes," in Henstock, 1975a. (University of Cardiff, Wales)

Hughes, David, and Clive Jones (1975), "Waste Not, Want Not," *New Scientist,* **65**(941), 705-708. (Department of Microbiology and Wolfson Laboratory for the Biology of Industry, University College, Cardiff, and Cardiff University Industry Centre, Wales)

Humber, N. (1975), "Waste Reduction and Resource Recovery—There's Room for Both," *Waste Age,* **6**(11), 38, 40, 41, 44.

Humphrey, Arthur E. (1966), "Starvation: Chemical Engineering Can Help Fight It," *Chemical Engineering,* **73**(18 July), 149-154. (School of Chemical Engineering, University of Pennsylvania)

Hunt, Robert G., and William E. Franklin (1973), "Environmental Effects of Recycling Paper," in AIChE, 1973, pp. 67-78; Midwest Research Institute Reprint MRI 1106.

Hunt, Robert G., and William E. Franklin (1975), "Resource and Environmental Profile Analysis...of Beer Containers," *Chem-Tech,* **5**(8), 474-481. (Franklin Associates, Prairie Village, Kansas)

ICI Australia Petrochemicals Ltd (1975), "Plastics Waste Recycling Process."

IFIAS (1974), *Guidelines for Energy Analysis,* Summary of Recommendations of Workshop, Guldsmedshyttan, Sweden, 26-30 August, International Federation of Institutes of Advanced Study.

IGT (1976), *Clean Fuels from Biomass, Sewage, Urban Refuse, and Agricultural Wastes,* Institute of Gas Technology Symposium, 27 January, Orlando, Florida.

IIT (1971), *Conversion of Polymer Wastes, A Multiclient Research Program Proposal,* No. 71-413C, IIT Research Institute, Chicago, Illinois.

Illner-Paine, O. (1970), "Reverse Osmosis: Cheaper Food, Purer Water," *Science Journal,* **6**(12), 53-55.

Imrie, Frazer (1975), "Single Cell Protein from Agricultural Wastes," *New Scientist,* **66**(950), 458-460. (Tate and Lyle, U.K.)

Imrie, Frazer K. E. (1976), "Foods from Fermentation Processes," in SCI, 1976b, pp. 584-588. (Talres Development Ltd., Reading, U.K.)

Imrie, F. K. E., and R. C. Rightelato (1976), "Production of Microbial Protein from Carbohydrate Wastes in Developing Countries," in Birch et al., 1976, pp. 79-97. (Tate and Lyle Ltd., Reading, U.K.)

Industrial Recovery (1976a), "Domestic Refuse as Fuel in Cement Making," **22**(5), 14, 16.

Industrial Recovery (1976b), "New Uses for Waste Glass," **22**(1), 18, 22.

Industrial Recovery (1976c), "Power-from-Refuse Plant," **22**(6), 12, 14.

Industrial Recovery (1976d), "Packaging—the Scope for Recycling," **22**(7), 12, 14.

Industrial Recovery (1976e), "Re-using Plastics Wastes," **22**(10), 8, 10.

Industrial Recovery (1976f), "Oil Water Separator," **22**(10), 12, 14.

Inglett, George E. (1973), Ed., *Processing Agricultural and Municipal Wastes,* Selected Papers from American Chemical Society Symposium, New York, 27-28 August 1972, Avi Publishing Co., Westport, Connecticut.

Inoue, Kimio (1974), "Method of Oil Resources Recovery by Pyrolysis of Scrap Tires," *Nenryo Oyobi Nensho,* **41**(11), 1001-1010; *Chemical Abstracts,* **83,** 32620. (Kobe Steel Ltd., Kobe, Japan)

Iron and Steel (1971), "Cryogenic Scrap Processing: Production of High Purity Steel Scrap," 346-348.

IRT(1974), *Tire Recycling and Reuse Incentives,* International Research and Technology Corp., Publication SW-32c, U.S. Environmental Protection Agency; U.S. NTIS PB-234,602.

ISIS (1973), *Specifications for Iron and Steel Scrap* (pamphlet), Institute of Scrap Iron and Steel, Washington, D.C.

ISIS (1974), *Railroad Freight Rates: Ferrous Scrap: An Equitable Solution* (pamphlet), Institute of Scrap Iron and Steel, Washington, D.C.

Ito, Kanichi, and Yoshio Hirayama (1975a), "Semi-Wet Selective Pulverizing System," *Resource Recovery and Conservation,* **1**, 45-53. (Ebara Manufacturing Co., Fujisawa, Japan)

Ito, Kanichi, and Yoshio Hirayama (1975b), "Resource Recovery from Municipal Refuse by Semi-Wet Selective Pulverizing System," in Kirov, 1975b, pp. 354-359. (Ebara Manufacturing Co., Fujisawa, Japan)

Ito, Kanichi, and Yoshio Hirayama (1976), "Two-Bed Pyrolysis System—Resource Recovery from Municipal Refuse," Ebara Manufacturing Co., Fujisawa, Japan.

IUPAC (1963), *Reuse of Water in Industry,* International Union of Pure and Applied Chemistry, Butterworth, London.

Ivanova, I.S., Yu. A. Gugnin, A. N. Aleksandrovich, G. I. Zheltova, N. P. Rushnov, and

M. V. Koperina (1975), "Important Raw Material Reserve," *Bumazhnaya Promyshlennost,* (10), 11-13 (Russian); *Chemical Abstracts,* **84**(2), 6723.

Jackson, D. V. (1972), "Metal Recovery from Effluents and Sludges," *Metal Finishing Journal,* **18**(211), 235, 238, 241, 242. (Warren Spring Laboratory, Stevenage, U.K.)

Jackson, D. V. (1973), "Reclamation of Non-ferrous Metals from Waste," *Process Technology International,* **18**(8/9), 329-330. (Warren Spring Laboratory, Stevenage, U.K.)

Jackson, Frederick R. (1974), *Energy from Solid Waste* (Pollution Technology Review No. 8, Energy Technology Review No. 1), Noyes Data Corporation, Park Ridge, New Jersey.

Jackson, Frederick R. (1975), *Recycling and Reclaiming of Municipal Solid Wastes* (Pollution Technology Review No. 17), Noyes Data Corporation, Park Ridge, New Jersey.

Jacques, Keith (1975), "Salt Solution," *Chemistry in Britain* **11**(1), 12-13. (University of Stirling, Scotland)

James, G. V. (1971), *Water Treatment,* Technical Press, London, 4th ed.

Jarvis, E. A. (1959), "Freezing Process to Reclaim Scrap Plastics," *U.S. Patent,* 2,879,005.

Jenkins, D. (1963), "Sewage Treatment," in Rainbow and Rose, Chapter 15, pp. 508-536. (SERL, University of California, Berkeley)

Jensen, James W., James L. Holman, and James B. Stephenson (1974), "Recycling and Disposal of Waste Plastics," in Yen, 1974a, Chapter 7, pp. 219-249. (Rolla Metallurgy Research Center, U.S. Bureau of Mines)

Jimeson, Robert M., and Roderick S. Spindt (1973), Eds., *Pollution Control and Energy Needs,* Symposium, 164th American Chemical Society Meeting, 1972, *Advances in Chemistry Series,* **127.**

Johnston, C. D. (1974), "Waste Glass as Coarse Aggregate for Concrete," *Journal of Testing and Evaluation,* **2**(5), 344-350. (Department of Civil Engineering, University of Calgary, Alberta)

Johnston, I. R. W. (1975), Ed., *Treatment, Recycle and Disposal of Wastes,* Proceedings, Third National Chemical Engineering Conference, Mildura, Victoria, Australia, 20-23 August.

Jones, D. T. (1976), "A Versatile Continuous Ion-Exchange Process for Protein Recovery," in Birch et al., 1976, pp. 242-255. (Development Division, Viscose Group Ltd., South Dock, Swansea, Wales, U.K.)

Jones, Robert W., and K. T. Chandy (1974), "Synthetic Plastics," in Kent, 1974, Chapter 10, pp. 238-300.

Joshi, Ramesh C., Donald M. Duncan, and Howard M. McMaster (1975), "New and Conventional Engineering Uses of Fly Ash," *American Society of Civil Engineers Proceedings,* **101,** 791-806 *(Transportation Engineering Journal,* TE4, No. 11730), (Woodward-Clyde Consultants, Central Region, Kansas City, Missouri)

Kaiser, Elmer R. (1969), "Incineration in Metal Salvage," in Corey, 1969, Chapter 8, pp. 227-238. (School of Engineering and Science, New York University.

Kakela, Peter (1975), "Railroading Scrap," *Environment,* **17**(2), 27-34. (Environments and People Program, Sangamon State University, Springfield, Illinois)

Kaplan, Robert S. (1974), "Bureau of Mines Research on the Recovery and Reuse of Post-Consumer Ferrous Scrap," *Transactions American Foundrymen's Society,* **82,** 487-494.

Kasper, William C. (1974), "Power from Trash," *Environment,* **16**(2), 34-48. (Office of Economic Research, New York State Public Service Commission)

Kaufman, James A., and Alvin H. Weiss (1975), *Solid Waste Conversion. Cellulose Liquefaction,* Worcester Polytechnic Institute Report 670/2-75-031 to Environmental Protection Agency, Washington, D.C.; U.S. NTIS PB-239,509.

Keating, K. B., and J. M. Williams (1976), "The Recovery of Soluble Copper from an Industrial Chemical Waste," *Resource Recovery and Conservation,* **2**(1), 39-55. (Engineering Research and Development Division, du Pont, Wilmington, Delaware)

Kehler, L. H., and G. L. Miles (1972), "The Nature and Disposal of Radioactive Wastes," in Kirov, 1972, pp. 129-135. (Australian Atomic Energy Commission)

Keller, Eugenia (1973), "Recycling: the Modern Day Phoenix," *Chemistry,* **46**(9), 22-26. (Managing Editor, *Chemistry*)

Kelley, Donald W. (1976), "Profitability of Industrial Waste Recovery," in SCI, 1976c, pp. 714-716. (Shirley Aldred Group, Sandy Lane, Worksop, Nottinghamshire, S80 3EY, U.K.)

Kemp, Clinton C., and George C. Szego (1974), "The Energy Plantation," Symposium on Energy Storage, 168th American Chemical Society National Meeting, Atlantic City, New Jersey, 12 September. (Intertechnology Corporation, Warrenton, Virginia 22186)

Kemp, Clinton C., George C. Szego, and Malcolm D. Fraser (1975), "The Energy Plantation," in APS, 1975, pp. 11-19. (Intertechnology Corporation, Warrenton, Virginia 22186)

Kenahan, Charles B. (1971), "Solid Waste: Resources Out of Place," *Environmental Science and Technology,* **5**(7), 594-600; reprinted in Giddings and Monroe, 1972, pp. 229-237; and in ACS, 1973, pp. 74-81. (U.S. Bureau of Mines)

Kenahan, C. B., R. S. Kaplan, J. T. Dunham, and D. G. Linnehan (1973), *Bureau of Mines Research Programs on Recycling and Disposal of Mineral-, Metal-, and Energy-based Wastes,* Bureau of Mines Information Circular 8595, U.S. Department of the Interior, Washington, D.C.; U.S. NTIS PB-227,476.

Kent, James A. (1974), Ed., *Riegel's Handbook of Industrial Chemistry,* Van Nostrand-Reinhold, New York.

Kenward, Michael (1975a), "Garbage: Slow Recovery," in "Energy File," *New Scientist,* **67**(967), 655.

Kenward, Michael (1975b), "Ousting Coals in Newcastle," in "Energy File," *New Scientist,* **68**(977), 549.

Kenward, Michael (1975c), "Incinerators. On Nashville's Skyline," in "Energy File," *New Scientist,* **65**(933), 215.

Kern, P. L., H. C. Peterson, and L. Mariani (1974), "Titanium Dioxide. Recycling of

Sulfuric Acid," *Information Chimie*, **129**, 99-101 (French); *Chemical Abstracts*, **81**(18), 111060. (New Jersey Zinc Co., New York)

Kerridge, David H. (1975). "Recent Advances in Molten Salts as Reaction Media," *Pure and Applied Chemistry*, **41**(3), 355-371.

Khazzoom, J. Daniel, Robert N. Anderson, and Wei-Yue Lim (1976), "Evaluation of Mineral Waste Utilization Processes and Strategies by Net Energy Analysis," in Aleshin, 1976, pp. 33-38. (Department of Operations Research and Institute for Energy Studies, Stanford University; Department of Materials Science, San Jose State University; Department of Applied Earth Sciences, Stanford University, California)

Kiefer, Irene (1973a), *Making Polyethylene More Disposable*, Solid Waste Management Report SW-14c.1, U.S. Environmental Protection Agency; condensation of Gutfreund, 1971.

Kiefer, I. (1973b), *The Salvage Industry; What It Is—How It Works*, Solid Waste Management Report SW-29c.1, U.S. Environmental Protection Agency.

Kihlberg, Reinhold (1972), "The Microbe as a Source of Food," *Annual Reviews of Microbiology*, **26**, 427-446. (Department of Applied Microbiology, Karolinska Institute, Stockholm, Sweden)

King, C. Judson (1971), *Separation Processes*, McGraw Hill (Chemical Engineering Series), New York.

Kinnersly, Pat (1973), "Recycling — Backwards to an Optimum," *New Civil Engineer*, special feature, August.

Kinney, A. M., Inc. (1974), *Franklin, Ohio's Solid Waste Disposal and Fiber Recovery Demonstration Plant; Final Report*, Publications SW-47d.1 and SW-47d.2, U.S. Environmental Protection Agency; U.S. NTIS PB-234,715 and PB-234,716.

Kirov, N. Y. (1972), Ed., *Solid Waste Treatment and Disposal* (1971 Australian Waste Disposal Conference Papers), Ann Arbor Science, Ann Arbor, Michigan. (Department of Fuel Technology, University of New South Wales, Australia)

Kirov, N. Y. (1975a), Ed., *Waste Management, Control, Recovery and Reuse* (Waste Management and Control Conference, 17-19 July 1974, Sydney, Australia), Ann Arbor Science, Ann Arbor, Michigan.

Kirov, N. Y. (1975b), Chairman, *Conversion of Refuse to Energy*, First International Conference and Technical Exhibition, Montreux, Switzerland, 3-5 November.

Kispert, R. G., S. E. Sadek, and D. L. Wise (1975), "An Economic Analysis of Fuel Gas Production from Solid Waste," *Resource Recovery and Conservation*, **1**, 95-109. (Dynatech Corp., Cambridge, Massachusetts)

Kispert, R. G., S. E. Sadek, and D. L. Wise (1976), "An Evaluation of Methane Production from Solid Waste," in RRC, 1976, pp. 245-255. (Dynatech Corp., Cambridge, Massachusetts)

Klass, Donald L. (1974), "Perpetual Methane Economy. Is it Possible?" *ChemTech*, **4**(3), 161-168. (Institute of Gas Technology, Chicago, Illinois)

Klass, Donald L. (1975), "Synthetic Crude Oil from Shale and Coal," *ChemTech*, **5**(8), 499-510. (Institute of Gas Technology, Chicago, Illinois)

Klass, Donald L., and Sambhunath Ghosh (1973), "Fuel Gas from Organic Wastes," *ChemTech,* **2**(11), 689. (Institute of Gas Technology, Chicago, Illinois)

Kleespies, Ernst K., J. P. Bennetts, and Thomas A. Henne (1970), "Gold Recovery from Scrap Electronic Solders by Fused Salt Electrolysis," *Journal of Metals,* **22**(1), 42-44. (Metallurgy Research Center, U.S. Bureau of Mines, Reno, Nevada)

Klein, S. A. (1972), "Anaerobic Digestion of Solid Wastes," *Compost Science,* **13**(1), 6-11. (University of California, Berkeley)

Klötgen, Gerhard E. (1974), "Entfernung von Lacken, Kunstoffen und Emails in oxidierenden Salzschmelzen" (Removal of Varnish, Plastics and Enamels in Oxidizing Fused Salts), *Chem.-Anlagen Verfahren,* (6), 44, 47, 48, 51; *Chemical Abstracts,* **82,** 21351.

Klopfenstein, Terry, and Walter Koers (1973), "Agricultural Cellulosic Wastes for Feed," in Inglett, 1973, Chapter 5, pp. 38-54. (Animal Science Department, University of Nebraska, Lincoln)

Klumb, David (1976a), "Union Electric Company's Solid Waste Utilization System," in RRC, 1976, pp. 225-233. (Union Electric Co., St. Louis, Missouri)

Klumb, David (1976b), "Utilization of Municipal Wastes in Coal Burning Power Plants," Fourth International Ash Utilization Symposium, 24-25 March, St. Louis, Missouri. (Union Electric Co., St. Louis, Missouri)

Klungness, John H. (1974), "Recycled Fiber Properties as Affected by Contaminants and Removal Processes," Forest Service Research Paper FPL 223, U.S. Department of Agriculture, Madison, Wisconsin.

Knights, Dan (1976), "Domestic Refuse Disposal via Cement Kilns," *Solid Wastes,* **66**(7), 320-326. (Associated Portland Cement Manufacturers Ltd., U.K.)

Kondo, Renichi, Masaki Daimon, Seishi Goto, Atsushi Nakamura, and Tadashi Kobayashi (1976), "Fuel Economized Ferrite Cement Made from Blastfurnace and Converter Slags," in Aleshin, 1976, pp. 329-340. (Tokyo Institute of Technology, Ookayama, Meguro-ku, Tokyo 152; Nihon Cement Co., Kiyosumi, Koto-ku, Tokyo 135; and Nippon Kokan Co., Ootemachi, Chiyoda-ku, Tokyo 100, Japan)

Kosaric, N. (1973), "Microbial Products from Food Industry Wastes," in Inglett, 1973, Chapter 13, pp. 143-160. (Chemical and Biochemical Engineering Department, University of Western Ontario, London, Ontario, Canada)

Koziorowski, B., and J. Kucharski (1972), *Industrial Waste Disposal,* transl. J. Bandrowski, ed. G. R. Nellist, Pergamon, Oxford.

Kremen, Seymour S. (1975), "Reverse Osmosis Makes High Quality Water Now," *Environmental Science and Technology,* **9**(4), 314-318. (ROGA Division, UOP, San Diego, California)

Kuester, James L., and Loren Lutes (1976), "Fuel and Feedstock from Refuse," *Environmental Science and Technology,* **10**(4), 339-344. (Arizona State University, Tempe, Arizona 85281; and Rice University, Houston, Texas 72001)

Kugelman, I. J. (1972), *Status of Advanced Waste Treatment,* U.S. Environmental Protection Agency, National Environmental Research Center, Cincinnati, Ohio; U.S. NTIS PB-213,819.

Kuhn, A. T. (1971a), ''A Review of the Electrolytic Recovery of Copper from Dilute Solutions,'' *Chemistry and Industry* (London), (18), 473-476. (Department of Chemistry and Applied Chemistry, University of Salford, Lancashire, U.K.)

Kuhn, A. T. (1971b), ''Electrochemical Techniques for Effluent Treatment,'' *Chemistry and Industry* (London), (34), 946-950. (Department of Chemistry and Applied Chemistry, University of Salford, Lancashire, U.K.)

Kunze, W. (1975), *Investigations of Light Aggregate Made from Sintered Washery Refuse,* Building Research Establishment (Watford, U.K.), Library Translation 1994, from Beton, 1974, (6), 217-222.

Lacy, William J. (1974), ''Industrial Wastewater Technology,'' in Kent, 1976, Chapter 26, pp. 823-873.

Lakshmanan, V. I., and G. J. Lawson (1973), ''Extraction of Cobalt by Kelex 100 and Kelex 100/Versatic 911 Mixtures,'' *Journal of Inorganic and Nuclear Chemistry,* **35**, 4285-4294.

Lakshmanan, V. I., and G. J. Lawson (1975), ''The Extraction of Copper from Aqueous Chloride Solutions with Lix-70 in Kerosene,'' *Journal of Inorganic and Nuclear Chemistry,* **37**, 207-209. (Wolfson Secondary Metals Research Group, University of Birmingham, U.K.)

Landau, Bernard (1974), *Raw Material Outlook Today and Tomorrow,* 82nd General Meeting, 23 May, American Iron and Steel Institute, published as pamphlet, Institute of Scrap Iron and Steel.

Lapp, Ralph E. (1973), *The Logarithmic Century,* Prentice-Hall, New Jersey.

Large, David B. (1973), Ed., *Hidden Waste: Potential for Energy Conservation,* Conservation Foundation, Washington, D.C.

Lash, Leslie D., and Edward G. Kominek (1975), ''Primary-Waste-Treatment Methods,'' *Chemical Engineering* (N. Y.), **82**(21), 49-61. (Eimco Division, Envirotech Corporation, Salt Lake City, Utah)

Lash, L. D., James Martin, and Bill Pude (1973), ''Recycle of Potato Processing Wastewater,'' in Complete WateReuse, 1973. (Eimco Division, Envirotech Corporation, Salt Lake City, Utah)

Lauber, Jack D., Frank W. Conley, and Robert D. Barshield (1973), ''Air Pollution Control of Aluminum and Copper Recycling Processes,'' *Pollution Engineering,* **5**(12), 23-26. (New York State Department of Environmental Conservation, Albany, N. Y.)

Laundrie, J. F., and J. H. Klungness (1973a), *Effective Dry Methods of Separating Thermoplastic Films from Wastepapers,* Forest Service Research Paper FPL 200, U.S. Department of Agriculture.

Laundrie, J. F., and J. H. Klungness (1973b), ''Dry Methods of Separating Plastic Films from Waste Paper,'' *Paper Trade Journal,* 5 February, 34-36. (Forest Products Laboratory, USDA, Madison, Wisconsin)

Lawrence, Alonzo Wm. (1971), ''Granular Activated Carbon Treatment of Primary and Chemically Treated Effluents,'' in ANERAC, 1971, pp. 65-80. (Department of Water Resources Engineering, Cornell University, Ithaca, New York)

Lawson, Gregory J. (1975), "Solvent Extraction of Metals from Chloride Solutions," *Journal of Applied Chemistry and Biotechnology,* **25**, 949-957; presented at the Solvent Extraction and Ion Exchange Group Symposium "Solvent Extraction and Ion Exchange in Hydrometallurgy," University of Leeds, 4-5 April 1974. (Department of Minerals Engineering, University of Birmingham, U.K.)

Layne, G. S., J. O. Huml, L. B. Bangs, and J. H. Meserve (1974), "Aluminum Extraction from Impure Sources by Vapor Transport with Magnesium Fluoride," in Aleshin, 1974, pp. 381-391. (Dow Chemical Co., Midland, Michigan)

League of Women Voters (1972), *Recycle; in Search of New Policies for Resource Recovery,* Publication No. 132, League of Women Voters of the United States.

Leake, Vance G., M. M. Fine, and Henry Dolezel (1974), "Separating Copper from Scrap by Preferential Melting: Laboratory and Economic Evaluation," in Aleshin, 1974, pp. 409-416. (U.S. Bureau of Mines, Twin Cities, Rolla, and Salt Lake City)

Leatherwood, J. M. (1973), *Utilization of Fibrous Wastes as Sources of Nutrients,* U.S. Environmental Protection Agency; U.S. NTIS PB-223, 625.

Lederman, Peter B. (1974), "The Challenge of Disposal of Plastic Packages," in Bruins, 1974, pp. 49-72. (Polytechnic Institute of Brooklyn, New York)

Leek, R. W. (1975), "The Use of Ferrous Scrap in Steelmaking," in Henstock, 1975a. (British Steel Corporation)

Leitz, Frank B. (1976), "Electrodialysis for Industrial Water Cleanup," *Environmental Science and Technology,* **10**(2), 136-139. (U.S. Bureau of Reclamation, Denver, Colorado)

Lessing, Lawrence (1973), "The Salt of the Earth Joins the War on Pollution," *Fortune,* July, 138-147.

Levy, S. J. (1975), *San Diego County Demonstrates Pyrolysis of Solid Waste to Recover Liquid Fuel, Metals, and Glass,* Publication SW-80d.2, U.S. Environmental Protection Agency.

Liebeskind, Judith E. (1973), "Pyrolysis for Solid Waste Management," *ChemTech,* **3,** 537-542.

Lightsey, George R. (1975), *Bark as Trickling-Filter Dewatering Medium for Pulp and Paper Mill Sludge,* U.S. NTIS PB-244,820. (Environmental Resource Center, Georgia Institute of Technology, Atlanta, Georgia)

Liles, K. J., and M. E. Tyrell (1976), *Waste Glass as a Raw Material for Lightweight Aggregate,* Bureau of Mines Report of Investigations 8104, U.S. Department of the Interior, Washington, D.C. (Tuscaloosa Metallurgy Research Laboratory, Alabama)

Lindberg, J. Johan, Vaino A. Era, and Tuure P. Jauhianen (1975), "Lignin as a Raw Material for Synthetic Polymers," in APS, 1975, pp. 269-275. (Department of Wood and Polymer Chemistry, University of Helsinki)

Lindstrom, Carl R. (1975), "The Clivus Multrum System—Composting of Toilet Waste, Food Waste, and Sludge Within the Household," *Water Pollution Control in Low Density Areas,* Proceedings, Rural Environmental Engineering Conference, 1973, ed. William J. Jewell, Rita Swan, University Press New England, Hanover, New Hampshire, pp. 429-444. (Statens Naturvardsverk, Solna, Sweden)

Lingle, S. (1976), Ed., "Resource Recovery Technology Update from U.S.E.P.A.: Status Report in Resource Recovery Technology: Demonstrating Resource Recovery," *Waste Age,* **7**(6), 19, 22, 26, 42, 44-46.

Little, Arthur D., Inc. (1976), *Analysis of Demand and Supply for Secondary Fiber in the U.S. Paper and Paperboard Industry,* Vol. 1, Sections 1-8, 10; Vol. 2, Section 9; Vol. 3, Appendices; Publications SW-115c.1, SW-115c.2, and SW-115c.3, U.S. Environmental Protection Agency; U.S. NTIS PB-250,798, PB-250,905, and PB-250,802.

Löf, George O. G., and Allen V. Kneese (1968), *The Economics of Water Utilization in the Beet Sugar Industry,* Resources for the Future, and John Hopkins University Press, Baltimore, Maryland.

Loehr, Raymond C. (1974), *Agricultural Waste Management: Problems, Processes, and Approaches,* Academic Press, New York. (Department of Agricultural Engineering, Cornell University, Ithaca, New York)

Loehr, Raymond C., T. B. S. Prakasam, E. G. Srinath, and V. D. Joo (1973), *Development and Demonstration of Nutrient Removal from Animal Wastes,* Report EPA-R2-73-095, Office of Research and Monitoring, U.S. Environmental Protection Agency. (College of Agriculture and Life Sciences, Cornell University, Ithaca, New York)

Loran, Bruno (1973), "The Earth Phosphate Cycle," *Journal of Environmental Systems,* **3**(1), 49-68. (Western Research Application Center, University of Southern California)

Loube, M. (1975), *Beverage Containers: The Vermont Experience,* Publication SW-139, U.S. Environmental Protection Agency.

Love, Sam (1974), "The Overconnected Society," *Futurist,* **8**(6), 293-295.

Lowe, Robert A. (1973), *Energy Recovery from Waste. Solid Waste as Supplementary Fuel in Power Plant Boilers,* Solid Waste Management Report SW-36d.ii, U.S. Environmental Protection Agency, Washington, D.C.

Lucas, Ted (1975), *How to Build a Solar Heater*, Ward Ritchie Press.

Lyons, L. A. (1972), "Salvaging and Utilisation of Metals," in Kirov, 1972, pp. 179-183. (Sims Consolidated Ltd., Australia)

Lyons, L. A., and C. S. Tonkin (1975), "Recycling of Metals from Scrapped Automobiles in Australia," in Kirov, 1975a, pp. 223-228. (Sims Consolidated Ltd., Australia)

Macadam, Walter K. (1976), "Design and Pollution Control Features of the Saugas, Massachusetts Steam Generating Refuse-Energy Plant," in RRC, 1976, pp. 235-243. (Wheelabrator-Frye Inc., New York)

McBain, David (1976), "Reverse Osmosis," *Chemistry in Britain,* **12**(9), 281-284. (Elga Products Ltd., Lane End, Buckinghamshire, HP14 3JH, U.K.)

McBee, W. C., and T. A. Sullivan (1976), "Utilization of Secondary Sulfur in Construction Materials," in Aleshin, 1976, pp. 39-52. (Boulder City Metallurgy Research Laboratory, U.S. Bureau of Mines)

McCaull, Julian (1974), "Back to Glass," *Environment,* **16**(1), 6-11.

McElroy, C.T. (1975), ''Emplacement, Disposal, and Utilisation of Colliery Waste,'' in Kirov, 1975a, pp. 171-177. (Clifford McElroy and Associates Pty Ltd.)

McEwen, L. (1976), *A Nationwide Survey of Resource Recovery Activities,* Report SW-142.1, U.S. Environmental Protection Agency.

McGough, Elizabeth (1972), ''Recycling Our Resources,'' *American Youth,* **13**(1), 18-21.

Machida, Hirokichi, and Hideyuki Fujita (1975), ''Recovery of Silver from Aqueous Solutions,'' *Japan. Kokai,* 75 18,318; *Chemical Abstracts,* **83**(12), 106178. (Konishiroku Photo Industry Co. Ltd.)

McIntyre, A. D., and M. M. Papic (1974), ''Pyrolysis of Municipal Solid Waste,'' *Canadian Journal of Chemical Engineering,* **52**(2), 263-272. (British Columbia Research, Vancouver)

Mackenzie, John D. (1973), ''Foamed Glass and Tiles from Waste Containers,'' in Albuquerque, 1973, pp. 291-303. (School of Engineering and Applied Sciences, University of California at Los Angeles)

MacLaren, D. D. (1975), ''Single-Cell Protein—An Overview,'' *ChemTech,* **5**(10), 594-597; presented at American Chemical Society Meeting, 10 April. (Exxon Corporation, New York)

MacLennan, D. G. (1974), ''Single Cell Protein—A Novel Protein Source with a Future for Australia,'' *Process and Chemical Engineering* (Australia), **27**(4), 13-17. (Department of Chemical Engineering, University of Sydney, Australia)

MacLennan, D. G. (1975), ''Single Cell Protein from Starch: A New Concept in Protein Production,'' Australian and New Zealand Association for the Advancement of Science Conference, Canberra, January. (Department of Chemical Engineering, University of Sydney, Australia)

McMillan, F. (1975), ''Cryogenic Tyre Granulation,''in Johnston, 1975, pp. T90-T91. (Commonwealth Industrial Gases Ltd, Australia)

McRoberts, T. S. (1973), ''Disposal of Plastics Waste,'' *Journal of Environmental Planning and Pollution Control,* **1**(4), 20-26. (QMC Industrial Research Unit, Queen Mary College, London, U.K.)

McRoberts, T. S. (1975), in *Nature,* 1975. (QMC Wolfson Recycle Unit, Queen Mary College, London, U.K.)

Madigan, D. C. (1971a), ''The Cultivation of Algae—A Review,'' *Bulletin of the Australian Mineral Development Laboratories,* (11), 73-97. (AMDEL, South Australia)

Madigan, D. C. (1971b), ''The Bacterial Sulphur Cycle as a Source of Sulphur,'' *Bulletin of the Australian Mineral Development Laboratories,* (12), 14-22. (AMDEL, South Australia)

Madigan, D. C. (1975), ''Treatment of Domestic Organic Refuse in the Adelaide Area,'' *Process and Chemical Engineering* (Australia), **28**(4) 33-36; reprinted from *Bulletin of the Australian Mineral Development Laboratories,* (18), 1974. (AMDEL, South Australia)

Mahoney, Lee R., Steven A. Weiner, and Fred C. Ferris (1974), ''Hydrolysis of

Polyurethane Foam Waste," *Environmental Science and Technology*, **8**(2), 135-139. (Chemistry Department, Ford Motor Co., Dearborn, Michigan)

Maizus, S. (1975), *Recycling of Waste Oils*, U.S. Environmental Protection Agency; U.S. NTIS PB-243,222. (National Oil Recovery Corporation)

Makar, H. V., R. S. Kaplan, and L. Janowski (1975), *Evaluation of Steel Made with Ferrous Fractions from Urban Refuse,* Bureau of Mines Report of Investigations 8037, U.S. Department of the Interior, Washington, D.C. (College Park Metallurgy Research Center)

Makino, Hiro, and R. Stephen Berry (1973), *Consumer Goods. A Thermodynamic Analysis of Packaging, Transport and Storage*, a study carried out for the Illinois Institute for Environmental Quality. (Department of Chemistry, University of Chicago)

Malcolm, Dorothea C. (1974), *Art from Recycled Materials*, Davis Publications, Worcester, Massachusetts.

Malina, Joseph F., Jr. (1970), "Water Renovation Processes and Reuse Systems," *Water Resource Management Series,* **1**, III, 1-20. (University of Texas, Austin)

Malisch, Ward R., Delbert E. Day, and Bobby G. Wixson (1970), "Use of Domestic Waste Glass as Aggregate in Bituminous Concrete," *Highway Research Record,* (307), 1-10. (University of Missouri, Rolla)

Malisch, Ward R., James T. Schneider, Bobby G. Wixson, and Delbert E. Day (1973a), "Laboratory and Field Experience with Asphaltic Cements Containing Glass Aggregates," in Albuquerque, 1973, pp. 35-72. (Departments of Civil Engineering, Environmental Health, and Ceramic Engineering, University of Missouri, Rolla)

Malisch, W. R., D. E. Day, and B. G. Wixson (1973b), *Use of Domestic Waste Glass for Urban Paving*, U.S. Environmental Protection Agency; U.S. NTIS PB-222,052.

Malisch, W. R., D. E. Day, and B. G. Wixson (1975), *Use of Domestic Waste Glass for Urban Paving, Summary Report,* U.S. Environmental Protection Agency; U.S. NTIS PB-242,536.

Mallan, G. M., and Titlow, E. I. (1976), "Energy and Resource Recovery from Solid Wastes," in RRC, 1976, pp. 207-216. (Occidental Research Corporation, La Verne, California)

Manahan, Stanley E. (1972), *Environmental Chemistry*, Willard Grant Press, Boston, Massachusetts.

Manchester, Alden C., and James G. Vertrees (1973), "Economic Issues in Management and Utilization of Waste," in Inglett, 1973, Chapter 2, pp. 6-12. (Economic Research Service, U.S. Department of Agriculture, Washington, D.C.)

Mandell, David A. (1974), *Thermal Power Plant Waste Heat Utilization,* Washington State University Bulletin 334, for Department of Lighting, City of Seattle. (Department of Mechanical Engineering, Washington State University, Pullman, Washington)

Mandels, Mary, and David Sternberg (1976), "Recent Advances in Cellulase Technology," *Journal of Fermentation Technology* (Japan), **54**(4), 267-286. (U.S. Army Natick Laboratories)

Maneval, David R. (1974), "Utilization of Coal Refuse for Highway Base or Subbase Material," in Aleshin, 1974, pp. 222-228. (Appalachian Region Commission, 1666 Connecticut Avenue N. W., Washington, D.C.)

Mantell, C. L. (1975), *Solid Wastes: Origin, Collection, Processing, and Disposal,* Wiley-Interscience, New York.

Mantle, E. C. (1976), "Advances in the Recovery of Waste Non-Ferrous Metals," in SCI, 1976c, pp. 716-720. (BNF Metals Technology Centre, Grove Laboratories, Denchworth Road, Wantage, Oxfordshire, OX12 9BJ, U.K.)

Marchant, B. D., and I. B. Cutler (1973), "Foamed Glass Insulation from Waste Glass," in Albuquerque, 1973, pp. 270-290. (Division of Materials Science and Engineering, University of Utah)

Mark, H. F. (1974), "Plastics and the Environment," *Proceedings of the Royal Australian Chemical Institute,* **41**(11), 271-272. (Polytechnic Institute of Brooklyn, New York 11201)

Markind, J., P. Minard, J. Neri, and R. Stana (1974), "Use of Reverse Osmosis for Concentrating Waste Cutting Oils," *American Institute of Chemical Engineers Symposium Series,* **70**(144), 157-162. (Westinghouse Electricity Corporation, Pittsburgh)

Marr, Harold E., III, Stephen L. Law, and William J. Campbell (1976), "Concentration and Source of Trace Elements in the Combustible Fraction of Urban Refuse," in Aleshin, 1976, pp. 251-252. (U.S. Bureau of Mines, College Park, Maryland)

Marsden, A. (1973), "Pulverisation and Tipping of Municipal Wastes" in CE, 1973, pp. 80-85. (Poole County Borough, Parkstone, Poole, Dorset)

Martin, C. B. (1972), "Oil Waste Reclaiming," in Kirov, 1972, pp. 197-198. (Martin's Refineries Pty Ltd., Australia)

Martin, D. F., L. S. Leung, and E. T. White (1975), "Recovery of Methyl Bromide from Wheat Fumigation by Adsorption," in Johnston, 1975, pp. T121-T124. (Department of Chemical Engineering, University of Queensland, Brisbane, Australia)

Masters, Gilbert M. (1974), *Introduction to Environmental Science and Technology,* Wiley, New York.

Mathew, K. C., S. A. A. Rizvi, S. P. Bhagat, and V. S. Bhaskar Rao (1974), "Isolation and Characterization of Wax from Decaffeinated Tea Waste," *Chemical Age* (India), **25**(6), 371-374.

Maydl, P., K. Ott, and R. M. E. Diamant (1974), "Using Slag from Refuse Incinerators as a Building Material," *Solid Wastes Management,* **64**(10), 604-608; reprinted from *Engineering* (Design Council). (Emphasis on West Germany)

Mayer, Albert (1976), "Some Aspects Concerning the Recirculation of City Water" (French), *Technique de l'Eau et de l'Assainissement,*349, 29-36.

ME (1976a), "A Plant to Make the Libyan Desert Bloom," *Municipal Engineering,* **153**(11), 406.

ME (1976b), "Coal/Refuse Mix to Fire Westbury Cement Kilns," *ibid.,* **153**(16), 591-592.

Meares, Patrick(1976), Ed., *Membrane Separation Processes,* Elsevier, Amsterdam.

Meers, John L. (1976), "Enzymes—Catalysts of the Future?" *Chemistry in Britain,* **12**(4), 115-118. (ICI Agricultural Division, Billingham, Cleveland TS23 1LB, U.K.)

Mehta, P. K., and N. Pitt (1976), "Energy and Industrial Materials from Crop Residues," *Resource Recovery and Conservation,* **2**(1), 23-38. (Department of Civil Engineering, University of California, Berkeley)

Meikle, Philip G. (1975), "Fly Ash," in Mantell, 1975, pp. 727-749. (National Ash Association, Inc., Washington, D.C.)

Meikle, A. B., and S. K. Nicol (1975), "Some Problems in the Recycling of Waste Ferruginous Dust from Iron and Steel Production," in Johnston, 1975, pp. T5-T6. (A. I. S., Port Kembla, N. S. W.; and B. H. P., Australia)

Meister, Erhard, and Norman R. Thompson (1976), "Physicochemical Methods for the Recovery of Protein from Waste Effluent of Potato Chip Processing," *Journal of Agricultural and Food Chemistry,* **24**(5), 919-923. (Department of Crop and Soil Sciences, Michigan State University, East Lansing, Michigan)

Melford, Sara Steck (1976), "Solid Waste Chemistry: A Translation of the Mineral Waste Utilization Symposia into a College Chemistry Course," in Aleshin, 1976, pp. 58-60. (DePaul University, Chicago, Illinois 60614)

Mellanby, Kenneth (1975), "Waste Line," *Nature,* **257**(5528), 639.

Menzel, Jürgen, Helmut Perkow, and Hansjörg Sinn (1973), "Recycling Plastics," *Chemistry and Industry* (London) (12), 570-573. (Chemical Institute of Hamburg University, 2 Hamsburg 13, Papendamm 6, West Germany)

Meyer, Judith G. (1972), "Renewing the Soil," *Environment,* **14**(2), 22-32.

Meyers, Sheldon (1976), "EPA and Municipal Resource Recovery," *NCRR Bulletin,* **6**(3), 62-65.

Miami University Pulp and Paper Foundation Conference (1971), *Fiber Recovery from Solid Waste—And Its Use,* 21-23 June, Oxford, Ohio.

Michaelis, Anthony R. (1975), "Fermentation, an Answer to the Solvent Crisis," *Chemistry in Britain,* **11**(2), 49. (Review of symposia on industrial fermentation, Octagon Group; proceedings available from Department of Extra-Mural Studies, University of Manchester, Manchester M13 9PL, U.K.)

Michell, A. J., W. E. Hillis, and J. E. Vaughan (1975), "Expeimental Mouldings from *Pinus radiata* Sawmill Residue," *CSIRO, Australia, Division of Chemical Technology Technical Papers,* No. 3, 1-23. (Divisions of Chemical Technology and Building Research, CSIRO, Melbourne, Australia)

Mihelich, Donald L. (1976), "Breakeven Economics of Resource Recovery Systems," in Aleshin, 1976, pp. 53-57. (Williams Brothers Urban Ore, Inc., 6600 South Yale Avenue, Tulsa, Oklahoma 74136)

Miles, J. E. P., and E. Douglas (1972), "Recovery of Non-ferrous Metals from Domestic Refuse," *Surveyor,* 8 December. (Warren Spring Laboratory, Stevenage, U.K.)

Milgrom, Jack (1972), *Incentives for Recycling and Reuse of Plastic,* Solid Waste Management Report SW-41c, U.S. Environmental Protection Agency, Washington, D.C.; U.S. NTIS PB-214,045; summary in Jackson, 1975. (Arthur D. Little, Inc., Cambridge, Massachusetts)

Milgrom, J. (1975), "The Present State of Recycling Thermoplastics—And Future Trends," *Solid Wastes,* **65**(11), 533-542; printed with acknowledgment to The Plastics and Rubber Institute, London. (Arthur D. Little, Inc., Cambridge, Massachusetts)

Millard, R. F. (1976), "Reclamation," *Solid Wastes,* **66**(7), 327-331. (Department of the Environment, U.K.)

Miller, M. W. (1961), *The Pfizer Handbook of Microbial Metabolities,* McGraw-Hill, New York.

Miller, G. Tyler (1971a), *Energetics, Kinetics and Life,* Wadsworth, Belmont, California.

Miller, Dwight L. (1971b), "Paper and Paperboard from Cellulosic Byproducts," presented at the Western Agricultural Experiment Station Collaborators Conference at the Western Marketing and Nutrition Research Division, USDA, Albany, California, 16 March. (Northern Regional Research Laboratory, U.S. Department of Agriculture, Peoria, Illinois)

Miller, Dwight L. (1973a), "Paper and Paperboard from Cellulosic Byproducts," in Inglett, 1973, Chapter 11, pp. 119-128. (Northern Regional Research Laboratory, USDA, Peoria, Illinois)

Miller, B. F. (1973b), *Biological Conversion of Animal Wastes to Nutrients,* U.S. Environmental Protection Agency; U.S. NTIS PB-221,171.

Miller Dwight L. (1975), "Annual Crops—A Renewable Source for Cellulose," in APS, 1975, pp. 21-28. (Northern Regional Research Laboratory, USDA, Peoria, Illinois)

Miller, Richard H., and Robert J. Collins (1974), "Waste Materials as Potential Replacements for Highway Aggregates," in Aleshin, 1974, pp. 50-61. (Valley Forge Laboratories, Inc., 6 Berkeley Rd., Devon, Pennsylvania)

Miller, Dwight L., and Ivan A. Wolff (1975), "Agricultural Fibers in the Paper Industry," in Mantell, 1975, pp. 525-547. (Northern Regional Research Center, USDA, Peoria, Illinois; and Eastern Regional Research Center, USDA, Philadelphia, Pennsylvania)

Millett, M. A. (1975), "Discussion of 'The Acid Hydrolysis of Refuse' " (Grethlein, 1975), *Biotechnology and Bioengineering Symposium,* **5** (Cellulose Chemical and Energy Resources), 319-320. (Forest Products Laboratory, Madison, Wisconsin)

Mills, G. Alex, and Brian M. Harney (1974), "Methanol—The 'New Fuel' from Coal," *ChemTech,* **4**(1), 26-31. (Division of Coal, Energy Research, U.S. Bureau of Mines)

Milton, John P. (1974), "Communities that Seek Peace with Nature," *Futurist,* **8**(6), 264-271.

Minnick, L. John (1970), "Lightweight Concrete Aggregate from Sintered Fly Ash," *Highway Research Record,* (307), 21-32. (G. and W. H. Corson, Inc., Plymouth Meeting, Pennsylvania)

Mitchell, Charles H. (1976), "Oil in the Hand," *Industrial Recovery,* **22**(1), 10, 12, 14. (Century Oils Ltd., Anglo Pennsylvanian Oil Co. Ltd., and Braybrooke Chemical Services Ltd., U.K.)

Mitra, Gautam, and Charles R. Wilke (1975), "Continuous Cellulase Production," *Biotechnology and Bioengineering,* **17**(1), 1-13. (Department of Chemical Engineering and Lawrence Berkeley Laboratory, University of California, Berkeley)

Miyazaki, Hajime, and Ichimariji Terada (1974), "Treatment of Waste Rind of Citrus Fruits and Extractions of the Components," *Shokuhin Kogyo,* **17**(20), 81-87 (Japanese); *Chemical Abstracts,* **82**(10), 64000. (Ube Industries Ltd., Ube, Japan)

Moats, E. R. (1976), "Goodyear Tire-Fired Boiler," in RRC, 1976, p. 315. (Goodyear Tire and Rubber Co., Akron, Ohio)

Moehle, Fred W. (1967), "Fly Ash Aids in Sludge Disposal," *Environmental Science and Technology,* **1**(5); reprinted in ACS, 1971, pp. 46-51. (Uniroyal, Inc., Mishawaka, Indianna 46544)

Mohaupt, A. A., and J. W. Koning, Jr. (1972), "A Practical Method for Recycling Wax-Treated Corrugated," *Boxboard Containers,* January, 60-62. (Forest Products Laboratory, USDA, Madison, Wisconsin)

Money, C. A., and U. Adminis (1974), "Recycling of Lime-Sulfide Unhairing Liquors. I. Small Scale Trials," *Journal of the Society of Leather Technologists and Chemists,* **58**(2), 35-40. (Division of Protein Chemistry, CSIRO, Parkville, Victoria 3052, Australia)

Monroe, Robert G. (1973), *Wastewater Treatment Studies in Aggregate and Concrete Production,* Report EPA-R2-73-003, Office of Research and Monitoring, U.S. Environmental Protection Agency, Washington, D.C. (Smith and Monroe and Gray Engineers, Inc., Lake Oswego, Oregon)

Moodie, S. P., and R. Hansen (1975), "Disposal of Solid Wastes from an Alumina Refinery," in Johnston, 1975, pp. T22-T25. (Department of Chemical Engineering, University of Queensland; and Queensland Alumina Ltd, Australia)

Moore, John W., and Moore, Elizabeth A, (1976), *Environmental Chemistry,* Academic Press, New York.

Morey, Brooker (1974), "Inorganic Resource Recovery and Solid Fuel Preparation from Municipal Trash," in Aleshin, 1974, pp. 85-94. (Garrett Research and Development Co., La Verne, California)

Morey, B., T. D. Griffin, A. K. Gupta, and I. J. T. Hopkins (1976), "Resource Recovery from Refuse," in Aleshin, 1976, pp. 184-194. (Occidental Research Corporation, La Verne, California)

Morris, David C., William R. Nelson, and Gerald O. Walraven (1972), *Recycle of Papermill Wastewaters and Application of Reverse Osmosis,* Water Pollution Control Research Series, 12040FUB01/72; *Chemical Abstracts,* **81,** 29295. (Green Bay Packaging, Inc., Green Bay, Wiconsin)

Morrison, E. D., and Chen-I Lu (1975), "Silver Recovery Process," *Research Disclosures,* **134,** 5-6.

Mother Earth News (1972), "Plowboy Interview: Ram Bux Singh," (18), November, 7-13.

MRW (1974), "Schh... on the One-Trip Argument," *Materials Reclamation Weekly,* **125**(46), 13.

MRW (1976a), "Chemicals Giant Comes into Plastic Recycling. First Reverzer Process for Britain," *ibid.,* **127**(5), 14; "Laporte Brings the First Reverzer to Britain," *ibid.,* **128**(16), 23-24.

MRW (1976b), "On the Road to Materials Recovery. Reclamation—Japan's Number One

Priority,'' *ibid.*, **127**(6), 22-23.

MRW (1976c), ''Polypropylene and Electrolysis Keys to Heavy Metal Recovery,'' *ibid.*, **127**(4), 22.

MRW (1976d), ''Metal Recovery Starts Tin Can Separation at Benwell,'' *ibid.*, **127**(14), 18, 20.

MRW (1976e), ''They Want to Turn the Desert Green—with Household Refuse,'' *ibid.*, **127**(11), 42.

MRW (1976f), ''Household Waste Will go to Cement Making,'' *ibid.*, **127**(15), 33.

MRW (1976g), ''Power From a Region's Refuse,'' *ibid.*, **127**(16), 32-33.

MRW (1976h), ''Ferrous and Fuel from Tyne and Wear's Refuse Reclamation Plant,'' *ibid.*, **127**(26), 16.

MRW (1976i), ''Fragmentiser Fever?'' *ibid.*, **127**(13), 22, 23, 26, 27.

MRW (1976j), ''Crashing Down,'' *ibid.*, **127**(26), 17.

MRW (1976k), ''Exploiting the Mine Above the Ground,'' *ibid.*, **128**(13), 22-24.

Müller, Ilse, and Dieter Schottelius (1975), ''Waste Exchange as a Solution to Industrial Waste Problems,'' *Israel Journal of Chemistry,* **14**, 226-233; from 25th IUPAC Congress, Jerusalem, July. (Verband der Chemischen Industrie, POB 119081, D-6000 Frankfurt/Main, FRG)

Muller, L. L. (1969), ''Yeast Products from Whey,'' *Process Biochemistry,* **4**(1), 21-23, 26. (CSIRO, Australia)

Murdoch, William W. (1971), Ed., *Environment: Resources, Pollution and Society,* Sinauer Associates, Stamford, Connecticut.

Myers, Gary C. (1971a), *Household Separation of Wastepaper: FPL Employee Survey,* Forest Service Research Paper FPL 159, U.S. Department of Agriculture.

Myers, Gary C. (1971b), ''What's in the Wastepaper Fiber Collected from Municipal Trash,'' *Paper Trade Journal,* 30 August, 32-34. (Forest Products Laboratory, USDA, Madison, Wisconsin)

Nagaya, Kiichi, and Akatsuki Adachi (1972), ''Separating Waste Plastics According to Specific Gravity Using Steam,'' *Japanese Kokai*, 72 44,545; *Chemical Abstracts,* **79,** 149238. (Hitachi Shipbuilding and Engineering Co. Ltd., Japan)

Nagiev, M. F. (1964), *The Theory of Recycle Processes in Chemical Engineering,* Transl. R. Hardbottle, Ed. R. M. Nedderman, Pergamon, Oxford.

Nakano, Junzo (1975), ''Trends of Utilization of Pulping Spent Liquor in Japan,'' in APS, 1975, pp. 85-91. (Department of Forest Products, University of Tokyo)

NASMI (1972), ''A Review of the Problems Affecting the Recycling of Selected Secondary Materials,'' National Association of Secondary Materials Industries, Inc., and Battelle Memorial Institute, in Stump, 1972, pp. 207-219; summary of USEPA, 1972a.

Nature (1975), ''Any Old Fe, Cu, Al, Pb, Zn, glass, paper,....?'' **253**(5488), 148-150.

NCRR (1972), *Paper,* Fact Sheet FS-1972-12, National Center for Resource Recovery, Washington, D.C.

NCRR (1973a), *Voluntary Separation of Refuse,* Fact Sheet NCF-19-01, National Center for Resource Recovery, Washington, D.C.

NCRR (1973b), *Shredders,* Fact Sheet NCF-18-01, National Center for Resource Recovery, Washington, D.C.

NCRR (1973c), *Pyrolysis,* Fact Sheet FS-9, National Center for Resource Recovery, Washington, D.C.

NCRR (1973d), *Plastics,* Fact Sheet NCF-14-01, National Center for Resource Recovery, Washington, D.C.

NCRR (1974a), "Magnetic Separation...A Basic Process," *NCRR Bulletin,* **4**(4), 2-6.

NCRR (1974b), "Solid Waste Shredders...A Survey," *ibid.,* **4**(4), 17-23; reprinted from *Waste Age,* June.

NCRR (1974c), *Resource Recovery from Municipal Solid Waste,* National Center for Resource Recovery, Inc., Lexington Books, Lexington, Massachusetts.

NCRR (1976), "State Initiatives," *NCRR Bulletin,* **6**(3), 66-72.

Neal, A. W. (1971), *Industrial Waste: Its Handling, Disposal, and Reuse,* Business Books, London.

Nemerow, Nelson L. (1963), *Theories and Practices of Industrial Waste Treatment,* Addison Wesley, Reading, Massachusetts.

Ness, Howard (1972), "Recycling as an Industry," *Environmental Science and Technology,* **6**(8), 700-704; summary of USEPA, 1972a. (NASMI, New York)

Netzer, A., and P. Wilkinson (1974), "Removal of Heavy Metals from Waste Water Utilizing Discarded Automotive Tires," *Water Pollution Research Canada,* **9,** 62-66. (Canada Center for Inland Waters, Burlington, Ontario)

Netzer, A., P. Wilkinson and S. Beszedits (1975), "Removal of Trace Metals from Waste Water by Treatment with Lime and Discarded Automotive Tires," *Water Research,* **8**(10), 813-817. (Canada Center for Inland Waters, Burlington, Ontario)

New, Colin (1974), "Can We Abolish Packaging?" *Ecologist,* **4**(6), 226-229.

Newell, John (1975a), "Poisons, and How to Dump Them," *Nature,* **255**(5504), 96.

Newell, John (1975b), in *Nature,* 1975.

Newell, Scott, and Michael A. Sandoval (1976), "An Urban Refuse Processing and Metal Reclamation Operation in Odessa, Texas," in Aleshin, 1976, pp. 146-152. (Newell Manufacturing Company, San Antonio, Texas)

New Scientist (1974a), "Sorting Out the Rubbish," **64**(928), 874.

New Scientist (1974b), "Problems with Glass Recycling...While Dispute Flares on Paper Recycling," **64**(923), 491.

New Scientist (1975a), "Food from Waste" (Report on Food from Waste Conference, Weybridge, Surrey, 1975), **66**(945), 134.

New Scientist (1975b), "Recycling Plant Awaits Treasury Approval," **68**(980), 697.

New Scientist (1975c), "Carrying the Can—To Be Recycled," **68**(976), 475.

New Scientist (1976a), "Burning Rubbish to Cement and Power," **70**(996), 132.

New Scientist (1976b), "Government Backs Two Waste Recycling Schemes," **69**(983), 107.

New Scientist (1976c), "Selling Recycled Tin, Glass, and Paper," **69**(984), 185.

New Scientist (1976d), "Scope for Retreading More of Britain's Waste Tyres," **69**(991), 569.

New Scientist (1976e), "Polymers for Leaching Metals from Solution," **71**(1013), 336.

New York State (1974), "Solid Waste as a Resource," Special Report, *NYS Environment,* October 1.

Nishimura, Yamaji, and Etsuo Inoue (1974), "Recovery of Chromium Ions from Electroplating Waste Solutions," *Japan.Kokai,* 74 32,825; *Chemical Abstracts,* **81**(12), 71892. (Daido Oxygen Co.)

Nordstrom, Robert P., Jr., (1974), "Ultrafiltration Removal of Soluble Oil," *Pollution Engineering,* **6**(10), 46-47. (Abcor, Inc.)

NSF (1975), "Cellulose as a Chemical and Energy Source," NSF Seminar, Berkeley, California, 25-27 June 1974; papers published in *Biotechnology and Bioengineering Symposia* (5), and Golueke, 1975.

Nuss, Gary R., William E. Franklin, David Hahlin, William Park, Michael Urie, and James Cross (1975), *Base Line Forecasts of Resource Recovery, 1972 to 1990; Final Report,* Publication SW-107c, U.S. Environmental Protection Agency; U.S. NTIS PB-245,924.

Nussbaum, Robert F. (1976), "Scrap—Demand versus Newly Available Supply: 1975-1985," in Aleshin, 1976, pp. 342-349. (A. T. Kearney, Inc., 100 S. Wacker Dr., Chicago, Illinois 60606)

O'Connor, G. V. (1971), "A New Approach to the Recovery of Metals from Plating Wastes," in ANERAC, 1971, pp. 28-46. (Monsanto Biodize Systems Inc., U.S.A)

Omori, Shinya (1975), "Reuse of Lime Cake in a Sugar Refinery," *Seito Gijutsu Kenkyukaishi,* **25,** 61-64 (Japanese); *Chemical Abstracts,* **83**(18), 149461. (Meiji Sugar Manuf. Co., Kawasaki, Japan)

Oosten, B. J. (1976), "Protein from Potato Starch Mill Effluent," in Birch et al., 1976, pp. 196-204. (Koninklijke Scholten-Honig Research NV, Foxhol (GR), The Netherlands)

Osell, Vernon A. (1973), "The Eidal Vertical Shaft Grinder," in Albuquerque, 1973, pp. 200-207. (Eidal International Corp., Albuquerque, New Mexico)

OST (1969), *Solid Waste Managment. A Comprehensive Assessment of Solid Waste Problems, Practices, and Needs,* Office of Science and Technology, Washington, D.C.

Ostrowski, E. J. (1971a), "Use of Ferrous Scrap from Municipal Waste Incinerators in Steelmaking," Paper EQC 28, Environmental Quality Conference, Extractive Industries, 7-9 June, Washington, D.C., American Institute of Mining, Metallurgical, and Petroleum Engineers. (National Steel Corp., Weirton, West Virginia)

Ostrowski, E. J. (1971b), "Recycling of Tin Free Steel Cans, Tin Cans and Scrap from Municipal Incinerator Residue," at 79th General Meeting, 26 May, New York, American Iron and Steel Institute. (National Steel Corporation, Weirton, West Virginia)

Ostrowski, E. J. (1973a), "Recycling of Ferrous Scrap from Solid Waste," in AIChE, 1973, pp. 37-50. (National Steel Corp., Weirton, West Virginia)

Ostrowski, E. J. (1973b), "Recycling of Ferrous Scrap from Incinerator Residue," *Secondary Raw Materials,* **11**(May), 37-46. (National Steel Corp., Weirton, West Virginia)

Ostrowski, Edward J. (1974), "The Bright Outlook for Recycling Ferrous Scrap from Solid Waste," American Institute of Chemical Engineers Conference, Salt Lake City, Utah, August. (National Steel Corp., Weirton, West Virginia)

Ostrowski, Edward J. (1975), "Bright Outlook for Recycling Ferrous Scrap from Solid Waste," *Recycling Today,* **13**(8), 66, 68. (National Steel Corp., Weirton West Virginia)

Oswald, William J., and Clarence G. Golueke (1960), "Biological Transformations of Solar Energy," *Advances in Applied Microbiology,* **2,** 223-262. (SERL, University of California, Berkeley)

Oswald, William J., and Clarence G. Golueke (1964), "Solar Power via a Botanical Process," *Mechanical Engineer,* **86**(2), 40-43. (SERL, University of California, Berkeley)

Owen, Paul H., and John P. Barry (1972), *Electrochemical Carbon Regeneration,* Environics Report, U.S. NTIS PB-239,156; *Chemical Abstracts,* **83,** 209176. (Environics, Huntington Beach, California)

Owen, T., and M. Bevis (1975), "Reclaimed Impact Polystyrene from Used Containers," *Polymer Age,* **6**(7-8), 212-213. (University of Liverpool, U.K.)

Packard, Vance Oakley (1960), *The Waste Makers,* David McKay Co., New York.

Palmer, C. (1975), "Resource Recovery from Municipal Solid Wastes in U.S.A.—A British View," *Solid Wastes,* **65**(3), 129-140. (Suffolk County Council, U.K.)

Park, W. R., and D. Bendersky (1973), "Analysis of Economics and Markets for Secondary Glass Products," in Albuquerque, 1973, p. 304; summary in Jackson, 1975. (Midwest Research Institute, Kansas City, Missouri)

Park, Charles F., and Margaret Cooper Freeman (1970), "Famine in Raw Materials by 2000 A.D.," *Science Journal,* **6**(5), 69-73 (Department of Mineral Engineering, Stanford University, California; and Freeman, Cooper and Co.)

Parker, A. J. (1972), "Solvation of Ions—Some Applications," *Proceedings of the Royal Australian Chemical Institute,* **39**(6), 163-170. (Research School of Chemistry, Australian National University, Canberra, Australia; currently Murdoch University, Western Australia)

Parker, A. J. (1973a), "Hydrometallurgy of Copper and Silver in Solvent Mixtures," *Search,* **4**(10), 426-432; Liversidge Lecture to Australian and New Zealand Association for the Advancement of Science, Congress, 13 August. (Research School of Chemistry, Australian National University, Canberra, Australia; currently Murdoch University, Western Australia)

Parker, C. A. (1973b), "Microbiological Means of Meeting the Increasing World Demand for Protein," Conference, *Global Impacts of Applied Microbiology, IV,* Sao Paulo, July. (Department of Soil Science and Plant Nutrition, University of Western Australia)

Patrick, Philip K. (1973), "Treatment and Disposal of Solid Waste—Resource Recovery Aspects," *Chemistry and Industry* (London), (12), 551-553. (Department of Public Health Engineering, Greater London Council, 10 Great George St., London SW1P 3AB, U.K.)

Pattengill, Maurice, and T. C. Shutt (1973), "Use of Ground Glass as a Pozzolan," in

Albuquerque, 1973, pp. 156-166. (School of Mines Research Institute, Colorado)

Pausacker, Ian (1975), *Is Recycling the Solution?* Patchwork Press, Melbourne, Australia; republished (1978) as *Recycling. Is It The Solution for Australia?* Penguin, Ringwood, Australia.

Pavoni, Joseph L., John E. Heer, Jr., and D. Joseph Hagert (1975), *Handbook of Solid Waste Disposal: Materials and Energy Recovery,* Van Nostrand-Reinhold, New York.

Pearl, Irwin A. (1968), "Waste Product Use Helps Paper Industry Control Pollution," *Environmental Science and Technology,* **2**(9), 676-681. (Institute of Paper Chemistry, Appleton, Wisconsin)

Pearson, D. (1972), "Hydrometallurgical Process for Recovering Copper from Bearing Scrap and Similar Copper Clad Materials," *Reclamation Industries,* July, 10-11. (Warren Spring Laboratory, Stevenage, U.K.)

Pearson, D. (1973), "Waste Reclamation 2: Economic Aspects," *Journal of Environmental Planning and Pollution Control,* **1**(2), 19-27. (Warren Spring Laboratory, Stevenage, U.K.)

Pearson, D., and M. Webb (1973), "The Salvage and Recycling of Useful Materials," in CE, 1973, pp. 89-93. (Warren Spring Laboratory, Stevenage, U.K.)

Peat, Derek (1974), "Telling it to the Dead Marines," *Current Affairs Bulletin* (Australia), **51**(5), 30-31. (Department of Adult Education, University of Sydney, Australia)

Pederson, H. (1975), "Salvage and Reuse of Textile Waste Material, with Special Reference to the Wool and Textile Industry," in Kirov, 1975a, pp. 179-182. (School of Chemical Engineering, University of New South Wales, Australia)

Peeler, J. P. K. (1975), "Advanced Wastewater Treatment: Technico-Economic Aspects," in Johnston, 1975, pp. T107-T109. (Division of Chemical Engineering, CSIRO, Clayton, Australia)

Perishing, D. W., and E. E. Berkau (1973), "The Chemistry of Nitrogen Oxides and Control Through Combustion Modifications," in Jimeson and Spindt, 1973, Chapter 19, pp. 218-240. (U.S. EPA, National Research Center, Research Triangle Park, North Carolina)

Perna, Angelo J. (1971), "Economics of Mechanically Aided Composting Process for Municipal Refuse," in ANERAC, 1971, pp. 113-123. (Newark College of Engineering, U.S.A.)

Perry, Harry (1974), "Coal Liquefaction," *American Institute of Physics Conference Proceedings,* **19,** 43-55. (Resources for the Future, Inc., Washington, D.C.)

Peterson J. R., T. M. McCaller, and G. E. Smith (1971), "Human and Animal Wastes as Fertilizers," in R. A. Olson, T. J. Army, J. J. Hanway, and V. J. Kilmer (editorial committee), *Fertilizer Technology and Use,* Soil Science Society of America, Madison, Wisconsin, 2nd ed.

Pfeffer, J. T. (1974), *Reclamation of Energy from Organic Waste,* U.S. Environmental Protection Agency; U.S. NTIS PB-231,176.

Pfeffer, John T., and Jon C. Liebman (1976), "Energy from Refuse by Bioconversion,

Fermentation, and Residue Disposal Processes,'' in RRC, 1976, pp. 295-313. (Department of Civil Engineering, University of Illinois, Urbana)

Philleo, Robert E. (1967), ''Fly Ash in Mass Concrete,'' in Faber et al., 1967, pp. 69-79. (Concrete Branch, Engineering Division Civil Works, Dept. of Army, Washington, D.C.)

Phillips, J. Craig (1973), ''Reuse Glass Aggregate in Portland Cement Concrete,'' in Albuquerque, 1973, pp. 138-153. (Riverside Chemical Co., Riverside, California)

Phoenix Quarterly (1976), ''Resource Recovery: The Vision and the Verities,'' **8**(1), 3-11.

Physics Education (1976), ''Sorting Mixed Metals by the Thermoelectric Effect,'' **11**(4), 290-291.

Pigott, Phillip G., E. G. Valdez, and K. C. Dean (1971), *Dry-Pressed Building Bricks from Copper Mill Tailings,* Bureau of Mines Report of Investigations 7537, U.S. Department of the Interior, Washington, D.C.

Pillai, S. C., E. G. Srinath, M. L. Mathur, P. M. N. Naidu, and P. G. Muthanna (1967), ''Activated Sludge as a Feed Supplement for Poultry,'' *Water and Waste Treatment,* **11**(7), 316-320, 322. (Department of Biochemistry, Indian Institute of Science, Bangalore; National Dairy Research Institute, Andugdi, Bangalore; and Department of Animal Husbandry and Veterinary Services, Government of Mysore, Bangalore, India)

Pincott, R. L. (1975), ''The Disposal of Tyres,'' in Kirov, 1975a, pp. 219-222. (R. L. and A. T. Pincott, Scrap Tyre Merchants, Australia)

Piper, Allan (1976), ''Flashes in Ashes,'' *Nature,* **263**(5573), 86-87.

Pirie, N. W. (1976), ''Food from Waste: Leaf Protein,'' in Birch et al., 1976, pp. 180-195. (Rothamstead Experimental Station, Harpenden, Herts, U.K.)

Pötschke, Herbert (1975), ''Recycling of Sewage in Industry,'' *Wasser, Luft, und Betrieb,* **19**(10), 569-574 (German).

Poll, Alan, and John Allen (1976), ''U.K. Waste Materials Exchange,'' in SCI, 1976a, p. 238. (Warren Spring Laboratory, Stevenage; and ICI Ltd., Higher Blackley, Manchester, U.K.)

Poole, Alan (1975), ''The Potential for Energy Recovery from Organic Wastes,'' in Williams, 1975, Chapter 6, pp. 219-308. (Institute Public Policy Alternatives, State University of New York, Albany, New York)

Porges, Nandor (1960), ''Newer Aspects of Waste Treatment,'' *Advances in Applied Microbiology,* **2,** 1. (Eastern Regional Research Laboratory, USDA, Philadelphia, Pennsylvania)

Porteous, Andrew (1969), ''The Recovery of Industrial Ethanol from Paper in Waste,'' *Chemistry and Industry* (London), 6 December, 1763-1770. (Department of Mechanical Engineering, University of Glasgow, Scotland)

Porteous, Andrew (1975a), ''The Economical Recovery of Fermentation Products from Cellulose Wastes via Acid Hydrolysis,'' in Johnston, 1975, pp. T86-T89. (Faculty of Technology, Open University, Milton Keynes, U.K.)

Porteous, Andrew (1975b), ''An Assessment of Energy Recovery Methods Applicable to

Domestic Refuse Disposal,'' *Resources Policy,* **1**(5), 284-294. (Faculty of Technology, Open University, Milton Keynes, U.K.)

Porteous, A. (1975c), ''Bulk Reduction by Incineration, Hydrolysis and Pyrolysis,'' in Henstock, 1975a. (Open University, U.K.)

Porter, John J., and Craig Brandon (1976), ''Zero Discharge as Exemplified by Textile Dyeing and Finishing,'' *ChemTech,* **6**(6), 402-407. (College of Industrial Management and Textile Science, Clemson University, Clemson, South Carolina 29631)

Potnis, S. P., and J. S. Aggarwal (1974), ''Cashewnut Shell Liquid. Raw Material for Useful Chemicals,'' *Journal of the Colour Society* (India), **13**(1), 10-12, 14. (Department of Chemical Technology, University, Bombay)

Povich, Michael J. (1976), ''Fuel Farming,'' *ChemTech,* **6**(7), 434-439. (Physical Chemistry Laboratory, Research and Development Center, General Electric Co., Schenectady, New York)

Powell, H. E., H. Fukubayashi, L. W. Higley, and L. L. Smith (1972a), *Recovery of Zinc, Copper, and Lead-Tin Mixtures from Brass Smelter Flue Dusts,* Bureau of Mines Report of Investigations 7637, U.S. Department of the Interior, Washington, D.C.

Powell, H. E., L. L. Smith, and A. A. Cochran (1972b), *Recovery of Phosphates and Metals from Phosphate Sludge by Solvent Extraction,* Bureau of Mines Report of Investigations 7662, U.S. Department of the Interior, Washington, D.C.

Powell, H. E., W. M. Dressel, and R. L. Crosby (1975), *Converting Stainless Steel Furnace Flue Dusts and Wastes to a Recyclable Alloy,* Bureau of Mines Report of Investigations 8039, U.S. Department of the Interior, Washington, D.C. (Rolla Metallurgy Research Center)

Power, G. P., and I. M. Ritchie (1975), ''The Electrochemistry of Metal Displacement (Cementation) Reactions,'' *Proceedings of the Royal Australian Chemical Institute,* **42**(2), 39-43. (Department of Inorganic and Physical Chemistry, University of Western Australia)

Powers, John A. (1974), ''Recovery of Sodium Tungstate from Scrap Tungsten Carbide,'' in Aleshin, 1974, pp. 377-380. (GTE Sylvania, Inc., Towanda, Pennsylvania 18848)

Prescott, James H. (1967), ''Composting Plant Converts Refuse into Organic Soil Conditioner,'' *Chemical Engineering,* **74**(23), 232-234.

Preston, G. T. (1976), ''Resource Recovery and Flash Pyrolysis of Municipal Refuse,'' in IGT, 1976, pp. 89-114; Occidental Research Corporation Reprint ORC 75-087.

Price, I. R. (1972), ''Decomposition of Organic Wastes from Domestic Refuse,'' in Kirov, 1972, pp. 47-49. (Mason Kockum Pty Ltd.)

Priestley, G. (1976), ''Algal Proteins,'' in Birch et al., 1976, pp. 114-138. (Western Biological Equipment Ltd., Sherborne, Dorset DT9 4RW, U.K.)

Pringle, Benjamin H. (1974), ''Water Reuse in the United States,'' Aerospace Medical Research Laboratory Report AMRL-TR-74-125, Paper No. 6, pp. 75-89, in Fifth Annual Conference on Environmental Toxicology, Aerospace Division, U.S. Air Force; U.S. NTIS-AD-A011856; *Chemical Abstracts,* **84,** 155288. (U.S. Environmental Protection Agency, Washington, D.C.)

Progress in Water Technology (from 1972), Series of Volumes of Conference Proceedings, International Association on Water Pollution Research Pergamon, Oxford.

Public Works (1976), "Solid Waste Resource Recovery: The 'Biogas' Concept," **107**(2), 71-75; summary of Ghosh and Klass, 1974.

Purcell, Arthur H., and Fred L. Smith (1976), "Energy and Environmental Impacts of Material Alternatives: An Assessment of Quantitative Understanding," *Resource Recovery and Conservation,* **2,** 93-102. (School of Engineering and Applied Science, George Washington University, Washington, D.C. 20052; and U.S. Environmental Protection Agency, Washington, D.C.)

Pyle, James L. (1974), *Chemistry and the Technological Backlash,* Prentice Hall, Englewood Cliffs, New Jersey, Chapter 6, "An Untapped Resource—Solid Waste."

Quang, Dang Vu, Georgio Carriero, Renato Schieppati, Andre Comte, and John W. Andrews (1975), "Experience with the French Petroleum Institute Propane Clarification Process in Re-refining Spent Crankcase Oils," *Journal of Environmental Sciences,* **18**(3), 18-20. (Institut Français du Pétrole, Rueil-Malmaison, France; Viscolube Italiano, Milano; Gargano Engineering, Novara, Italy; and Institut Français du Pétrole, New York)

Raask, E., and M. C. Bhaskar (1975), "Pozzolanic Activity of Pulverized Fuel Ash," *Cement and Concrete Research,* **5**(4), 363-375. (Central Electricity Research Laboratory, Leatherhead, Surrey, U.K.)

Rainbow, C., and A. H. Rose (1963), *Biochemistry of Industrial Micro-organisms,* Academic Press, London.

Randall, J. M., E. Hautala, and A. C. Waiss, Jr., (1974), "Removal and Recycling of Heavy Metal Ions from Mining and Industrial Waste Streams with Agricultural Byproducts," in Aleshin, 1974, pp. 329-334. (Western Regional Research Laboratory, U.S. Department of Agriculture, Berkeley, California)

Rao, S. Sathyanarayana, and S. C. Pillai (1962), "Recent Work on the Utilization of Activated Sludge," *Journal of Scientific and Industrial Research* (India), **21A,** January, 33-35. (Department of Biochemistry, Indian Institute of Science, Bangalore)

RAPRA (1976), *A Study of the Reclamation and Reuse of Waste Tyres* (for the U.K. Department of Industry), Rubber and Plastics Research Association, Shrewsbury, U.K.

Rapson, W. H. (1976), "Pulp and Paper Technology: The Closed-Cycle Bleached Kraft Pulp Mill," *Chemical Engineering Process,* **72**(6), 68-71. (University of Toronto, Canada)

Rasmuson, Anders S. H. (1973), "What About Wastes and How to Reuse Them," *Environmental Engineering,* 19-31. (Department of Chemical Engineering, Royal Institute of Technology, Stockholm)

Rath, A. E. (1976), "A Solution to Plastic Waste Problems?" *Solid Wastes,* **66**(10), 521-522. (Reclamat International Ltd., U.K.)

Ratledge, C. (1976), "Microbial Production of Oils and Fats," in Birch et al., 1976, pp. 98-113. (Department of Biochemistry, University of Hull, U.K.)

Ratter, B. G. (1975), "Waste Management in the Lead-Zinc Industry," in Kirov, 1975a, pp. 167-170. (School of Chemical Engineering, University of New South Wales, Australia)

Rees, G. J. (1976), "Recovery of Diffusion Pump Fluids," *Water and Waste Treatment,* **19**(6), 26. (Department of Chemical Engineering, Glamorgan Polytechnic, Treforest, Wales)

Regan, William J., III (1972), "An Approach to Ferrous Solid Waste," in Stump, 1972, pp. 221-235. (Battelle Memorial Institute, Columbus, Ohio)

Regan W. J., R. W. James, and T. J. McLeer (1972), *Identification of Opportunities for Increased Recycling of Ferrous Solid Waste,* Publication SW-45d, U.S. Environmental Protection Agency; U.S. NTIS PB-213,577.

Reimers, G. W., S. A. Rholl, and H. B. Dahlby (1976), "Density Separation of Nonferrous Scrap Metals with Magnetic Fluids,"in Aleshin, 1976, pp. 371-376. (Twin Cities Metallurgy Research Center, U.S. Bureau of Mines)

Reinhardt, Hans (1975), "Solvent Extraction for Recovery of Metal Waste," *Chemistry and Industry* (London), (5), 210-213. (MX Processor AB, Goteborg, Sweden)

Rembaum, A. (1974), "Flocculation of Model Waste Particles by Means of Ionene Polymers," in Yen, 1974a, Chapter 9, pp. 261-300. (Jet Propulsion Laboratory, California Institute of Technology, Pasadena)

Rex, Mervyn J. (1975), "Choosing Equipment for Process Energy Recovery," *Chemical Engineering,* **82**(16), 98-102; adapted from "Equipment for Power Recovery," presented at meeting "Energy Recovery in Process Plants," Institute of Mechanical Engineers. (Davy Powergas Ltd., 8 Baker St., London W1M 1DA)

Rickles, Robert N. (1965), "Waste Recovery and Pollution Abatement," *Chemical Engineering,* **72**(20), 133-152. (Dorr-Oliver, Inc.)

Ridge, M. J. (1975), "Chemical Gypsum," in Johnston, 1975, pp. T57-T58. (CSIRO Division of Building Research, Australia)

Rightelato, R. C., F. K. E. Imrie, and A. J. Vlitos (1976), "Production of Single Cell Protein from Agricultural and Food Processing Wastes," in RRC, 1976, pp. 257-269. (Tate and Lyle, Reading, U.K.)

Robinson, Donald J., Harold E. Weisberg, Glenn I. Chase, Kenneth R. Libby, Jr., and James L. Capper (1974), *An Ion-Exchange Process for Recovery of Chromate from Pigment Manufacturing,* U.S. EPA, Office of Research and Development, Report EPA 670/2-74-044; U.S. NTIS PB-233,641. (Mineral Pigment Corporation, Beltsville, Maryland)

Robson, G. H. (1972), "Salvaging and Reuse of Waste Glass and Glass Bottles," in Kirov, 1972, pp. 185-186.

Rogers, Charles J. (1976), "Problems and Potential Associated with the Production of Protein from Cellulosic Wastes," in RRC, 1976, pp. 271-277. (Solid and Hazardous Waste Research Laboratory, EPA, Cincinnati, Ohio)

Rogers, Charles J., and Donald Spino (1973), "Microbial Protein Production from Cellulosic Wastes," in Inglett, 1973, Chapter 14, pp. 161-171. (Solid Waste Research Laboratory, EPA, Cincinnati, Ohio)

Rogers, Charles J., P. V. Scarpino, Emile Coleman, Donald F. Spino, Thomas C. Purcell, and Pasquale V. Scarpino (1972), "Production of Fungal Protein from Cellulose and Waste Cellulose," *Environmental Science and Technology*, **6**(8), 715-719. (U.S. EPA, Cincinnati, Ohio; and Department of Civil Engineering, University of Cincinnati)

Rolfe, E. J. (1976), "Food From Waste in the Present World Situation," in Birch et al., 1976, pp. 1-7. (National College of Food Technology, University of Reading, U.K.)

Rook, J. F. (1976), "Feed from Waste," in SCI, 1976b, pp. 581-584. (Hannah Research Institute, Ayr, Scotland KA6 5HL)

Roper, G. H. (1972), "Utilisation of Biological Wastes," in Kirov, 1972, pp. 157-159. (Department of Biological Process Engineering, University of New South Wales, Australia)

Rose, David J., John H. Gibbons, and William Fulkerson (1972), "Physics Looks at Waste Management," *Physics Today*, **25**, 32-41. (Department of Nuclear Engineering, Massachusetts Institute of Technology; and Oak Ridge National Laboratory, Tennessee)

Rosen, Irving N., and June B. James (1974), "The New Game in Town: Materials Utilization: Everyone Wins; No One Loses," in Aleshin, 1974, pp. 150-161. (INS Equipment Co. Inc., 1329 William St., Buffalo; and Erie County Environmental Management Council, 831 West Ferry, Buffalo, New York)

Rosich, Ronald S. (1975a), "Resource Recovery from Solid Wastes," *Search* (Journal of the Australian and New Zealand Association for the Advancement of Science), **6**, 120-126. (School of Applied Sciences, Canberra College of Advanced Education, Australia)

Rosich, Ronald S. (1975b), "Resource Recovery in Solid Waste Management," in Kirov, 1975a, pp. 191-197. (School of Applied Sciences, Canberra College of Advanced Education, Australia)

Ross, Richard D. (1968), Ed., *Industrial Waste Disposal*, Van Nostrand-Reinhold, New York.

RRC (1976), Symposium, "Energy Recovery from Solid Waste," College Park, Maryland, 13-14 March, 1975, *Resource Recovery and Conservation*, **1**(3).

Russell, C. S. (1971), "Models for Investigation of Industrial Responses to Residuals Management Action," *Swedish Journal of Economics*, **73**(1), 134-156; reprinted in Peter Bohm and Allen V. Kneese, Eds., *The Economics of Environment: Papers from Four Nations*, Macmillian, London, 1971.

Russell, Clifford S. (1973), *Residuals Management in Industry: A Case Study of Petroleum Refining*, Resources for the Future and John Hopkins University Press, Baltimore, Maryland.

Russell, E. R. (1975), *Removal of Mercury from Aqueous Solutions by Shredded Rubber*, du Pont Report DP-1395 for U.S. Energy Research and Development Administration. (du Pont, Aiken, South Carolina)

Russell R. R., and J. A. Mraz (1974), "Hydrochloric Acid Recovery from Chlorinated Organic Wastes," *Industrial Process Design for Pollution Control, Proceedings*

American Institute of Chemical Engineers Workshop, **5,** 38-45. (Carbon Products Division, Union Carbide, Cleveland, Ohio)

Russell, T. W. F., and M. W. Swartzlander (1976), "The Recycling Index," *ChemTech,* **6**(1), 32-37. (Department of Chemical Engineering, University of Delaware, Newark, Delaware 19711).

Russell, Clifford S., and William J. Vaughan (1974), "A Linear Programming Model of Residuals Management for Integrated Iron and Steel Production," *Journal of Environmental Economics and Management,* **1**(1), 17-42. (Resources for the Future, Inc., Washington, D.C.)

Saito [Saitoh], Kozo (1975), "Separation Technology of Waste Plastics," *Hyomen,* **13**(6), 344-353 (Japanese), *Chemical Abstracts,* **84,** 111114. (Mitsui Mining and Smelting Co., Tokyo)

Saito [Saitoh], Kozo, and Sumio Izumi (1976), "Separation of Strongly Hydrophobic Plastics from a Plastics Mixture," *German Offen.,* 2,535,502, 26 February; *Chemical Abstracts,* **84,** 165537. (Mitsui Mining and Smelting Co., Tokyo)

Saitoh, Kozo, Ikuo Nagano, and Sumio Izumi (1976), "New Separation Technique for Waste Plastics," in Aleshin, 1976, pp. 322-328; *Resource Recovery and Conservation,* **2,** 127-145. (Mitsui Mining and Smelting Co., Tokyo)

Sammon, David C., and Brian Stringer (1975), "The Application of Membrane Processes in the Treatment of Sewage," *Process Biochemistry,* **10**(2), 4-12. (Atomic Energy Research Establishment, Harwell, U.K.)

Sandscheper, Günter (1975), "Sun Shines on Family of Four," *New Scientist,* **67**(962), 382-383.

Sanks, Robert L., and Takashi Asano (1976), *Land Treatment and Disposal of Municipal and Industrial Wastewater,* Ann Arbor Science, Ann Arbor, Michigan.

San Martin, Robert L. (1975), "Solar Heating and Cooling at New Mexico State University," in ASME, 1975, pp. 27-39.

Savage, G., and G. J. Trezek (1976), "Screening Shredded Municipal Solid Waste," *Compost Science,* **17**(1), 7-11. (Department of Mechanical Engineering, University of California, Berkeley)

Savage, George, Luis F. Diaz, and G. J. Trezek (1975), "The Cal Recovery System: A Resource Recovery System for Dealing with the Problems of Solid Waste Management," *Compost Science,* **16**(5). (Department of Mechanical Engineering, University of California, Berkeley)

Sawyer, James W., Jr., (1974), *Automotive Scrap Recycling,* Resources for the Future and John Hopkins University Press, Baltimore, Maryland.

Scammell, G. W. (1975), "Anaerobic Treatment of Industrial Wastes," *Process Biochemistry,* **10**(8), 34-36. (Biomechanics Ltd., Kent, U.K.)

Schickler, W. J., I. L. Gomez, and F. Metz (1973), "The Technical Aspects of Refilling and Recycling Lopac Containers," in *Environmental Impact, Nitrile Barrier Containers: LOPAC,* Proceedings of Symposium, ed. Ferdinand Decatur Wharton, Jr., pp. 130-137. (Lopac Container Group, Monsanto Co., St. Louis, Missouri)

Schlesinger, M. D., W. S. Sanner, and D. E. Wolfson (1973), "Energy from the

Pyrolysis of Agricultural Wastes,'' in Inglett, 1973, Chapter 9, pp. 93-100. (Pittsburgh Energy Research Center, U.S. Bureau of Mines)

Schlömann, Ernst (1975), ''Separation of Nonmagnetic Metals from Solid Waste by Permanent Magnets, I. Theory. II. Experiments on Circular Disks,'' *Journal of Applied Physics,* **46**(11), 5012-5021, 5022-5029. (Raytheon Research Division, Waltham, Massachusetts)

Schloemann, E. (1976), ''A Rotary-Drum Metal Separator Using Permanent Magnets,'' *Resource Recovery and Conservation,* **2,** 147-158. (Raytheon Research Division, Waltham, Massachusetts)

Schmidt, Curtis J., Irwin Kugelman, and Ernest V. Clements, III (1975), ''Municipal Waste Water Reuse in the U.S.,'' *Journal Water Pollution Control Federation,* **47**(9), 2229-2245. (SCS Engineers, Long Beach, California; and Advanced Waste Treatment Research Laboratory, National Experimental Research Center, EPA, Cincinnati, Ohio)

Scholes, Stafford (1974), ''Utilization of Industrial Wastes in the Production of Glass-ceramics,'' in Aleshin, 1974, pp. 316-328. (Advanced Materials Consultancy, Durham, England)

Schroeder, Henry A. (1971), *Pollution, Profits and Progress,* Stephen Green Press, Battleboro, Vermont.

Schulz, Helmut W. (1975), ''Cost/Benefits of Solid Waste Reuse,'' *Environmental Science and Technology,* **9**(5), 423-427. (Columbia University, New York 10027)

SCI (1976a), ''Economic Aspects of the Conservation of Resources in the Chemical and Allied Industries,''Symposium, Society of Chemical Industry, Manchester, U.K.; *Chemistry and Industry* (London), (6).

SCI (1976b), ''Novel Feeds and Foods,'' Annual Meeting, Society of Chemical Industry, 29 June, Stirling, U.K., *Chemistry and Industry* (London), (14).

SCI(1976c), ''Recent Advances in the Recovery of Useful Materials from Industrial Waste,'' Papers presented to the London Section of the Society of Chemical Industry, 2 March, *Chemistry and Industry* (London), (17), 709-729.

Scott, Pickett (1973), ''Terrazzo and Other Glass Products in Existing Buildings,'' in Albuquerque, 1973, pp. 131-136. (Glass Containers Corp., Fullerton, California)

Scott, Gerald (1976), ''Some Chemical Problems in the Recycling of Plastics,''*Resource Recovery and Conservation,* **1,** 381-395. (University of Aston in Birmingham, England)

Scott, D. S., and H. Horlings (1974), ''Production of Iron Salts from Waste Materials for Use in Phosphate Removal, '' in Aleshin, 1974, pp. 364-375. (Department of Chemical Engineering, University of Waterloo, Ontario)

SCS Engineers, Inc. (1974), *Analysis of Source Separate Collection of Recyclable Solid Waste.* [Vol. 1] *Separate Collection Studies;* [Vol. 2] *Collection Center Studies,* Publications SW-95c.1, SW-95c.2, U.S. Environmental Protection Agency: U.S. NTIS PB-239,775, PB-239,776.

Seaborg, Glenn T. (1974), ''The Recycle Society of Tomorrow,'' *Futurist,* **8**(3), 108-112, 114-115. (Department of Chemistry, University of California, Berkeley)

Seaborg, Glenn T. (1975a), "Opportunities in Today's Energy Milieu,"*Futurist,* **9**(1), 22. (Department of Chemistry, University of California, Berkeley)

Seaborg, Glenn T. (1975b), "Toward a Recycle Society," *Conference on Facing a World of Scarce Resources,* Los Angeles, March.

Seal, K. J., and H. O. W. Eggins (1976), "The Upgrading of Agricultural Wastes by Thermophilic Fungi," in Birch et al., 1976, pp. 58-78. (Biodeterioration Information Centre, University of Aston in Birmingham, U.K.)

Sealy, Gary D. (1975), "Magnets as Applied to Ferrous Scrap Recovery," *Recycling Today,* **13**(8), 90, 92, 94, 122. (Separation Division, Eriez Magnets)

Sear, Derek W. (1975), "Treatment of Nonplastic Rubber Pieces with Crosslinked Structure," *Fr. Demande,* 2,265,530; *Chemical Abstracts,* **84,** 165987.

Search, W. J., and T. E. Ctvrtnicek (1976), "Resource Recovery Systems for Non-recappable Rubber Tires," *Resource Recovery and Conservation,* **2,** 159-170. (Monsanto Research Corporation, Dayton, Ohio)

Seattle (1975), *Final Environmental Impact Statement on Solid Waste of Methanol/ Ammonia Conversion Plant,* City of Seattle, Washington.

Sebastian, Frank P. (1970), "Tahoe and Windhoek: Promise and Proof of Clean Water," *Chemical Engineering Progress Symposium Series,* **67**(107), 410-412. (Envirotech Corp., Palo Alto, California)

Sebastian, Frank P. (1974), "Purified Wastewater—The Untapped Water Resource." *Journal Water Pollution Control Federation,* **46**(2), 239-246. (Envirotech Corporation, Menlo Park, California)

Sebastian F. P., and M. C. Isheim (1970), "Advances in Incineration and Resource Reclamation," *Proceedings 4th National Incinerator Conference,* 71-78. (Envirotech Corp., Brisbane, California)

Seldman, Neil N. (1976), "High Technology Recycling: Costly and Still Wasteful," *Compost Science,* **17**(2), 28-29; excerpted from Seldman (below). (Institute for Local Self-Reliance, Washington, D.C.)

Seldman, Neil N., *Garbage in America,* Institute for Local Self-Reliance, Washington, D.C.

Senturia, Stephen D. (1974), "Automated Recycling," *Computers and People,* **23**(2), 8-10. (Massachusetts Institute of Technology, Cambridge, Massachusetts)

Senturia, Stephen D., David Gordon Wilson, P. Frank Winkler, Kendall H. Lewis, and Hubert Hibberd (1971), "New Sensors for the Automatic Sorting of Municipal Solid Waste," *Compost Science,* **12**(5). (Departments of Electrical Engineering and Mechanical Engineering, Massachusetts Institute of Technology, Cambridge; and Middlebury College, Vermont)

Seymour, Raymond B., and G. Allan Stahl (1976), "Separation of Waste Plastics: An Experiment in Solvent Fractionation," *Journal of Chemical Education* **53**(10), 653. (University of Houston, Texas)

Shafizadeh, Fred (1971), "Thermal Behavior of Carbohydrates," *Journal of Polymer Science, Part C,* **36,** 21-51. (Wood Chemistry Laboratory, University of Montana, Missoula)

Shafizadeh, Fred (1975), "Industrial Pyrolysis of Cellulosic Materials," in APS, 1975. (Wood Chemistry Laboratory, University of Montana, Missoula)

Shafizadeh, Fred, and Y. L. Fu (1973), "Pyrolysis of Cellulose," *Carbohydrate Research,* **29,** 113-122. (Wood Chemistry Laboratory, University of Montana, Missoula)

Shafizadeh, Fred, Craig McIntyre, Hans Lundstrom, and Yun-Lung Fu (1973), "Chemical Conversion of Wood and Cellulosic Wastes," *Proceedings of the Montana Academy of Sciences,* **33,** 65-96. (Wood Chemistry Laboratory, University of Montana, Missoula)

Shale, C. C. (1973), "Ammonia Injection: A Route to Clean Stacks," in Jimeson and Spindt, 1973, Chapter 17, pp. 195-205. (Morgantown Energy Research Center, U.S. Bureau of Mines)

Shannon, Earl E., and Peter J. A. Fowlie (1974), "The Utility of Selected Industrial Wastes for Phosphorus Removal," in Aleshin, 1974, pp. 335-345. (Canada Centre for Inland Waters, Wastewater Technology Centre, Department of the Environment, Burlington, Ontario)

Sheehan, Robert G. (1975), "Energy from Solid Waste: Appraisal of Alternatives," in 1975 Joint IEEE/ASME/ASCE Power Generation Technical Conference, Portland, Oregon, 28 September-1 October. (City of Seattle, Washington)

Sheehan, Robert G., and Richard Corlett (1975), "Methanol or Ammonia Production from Solid Waste by the City of Seattle," 169th American Chemical Society Meeting, Philadelphia, Pennsylvania, April. (Seattle City Light, Washington 98104; and Mathematical Sciences Northwest, Inc., Bellevue, Washington 98004)

Shelley, Steven V. (1976), "Recycling: A Contribution to the Environment," *Packaging Technology,* **22**(142), 4, 16.

Sheng, Henry P., and Harvey Alter (1975), "Energy Recovery from Municipal Solid Waste and Method of Comparing Refuse-derived Fuels," *Resource Recovery and Conservation,* **1,** 85-93. (Youngstown State University, Ohio 44555; and NCRR, Washington, D.C.)

Sherlock, Derek (1976), "Water as a Raw Material," in SCI, 1976a, pp. 232-233. (North West Water Authority, Dawson House, Liverpool Rd., Gt Sankey, Warrington, Lancashire, U.K.)

Sherwood, Robert J., and Frank P. Sebastian (1970), "Chemicals—Their Use and Reuse in Advanced Wastewater Treatment," at American Institute of Chemical Engineers Convention, 1 December, Chicago, Illinois. (Envirotech Corporation)

Showyin, L. (1972), "The Problems Associated with the Disposal and/or Re-use of Plastic Containers," in Kirov, 1972, pp. 187-189. (Reckitt and Colman Pty Ltd, Australia)

Sias, D. R., and T. A. Nevin 1973), "Experimental Hydroponic Gardening with Municipal Waste-Water," *Bulletin of Environmental Contamination and Toxicology,* **10**(5), 272-278. (Department of Oceanography, Florida Institute of Technology, Melbourne, Florida 32901)

Singh, Narinder, and S. B. Mathur (1973), "Survey of Indigenous Coals, Fly Ashes, and Flue Dusts as a Potential Source of Germanium," *NMR Technical Journal,* **15**(2),

42-48; *Chemical Abstracts,* **82**(18), 113927. (National Metall. Laboratory, Jamshedpur, India)

Sittig, Marshall (1975), *Resource Recovery and Recycling Handbook of Industrial Wastes,* Noyes Data Corporation, Park Ridge, New Jersey.

Skinner, E. C. (1972), "Treatment of Effluents by Centrifuges," in Kirov, 1972, pp. 151-155. (Alfa-Laval Pty Ltd.)

Skinner, Karen Joy (1975), "Single-Cell Protein Moves Toward Market," *Chemical and Engineering News,* **53**(18), 24-26.

Skitt, John (1972), *Disposal of Refuse and Other Waste,* Charles Knight, London.

Skitt, John (1973), "Waste Disposal—A General Review," in CE, 1973, pp. 55-60. (Cleansing and Transport Department, Booth Street, City of Stoke-on-Trent, U.K.)

Skogman, H. (1976), "Production of Symba-Yeast from Potato Wastes," in Birch et al., 1976, pp. 167-179. (AB Sorigona, Staffanstorp, Sweden)

Sleppy, William C., and Robert H. Goheen (1975), "Electrolytic Recovery of Metallic Gallium," *U.S. Patent,* 3,904,497; *Chemical Abstracts,* **83,** 185509. (Aluminum Company of America)

Sloneker, J. H., R. W. Jones, H. L. Griffin, K. Eskins, B. L. Bucher, and G. E. Inglett (1973), "Processing Animal Wastes for Feed and Industrial Products," in Inglett, 1973, Chapter 3, pp. 13-28. (Northern Regional Research Laboratory, U.S. Department of Agriculture, Peoria, Illinois)

Smith, L. W. (1973), "Nutritive Evaluations of Animal Manures," in Inglett, 1973, Chapter 6, pp. 55-74. (Agricultural Environmental Quality Institute, U.S. Department of Agriculture, Beltsville, Maryland)

Smith, B. R. (1975), "Membrane Processes—The State of the Art," in Johnston, 1975, pp. T187-T189. (Division of Chemical Engineering, CSIRO, Clayton, Australia)

Smith, Frank Austin (1976), "Quantity and Composition of Post-Consumer Solid Waste: Material Flow Estimates for 1973 and Baseline Future Projections," *Waste Age,* **7**(4), 2, 6-8, 10; summary of USEPA, 1975. (Resource Recovery Division, Office of Solid Waste Management Programs, U.S. Environmental Protection Agency)

Snyder, M. Jack (1964), "Properties and Uses of Fly Ash," *Battelle Technical Review,* **13**(2), 14-18. (Battelle Columbus Laboratories, Ohio)

Snyder, M. Jack (1967), "Fly Ash: Specifications, Limitations, and Restrictions," in Faber et al., 1967, pp. 37-45. (Battelle Columbus Laboratories, Ohio)

Sobkowicz, G. (1976), "Yeast from Molasses," in Birch et al., 1976, pp. 42-57. (Institute of Storage and Food Technology, Wroclaw, Poland)

Solid Wastes (1976a), "Go-ahead for Two Waste Sorting Plants," **66**(3), 139-141.

Solid Wastes (1976b), "Tin Can Recycling Plant Starts Up," **66**(7), 319.

Solid Wastes (1976c), "Dano Refuse Pulverization Plant for Buckinghamshire," **66**(3), 142-143.

Solt, G. S. (1976), "Electrodialysis," in Mears, 1976, Chapter 6, pp. 229-260.

Sommer, Edward J., Jr., and Garry R. Kenny (1974), "An Electromagnetic System for Dry Recovery of Nonferrous Metals from Shredded Municipal Solid Waste," in Aleshin, 1974, pp. 77-84. (Magnetic Separation Systems, Inc., 1612 18th Avenue, South Nashville, Tennessee)

Sorrentino, John A., Jr., and Andrew B. Whinston (1975), "The Economic Implications of Recycling Exhaustible Natural Resources: The Case for Crude Oil," *Ecological Modelling,* **1,** 219-233. (Department of Economics, Temple University, Philadelphia, Pennsylvania; and Department of Economics, Krannert Graduate School, Purdue University, West Lafayette, Indiana)

Spano, L. A., J. Medeiros, and M. Mandels (1976), "Enzymatic Hydrolysis of Cellulosic Wastes to Glucose," in RRC, 1976, pp. 279-294. (U.S. Army Natick Laboratories)

Spedding, P. L. (1974), "Processing Lead Recovery Furnace Slag," *Chemical Processing* (London), **20**(3), 13-15, 17. (Department of Chemical and Materials Engineering, University of Auckland, New Zealand)

Spedding, C. R. W., and J. M. Walsingham (1975), "Energy Use in Agricultural Systems," *Span,* **18**(1), 7-9. (Department of Agriculture and Horticulture, University of Reading; and The Grassland Research Institute, Hurley, Berkshire, U.K.)

Spencer, David B., and Ernst Schlömann (1975), "Recovery of Non-ferrous Metals by Means of Permanent Magnets," *Resource Recovery and Conservation,* **1,** 151-165. (Raytheon Service Co., Burlington, Massachusetts, and Raytheon Research Division, Waltham, Massachusetts)

Spendlove, Max J. (1976), *Recycling Trends in the United States: A Review,* Bureau of Mines Information Circular 8711, U.S. Department of the Interior, Washington, D.C.; reviewed in MRW, 1976k. (College Park Metallurgy Research Center)

Sperber, Robert J. (1975), personal communication.

Sperber, Robert J., and S. L. Rosen (1974), "Reuse of Polymer Waste," *Polymer-Plastics Technology and Engineering,* **3** (2), 215-239. (Department of Chemical Engineering, Carnegie-Mellon University, Pittsburgh, Pennsylvania)

Sperber, Robert J., and S. L. Rosen (1975), "Recycling of Thermoplastic Waste. Phase Equilibrium in Polystyrene," Society of Plastics Engineers, *Technical Papers*, **21,** 521-524; *Chemical Abstracts,* **83,** 102803.

Spittle, L. F. (1975), "Resource Recovery: The Magnetic Separation and Recycling of Steel Cans Collected from Solid Wastes," in Johnston, 1975, pp. T13-T15. (Steel Can Group, Melbourne, Australia)

Spofford, Walter O., Jr. (1971), "Solid Residuals Management: Some Economic Considerations," *Natural Resources Journal,* **11**, 561-589. (Resources for the Future, Washington, D.C.)

Stanier, R. Y., M. Doudoroff, and E. A. Adelberg (1971), *General Microbiology,* Macmillan, London, 3rd ed.

Staudinger, J. J. P. (1970), *Plastic Waste and Litter,* Society of Chemical Industry Monograph No. 35.

Staudinger, J. J. P. (1974), Ed., *Plastics and the Environment,* Hutchinson, London, in association with the British Plastics Federation; previously published (1973) as separate booklets; contributions by Evans, Fergusson, Flintoff and Higginson.

Stauffer, James M. (1975), "Newsprint, and Economical Filtration Medium for Waste Water," *Filtr. Sep.-Multi-Ind. Technol., Reprints of Papers, Filtration Society Conference,* 18-22; *Chemical Abstracts*, **85,** 9946.

Stearns, S. Russell (1973), "Glass Aggregate in Concrete," in Albuquerque, 1973, pp.

168-185. (Thayer School of Engineering, Dartmouth College, Hanover, New Hampshire)

Steffgen, Fred W. (1974), ''Clean Fuels from Solid Organic Wastes,'' in Aleshin, 1974, pp. 13-21. (Pittsburgh Energy Research Center, U.S. Bureau of Mines)

Steinberg, Meyer, and Morris Beller (1974), ''Glass-Polymer Composites for Sewer Pipe Construction'', in Aleshin, 1974, pp. 162-173. (Department of Applied Science, Brookhaven National Laboratory, Upton, New York)

Stephens, G. K. (1976), ''Desalination by Thermally Regenerable Ion Exchange. Part II. Applications,'' *Proceedings of the Royal Chemical Institute,* **43** (11), 350-354. (ICI Australia Ltd., Melbourne)

Stephenson, Junius W., R. K. Hampton, E. R. Kaiser, and C. O. Velzy (1975), *Incinerator and Solid Waste Technology,* The American Society of Mechanical Engineers, New York.

Stevens, Leonard A. (1967), ''The Town That Launders Water,'' *Reader's Digest,* May, 92-99. (Condensed from *National Civic Review.*)

Stevens, Bruce W., and Jeffrey W. Kerner (1975), ''Recovering Organic Materials from Wastewater,'' *Chemical Engineering* (N.Y.), **82** (3), 84, 86-87. (Rohm and Haas Co.; and Alberts and Associates)

Stevenson, M. K., J. O. Leckie, and R. Eliassen (1973), *Preparation and Evaluation of Activated Carbon Produced from Municipal Refuse,* U.S. Environmental Protection Agency; U.S. NTIS PB-221,172.

Stewart, Robert F., and William L. Farrier, Jr. (1967), ''Nuclear Measurement of Carbon in Fly Ash,'' in Faber et al., 1967, pp. 262-270. (Morganton Coal Research Center, U.S. Bureau of Mines, West Virginia)

Stock, R. (1975) ''Textile Recycling,'' in Henstock, 1975a. (Trent Polytechnic, U.K.)

Stoia, John, and Anil K. Chatterjee (1972), ''An Advanced Process for the Thermal Reduction of Solid Waste: the Torrax Solid Waste Conversion System,'' in Stump, 1972, pp. 109-127. (Torrax Systems, Inc., North Tonawanda, New York)

Stuck, James D. (1973), ''Enzymatic Hydrolysis of Pure and Waste Cellulose,'' State University of New York, Buffalo, *Dissertation Abstracts International B,* 1973, **34** (6), 2597.

Stumm, Werner, and Joan Davis (1974), ''Kann Recycling die Umweltsbeeinträchtigung vermindern?'' *Brennpunkte* (2), 29-41. (Eidgenössische Technische Hochschule, Eidgenössischen Anstalt fur Wasserversorgung, Abwassereinigung and Gewässerschutz (EAWAG), 8600 Dübendorf, Switzerland)

Stump, Patricial L. (1972), compiler, *Solid Waste Demonstration Projects,* Symposium Proceedings, Cincinnati, May 4-6, 1971, U.S. Environmental Protection Agency Report SW-4p, U.S. Government Printing Office, Washington, D.C.

Su, Tah-Mun, and Irene Paulavicius (1975), ''Enzymatic Saccharification of Cellulose by Thermophilic Actinomyces,'' in APS, 1975, pp. 221-236, (Physical Chemistry Laboratory, General Electric Co., Schenectady, New York)

Sullivan, P. M., and Harry V. Makar (1974), ''Bureau of Mines Process for Recovering Resources from Raw Refuse,'' in Aleshin, 1974, pp. 128-141. (College Park Metallurgy Research Center, U.S. Bureau of Mines, Maryland)

Sullivan, P. M., and H. V. Makar (1976), "Quality of Products from Bureau of Mines Resource Recovery Systems and Suitability for Recycling," in Aleshin, 1976, pp. 223-233. (College Park Metallurgy Research Center, U.S. Bureau of Mines, Maryland)

Sullivan, T. A., and W. C. McBee (1974), "Sulfur Utilization in Pollution Abatement," in Aleshin, 1974, pp. 245-254. (Boulder City Metallurgy Research Laboratory, U.S. Bureau of Mines, Nevada)

Sullivan, P. M., and M. H. Stanczyk (1971), *Economics of Recycling Metals and Minerals from Urban Refuse,* Bureau of Mines Solid Waste Research Program Technical Progress Report 33, U.S. Department of Interior, Washington, D.C.; U.S. NTIS PB-200,052; summary in Jackson, 1975.

Summers, W. K., and Zane Spiegel (1974), *Ground Water Pollution: A Bibliography,* Ann Arbor Science, Ann Arbor, Michigan.

Surfleet, B. (1970), "The Electrochemical Treatment of Industrial Effluents" *Electronics and Power,* November, 411-413.

Surveyor (1976), "Plan to Use Refuse to Fuel Cement Kilns," **147** (4373), 7.

Sussman, D. B. (1975), *Baltimore Demonstrates Gas Pyrolysis, Resource Recovery from Solid Waste,* Report SW-75d.i, U.S. Environmental Protection Agency.

Sutterfield, G. W. (1974), *Refuse as a Supplementary Fuel for Power Plants—November 1973 through March 1974, Interim Progress Report,* U.S. Environmental Protection Agency, Report SW-36d.iii.

Sutterfield, G. Wayne, and F. Ed Wisely (1972), "Refuse as Supplementary Fuel for Power Plants," in Stump, 1972; pp. 129-147. (City of St. Louis, Missouri; and Horner and Shifrin, Inc., St. Louis, Missouri)

Sutterfield, G. W., Earl K. Dille, and F. E. Wisely (1972), "Refuse for Fuel to Generate Power," *Energy and Environmental Engineering, Proceedings,* (10).

Sutterfield, G. Wayne, D. L. Klumb, and F. E. Wisely (1974), "Municipal Refuse. Fuel for Electric Utility Boiler," *American Institute of Chemical Engineers Symposium Series,* **70** (137) *(Recent Advances in Air Pollution Control),* 485-488. (City of St. Louis, Missouri; Union Electric Co., St. Louis; and Horner and Shifrin, Inc., St. Louis, Missouri)

Swager, William L. (1967), "Materials," *Science Journal,* **3** (10), 107-112. (Battelle Memorial Institute, Department of Economics and Information Research, Columbus, Ohio)

Szego, George C., and Clinton C. Kemp (1973), "Energy Forests and Fuel Plantations," *ChemTech,* **3,** 275-284. (Intertechnology Corporation)

Szekely, Julian, Jack J. Fritz, and Frank A. Berczynski (1975), "The Andco-Torrax Slagging Pyrolysis Solid Waste Disposal System," in Johnston, 1975, pp. T98--T100. (Department of Chemical Engineering, State University of New York at Buffalo; and Andco Incorporated, Buffalo, New York)

Szilard, Jules A. (1973), *Reclaiming Rubber and Other Polymers,* (Chemical Technology Review No. 5), Noyes Data Corporation, Park Ridge, New Jersey; summary in Jackson, 1975.

Szpindler, G. à Donau, and P. L. Waters (1976), "Heat Recovery in the Fluidised Incineration of Industrial Wastes," *Process and Chemical Engineering* (Australia),

29 (7), 19-20 (a synopsis of a paper presented in Kirov, 1975b). (CSIRO Division of Process Technology, North Ryde, N.S.W., Australia)

Tannenbaum, Steven R., and Richard I. Mateles (1968), "Single Cell Protein," *Science Journal,* **4** (5), 87-92. (Department of Nutrition and Food Science, Massachusetts Institute of Technology).

Tannenbaum, S. R., and G. W. Pace (1976), "Food from Waste: An Overview," in Birch et al., 1976, pp. 8-22. (Department of Nutrition and Food Science, Massachusetts Institute of Technology; and Tate and Lyle Ltd., Reading, U.K.)

Tauber, E. (1966), "Basalt Makes Inexpensive Stoneware and Architectural Shapes," *Ceramic Industry Magazine,* **86**(4), 122-125, 152-153. (CSIRO Division of Building Research, Melbourne, Australia)

Tauber, E., and D. N. Crook (1965), "Bonding Lightweight Aggregate with Glass," *British Clayworker,* **74** (873), 42-43. (CSIRO Division of Building Research, Melbourne, Australia)

Tauber, E., and M. J. Murray (1968), "Coloured Stone for Exposed Aggregate Panels," *Constructional Review,* **41** (7), 24. (CSIRO Division of Building Research, Melbourne, Australia)

Tauber, E., and M. J. Murray (1969), "Low Cost Glass Enamel Composite Makes Attractive Concrete Decoration," *Ceramic Industry Magazine,* **86,** May. (CSIRO Division of Building Research, Melbourne, Australia)

Tauber, E., and M. J. Murray (1971), "Glass-Clay Mosaics and Tiles," *Journal of the Australian Ceramic Society,* **7** (2), 47-51. (CSIRO Division of Building Research, Melbourne, Australia)

Tauber, E., R. K. Hill, D. N. Cook, and M. J. Murray (1970), "Ilmenite Ceramics," *Journal of the Australian Ceramic Society,* **6** (1). (CSIRO Division of Building Research, Melbourne, Australia)

Tauber, E., R. K. Hill, D. N. Crook, and M. J. Murray (1971), "Red Mud Residues from Alumina Production as a Raw Material for Heavy Clay Products," *Journal of the Australian Ceramic Society,* **7** (1), 12-17. (CSIRO Division of Building Research, Melbourne, Australia)

Taylor, Lynn J. (1973), "Polymer Degradation: Some Positive Aspects," *ChemTech,* **3,** 552-559. (Owens-Illinois Technical Center, Toledo, Ohio)

Taylor, O. R. (1975a), "Collection and Sale of Recyclable Goods from Households," in Kirov, 1975a, pp. 199-201. (The Smith Family, Sydney, Australia)

Taylor, G. T. (1975b), "The Formation of Methane by Bacteria," *Process Biochemistry,* **10** (8), 29-31, 33. (Microbiology Department, Queen Elizabeth College, London)

Taylor, J. C. (1975c), "Continuous Regeneration of Hydrochloric Acid Pickle Liquors," in Johnston, 1975, pp. T10-T12. (Taylor, Lurgi, Australia, Pty. Ltd., Melbourne)

Tebbut, T. H. Y. (1973), *Water Science and Technology,* Barnes and Noble, New York.

Technology Review (1971), "Mining the Dumps: The Economics, the Technology, and the Steelworks," **73** (9), 57-58.

Tegart, W. J. McG., and G. Mowat (1975), "Recent Examples of By-Product Utilization by B.H.P.," in Johnston, 1975, pp. T2-T4. (B.H.P., Australia)

Teknekron (1973), *A Technical and Economic Study of Waste Oil Recovery,* Parts I, II, and III, Reports SW-90c.1, SW-90c.2, and SW-90c.3, U.S. Environmental Protection Agency, Washington, D.C.; U.S. NTIS PB-237,618, PB-237,619, and PB-237,620. (Teknekron, Inc., Berkeley, California; and the Institute of Public Administration)

Teknekron (1975), *A Technical and Economic Study of Waste Oil Recovery,* Parts IV-VI, Report SW-90c.4 to Office of Solid Waste Management Programs, U.S. Environmental Protection Agency, Washington, D.C.; U.S. NTIS PB-251,716. (Teknekron, Inc., Berkeley, California; and the Institute of Public Administration)

Teskey, B. J. E., and K. R. Wilson (1975), "Tire Fabric Waste as Mulch for Fruit Trees," *Journal of American Society for Horticultural Science,* **100** (2), 153-157. (University of Guelph, Ontario)

Testin, Robert F. (1971), "Recycling of Used Aluminum Products," in ANERAC, 1971, pp. 12-27. (Reynolds Metal Company, U.S.A.)

Theilig, Gerhard, Gerhard Müller, and Kurt Bodenbenner (1975), "Re-Use and Disposal of Waste Acids," *Israel Journal of Chemistry,* **14,** 234-243; 25th IUPAC Congress, Jerusalem, July. (Hoechst A.G, West Germany)

Thomas, Christine (1974), *Material Gains: Reclamation, Recycling and Re-Use,* Friends of the Earth Ltd., London.

Thygeson, J. R., and E. D. Grossmann (1975), "Recycling Cow Manure—A System Analysis," at Second International Conference on Environmental Problems of the Extractive Industries, 18 June. (Department of Chemical Engineering, Drexel University, Philadelphia, Pennsylvania)

Thygeson, J.R., E. D. Grossmann, and Joseph MacArthur (1971), "Through-Circulation Drying of Manure in Superheated Steam," *Proceedings of the International Symposium on Livestock Wastes,* American Society of Agricultural Engineers, pp. 185-189. (Department of Chemical Engineering, Drexel University, Philadelphia, Pennsylvania)

Tillman, David A. (1975), "Fuels from Recycling Systems," *Environmental Science and Technology,* **9** (5), 418-422. (Materials Associates, Inc., Washington, D.C. 20037)

Timpe, W. G., E. Lang, and R. L. Miller (1973), *Kraft Pulping Effluent Treatment and Reuse—State of the Art,* Report EPA-R2-73-164, Office of Research and Monitoring, U.S. Environmental Protection Agency, Washington, D.C. (St. Regis Paper Co., Research and Development Center, Pensacola, Florida)

Tinker, Jon (1976), "Sump Oil: A Million Wasted Quid," *New Scientist,* **69** (989), 435-436.

Toffler, Alvin (1972), Ed., *The Futurists,* Random House, New York.

Toyabe, Yutaka, Genji Matsumoto, and Hitoshi Kishikawa (1974), "Manufacturing Ceramic Goods Out of Mining Wastes," in Aleshin, 1974, pp. 240-244. (Chemical Consulting Engineers, 17-28, Shimouma-6, Setagaya-Ku, Tokyo; and Arita Bussan Co. Ltd., Arita-Cho, Saga Prefecture, Tokyo)

Treacher, Sydney (1973), "Recycling—A Positive Approach to Used Solvent Handling," *Environmental Pollution Management,* **3** (4), 203, 205. (Croftshaw (Solvents) Ltd., London, SE6 1DG)

Trevelyan, W. E. (1975), "Renewable Fuels: Ethanol Produced by Fermentation," *Tropical Science,* **17** (1), 1-13. (Tropical Products Institute, 56/62 Gray's Inn Road, London, WClX 8LU)

Trevitt, E. W. (1976), "Recycling of Mixed Plastics Waste," *Polymers Paint and Colour Journal,* March; in condensed form in *Industrial Recovery,* **22** (4), 12, 14. (Edwin G. Fisher Associates, Consulting Engineers)

Trezek, G. J., and G. Savage (1976), "MSW Component Size Distributions Obtained from the Cal Resource Recovery System," *Resource Recovery and Conservation,* **2** (1), 67-77, (Department of Mechanical Engineering, University of California, Berkeley)

Tron, A. R. (1976), *A Quantitative Assessment of Old and New Copper Scrap Collected in the U.K.,* Warren Spring Laboratory Publication LR 222 (MR), Department of Industry, U.K.

Troth, S. J. (1973), "Composting Agricultural and Industrial Organic Wastes," in Inglett, 1973, Chapter 15, pp. 172-182. (Soils and Crops Department, Rutgers State University of New Jersey, New Brunswick)

Truby, Randolph L., and James H. Sleigh (1974a), "Cleaning Water by Reverse Osmosis...An Overview of System Requirements and Capabilities," *Plant Engineering,* October 31. (Roga Systems Division of UOP, San Diego, California)

Truby, Randolph L., and James H. Sleigh (1974b), "Cleaning Water by Reverse Osmosis...Membranes Design Parameters, Costs," *Plant Engineering,* December 12. (Roga Systems Division of UOP, San Diego, California)

Trussell, R. Rhodes (1975), "The Use of Renovated Water in Cooling Towers." in Johnston, 1975, pp. T51-T53. (James M. Montgomery, Consulting Engineers, Pasadena, California)

Tsuchiya, Harufumi (1976), "Recovery of the Mist Generated from Aqueous Coating Compositions," *Japanese Kokai* 76 20,233; 76 20,234; *Chemical Abstracts,* **85** (2), 7373, 7374. (Dai Nippon Toryo Co. Ltd.)

Tsutsumi, Shigeru (1974), "Recycling of Waste Plastics," *Chemical Economy and Engineering Review,* **6** (1), 7-10, 20. (Osaka University Japan)

Tudge, Colin (1975), "Why Turn Waste into Protein," *New Scientist,* **66,** (945), 138-139. (Science Editor, *World Medicine*)

Tunnah B. G., A. Hakki, and R. J. Leonard (1974), *Where the Boilers Are; A Survey of Electric Utility Boilers with Potential Capacity for Burning Solid Waste as Fuel,* Publication SW-88c, U.S. Environmental Protection Agency; U.S. NTIS PB-239,392. (Gordian Associates, Inc.)

Twidwell, Larry G., Jin-Rong Hwang, and Ralph E. Dufresne (1976), "Industrial Waste Disposal. Excess Sulfuric Acid Neutralization with Copper Smelter Slag," *Environmental Science and Technology,* **10** (7), 687-691. (Metallurgy-Mineral Processing Department, Montana College of Mineral Science and Technology, Butte)

Ungewitter, C. (1938), *Verwertung des Wertlosen* (Utilization of the Worthless), Wilhelm Limpert-Verlag, Berlin.

Updegraff, David M. (1971), "Utilization of Cellulose from Waste Paper by *Myrothecium verrucaria,'' Biotechnology and Bioengineering,* **13,** 77-97. (Denver Research Institute, University of Denver, Colarado)

Upsher, F. J. (1976), ''Microbial Attack on Materials,'' *Proceedings of the Royal Australian Chemical Institute,* **43** (6), 173-176. (Materials Research Laboratories, Australian Defence Scientific Service, Maribyrong, Victoria)

U.S. Congress (1972), *The Economics of Recycling Waste Materials,* U.S. Congress Joint Economic Committee, Subcommittee on Fiscal Policy, Hearings, 92nd Congress, First session, 8-9 November, 1971, U.S. Government Printing Office, Washington, D.C.

USDI (1972), *First Annual Report of the Secretary of the Interior Under the Mining and Minerals Policy Act of 1970,* U.S. Department of the Interior, Washington, D.C.

USEPA (1971), *Zinc Precipitation and Recovery from Viscose Rayon Waste Water,* Report 12090 ESG 01/71, Water Quality Office, U.S. Environmental Protection Agency, Washington, D.C. (American Enka Company, Enka, North Carolina 28728)

USEPA (1972a), *Study to Identify Opportunities for Increased Solid Waste Utilization,* U.S. Environmental Protection Agency Report SW-40d, U.S. NTIS PB-212,729, PB-212,730; summaries in Jackson, 1975; NASMI, 1972; Ness, 1972.

USEPA (1972b), *Metropolitan Housewives' Attitudes Towards Solid Waste Disposal,* Report EPA-R5-72-003, Office of Research and Monitoring, Environmental Protection Agency, Washington, D.C. (National Analysts, Inc., Philadelphia, Pennsylvania)

USEPA (1972c), *The Automobile Cycle; an Environmental and Resource Reclamation Problem,* Federal Solid Waste Management Program Publication SW-80 ts.1, U.S. Environmental Protection Agency.

USEPA (1974a), *Resource Recovery and Source Production, First Report to Congress,* Office of Solid Waste Management Programs, U.S. Environmental Protection Agency, Report SW-118, Washington, D.C., 3rd ed.

USEPA (1974b), *Resource Recovery and Source Reduction, Second Report to Congress,* Office of Solid Waste Management Programs, U.S. Environmental Protection Agency, Report SW-122, Washington, D.C.

USEPA (1974c), *Solid Waste Managements: Available Information Materials,* Solid Waste Management Report SW-58.21, U.S. Environmental Protection Agency.

USEPA (1974d), *Recycling and the Consumer; Solid Waste Management,* Publication SW-117, U.S. Environmental Protection Agency.

USEPA (1975), *Resource Recovery and Waste Reduction, Third Report to Congress,* Office of Solid Waste Management Programs, Report SW-161, U.S. Environmental Protection Agency, Washington, D.C.: summary in Smith, 1976.

USEPA (1976a), ''Guidelines for Procurement of Products that Contain Recycled Material,'' U.S. Environmental Protection Agency, *Federal Register,* **41**(10), 2356-2358.

USEPA (1976b), ''Source Separation for Materials Recovery; Guidelines,'' U.S. Environmental Protection Agency, *Federal Register,* **41**(80), 16950-16956.

USEPA (1976c), *Solid Waste Management: Available Information Materials* (Total

Listing, 1966-1976), compiler Julie L. Larsen, Office of Solid Waste Management Programs Report SW-58.26, U.S. Environmental Protection Agency.

Valdez, E. G. (1976), "Separation of Plastics from Automobile Scrap," in Aleshin, 1976, pp. 386-392. (Salt Lake City Metallurgy Research Center, U.S. Bureau of Mines)

Valdez, E. G., and K. C. Dean (1975), *Experiments in Treating Zinc-Lead Dusts from Iron Foundries,* Bureau of Mines Report of Investigations 8000, U.S. Department of the Interior, Washington, D.C. (Salt Lake City Metallurgy Research Center)

Valdez, E. G., K. C. Dean, J. H. Bilbrey, Jr., and L. R. Mahoney (1975), *Recovering Polyurethane Foam and Other Plastics from Auto-Shredder Reject,* Bureau of Mines Report of Investigations 8091, U.S. Department of the Interior, Washington, D.C. (Salt Lake City Metallurgy Research Center; and Ford Motor Co., Dearborn, Michigan)

Valdmaa, Kalju (1973), "Composting of Wastes," in *Environmental Engineering,* ed. Goesta Lindner and Klas Nyberg, Reidel, Holland, pp. 445-449. (Agriculture College of Sweden, Uppsala)

Vale, Brenda, and Robert Vale (1975), *The Autonomous House. Design and Planning for Self-Sufficiency,* Thames and Hudson, London.

van Dam, André (1975), "The Limits to Waste," *Futurist,* **9**(1), 18-21. (Cerrito 866, Buenos Aires, Argentina)

van den Broek, E., and N. Y. Kirov (1972), "The Characterisation of Municipal Solid Waste," in Kirov, 1972, pp. 23-29. (Department of Fuel Technology, University of New South Wales, Australia)

Van der Molen, Robert H. (1973), "Energy from Municipal Refuse through Fluidized Combustion: the CPU-400 Pilot Plant," at 66th Annual Meeting, American Institute of Chemical Engineers, 11-15 November, Philadelphia, Pennsylvania. (Combustion Power Co., Inc., Menlo Park, California)

Van Eeden, W. N. (1975), "Power Station Effluent Control and the Reuse of Ash Water for Cooling Water Treatment," *Water Pollution Control,* **74**(2), 211-215. (Electricity Supply Commission, South Africa)

Varjavandi, J. J. (1975), "Management of Municipal Solid Waste," in Johnston, 1975, pp. R13-R23. (Crooks Michell Peacock Stewart Pty Ltd., Chatswood, NSW, Australia)

Varjavandi, J. J., and T. J. Fischof (1975), "A Survey of Community Solid Waste Practices in Australia," in Kirov, 1975a, pp. 65-71. (A.C.I. Ltd., Australia)

Verami, Kh., and I. Mladenov (1972), "Possibilities of Using Polyacrylonitrile Wastes by Hydrolysis," *Godishnik na Visshiya Khimiko-Tekhnologicheski Institut, Burgas, Bulgaria,* **9**(9), 187-197 (Bulgarian), published 1973; *Chemical Abstracts,* **83,** 29612.

Videla, H. A., and A. J. Arvia (1975), "The Response of a Bioelectrochemical Cell with *Saccharomyces cerevisiae* Metabolizing Glucose under Various Fermentation Conditions," *Biotechnology and Bioengineering,* **17**(10), 1529-1543. (University of La Plata, Argentina)

Vincent, W. A. (1969), "Algae for Food and Feed, *Process Biochemistry,* **4**(June), 45-47. (Battelle Institute, Geneva)

Viscomi, Vincent (1976), ''Feasibility Study for Burning Refuse-derived Fuel in the District of Columbia by Potomac Electric Power Co.,'' in RRC, 1976, pp. 217-224. (Lafayette College, Easton, Pennsylvania 18042)

Wade, Charles G. (1976), *Contemporary Chemistry: Science, Energy, and Environmental Change,* Macmillan, New York.

Waggoner, Don (1976), ''The Oregon Bottle Bill—What It Means to Recycling,'' *Compost Science,* **17**(4), 10-13. (Past-President, Oregon Enviromental Council)

Walker, Colin (1971), *Environmental Pollution by Chemicals,* Hutchinson Biological Monograph, Hutchinson Educational, London.

Wang, Lawrence K., and John Y. Yang (1975), ''Total Waste Recycle System for Water Purification Plant Using Alum as Primary Coagulant,'' *Resource Recovery and Conservation,* **1,** 67-84. (Department of Chemical and Environmental Engineering Systems Department, Calspan Corporation, Buffalo, New York)

Ward, Benjamin F. (1975), ''Tall Oil Chemicals from a Natural, Renewable Source,'' in APS, 1975, pp. 329-334. (Westvaco Corporation, Charleston Heights, South Carolina)

Wary, John, and Robert B. Davis (1976), ''Cryopulverizing,'' *ChemTech,* **6**(3), 200-203. (Linde Division, Union Carbide Co., New York)

Wayne, T. J., and A. J. Perna (1971), ''Effects, Recovery, and Reuse of Oil from Aqueous Environments,'' in ANERAC, 1971, pp. 232-243. (Worthington Corporation; and Newark College of Engineering, U.S.A.)

Weber, Walter J., Jr. (1972), *Physicochemical Processes for Water Quality Control,* Wiley-Interscience, New York.

Wechsler, R. (1975), ''Direct Reuse of Municipal Waste Water,'' in Johnston, 1975, pp. T110-T112. (Keurinsinstitut voor Waterleidingartikelen, Rijswijk, The Netherlands)

Weeden, C. (1975), ''Glass Containers in the Environment,'' in Henstock, 1975a. (Glass Manufacturers' Federation, U.K.)

Weinstein, N. J. (1974), *Waste Oil Recycling and Disposal,* U.S. Environmental Protection Agency; U.S. NTIS PB-235,857.

Weinstein, Norman J., and Richard F. Toro (1976), *Thermal Processing of Municipal Solid Waste for Resouce and Energy Recovery,* Ann Arbor Science, Ann Arbor, Michigan.

Weiss, Donald (1976), ''Resource-Full Australia,'' *Chemistry in Britain,* **12**(1), 8-15. (Division of Chemical Technology, CSIRO, Australia)

Weiss, D. E., B. A. Bolto, A. S. Macpherson, R. McNeill, R. Siudak, and D. Willis (1966), ''Thermally Regenerated Ion Exchange Process—An Aid to Water Management,'' *Journal Water Pollution Control Federation,* **38**(11), 1782-1804. (CSIRO Chemical Research Laboratories, Melbourne, Australia)

Wender, Irving, Fred W. Steffgen, and Paul M. Yavorsky (1974), ''Clean Liquid and Gaseous Fuels from Organic Solid Waste,'' in Yen, 1974a, Chapter 2, pp. 43-99. (Pittsburgh Energy Research Center, U.S. Bureau of Mines)

Westerhoff, Garret P. (1973), ''Alum Recycling: An Idea Whose Time Has Come?'' *Water and Wastes Engineering,* **10**(12), 28-31, 48. (Malcolm Pirnie, Inc., Paramus, New Jersey)

Wharton, F. D., Jr., and J. Kenneth Craver (1975), ''Technology Assessment in Product Development. A Case History,'' *ChemTech,* **5**(9), 547-551. (Monsanto, U.S.A.)

Wheeler, Tone (1975), ''The Autonomous House at Sydney University,'' *Architecture in Australia,* April, 74-77.

Whisman, M. L., J. W. Goetzinger, and F. O. Cotton (1974), *Waste Lubricating Oil Research* (A Comparison of Bench-Test Properties of Re-refined and Virgin Lubricating Oils), Bureau of Mines Report of Investigations 7973, U.S. Department of Interior, Washington, D.C.

White, Jonathan W. (1973), ''Processing Fruit and Vegetable Wastes,'' in Inglett, 1973, Chapter 12, pp. 129-142. (Eastern Regional Research Laboratory, U.S. Department of Agriculture, Philadelphia)

Wiley, John S., F. E. Gartrell, and H. Gray Smith (1966), ''Concept and Design of a 3-Way Composting Project,'' *Compost Science,* **7**(2), 11-14. (Office of Solid Wastes, Public Health Service, Chattanooga, Tennessee; and Tennessee Valley Authority)

Wiley, Averill J., George A. Dubey, and I. K. Bansal (1972), ''Reverse Osmosis Concentration of Dilute Pulp and Paper Effluents,'' Report 12040EEL 02/72, Office of Research and Monitoring, U.S. Environmental Protection Agency, Washington, D.C. (Pulp Manufacturers Research League and the Institute of Paper Chemistry, 1043 E. South Piver Street, Appleton, Wisconsin 54911)

Wilkinson, J. F., and A. H. Rose (1963), ''Fermentation Processes,'' in Rainbow and Rose, 1963, Chapter 11, pp. 379-414. (Department of Bacteriology, University of Edinburgh, Scotland; and Department of Microbiology, University of Newcastle-upon-Tyne, England)

Williams, Robert H. (1975), *The Energy Conservation Papers,* Ballinger, Cambridge, Massachusetts.

Willson, R. Thomas (1974), ''Urban Refuse: New Source for Energy and Steel,'' *Professional Engineer,* **44**(11), 20-23. (American Iron and Steel Institute)

Wilson, David Gordon (1974a), ''Review of Advanced Solid-Waste Processing Technology,'' in American Institute of Chemical Engineers Symposium on Solid Waste Management, 4 June, Paper 40a, *Solid Wastes Management,* **64**(10), 581-602. (Department of Mechanical Engineering, Massachusetts Institute of Technology)

Wilson, David Gordon (1974b), ''Universities Attack Solid Waste,'' *NCRR Bulletin,* **4**(4), 12-16.

Wilson, David Gordon (1975), ''The Resource Potential of Demolition Debris in the United States,'' *Resource Recovery and Conservation,* **1,** 129-140. (Massachusetts Institute of Technology)

Wilson, D. A. (1976), ''A New Sulfur Dioxide-Free Process for Recovering Lead from Battery Scrap,'' in Aleshin, 1976, pp. 393-397. (College Park Metallurgy Research Center, U.S. Bureau of Mines)

Wilson, E. Milton, and Harry M. Freeman (1976), ''Processing Energy from Wastes,'' *Environmental Science and Technology,* **10**(5), 430-435. (The Ralph M. Parsons Co., Pasadena, California 91124; and Industrial Environmental Research Laboratory, U.S. EPA, Cincinnati, Ohio)

Wilson, David Gordon, and Stephen D. Senturia (1974), "Design and Performance of the M.I.T. Process for Separating Mixed Municipal Refuse," in Aleshin, 1974, pp. 117-127. (Massachusetts Institute of Technology)

Wilson, David Gordon, and Ora E. Smith (1972), "How to Reclaim Goods from Wastes," *Technology Review,* **74**(6). (Department of Mechanical Engineering, Massachusetts Institute of Technology)

Wilson, Maurice J., and David W. Swindle (1976), "The Markets for and the Economics of Heat Energy from Solid Waste Incineration," in RRC, 1976, pp. 197-206. (I.C. Thomasson and Associates, Inc., Nashville, Tennessee 37204)

Wilson, David Gordon, Patricia Foley, Richard Wiesman, and Stamatia Frondistou-Yannas (1976), "Demolition Debris: Quantities, Composition, and Possibilities for Recycling," in Aleshin, 1976, pp. 8-15. (Massachusetts Institute of Technology)

Winkler, P. Frank, and David G. Wilson (1973), " Size Characteristics of Municipal Solid Waste," *Compost Science,* **14**(5). (Department of Physics, Middlebury College, Middlebury, Vermont; and Department of Mechanical Engineering, Massachusetts Institute of Technology)

Wise, D. L., S. E. Sadek, R. G. Kispert, L. C. Anderson, and D. H. Walker (1975), "Fuel Gas Production for Solid Waste," *Biotechnology and Bioengineering Symposia,* **5** (Cellulose Chemical and Energy Resources), 285-301. (Dynatech Research and Development Co., Cambridge, Massachusetts)

Wisely, F. E., G. W. Sutterfield, and D. L. Klumb (1974), "St. Louis Energy Recovery Project," in Aleshin, 1974, pp. 191-195. (Horner and Shifrin, Inc., St. Louis; City of St. Louis; and Union Electric Co., St. Louis, Missouri)

Wiser, Wendell H., and Larry L. Anderson (1975), "Transformation of Solids to Liquid Fuels," *Annual Review of Physical Chemistry,* **26,** 339-357. (Department of Fuels Engineering, University of Utah, Salt Lake City, Utah 84112)

Witwer, J. G., K. K. Ushiba, and K. T. Semrau (1976), "Energy Conservation with LNG Cold," *Chemical Engineering Progress,* **72**(1), 50-55. (Stanford Research Institute, Menlo Park, California)

Wolk, R. H., and C. A. Battista (1973), *Study of the Technical and Economic Feasibility of a Hydrogenation Process for Utilization of Waste Rubber,* U.S. Environmental Protection Agency; U.S. NTIS PB-222,694.

Wolverton, Bill, and Rebecca C. McDonald (1976), "Don't Waste Waterweeds," *New Scientist,* **71**(1013), 318-320. (National Space Technology Laboratories, U.S. National Aeronautics and Space Administration, Bay St. Louis, Mississippi)

Wood, Richard (1975), "Utilization of Sewage Sludge," *Effluent Water Treatment Journal,* **15**(9), 455-457, 459-461.

Worgan, J. T. (1976), "Wastes from Crop Plants as Raw Materials for Conversion by Fungi to Food or Livestock Feed," in Birch et al., 1976, pp. 23-41. (National College of Food Technology, University of Reading, U.K.)

Wright, J. (1971), "Electrochemical Aspects of Sulphide Hydrometallurgy," *Bulletin of the Australian Mineral Development Laboratories,* (12), October, 47-73.

WSL (1972a), *Treatment of Domestic Wastes,* Information Sheet 72 MHL, Warren Spring Laboratory, Stevenage, U.K.

WSL (1972b), *Treatment of Industrial Wastes,* Information Sheet 72 MHM, Warren Spring Laboratory, Stevenage, U.K.

WSL (1975a), *Acid Mine Drainage,* Publication 2ME/75F/10C, Warren Spring Laboratory, Stevenage, U.K.

WSL (1975b), *Microbial Processes Associated with Minerals and Metals,* Publication 8ME/75F/10C, Warren Spring Laboratory, Stevenage, U.K.

WSL (1975c), *Pyrolysis—Some Questions Answered,* Publication 2MR/75F/37C, Warren Spring Laboratory, Stevenage, U.K.

WSL (1975d), *Developments in the Pyrolysis of Domestic Refuse,* Publication 1MR/75F/27C, Warren Spring Laboratory, Stevenage, U.K.

WSL (1975e), *Recovery of Re-usable Materials from Domestic Refuse,* Publication 3MR/75F/22C, Warren Spring Laboratory, Stevenage, U.K.

WSL (1975f), *Waste Recovery Centres—A Feasibility Study,* Publication 5MR/75F/37C, Warren Spring Laboratory, Stevenage, U.K.

WSL (1975g), *Bibliography of WSL Publications on Waste Recovery,* Warren Spring Laboratory, Stevenage, U.K. (publications since 1969).

WWE (1973), "Algae Can be Profitable," *Water and Wastes Engineering,* **10**(2), 39.

WWT (1975a), "High Throughput Filter for Solid Recovery," *Water Waste Treatment,* **18**(6), 14.

WWT (1975b), "Using Water Coolers Intelligently," *ibid.,* **18**(6), 16.

WWT (1975c), "Recovery of Protein by Ion Exchange," *ibid.,* **18**(6), 22.

WWT (1975d), "Recycling Old Motor Tyres," *ibid.,* **18**(6), 24.

WWT (1975e), "Submerged Combustion Techniques in Recycling," *ibid.,* **18**(6), 26.

WWT (1975f), "Reclaiming Antifreeze—A New Venture," *ibid.,* **18**(3), 19.

WWT (1975g), "Chromatic Acid Recovery," *ibid.,* **18**(6), 20.

WWT (1975h), "A Successful Chemical Regeneration Process," *ibid.,* **18**(6), 18-19.

WWT (1976a), "German Ultrafiltration Plant," *ibid.,* **19**(6), 28.

WWT (1976b), "Magnetic Filtration Attracts Industry," *ibid.,* **19**(6), 28.

WWT (1976c), "The Recovery of Metals from Effluent," *ibid.,* **19**(6), 32.

WWT (1976d), "Flux Recovery Plant to be Marketed," *ibid.,* **19**(6), 34.

WWT (1976e), "The Theory and Practice of Ion Exchange," (Report of the International Ion Exchange Conference, Cambridge, 25-30 July) *ibid.,* **19**(8), 24.

WWT (1976f), "Integral Recycling of Fibreboard Production," *ibid.,* **19**(7), 40.

Wylie, John C. (1955), *Fertility from Town Waste,* Faber and Faber, U.K.

Yamazaki, Daizo, Yoshiaki Mizumoto, Shigeo Hasegawa, and Iwao Tsukuda (1974), "Recycling Polymers," *Japanese Kokai,* 74 34,576; *Chemical Abstracts,* **81,**78881. (Mitsubishi Heavy Industries Ltd., Japan)

Yen, T. F. (1974a), Ed., *Recycling and Disposal of Solid Wastes,—Industrial, Agricultural, Domestic,* Ann Arbor Science, Ann Arbor, Michigan.

Yen, T. F. (1974b), "Biodeterioration and Biodisintegration" in Yen 1974a, Chapter 1, pp. 1-41. (University of Southern California, Los Angeles)

Yosim, S. J., L. F. Grantham, D. E. McKenzie, and G. C. Stegman (1973), "The Chemistry of the Molten Carbonate Process for Sulfur Oxides Removal from Stack Gases," in Jimeson and Spindt, 1973, Chapter 15, pp. 174-194. (Atomics International, Canoga Park, California; and Consolidated Edison Co., New York)

Young, Richard A., and Ian O. Lisk (1976), "Paper Mill Incinerator Solves Town Waste Problem and Provides Energy," *Pollution Engineering,* **8**(8), 40.

Yudkin, J. (1972), "The Psycho-Sociological Problems Involved in Reactions to New Foods," in Gounelle de Pontanel, 1972, pp. 149-163.

Zandi, Iraj, and John A. Hayden (1969), "Are Pipelines the Answer to Waste Collection Dilemma?" *Environmental Science and Technology,* **3**(9); reprinted in ACS, 1971, pp. 52-57. (Towne School of Civil and Mechanical Engineering, University of Pennsylvania, Philadelphia, Pennsylvania 19104)

Zerlaut, G. A., and A. M. Stake (1974), "Chemical Aspects of Plastic Waste Management," in Yen 1974a, Chapter 5, pp. 175-184. (IIT Research Institute, Chicago, Illinois)

Zimmer, F. V. (1967), "Problems in Fly Ash Marketing," in Faber et al., 1967, pp. 58-68. (Detroit Edison Co., Detroit, Michigan)

INDEX

g *refers to glossary entry*; t *refers to table entry*; f *refers to figure.*

Abandoned consumer goods, *see* Automobiles; Litter; Postconsumer waste
Abrasives, glass, 201
  woody wastes, 227
ABS plastic, 119, 287g
Absorption, gases, 50, 201-207
  by woody wastes, 227
Acacia mollisima, 236
Accelerometer, 32, 53, 287g
Accommodation, *see* Autonomous house; Building
Acetaldehyde, 236
Acetic acid, in microbiological processes, 96, 98t
  separation, by active carbon, 55
    by ion exchange, 68
  from wood, 86, 236
Acetone, from cellulose, 236, 237t
  in cryogenic embrittlement, 63
  from microbiological processes, 98t, 236
  in solvent extraction, 50
Acetonitrile, in solvent extraction, 51, 93
Acetylene, 194
Acid, metal dissolution, 94
  metal pickling, 194-195, 196-198
  waste neutralization, 190, 199, 201-205
  waste recovery, 193-194
  see also Acetic; Butyric; Carboxylic, Chromic; Formic; Hydrochloric; Lactic, Nitric; Oxalic; Propionic; Succinic; Sulfuric; Tartaric
Acid leach mine waters, uranium recovery, 68
Acidity, in precipitation, 66, 188
Acrylonitrile, 74
Acrylonitrile-butadiene-styrene (ABS) plastic, 119, 287g
Actinomyces, 287g
  in composting, 101
  single cell protein, 234
Activated carbon, *see* Activated charcoal
Activated charcoal, 54-55, 251, 287g
  advanced water treatment, 209-211
  from cellulose, 236
  from coal, 240
  gas recovery, 207, 214
  in hydrometallurgy, 19f
  from lignin, 236
  organics recovery, 188
  regeneration, 54, 72, 212, 214, 251-252
  solvent adsorption, 207, 214
  sulfur oxides recovery, 202-205
  from textiles, 225
  from urban waste, 145
  water treatment, 209, 211
Activated sludge, 148, 287g
Addition polymers, 73-75, 287g
Adhesives, from bark, 226, 236
  solvents, 214-229
  on waste paper, 229-230
Adsorbates, 287g
Adsorbents, 54, 287g
  synthetic, 55
  waste materials, 55
Adsorption, 54-55, 287g
  acid recovery, 193, 203-204
  see also Activated charcoal
Advanced water treatment, 41, 58, 65, 209-211
Advertizing, 28, 117
Aerobic process, 23, 96, 128, 148, 287g
Aerosols, 288g
  propellants, 205
  from pyrometallurgy, 190
Africa, SCP use, 235
Aggregates, 288g
  from wastes, 17, 178-184, 201
  water recycle in production, 212

Aging of precipitate, 41
Agricultural wastes, 9t, 175, 225-227
  anaerobic digestion, 104-105
  animal feed, 233
  composting, 100, 225
  energy source, 237-238
  heat content, 131t
  hydrolysis, 231-232
  as metal ion scavengers, 69, 189, 226, 236
  microbiological processing, 103, 231-235
  papermaking, 227, 229
Agriculture, energy, 248
  plastics use, 216
  water reuse, 210
  see also Aquaculture; Feedlot wastes, Fertilizer; Hydroponics; Intensive farming; Spray irrigation
Air, in ammonia synthesis, 145
Air blast separation, 31, 35-36, 38
Air classification, 48-50, 53, 124-125, 288g
  urban waste, *see* Urban waste
Air conditioning, 169, 173, 249
Air elutriation, *see* Air classification
Air flotation, *see* Froth flotation
Air pollution, 24, 133, 201-207
Airships, 18
Alcohols, from fermentation, 96-99, 102
  mixtures with water, 41, 221
  from plastics manufacture, 187
  from pyrolysis, 84
  solvents, 214
  from sulfite waste liquors, 192
  see also Butanediol; Ethanol; Methanol; Polyalcohols
Algae, 95, 233, 235, 288g
  alum effect, 235
  Chlorella, 234, 235
  energy storage, 106
  lipid content, 234
  nutrients, 211
  on sewage sludge ponds, 148, 235
  single cell protein, 131, 233
  Spirulina maxima, 235
Aliphatic, 288g
Aliphatic acids, flotation agents, 66
  in microbiological processes, 97-99, 101-102
  from partial oxidation, 91
  polymerization, 74
  from sulfite waste liquor, 192
  waxes, 223, 227
  see also Acetic; Butyric; Formic; Propionic; Succinic; Tartaric
Alkali, from fly ash, 189
  in liquefaction process, 90
  metal precipitation, 66, 188
  nickel alloy pickling, 198
  oxides from cement manufacture, 185, 190
  regeneration of ion exchange resin, 68
  waste neutralization, 190, 199, 203-205
Alkali carbonates, *see* Lithium carbonate; Molten salts; Potassium carbonate; Sodium carbonate
Alkali halides, *see* Lithium chloride; Molten salts; Potassium chloride; Sodium chloride
Alkanes, *see* Hydrocarbons
Alkyl halides, 206
Alkyl sulfonates, 209
Alloy, *see* Brass; Ferrous metal; Stainless Steel, Superalloy
Alternative technology, 10-12, 91, 103, 147
Altoona, Pennsylvania, 129
Alum, 288g
  recovery, 212, 235
  toxic effect on algae, 235
  in wastewater treatment, 189, 211
Alumina, adsorbent, 54, 202
  in aluminum manufacture, 25t, 181, 202
  from bauxite, 25t, 181
  in blast furnace slag, 180t
  in cement manufacture, 185-186
  from clay, 186
  in fly ash, 183t
Alumino-silicates, 179-180, 183, 288g
Aluminum, 253-254
  in automobiles, 49, 61t, 150, 153
  from cementation of copper and silver, 72
  energy content, 25t, 216
  metallurgy, 24-25, 72-73, 187, 190, 202
  packaging, 117
  radiator scrap, 49
  recycle, 8t, 28, 119, 215
  recycle literature, 252-254
  removal from water by fly ash, 189
  resources, 16, 250
  separation, by cryogenic embrittlement, 63, 216
    by density methods, 37

by eddy current methods, 35
by electrochemical processes, 70
by selective melting, 61t
in steel, 61, 151
in urban waste, 117, 216
wiring, 22
Aluminum cans, 118-121
Aluminum chloride, 72
Aluminum fluoride, 72-73, 187, 202
Aluminium hydroxyphosphate, 211
Aluminum magnet, 35
Aluminum nitride, 73
Aluminum oxide, *see* Alumina
Aluminum phosphate, 211
Aluminum salts, coagulants, 227
in concrete, 181
see also specific salts
Aluminum sulfate, *see* Alum
Alwatech process, 227
American Chemical Society, 30, 213
American Gas Association, 172
Ames, Iowa, 134, 158
Amines, flotation agents, 66
isocyanate manufacture, 206
from microbiological processes, 98-99, 103, 236
polymerization, 74
solvent extraction agents, 51
Amino acids, 75, 233-236, 288g. *See also* Protein
Ammonia, from domestic waste, 92, 139, 145, 167
from microbiological processes, 98
recovery, 202
sulfur oxides recovery, 203-205
Ammonia fertilizer, from wastes, 92, 150, 193, 230, 239
Ammonia-soda (Solvay) process, 200
Ammonium bisulfite, 192-193, 204-205
Ammonium fluoride, 56
Ammonium ion, in ion exchange, 68
Ammonium sulfate, 150, 203
Ammonium sulfite, 203-204
Amsterdam, 133
Andco-Torrax, 137, 171
Anaerobic, 288g
Anaerobic digestion, 12, 96, 99, 104-105, 233, 288g
of sewage, 148
of urban waste, 14, 129-130
Anhydrite, 191
Animal feed, 233
anaerobic digestion, 233
from compost, 100, 233
from crop wastes, 225
from food processing wastes, 175, 227, 233
from manure, 64, 233
from paper, 233
pellets, 233
from sewage, 148
single cell protein, 106, 233, 234
from sulfite waste liquor, 192, 233, 234
from tanning wastes, 227
from urban waste, 130-131
from wood wastes, 233
Anions, 288g. *See also* Ions
Annealing, 288g
Anorthosite, 25t
Anthracite, 88t. *See also* Coal
Antibiotics, in composting, 101
Antifreeze, 214
Antimony, 8t, 215, 229
Antioxidants, for polymers, 77-78
Aquaculture, 106-107, 148, 249, 288g
Aqueous solutions, *see* Brine; Desalination; Hydrometallurgy; Water
Aquifer, 288g. *See also* Water
Arabinose, 97
"Ark," autonomous living, 11t
Aromatic liquids, 221, 223
Arsenic, separation by electrochemical processes, 71
Art, from recycled materials, 10
Asbestos, 54
mine tailings, 187
in water, 209
Ash, from urban waste and coal, 134t, 138
in urban waste, 114t
utilization, 133, 176, 178, 181
see also Fly ash
Aspergillus fumigatus, 235
Aspergillus niger, 235
Aspergillus oryzae, 235
Asphalt, 223, 239
microbiological decomposition, 99
rubberized, 17, 156, 182
Associated Portland Cement Manufacturers, 185
Atmosphere, 3, 251

Atmospheric pollution, 24, 133, 201-207
Atom, 2, 4, 27
Auckland, New Zealand, 158
Australia, ash utilization, 178
  automobile composition, 153
  beverage containers, 120
  coal waste utilization, 178
  construction materials from waste, 179, 181
  packaging, 117
  plastics wastes, 216
  sawdust utilization, 184
  urban waste, 116, 164
  see also CSIRO
Australian Conservation Foundation, 117
Autoclave, 288g. *See also* Reactors
Automated processes, 31-32, 37, 126-127
Automobiles, baling, 150-151
  composition, 152-153
  dismantling, 150-151
  energy costs, 154, 245-247
  incineration, 151
  manufacture, 28, 57, 150-153
  oil, 239
  plastic recovery, 17-18, 153, 216, 219, 222
  representative, 153
  resources, 215
  scrapped, 150-154
  selective dissolution, 71
  shredders, 18, 150-151
  shredder scrap nonmagnetic residue, 63
  shredder scrap separation, by density, 43, 151
    by fluidized bed, 49
    by hydrometallurgy, 94, 151
    by magnetism, 151
    by pyrometallurgy, 61, 94
    by selective melting, 61, 151
  shredding with cryogenic embrittlement, 63
  steel recycling, 22, 61
  thermodynamic potential, 245-247
  tires, *see* Rubber tires
  wiring, 22, 151
  wreckers, 150
Autonomous house, 11
Autotrophic, 96, 288g
AWT (advanced water treatment), 41, 58, 65, 209-211
Aztecs, SCP use, 235

Back end recovery, 123, 138, 288g
Bacteria, 95-109, 233-235, 288g
  cellulolytic, 98
  Cellulomonas, 235
  Cellulomonas flavigena, 98, 235
  chlorine killing rate, 211
  single cell protein, 234
  Thiobacillus ferrooxidans, 107-108
Bagasse, 227, 229, 232, 234-235, 289g
Bakers' yeast, 96, 102
Ballistic separator, 53-54, 289g
Baltimore, Maryland, 168
Band filter, 40
Bangkok, 128
Barium chloride, 60
Bark, 226
  as metal ion scavenger, 69, 189, 226
  pyrolysis, 88
Barytes, fluidized bed, 49t
Basalt, in stoneware, 186
Base, *see* Alkali
Basic oxygen furnace, 154
Battelle Memorial Institute, 224, 228
Battelle Northwest, 158
Battery, lead and mercury wastes, 189
Bauxite, 25t, 181
Bearing scrap, 94
Beet sugar, 14, 212
Benghazi, Libya, 128
Benzene, 29
Beta pinene, 236
Beverage containers, 27, 118-120, 199
  degradable, 80
  deposits, 27, 118-120
  energy costs, 119-120
  glass, 3, 24, 112, 117-121, 199-201
  legislation, 27, 120
  litter, 120, 122, 201
  non-returnable, 27, 118-120
  plastic, 75, 119, 218, 220
  returnable, 27, 118-120, 201, 218-219
Bibliographies, 10, 300-377
  elements, 250-263
  reviews, 264-269
  water, 213-214
Bicarbonate, in liquefaction process, 90
"Biocel" process, 230
"Bio-cyclic" plastics, 108

Biodegradation, 95, 209
hydrocarbons, 99, 193, 239
methane, 239
plastics, 80, 108, 128, 217, 251
Bioenergy (TM) process, 105
Biogas® process, 130
Biological fuel cell, 238
Biological oxygen demand, 208
anaerobic digestion, 105
dairy wastes, 23
hydrolyzed urban waste residue, 231
paper pulp mill wastes, 69, 192-231
potato wastes, 212
"Bioplex," 12
"Biosphere," autonomous house, 11t
"Bio-stabiliser" system, 129, 158
Biosynthesis, 95
Biotechnic community, 11t
Bituminous coal, 88t. *See also* Coal
Bituminous concrete, *see* Asphalt
Black Clawson, 123, 125-126, 146, 158-162, 230
Black liquor, 84, 192. *See also* Pulp and paper mill wastes
Bladder stones, 234
Blast furnace, copper, 190
steel, 13, 17, 18, 84, 136, 145
Blast furnace slag, aggregate, 17, 18, 179
composition, 180t
copper, 190
lightweight aggregate, 183
pozzolan, 185
Bond breaking, in polymers, 75, 80-81
Boston, Massachusetts, 14, 133, 171
Bottles, *see* Beverage containers
Bounce separation, 52
Bound metal, in ion exchange, 67
Brackish water, 56-58, 66, 213
Brass, 60, 150, 215
in automobiles, 153t
etching solutions, 197
pickling, 197
pyrometallurgy, 190
red, 150
separation, 18, 35t, 60
Bread-making, 96, 102, 207
Breakwaters, of tires, 155
Brewer's yeast, 96, 102
Brewery wastes, 233
Bricks, facing, 186
from fly ash, 184, 186
manganese dioxide, 186
from sawdust, 184, 227
from wastes, 181-183, 221
Brine, evaporation of water, 64
purification for reuse, 45, 187
recovery of chemicals, 54
regeneration of ion exchange resin, 68
Britain, *see* United Kingdom
Brittleness, cryogenic embrittlement, 18, 62-63, 151, 157, 216, 219
plastics, 81
Bromine, 432
Bronze, 215
Brooklyn, New York, 129
Brown coal, 131t. *See also* Coal
Bubble aggregates, 184
Bubbles, *see* Froth flotation; Electroflotation
Buffered, 289g
Building, 3. *See also* Autonomous house
Building board, adhesive, 226, 236
agricultural wastes, 229
gypsum, 191
mixed waste, 221
plastic, 221
sugarcane bagasse, 227
wood, 226, 236
Building demolition wastes, 179, 226
Building materials from wastes, 17, 178-187, 192
paper, 229
plastics, 216, 220-221
textiles, 225
tolerance limits, 179
wood, 226-227
Building Research Establishment, 178-179
Bureau of Mines, U.S., 6
ash, 178
automobiles, 151-153
cryogenic techniques, 63
electrochemical machining wastes, 199
mining wastes, 181
molten salts, 60
plastics separation, 219
plating wastes, 199
sulfur dioxide, 202
tires, 156
urban wastes, 89-90, 125, 139, 140, 143, 162-163, 168, 230

Butadiene, 75, 237t
Butanediol, 98t
Butanol, 98t, 236, 237t
Butylene, 237t
Butyric acid, 98t, 236

Cable scrap, 18, 22, 44, 53, 63
Cadmium, 254
in copper, 23, 216
environmental effect, 23, 216
metallurgy, 34
plating wastes, 198-199
recycle literature, 252-254
removal by fly ash, 189
separation by electrochemical processes, 71
Calcination, 84, 191, 212, 289g
Calcium, recycle, 107
Calcium bisulfite, 192
Calcium carbonate, aggregate, 181
alumina production, 186
blast furnaces, 179
cement manufacture, 185-186
foamed glass, 187
glass manufacture, 199-200
sulfur oxides recovery, 203
Calcium chloride, 41, 60, 196
Calcium fluoride, 191
Calcium hydroxide, as adsorber, 55
in aggregates, 181
for iron precipitation, 196
precipitation of metals, 188, 196
reactions with pozzolans, 185
recovery, 212
sulfur oxides recovery, 203-205
wastewater treatment, 189, 211
woodpulping, 192
Calcium oxide, in blast furnace slag, 180t
in cement manufacture, 185
in fly ash, 183t, 189
Calcium phosphate, 211
Calcium salts, coagulants, 227. *See also specific salts*
Calcium sulfate, *see* Anhydrite; Gypsum; Plaster-of-Paris
Calcium sulfite, 203-205
Calorific value, 131-132, 134t, 140t, 144t, 156t
Cal Recovery System, 126, 230
Canada, beverage container legislation, 120
climate and concrete, 184
Department of Energy, Mines and Resources, 181
highway construction, 178-179
Trudeau government, 10
Candida utilis, 234
Cans, cryogenic embrittlement, 62. *See also* Aluminium cans; Steel cans; Tinned steel cans
Canterbury, Australia, 164
Capital, for recycling plants, 14
Carbohydrate, 225-238, 289g
from cellulose, 78-79, 98, 102, 143, 231-232
fermentation, 227, 231-232, 236
in microbiological processes, 96-107, 236
as polymer, 75
polymer production, 236-237
separation, 226-229
by crystallization, 227
by membrane processes, 226
from urban waste hydrolysis, 143, 231
from wood pulping liquor, 192
Carbon, 250-252
active, *see* Activated charcoal
oxidation, 82-83
plastic reinforcing, 222
recycle, 4, 250-252
reduction, 82-83
see also Activated charcoal; Carbon black; Coke; Graphite
Carbonates, molten, 93, 203-205
Carbon black, tires, 88, 155-156, 251
Carbon cycle, 3, 237, 250-252
Carbon dioxide, 250-251
dry ice, 63
in foaming, 187
greenhouse effect, 202, 251
in microbiological processes, 96-97, 104, 251
from pyrolysis and incineration, 82-91, 202
Carbon disulfide, 207
Carbonization, *see* Pyrolysis
Carbon monoxide, from carbon dioxide, 251
from pyrolysis and incineration, 83-91
reaction with carbon compounds, 85-91
see also Synthesis gas

Carbon: nitrogen ratio, for microorganisms, 99, 130, 234
Carbon tetrachloride, 50
Carbonyl chloride, 206
Carborundum Co., 172
Carbothermic reduction, 94, 190, 289g
Carboxylic acids, from partial oxidation, 91
  from pyrolysis, 84
  flotation agents, 66
  in microbiological processes, 96-99, 101-102
  from plastics manufacture, 187
  polymerization, 74
  from pyrolysis, 84
  solvents for polyurethane, 221
  from wood pulping, 192
  see also Acetic; Butyric; Carboxylic; Formic; Lactic; Oxalic; Propionic; Succinic; Tartaric
Carpets, 225
Carpet underlay, 62, 157
Casa del Sol, 11t
Casting, 289g
  ceramics, 183
Cast iron, 150, 153
Catalysis, oxidation, 65
Catalyst, 289g
  metal, 215
Cation, 289g. *See also* Ions
Cattle feed, *see* Animal feed
Caustic soda, *see* Alkali; Sodium hydroxide
Cells, microbiological, 95
Cellulase, 98, 103, 231-232, 289g
Cellulolytic, 289g. *See also* Cellulose, chemical hydrolysis
Cellulomonas, 234-235
Cellulomonas flavigena, 98, 235
Cellulose, 78-79, 225-238, 289g
  activated charcoal, 236
  agricultural, 175, 225, 232
  bonds with synthetic fibers, 81, 221
  chemical hydrolysis, 78-79, 143, 231
  empirical composition, 88t, 116
  energy storage, 105-107, 237-238
  enzymatic hydrolysis, 98, 103, 231
  extraction, 226-227
  filter aid, 54-55
  heat content, 131-132
  ion exchange resin, 68-69
  microbiological degradation, 98, 103, 231
  microwave irradiation, 232
  native, 231-232
  oxidation, 79, 85-91
  photochemical degradation, 232, 235
  plastics feedstock, 236-237
  pyrolysis, 55, 84-91
  screening, 40, 125
  separation, 226-227
  single cell protein, *see* Single cell protein
  solar energy storage, 105-107, 237
  solvent, 229
  sources, 225-226
  structure, 78, 97
  in urban waste, 116, 125, 160, 226
  see also Paper; Pulp and paper mill wastes
Cellulose acetate, 57-58
Cement, 185-186, 289g
  kiln exhaust gases, 181, 185, 190
  kiln fuel, 134, 185-186
  low alkali, 190
  manufacture, 134, 181, 185-186, 190
  setting retarder, 191
  silica ash from rice hulls, 227
  for waste treatment, 82, 182-183
Cementation, 72, 289g
  copper, 72, 188
Centrifugation, 44-45, 66, 290g
  of fats, 227
  of protein, 227
  of yeast, 102
  vacuum, 62
Ceramics, 290g
  from waste, 180-183, 186
Cereal food, 232
Chad, SCP use, 235
Chahroudi, Day, 11t
Chain breaking, polymers, 73-81
Chain initiation, 73-74, 77, 81. *See also* Free radicals
Chain propagation, 74, 77-78, 80
Charcoal, active, *see* Activated charcoal
Charleston, West Virginia, 167
Cheese whey, 57-58, 226, 232, 234
Chelation, by resins, 68
Chemfix®, 82
Chemical analysis, wastes and fuels, 87-89, 116
Chemical engineering, 12, 30

Chemical formula, wastes and fuels, 87-89, 116
Chemical oxygen demand, 208
Chemical precipitation, *see* Cementation; Precipitation
Chemical processes, 30, 65-94
Chemical properties, 32
Chemical reactions, 30, 65-94
Chemical separation, 18, 30, 65-73, 218
Chemical values, 14-17
Chemical vapor transport, 72-73
Chemical wastes, *see* Industrial wastes
Chemoautotroph, 96, 290g
Chemoheterotroph, 96, 290g
Chicago Northwest incinerator, 133
Chickens, 105, 130, 131, 148, 235
China, Padi Straw mushroom, 235
  stoneware, 186
China clay, sand wastes, 186
Chitin, 75
Chlorella, 234, 235
Chlorinated hydrocarbons, 193-194, 205-206, 214, 221
Chlorine, from hydrogen chloride, 193, 206
  in hydrometallurgy, 19f
  oxidation of ferrous ions, 189
  from poly(vinyl chloride), 83
  recovery, 83-84
  recycle literature, 263
  water treatment, 211
Chlorofluorocarbons, 205
Chocolate wastes, 227, 233
Chromate, 68
Chrome liquors, in tanning, 187
Chromic acid, 197-198
Chromium, 254
  in electrochemical machining wastes, 199
  plating wastes, 198
  recycle literature, 252-254
  removal by fly ash, 189
  resources, 8t, 16f
  separation, by electrochemical processes, 70-71
    by hydrometallurgy, 18, 19f-21f, 50-51, 188, 197-198
    by ion exchange, 68
    by precipitation, 188-189
  in steel, 61, 151
  in tanning, 187
Chromium trioxide, 197-198
Cinders, *see* Ash; Fly ash
Citrate buffer, 202-203
Citrus peelings, pectin extraction, 227
Clam industry, 248-249
Clarke level, 246-247
Classification, 290g. *See also* Air classification; Separation; Water classification
Clay, in alumina manufacture, 25t, 186
  china, 186
Cleaning cloth, 224. *See also* Textiles
Climate, temperature fluctuations, 17, 184
Clinker, 49, 91, 136, 138, 181, 183, 290g
"Clivus Multrum," 11t, 148-149
Closed cycle, biological, 106
  drying, 64
  living, 10-12, 147
  water, 207-208, 212
Clothes, 225
Coagulation, fats, 227
  protein, 68, 227
  pulpmill wastes, 193
Coal, 238-240
  dust pelletization, 239
  empirical composition, 88t
  fly ash, 183-184
  gasification, 239-240
  heat content, 131t, 134t
  liquefaction, 141
  methanol production, 240
  pyrolysis, 55, 84
  resources, 16f, 240
  substitute, 124, 134-136, 146, 172-173
  sulfur content, 134t, 192, 239
  tar, 239
  technology research, 249
  town gas, 84, 239
  waste incineration, 83
  wastes, 8, 180-182, 184, 239
Coalescing media, 55
Coal gas, 84, 239
Coal shale, 180-181, 184, 290g
Coal washing wastes, 181
Coatings, *see* Surface coatings
Cobalt, 254-255
  catalyst, 215
  electrochemical machining wastes, 199
  recycle literature, 252-255
  resources, 16f

separation, by electrochemical processes, 70-71, 93
by hydrometallurgy, 18, 19f-20f, 34, 50-51, 188, 199
by precipitation, 188, 199
Cobalt-60, 106. *See also* Ionizing radiation
Coconut husks, as metal ion scavengers, 69, 189
Coconut shells, for producer gas, 91
Coding for separation, 18, 31, 37
Coke, from coal, 84
as absorber, 54
Cola bottle, 12. *See also* Beverage containers
Cold, source of, 63
Collection, 17-18
Colliery spoit, 180-182, 184
Colloidal state, 290g
cellulose, 78
separation, 41, 46, 55
Colloid mill, 69-70
Color separation, 36-38, 200
Combustion, submerged, 41. *See also* Incineration; Oxidation
Combustion Power Co., 134, 165
Commercial waste, 5, 9t, 110
Communications cable, *see* Cable; Copper
Compatibilizer, plastic, 221
Complex (chemical), 290g
cyanide, 198
formation, 50-51, 55, 68
resin, 68
Composition, urban waste, 87-88, 114-116, 138, 162
Compost, 99-101, 233
as animal feed, 100, 233
"Dano" process, 129, 158
as fuel, 100, 138
mushrooms, 235
see also Agricultural waste; Domestic waste; Industrial waste; Soil conditioner; Urban waste
Computer, in item identification, 31-32, 37-39
simulation, 146
Concentration, of solutions, 108. *See also* Distillation; Drying; Evaporation
Concentration differences, in dialysis, 55-56
Concentration polarization, 69
Concrete, 179-183, 185-186
aggregates, *see* Aggregates
frost damage, 184
incorporation of wastes, 179-183, 221
lightweight, 183-185
tensile stress reduction, 185
thermal conductivity, 181-182
water recycle in production, 212
Condensation, 61-62
Condensation polymers, 73-75, 222, 290g
Conductivity, *see* Electrical; Thermal
Confectionery wastes, 227
Conservation, 3, 10-12, 15-17, 243-244. *See also entries under* Environmental
Conserver society, 10
Construction materials from wastes, 17, 178-187, 192, 201, 220-221, 226-227, 229
tolerance limits, 179
see also specific materials and wastes
Construction metals, 252-261
Consumer goods, plastic, 216
Consumer habits, beverage containers, 118
Consumer society, 10
Consumer waste, *see* Postconsumer waste
Containers, *see* Aluminum Cans; Beverage containers; Glass bottles; Packaging; Plastic bottles; Steel cans; Tinned steel cans
Continuous process, 61, 62
Controlled potential electrolysis, 69, 188
Controlled release pesticides, 227
Cooling water, *see* Water, cooling
Copolymerization, 75, 81, 222. *See also* Crosslinking
Copper, 255-256
blast furnace slag, 190
catalyst, 215
from flue dusts, 190
fluidized bed, 49t
Copper, in automobiles, 49, 61, 150-151, 153t
metallurgy, 23, 25, 202
microbiology, 108
mine runoff water, 295
mine tailings, 17, 189
pickling, 197
plating wastes, 198
pollutant, 24
precipitation, 188

radiator scrap, 49
recycle, 8t, 22, 23, 215
recycle literature, 252-253, 255-256
removal from water by fly ash, 189
resources, 16, 216
separation, by ballistic methods, 53
by carbothermic reduction, 190
by cementation, 72, 188, 190
by cryogenic embrittlement, 63
by eddy current methods, 35t
by electrochemical processes, 69-71, 188-189, 197
by hydrometallurgy, 51, 93, 188, 197
by ion exchange, 67-68
by pyrometallurgy, 94, 190
by selective melting, 60-61
by solvent extraction, 50-51, 93
in steel, 60-61, 151
Copper oxide, 203t
Copper sulfide, microbiological leaching, 108
Coprecipitation, 34
Corncobs, 227, 233
Corrosion protection, 196
Costs, *see* Economics
Cotton, effect of fungi, 103, 232
reuse, 224
wastes, 209, 236
see also Textiles
Cotton hulls, hydrolysis, 78-79
Cotton mill wastes, 209
Covalent materials, solubility, 50
Crankcase oil, *see* Oil
Crop wastes, 225
incineration, 83
leaf protein, 227
pesticides, 226
Cross-linking, of polymer chains, 76, 81
Crushing, *see* Shredding
Cryogenic, 290g
embrittlement, 18, 62-63, 157, 216, 219
Cryolite, 187, 202
Cryopulverizing, 18, 62-63, 151, 157, 216, 219
Crystallization, 41, 62
ferrous sulfate, 195
ice, 62
trisodium phosphate, 51
CSIRO Division of Building Research, 179, 181, 184, 191
Cullet, *see* Glass
Culm, 290g
Culm banks, 180
Cuprammonium waste, 68
Cyanate, 198
Cyanides, photographic waste recovery, 66
plating wastes, 198-199
water pollutants, 208
Cyclohexanone, 51
Cyclone, 45. *See also* Air classification

Dairy wastes, 23, 57-58, 226-227, 233-234
Dams, concrete, 185
"Dano" composting process, 129, 158
Deacon process, 194
Decorative surface treatment, *see* Surface coatings
Definitions, 3-6, 287-289
Defogging, 249
Degradable plastics, 80, 108, 128, 217, 251
Degradation, *see* Biodegradation; Photodegradation
Dehydration, 30, 64, 187. *See also* Calcination
De-inking waste paper, 23, 228-229
Delaware, 166
De Matteo, 171
Demolition wastes, 22, 179, 226
Demonstration projects, 157-174
Denmark, composting, 128
Density, 30, 39-50
fluidized bed, 76t
plastics, 42t
variation in ferrohydrodynamics, 34
Density-conductivity ratio, 34-35
Density difference separation methods, 39-50, 125, 200, 218
Density-drag separation methods, 48-50, 52-54
Denver Regional Council, 112
Department of Agriculture (U.S.), 121, 125, 228
Department of Commerce (U.S.), 6, 154
Department of Energy (U.K.), 249
Department of the Interior (U.S.), 6. *See also* Bureau of Mines
De Paul University, Chicago, 12
Depolymerization, 73, 75-81, 222

Deposits on beverage containers, *see* Beverage containers
Desalination, 56-60, 66-69, 133, 213, 249. *See also* Brine
Design, for recycle, 12, 18, 22, 28, 186
Destructive distillation, *see* Pyrolysis
Detergents, biodegradation, 209
  embrittlement of plastics, 81
Devitrified glass, 180
Devulcanizing, 155
Dewatering, sludge, 187
Dialdehyde cellulose, 79
Dialysis, 55-56, 290g
  acid recovery, 193
Diatomaceous earth, 18, 32, 54, 290g
Dicalcium silicate, 179
Dicarboxyl cellulose, 79
Dichloroethane, 193-194
Dichromate, 197
Dielectric fluid, 41
Diet, 232-235
Diffusion, in dialysis, 55
Digesters, *see* Reactors
Dilute solutions, electrochemical processes, 69-71
  ion exchange, 66
Dilution, 22, 243
Dimethyl formamide, 221, 229
Dimethyl sulfide, 236
Dimethyl sulfoxide, 229, 236
Dinitrogen tetroxide, 79, 229
Dipolar cell, 69
Disaccharides, 97, 290g. *See also* Lactose; Sucrose; Sugar
Discarded consumer goods, *see* Automobiles; Litter; Postconsumer waste
Disease organisms, in agricultural wastes, 225
  in anerobic digestion, 104, 130
  in animal feed, 130
  in composting, 101-102
Disorder, 30-31, 241
Disposable containers, *see* Beverage containers; Packaging
Dissolution, of precipitate, 66
  selective, 71, 188, 213, 214
Distillation, 61-62, 72, 290g
Dolomite, 180, 230
Domestic heating, 133, 249
Domestic recycling, 224
Domestic waste, 5, 110
  composting, 11t
  empirical composition, 88t
  recycle rates, 110
  in U.K., 110, 114
  in U.S., 9t, 110
  see also Urban Waste
Doubling times, 15
Drainage, 187
Drainage tiles, 63, 181
Dredge spoil, 178
Drexel University, Philadelphia, 64
Dross, 290g. *See also* Molten metal; Slag; Smelting
Dry ice, 63
Drying, 30, 64, 187
Dry screening, 32, 39-40, 125, 231
Ductility, 62-63, 151
Dumps, *see* Landfill
Durability, 28-29, 154, 245
Dust extraction, 39-40, 187
  from incinerator gases, 35-36, 133, 190, 224
  from smelter flues, 70, 190
Dye, from coal, 239
  recycle, 58
Dynatech Co., 130

Ebara Co., Japan, 165
Ecological systems, 10, 102, 105-107, 147, 243, 291g
"Ecologist," community, 11t
Ecology Action, Berkeley, 121
Ecology, Inc., 129
Economics, 5, 10, 13-14, 178
  anaerobic digestion, 105
  cellulose hydrolysis, 232
  metal sludges, 189
  paper recycle, 228-229
  plastics, 237
  rubber, 156
  solvent recovery, 214
  urban waste recovery, 110, 143-147, 201, 231
  used oil recycling, 239
Ecosystems, 10, 102, 105-107, 147, 243
Eddy currents, 34, 61, 291g
Edmonton, London, 133, 165
Education, effect on recycling participation, 121

Effluent, *see* Air pollution; Furnace effluent gases; Water, pollution
Electrical coalescence, 41
Electrical conductivity, 30, 35-36
Electrical conductivity-density ratio, 34-35
Electrical discharge machining, 199
Electrical equipment, 18, 60, 63, 150
Electrical properties, 32-39
Electric arc furnace, 61, 94, 190
Electric charge filter, *see* Electrodialysis
Electric induction furnace, 61, 94, 154
Electric motors, 18, 60, 63, 150
Electric power generation, 93, 133-136, 146, 213, 248-249
Electrochemical machining wastes, 94, 188, 199
Electrochemical processes, *see* Electrolysis; Electroplating
Electrochemical regeneration of active charcoal, 54, 72
Electrode potential, 69-72
Electrode processes, 69-72
Electrodialysis, 56, 291g
  acid recovery, 193
  water treatment, 209, 211, 213
Electroflocculation, 46
Electroflotation, 46
Electrolysis, 69-72, 93
  biological fuel cell, 238
  chlorine, 194, 206
  copper, 188-189, 197-198
  cyanide, 198
  flow cells, 69, 198
  fluidized bed, 69, 198
Electromagnetic radiation, 36-39, 107, 291g. *See also* Ionizing radiation; Infrared radiation; Microwave radiation; Radioactivity; Ultraviolet radiation; Visible radiation
Electromagnetic separation, 32-39
Electromotive force, in electrodialysis, 56
  in electrolysis, 69-71
Electronics wastes, 70
Electron transfer, 82
Electroplating wastes, 189, 197-199
  separation, by ion exchange, 67-68
    by reverse osmosis, 58
Electrorecovery, *see* Electrolysis
Electroslag melting, 61
Electrostatic precipitators, 35-36, 133, 183
Electrostatic separation, 35-36, 219
Electrowinning, *see* Electrolysis
Elements, 2, 4, 27, 250-263, 291g
Elutriation, 41, 43, 291g
Embden-Meyerhof pathway, 97-98
Embrittlement, cryogenic, 18, 62-63, 151, 157
  plastics, 18, 63, 81
Empirical formulae, of wastes and fuels, 87-89, 116
Employment, 27
Empresa Nacional ADARO de Investigaciones Mineras, 168
Emulsion, 41, 57
Encapsulation, of wastes, 18, 81
Endothermic reactions, 87, 242-243, 291g
Energy, 1, 24-26, 242-249
  in agriculture, 248
  in automobiles, 154, 245-247
  "crisis," 241, 245
  efficiency, 248-249
  environmental clean-up costs, 245
  in environmental protection, 245
  in glass manufacture, 120, 201, 248
  low grade, 67, 248-249
  in manufacture, 28, 245-248
  in metal production, 24-25
  in nonferrous metals, 248
  from organic waste, 131-145, 184, 185, 192
  in packaging, 118-120, 248
  in paper recycling, 248
  in separation, 25-26, 243
  solar *see* Solar Energy
  in steel, 248
  thermal utilization, 67, 248-249
  in urban waste, 131-145, 248
  in water recycle, 248
Energy farming, 105, 237-238
Energy recovery incineration and pyrolysis, 83-93, 132-145
Energy use, growth, 15
Enthalpy changes, 242-245
  in incineration and pyrolysis, 86-87
  see also Heat content
Entropy, 10, 22, 30-31, 241-245, 291g
  "crisis," 241
  pollution, 245
Environmental "crisis," 241

Environmental impact report, 23
Environmental protection, 22-24
  automobiles, 150-157
  ecological systems, 10, 102, 105, 147, 243
  energy cost, 245
  glass, 200-201
  legislation, 22-24, 111
  litter, *see* Litter
  plastics, 217
  pollution, *see* Pollution
  solvents, 214
Environmental Protection Agency, 24
  automobiles, 154
  paper, 228
  urban waste demonstration projects, 157-173
  waste generation rates, 110
  waste utilization, 224
Enzymes, 291g
  in cellulose hydrolysis, 98, 103, 231-232
  production, 103
  protein disintegration, 230
Equilibrium, acid dependence in ion exchange, 67-68
  temperature dependence in ion exchange, 67
  thermodynamic, 244
Erie County, New York, 171-172
Ethane, 89, 140. *See also* Hydrocarbons
Ethanol, from cellulose pyrolysis, 86
  in microbiological processes, 96-98, 102, 143
  recovery, 55
  from urban waste, 143, 147, 162, 231
  from wood sugars, 192, 236-237
Ethyl acetate, 229
Ethyl alcohol, *see* Ethanol
Ethyl benzene, 221
Ethylene, copolymer, 182
  from cellulose, 237
  polymers, *see* Polyethylene
  from pyrolysis, 223
  recycling index, 29
  vinyl chloride manufacture, 193
Ethylene glycol, antifreeze, 214
  polymer solvent, 80, 222
  recovery, 214
Ethyl vanillate, 236
Ethyl, *see* Acetylene
Europe, composting, 128-129
  incineration, 132-133
  pulp and paper industry wastes, 193
Eutectic freezing technique, 62, 210, 291g
Eutrophication, 211
Evaporation, solar, 64
  of solvent, 40-41, 61-62, 64, 221
  see also Distillation; Drying
Exchange reactions, 67-73
Exothermic reactions, 86-87, 185, 242-243, 291g
Exponential, 1, 15, 291g
Extended surface electrolysis cell, 69
Extruded mineral papers, 193
Extrusion, plastics, 63, 81, 220

Fabrics, *see* Textiles
Facing bricks, *see* Surface coatings
Facultative anaerobe, 96
Fairfield-Hardy composting process, 128
Fats, 291g
  microbiological decomposition, 99
  microbiological synthesis, 234
  separation, 46, 227
Federal waste, 9t
Feedlot surfaces, 157
Feedlot wastes, 292g
  fiberboard, 229
  microbiological processing, 103, 105, 232, 233
Fermentation, 95-107, 126-130, 147-149, 292g
  carbohydrate, 143, 146, 231
  controlled, 96, 102
Ferric chloride, 84
Ferric hydroxide, 57, 188-189, 196
Ferric oxide, *see* Iron oxides
Ferricyanide, 66
Ferrocyanide, 66
Ferrofluid, 34
Ferrohydrodynamics, 34
Ferromagnetism, 18, 22, 32-35, 292g
Ferrosilicon, fluidized bed, 49
  heavy media, 41
Ferrous can scrap, 27, 94
Ferrous chloride, 195
Ferrous hydroxide, 34
Ferrous ions, oxidation to ferric, 189, 195
Ferrous metal, destructive oxidation, 84

from high temperature incineration and pyrolysis, 136, 143
recycle, 8t, 215
recycle literature, 252-253, 256-258
in urban waste, 114-117, 123, 138, 164
Ferrous-nonferrous metal separation, by cryogenic embrittlement, 63
by selective melting, 60
by solvent extraction, 50-51
Ferrous oxide, *see* Iron oxides
Ferrous pickling solution wastes, 51, 57, 194-196
Ferrous scrap, *see* Ferrous metal
Ferrous sulfate, 194-195
Ferrous sulfide, 181
Fertilizer, ammonia, 92, 150, 193, 230, 239
anaerobic digester sludge, 104-105
cement dust, 190
slow release, 193, 236
sulfite waste liquor, 192
tannery wastes, 227
urban waste compost, 128-129
waste sulfuric acid utilization, 193
zinc sulfate, 190
Fescue, 106
Fiber, asbestos, 54, 209
paper, *see* Paper
plastic, 81, 221
synthetic papermaking, 193
Fiberboard, *see* Building board
Fiber fluff, from tires, 62, 156
Fiberglass, from cullet, 200
insulation, 187
in tires, 155
"Fibreclaim," 125-126, 146, 158-162
Fibrous asbestos, 54, 209
Fillers, in plastics, 182, 223
Film spools, recycle, 220
Filter aids, 18, 32, 54-55, 187, 240
Filtration, 39-41, 209
chemical precipitates, 66, 188
electrochemical machining sludges, 199
rates, 41, 54, 188
sand beds, 55
Finance, 14. *See also* Economics
Fish farming, 106-107, 148, 248
Fissionable material, 240
Fixed bed reactors, 92
Flame photometry, 38
Float-sink, *see* Density difference separation methods
Flocculation, 41, 209, 292g
Flooring paper, 229
Flotation, agents, 66
froth, 46, 180, 200, 227
precipitate, 66
Flow cells, electrolytic, 69-70, 198
Flow diagrams, 4
Flue dusts, from incinerators, *see* Fly ash
from smelters, 70, 190
Fluidized beds, 48, 69, 92, 292g
in adsorption, 54, 203-204
in air separation, 48-50
densities, 49t
in electrochemical processes, 69, 198
in incineration, 92-93, 133, 165
in ion exchange, 66-68
in pyrolysis, 139, 166-167, 223
in separation, 48-50
Fluoride, 56, 202
Fluorine, 263
Fluoroanhydrite, 191
Flux, 292g
in aluminum resmelting, 187, 190
cullet, 120, 176, 200
Fly ash, 292g
in bricks, 184, 186
in cement, 185
coal furnace, 183
composition, 183
in construction, 17, 183-185
filter aid, 187
in flocculation, 41
pelletization, 184
pozzolan, 184-185
in road construction, 178-179
sewage treatment, 149
silicate treatment, 82
urban waste incinerator, 35, 113t, 133, 190, 224
water treatment, 212
Foamed glass, 187
Foam fractionation, 55
Foaming, 55
Food, 232-235
from waste, *see* Single cell protein
Food processing wastes, 45, 46, 57-58, 175, 226-227, 233
Forest, 3, 107

Forest Products Laboratory, USDA, 121, 125, 228
Forestry wastes, 226
  energy source, 238
  heat content, 131t
  as metal ion scavengers, 69
  microbiological processes, 235
  see also Cellulose; Wood
Formaldehyde, adhesives from bark, 226, 236
  cellulose solvent, 229
  from cellulose pyrolysis, 86
  from plastics manufacture, 187
  sterilization of microbiological media, 102
Formate, in liquefaction process, 90
Formic acid, from wood, 86, 236
  in microbiological processes, 98t
Fossil fuels, *see* Coal; Natural gas; Petroleum
Fouling, of filtration media, 60
  of ion exchange resin, 68
Fractionation, 61-62
Franklin, Ohio, 126, 158-162
Franklin Institute, Philadelphia, 125
Free energy change, 244
Free radicals, 292g
  in depolymerization, 75-77, 81, 222
  in polymerization, 74
  in polymer oxidation, 77-78
Freeze-drying, 62
Freezing, 62
Freight rates, 14, 220, 223. *See also* Transportation
Frictional separation methods, 53
Front end recovery, 123, 138, 169, 292g
Frost damage, in concrete, 184
Froth flotation, 45-46, 66, 180, 200, 227, 292g
Fructose, 97
Fruit wastes, 227, 233. *See also* Agricultural wastes
Fuel, from organic waste, 131-145, 184, 192
  heat content, 131t
  hydrogen/carbon ratio, 87-89
Fuel cell, 292g
  biological, 238
  hydrogen, 1, 89
Fuel farming, 105, 237-238
Fumes, 292g
  from pyrometallurgy, 190
Fumigation, chemical recovery, 54-55
Fungi, 95, 232-235, 292g
  Aspergillus fumigatus, 235
  Aspergillus niger, 235
  Aspergillus oryzae, 235
  feed-lot wastes, 103
  Fusarium, 235
  Fusarium semitectum, 235
  mushrooms, 235
  Myrothecium verrucaria, 235
  Padi Straw mushrooms, 235
  single cell protein, 233-235
  Trichoderma lignorum, 235
  Trichoderma viride, 103, 232, 235
Furfural, from hemicellulose, 236
  from pyrolysis, 84
Furnaces, clinker, *see* Clinker
  effluent gases, 65, 181, 193, 201-206
  industrial, 60, 93, 134, 154, 190
  power station, 134-137, 183
  steel, 13-14, 60, 94, 136, 145, 151, 154, 190
  see also Blast furnaces; Cement kilns; Reactors
Fusarium, 235
Fusarium semitectum, 235
Fused salts, *see* Molten salts

Galactose, 232, 292g
Galena, 43
Gallium, recovery by electrolysis, 71
  recycle literature, 261
Gamma radiation, 106, 240
Garbage, definition, 5, 293g. *See also* Domestic waste; Urban waste
Garbage grinders, 116, 130
Garden utensils, recycled plastic, 220
Garrett flash pyrolysis, 169-171
Gases, 201-207
  absorption, 50, 201-205
  dust extraction, 35-36, 40, 201
  recovery, 201-205
Gasification, 86-87, 91, 140, 144t
  coal, 239
  see also Pyrolysis; Urban waste
Gas-liquid solubility, 50
Gasoline, *see* Hydrocarbons
  tetraethyl lead, 215
Gas turbine, incineration gases, 134, 164-165

General Electric Co., 171
Genetic experiments, in microbiology, 103
Germanium, 262
Germany, waste heat utilization, 249
Gibbs free energy change, 244-245
Glasgow, Scotland, 133
"Glascrete," 182
"Glasphalt," 182, 293g
Glass, 199-201
  in asphalt, 182, 201
  in bricks, 186
  in ceramics, 186
  in concrete, 182-183
  cullet, 120, 176, 200, 290g
  disposal, 201
  foamed, 187
  in lightweight aggregate, 184
  manufacture, 200
  packaging, *see* Beverage containers
  pozzolan, 185
  recycle, 200
  remelting, 120, 176, 200
  secondary markets, 20
  separation, 32, 37, 52-53, 200
  in terrazzo, 186
  in urban waste, 32, 52-53, 114-115, 123, 138t, 160, 181-182, 187, 200
Glass bottles, *see* Beverage containers
Glass-ceramic, 277, 289, 293g
Glass Container Manufacturers' Association, 160
Glass etching wastes, 56
Glass-polymer composite, 182
Glass transition temperature, 17, 62-63
Glassy slag, 180-181, 185-186
Glossary, 287-299
Gluconic acid, 97
Glucose, 97, 293g
  from cellulose, 78-79, 143, 231-232
  from lactose, 232
Glucuronic acid, 97
Gluten recovery, 57
Glycerol, 97, 99
Glycine, from cellulose, 80, 236
Glyoxal, from cellulose pyrolysis, 86
Glyoxylic acid, from cellulose oxidation, 79
Gold, microbiological leaching, 108
  recycle, 8
  recycle literature, 261
  resources, 16f
  separation by electrochemical processes, 71, 93
Graft copolymerization, 81, 222
Granulation, *see* Shredding
Graphite, 73, 251
Grass, 106
Grass seeding mixtures, 17
Gravity, 34, 39-50. *See also* Centrifugation; Sedimentation
Greenhouse effect, 251
Grinding, *see* Shredding
Ground glass, pozzolan, 185
  soil conditioner, 201
Ground stabilization, with waste materials, 17, 179
Ground water contamination, 82, 187
Groundwood pulp, 192, 228, 232. *See also* Pulp and papermill wastes
Groveton, New Hampshire, 165-166
Gunmetal, 150
Gypsum, 190-192
  aggregate, 181
  from phosphoric acid manufacture, 190-191
  from sulfur oxides, 192, 203-205
  incorporation of coal dust, 181
  in natural waters, 207
  production, 191
  wallboard, 191

Halides, molten, 52, 60, 71, 73, 93
Halogens, 201, 263
Handcrafts, 10, 91, 103, 147
Hand-sorting, 31, 150, 219
Hard water, 66. *See also* Water, desalination; Water, purification
Heat, dissipation, 243
Heat capacity, glasses, 182
Heat content, 131-132, 134t, 140t, 144t, 156t
Heat recovery incineration and pyrolysis, 83-93, 132-145
Heat-sensitive material, 62, 214
Heat sterilization, 102
Heavy media slurries, 43-45
Heavy metals, in compost, 129
  environmental effect, 22-24, 216
  in incinerator fly ash, 133
  in pickling wastes, 189, 196

in recycled metals, 216
in water, 207
removal by fly ash, 189
removal by natural wastes, 69, 189, 226, 236
removal by rubber waste, 155
in sewage, 149
smelting wastes, 190
in urban waste, 23-24, 133
see also under specific metals
Helium, resources, 3, 16f, 252
Hematite, 25t
Hemicellulose, 97, 192, 232, 236, 293g
Hempstead, Long Island, 162
Herbicides, in urban waste, 129
Hercules urban waste system, 166-167
Hexane, 62. *See also* Hydrocarbons
Hexosan, *see* Hemicellulose
Hexoses, 97, 293g
HGMS, 32-33, 210
Hierachy of systems, 2-3, 241, 246, 250
High density polyethylene, *see* Polyethylene
High gradient magnetic separation, 32-33, 210
High melting point materials, 17, 60, 72-73, 136
High temperature gasification, 91
High temperature incineration, 136-137, 181, 201
High temperature reactors, *see* Electric arc furnace; Electric induction furnace; Furnaces; Molten metals; Molten salts; Smelting
High voltage separation, 35-37, 219
Highway construction, 178-179, 201, 221
Highway litter, 120, 122
Highway Materials Laboratory, 178-179
Highway Research Board, 122
History, 15
Holland, compost, 128
energy from wastes, 249
Home scrap, 5, 293g
plastic, 218
Homolactic fermentation, 97
Horner and Shifrin, 134-136, 146, 172-173
Hot pressing, 220. *See also* Molding
Houses, *see* Autonomous house; Buildings
House heating, 133, 249
Household utensils, recycled plastic, 220
Housewives, recycling attitudes, 121
HTI, 136-137, 181, 201
Humus, *see* Soil conditioner
Hydrapulping, 40, 123-125, 138, 158-162, 165, 230, 293g
"Hydraposal," 123, 125, 146, 158-162, 230
Hydraulic separation process, 41-48, 123-125, 138, 158-162, 165, 230
Hydrocarbons, 50, 238-240
alkanes, 89, 140, 238
biodegradation, 99, 193, 239
chemicals from petroleum, *see* Petrochemicals
chlorination, 193-194, 205-206
coal, *see* Coal
contaminants in containers, 219
cracking, 238
crude oil, *see* Petroleum
grafted onto polymer fiber, 81
from hydrogasification, 140-141
methane, *see* Methane
microbiological decomposition, 99, 193, 239
natural gas, *see* Natural gas
paraffins, 89, 140, 238
petroleum, *see* Petroleum
polymeric, 73-81
from pyrolysis, 86-91
solvents for plastics, 221
sulfur impurities, 134t, 192, 238
Hydrochloric acid, in hydrometallurgy, 51
recovery, 193
recycling index, 29
uses, 206
see also Hydrogen chloride
Hydrofluoric acid, manufacture, 191. *See also* Hydrogen fluoride
Hydrogasification, 87-89, 293g
of urban waste, 140-141, 144t
Hydrogen, chemical feedstock, 92, 145
from hydrogasification, 140
in microbiological processes, 96
from pyrolysis, 84-88, 145
reaction with carbon compounds, 83, 86-89
recycle, 1
reduction of metal hydroxides, 188
resources, 3
in sulfur recovery, 204
see also Synthesis gas

Hydrogen chloride, from poly(vinyl chloride), 83, 133, 139, 223
generation of chlorine, 206
recovery, 193, 205
uses, 206
see also Hydrochloric acid
Hydrogen cyanide, from polyurethane, 133
Hydrogen fluoride, recovery, 56, 202. *See also* Hydrofluoric acid
Hydrogen peroxide, free radical initiator, 74
in de-inking, 230
Hydrogen sulfide, in sulfur recovery, 202-205
Hydrolysis, 78, 231-232, 293g
of cellulose, 231-232
enzymatic, 98, 103, 231
of polymers, 78-80, 222
of urban waste, *see* Urban waste
Hydrometallurgy, 50-51, 65, 93, 188, 293g
in superalloy waste recovery, 18-21
see also Solvent Extraction
Hydroperoxide, 78, 108
Hydrophilic, 293g. *See also* Surface properties
Hydrophobic, 293g. *See also* Surface properties
Hydroponics, 106-107, 148, 210, 293g
Hydropulping, 40, 123-125, 138, 158-162, 165, 230, 293g
Hydroxide, *see* Alkali; Metal hydroxide; Sodium hydroxide
Hydroxyapatite, 211
Hydroxyl radical, 74
Hyperfiltration, *see* Reverse Osmosis

Ice-cream cone, 12
Ice crystallization, 62
Identification, 18, 22, 30-39
IIT Research Institute, Chicago, 179
Ilmenite, 16, 25t, 181, 195
Impact sensor, 32, 53
Incineration, 293g
clinker, 49, 136, 138, 181
energy recovery from waste, 83, 93, 132-138
enthalpy changes, 86-87
flue dusts, 35, 133, 190, 224
high temperature, 136-137, 181
plastics, 83-84, 118, 133, 217, 223
urban waste, *see* Urban waste
vortex, 93
wood pulping liquor, 192
Induction furnace, 61, 94, 154
Industrial Revenue Bonds, 14
Industrial wastes, 7t, 9t, 175-240
anaerobic digestion, 105
composting, 100
liquid, 68
Inertial separator, 52-54
Information, loss, 243
Infrared radiation, 31-32, 37-39
In-house scrap, 6, 293g
Injection molding, 220-221. *See also* Molding
Inks, 23-24. *See also* Paper, de-inking
In-line process, 61, 62
Inorganic microbiological recycling, 107-108
Inorganic polymerization, 81-82
In-plant water recycle, 207-208, 212
Insecticides, from lignin, 236
incineration, 83
Insects, 75, 102
Institute for Local Self Reliance, 11t
Institute of Gas Technology, 130
Insulation, 187
cryogenic embrittlement of cable, 63
glasses, 181-182, 187
paper, 229
vitreous slag, 181-182
Intensive farming, 175. *See also* Feed-lot wastes
International Association on Water Pollution Research, 213
International Energy Agency, 249
Iodine, 54, 263
Ion exchange, 66-69, 294g
acid recovery, 193, 198
lignin derivatives, 236
liquid, 68
membranes, 56
protein recovery, 68, 227
resin, 66-69
sewage treatment, 68, 150
water treatment, 209, 211, 213
Ion exclusion, 68
Ion flotation, 55
Ionization, 294g
by plasma, 94

Leaching, 294g
  microbiological, 107-108
  silicate treatment, 82
  from wastes, 187
Lead, 158-259
  in automobiles, 150-152, 153t
  battery wastes, 189
  from flue dusts, 189-190
  fluidized bed, 49t
  liquid electrode, 93
  metallurgy, 25t, 190, 202
  microbiology, 108
  pollutant, 24, 129, 216
  pyrolysis medium, 93
  recycle, 8t, 24, 215
  recycle literature, 252-254, 258-259
  removal by fly ash, 189
  removal by natural organic wastes, 189
  resources, 16f, 216
  salts in concrete, 181
  separation, by ballistic method, 53
    by eddy current method, 35t
  in steel, 151
  see also Heavy metals
Leaf protein, 227
Legislation, beverage containers, 27, 120
  recycled plastic, 220
  U.S., 6, 24, 112
Less-common metals, 261-262
Liège, Belgium, 63
Lifestyle, 10-11, 28, 122, 147
Lightweight construction materials, 183-187
  from blast furnace slag, 183-185
  from boiler slag, 183
  from fly ash, 182
  from glass, 184
  from organic wastes, 184
  from plastics, 184
  from shale and slate, 134, 184
Lignin, 294g
  in aggregates, 184
  microbiological processes, 99, 102, 104, 231-232
  in plastics, 182
  recovery from pulp mill wastes, 57, 192, 236
  source of organic chemicals, 192
  structure, 75, 99
Ligninolytic, 294g. *See also* Lignin, microbiological processes
Lignin sulfonates, 57, 184, 192-193, 227, 236
Lignite, *see* Coal
Lignocellulose, in microbiological processes, 104, 231-232
  separation, 79
Lignosulfonates, 57, 184, 192, 227, 236, 294g
Lime, *see* Calcium oxide; Calcium hydroxide
Lime coke, in sugar refining, 84
Lime sludge, 181, 188
Limestone, *see* Calcium carbonate
Linde Division, Union Carbide, 136, 143, 146, 167
Lipids, 99, 234, 294g
Liquefaction, of solid fuels, 89-90, 294g
  of urban waste, *see* Urban waste
Liquefied natural gas, 63
Liquid-gas solubility, 50
Liquid-liquid extraction, *see* Solvent extraction
Liquid-liquid solubility, 50, 61
Liquid nitrogen, 62-63
Liquids, organic, 50, 214
Liquid-solid solubility, 50
  temperature dependence, 188, 195
Literature of recycling, 10-11, 264-269, 300-377
  elements, 250-263
  reviews, 264-269
  water, 213
Lithium, electron, transfer, 82
Lithium carbonate, 205
Lithium chloride, 71
Litter, 108, 120, 122, 154, 201
  degradation, 80, 108-109
  highway, 120, 122
Local authority, 110, 112, 120, 122, 146
London (U.K.), Edmonton incinerator, 133, 165
Loops, recycling, 4
Low concentrations, *see* Dilute solutions
Low density polyethylene, *see* Polyethylene
Lowell, Massachusetts, 167-168
Low-grade heat, 67, 248-249
Low-iron nickel alloys, 199
Low technology processes, 10-12, 91, 103, 147

Lubricants, from fat wastes, 227. *See also* Oil
Lumber, *see* Cellulose; Sawdust; Wood

Machining wastes, 7t. *See also* Electrodal discharge machining; Electrochemical machining wastes
Macroporous ion exchange resin, 68
Madison, Wisconsin, 115, 121, 125
Madrid, Spain, 116, 168
Magnesia, *see* Magnesium oxide
Magnesium, for cementation of copper and silver, 72
  metallurgy, 25t, 34
  resources, 16
Magnesium carbonate, 180
Magnesium fluoride, 73
Magnesium oxide, in blast furnace slag, 180t
  in fly ash, 183t
  sulfur oxides recovery, 203t
Magnetic field, 32-35
  non-homogeneous, 34
Magnetic polymer beads, 18, 22, 32
Magnetic separation, 18, 22, 31-35
Magnetic susceptibility, 32, 294g
Magnetite, 34, 41, 210
  fluidized bed, 49
Malleability, *see* Cryogenic embrittlement
Malmo, Sweden, 249
Maltose, 97
Manganese, 259
  in bricks, 186
  metallurgy, 20f-21f, 34
  microbiology, 108
  recycle literature, 252-254, 259
  removal by fly ash, 189
  resources, 16f
Manganese dioxide, 186, 203t
Manganese sulfate, 203t
Man-made mines, 111
Mannitol, 97
Manual separation, 31
Manufacturing wastes, *see* Industrial wastes
Manure, animal feed, 64, 233
  drying, 64
  liquefaction, 90
  pyrolysis, 88
Massachusetts Institute of Technology, 11t, 31, 37, 123, 126, 130
Mass concrete, 185
Mastication, 81
Material specifications, 26-27, 63, 179
Materials utilization industry, 12, 147
Matter, conservation, 2
Maxwell demon, 31
McMaster University, Canada, 179
Meat-packing wastes, 46, 227
Mechanical separation, 31-32
Melting, 60-61. *See also* Molten metals; Molten salts
Membrane separation processes, 55-60
  dialysis, 55
  electrodialysis, 56
  reverse osmosis, 56, 58, 226-227
  ultrafiltration, 41, 57, 226-227
Menlo Park, California, 164
Mercury, battery wastes, 189
  microbiology, 108
  recycle, 8t
  recycle literature, 261-262
  removal by natural organic wastes, 189
  resources, 16
  separation, by hydrometallurgy, 34, 51
    by ion exchange, 67
  water pollutant, 208
  see also Heavy metals
Mesophilic organisms, 101, 104-105, 294g
Metabolites, 95, 294g
Metal detector, 32
Metal displacement, 69-72
Metal finishing wastes, 188-189, 194-199
  electroplating, 58, 67, 197-199
  hydrometallurgy, 51, 65, 93
  pickling solutions, 29, 51, 57, 70, 194-199
  pyrometallurgy, 61, 94
Metal hydroxide, dissolution, 188
  precipitation, 188
  reduction, 188
Metallurgy, 25, 50-51
  hydrometallurgy, 18-21, 51, 65, 93, 188
  pyrometallurgy, 61, 94, 190, 202, 216
Metal oxidation, 82, 84
  prevention, 60
Metal plating, 252-261
  electroplating, 58, 67-69
  wastes, *see* Metal finishing wastes
Metal powder, by electrochemical process, 70-72
Metals, 215
  heavy, *see* Heavy metals

less common, 261-262
microbiological recycling, 107-108
molten, *see* Molten metals
packaging, 117
recycle literature, 252-262
scrap, 31, 199, 215
separation, by cementation, 72
by chemical processes, 65, 93, 188-189
by eddy current methods, 34
by electrochemical processes, 69-72, 93, 188-189, 197-198
by hydrometallurgy, 50-52, 65, 93, 188
by magnetic methods, 31-35
by precipitation, 66, 188
by pyrometallurgy, 61, 94, 190
by selective melting, 60
by solvent extraction, 50-52, 68
by thermoelectric method, 38
solubility, 50-52
structural, 252-261
surface cleaning, *see* Pickling
in urban waste, 32, 113-115
Methanation, 86-89
Methane, biodegradation, 239
from hydrogasification, 140
in microbiological processes, 96, 102, 104
from pyrolysis, 83-89
sulfur reduction, 202
see also Hydrocarbons
Methanogenic microorganisms, 96, 104
Methanol, from coal, 239-240
as cryogenic liquid, 63
in microbiological processes, 96, 102
from waste materials, 92, 145, 167
Methyl alcohol, *see* Methanol
Methyl bromide, 55
2-Methyl-1,3-butadiene, 73
Methylene iodide, 41
Methyl methacrylate, 76, 182
Mexico, SCP use, 235
Microbial proteins, *see* Single cell protein
Microbiological processes, 23, 95-109, 126-130, 147-149, 209, 231
Microorganisms, 95-109
algae, 95, 106-107, 131, 148, 211, 233, 235
bacteria, 95-96, 98-99, 211, 233-235
fungi, 95, 103, 231, 233, 235
yeasts, 95-96, 103, 148, 233-234
Microstraining, *see* Ultrafiltration
Microwave radiation, 36
cellulose degradation, 232
polymer degradation, 81, 223
Milk bottles, polyethylene, 63
Milling, *see* Shredding
Mineral jig, 44
Mineral papers, 193
Mineral technology, applied to recycling, 111
Mineral wastes, 9t. *See also* Mining wastes
Mineral Waste Utilization Symposia, 176, 179
Mineral wool, 187
Mine waters, ion exchange, 68
uranium recovery, 68
Miniaturization, 22
Mining, 6
Mining wastes, 9t, 23, 176, 181
asbestos, 187
coal, *see* Coal
copper, 17, 186
iron, 17
oil shale, 181
Minnesota, 11t
Mixed metal salt, 188
Mixing, 10, 31, 243
Models, 2, 12
Molasses, 234. *See also* Sugar industry wastes
Molding, bricks, 182-183
plastics, 27, 63, 75, 217, 220-221
Molds, 95, 294g
Molecular mass, of polymers, 76-77
ultrafiltration, 57
Molecular size, in dialysis, 55
polymers, 77
Molecular weight, of polymers, 76-77
Molten metals, in incineration, 133
in pyrolysis, 93, 139
Molten salts, in aluminum industry, 73, 202
in electrochemical processes, 71
in electroslag melting, 61
in high temperature reactors, 93, 156
in incineration, 133
in oxidation, 93
in pyrolysis, 139, 223
in selective melting, 60
in solvent extraction, 52
in sulfur oxides recovery, 203-205
Molybdate, 54

Molybdenum, 259
carbon adsorption of molybdate, 54
catalyst, 215
in steel, 151
recycle literature, 252-254, 259
resources, 16f
separation by hydrometallurgy, 18-19, 51, 199
Monomer, 73, 294g
recovery from polymers, 76
Monosaccharides, 294g. *See also* Fructose; Glucose; Hexoses; Pentoses; Sugar
Monsanto Enviro-Chem, 168-169
Motor car, *see* Automobile
Mulch, *see* Soil conditioner
Municipal authority, 110-112, 120, 122-123, 146-147
Municipal waste, 5, 9t, 294g. *See also* Urban waste
Municipal wastewater, *see* Water
Murdoch University, Western Australia, 12
Museums, 10
Mushrooms, 235
Padi Straw, 235
Mustard seed residue, 227, 233
Myrothecium verrucaria, 235

Nashville, Tennessee, 169
National Association of Recycling Industries, 215
National Association of Secondary Material Industries, 224
National Center for Resource Recovery, 112, 114, 125, 169
National Coal Board (U.K.), 249
National Research Council (U.S.), 120
Natural gas, heat content, 131t
resources, 16f, 239-240
source of "cold," 63
Natural polymers, 57, 75, 98-99. *See also* Cellulose; Chitin; Lignin; Protein
Nature, 3, 10, 22
Neoprene, cryogenic embrittlement, 63
Netherlands, compost, 128
energy from wastes, 249
Neutralization, acid and alkaline wastes, 190, 199, 201-205
New Alchemy Institute, 11t
New Mexico, secondary glass utilization conference, 200
New Mexico State University, 11t
New Orleans, Louisiana, 169
Newsprint, *see* Paper
New York City, 112
New York State, 112, 172
Nickel, 259-260
in automobiles, 150
catalyst, 215
electrochemical machining wastes, 199
low-iron alloys, 199
pickling, 198
plating wastes, 198
recycle, 8t, 215
recycle literature, 252-254, 259-260
removal by fly ash, 189
resources, 16f
separation, by carbothermic reduction, 190
by electrochemical processes, 71, 93
by hydrometallurgy, 18, 19f-21f, 34, 51, 93, 188-189, 197-198
by ion exchange, 67-68
by precipitation, 34, 188-189, 197-198
by pyrometallurgy, 94, 190
by solvent extraction, 18, 19f-21f, 34, 51, 93, 197-198
in steel, 151
Nickel sulfide, microbiological leaching, 108
Niger, SCP use, 235
Niobium, in electrochemical machining wastes, 199
Nitrates, molten, 93
Nitric acid, from nitrogen oxides, 206
in plastics oxidation, 109
Nitric oxide, 206
Nitrogen, ammonia synthesis, 145
in composting, 99
liquid, 62-63
plant nutrient, 211
thermal fixation, 206
Nitrogen: carbon ratio, for microorganisms, 99, 130, 234
Nitrogen content, of bark, 69
Nitrogen cycle, 3
Nitrogen dioxide, 206
Nitrogen oxides, 202, 206
Nitrophenol, 54
Noble metals, separation by ion exchange, 67

Non-aqueous solvents, 50, 93
Non-conductors, separation, 35
Nonferrous-ferrous metal separation, *see* Ferrous-nonferrous metal separation
Nonferrous metals, energy costs, 248
  in urban waste, 114-115, 117, 138
  recycle, 215
  separation, by eddy currents, 34
    by float-sink methods, 43
    by incineration, 83
    by selective melting, 60-61
Nonmetals separation, 35, 43
Non-returnable bottles, 27, 118-120
Norfolk Naval Station, Virginia, 133
Noyes Data Corporation, 112
Nuclear energy research, 249
Nuclear fusion torch, 94, 126, 247f
Nuclear wastes, 240
Nucleic acids, 295g
  in single cell protein, 233
Nutrients, plant, 106-107, 189, 196, 211-212. *See also* Aquaculture; Microbiological processes; Solar energy; Photosynthetic storage
Nut shells, 55, 227
Nut wastes, 227
Nylon, 74
  solvent, 221
  in tires, 155
  see also Polyamides

Oak Ridge National Laboratory, 24
Obligate anaerobe, 96, 295g
Obsolescence, 28
Obsolete scrap, 6, 8, 110, 215, 218, 295g
Occidental flash pyrolysis, 169-171
Odessa, Texas, 128
OECD, 249
Office of Science and Technology, 6, 9
Oil, additives, 239
  biodegradation, 99, 193, 239
  from cellulose, 238
  coal dust pelletization, 239
  crankcase, 239
  crisis, 17
  crude, *see* Petroleum
  cutting, 239
  energy factors in recycling, 248
  lubricating usage, 239
  microbiological decomposition, 99, 193, 239
  recycling, 14, 182, 239
  refinery wastes, 227
  rerefining, 239
  reverse osmosis, 239
  separation, 41, 57-60
  spills, 226, 236, 239
  ultrafiltration, 239
  waste, 239
  water-emulsified, 239
  water-soluble, 239
Oil shale, 156, 181
Oil-water separation, 41, 57-60
Open hearth smelting, 154, 295g
Optical methods of separation, 36-37, 200
Order and disorder, 30-31, 241-243
Oregon, 27, 120
Ores, microbiological leaching, 107-108
  tailings, 17
  thermodynamic potential, 245
Organic acids, *see* Carboxylic acids
Organic liquids, 214
Organic waste, as fuel, 23-24, 132-145, 184, 185, 192. *See also* Urban waste
Organization for Economic Cooperation and Development, 249
Organizations, 270-286
Oriental aquaculture, 106-107, 148
Oscillating stratification, 44
Osmosis, 58-59
"Ouroboros," autonomous house, 11t
Oxalic acid, from lignin, 236
Oxidation, 82-88, 91, 295g
  of metals, 84
  in microbiological systems, 96-97, 232
  in molten salts, 93
  of polymers, 27, 77, 80, 220
  of sulfur dioxide, 65, 192, 193, 203-205
  in water purification, 209
  see also Incineration
Oxide scale removal, *see* Pickling
Oxidizable material, 62
Oxygen, as reagent, 83
  recycle, 1
  in urban waste pyrolysis, 136-137, 167
Oxygen cycle, 3
Ozone, 203t, 209

Packaging, 110, 116-120
  aluminum, 117-119

cost, 117-119
energy costs, 117-119
glass, 3, 24, 112, 117-121, 199-201
paper, 117
plastics, 75, 117-118, 216-219
see also Beverage containers
Packed beds, in drying, 64
in electrochemical processes, 69
Paint, wastes, 57
with glass, 201
see also Surface coatings
Pallets, recycled plastic, 220
Paper, additives, 23, 228-229
from agricultural wastes, 227-229
chemical products, 236-237
de-inking, 23, 228-230
economics of recycling, 236-237
empirical composition, 87-88
energy costs, 248
enzymatic hydrolysis, 96-97, 103, 231-232
fiber classification, 228
fiber from urban waste, 113-115, 125-126, 146, 158-162, 228-231
as filter, 55
grading, 229, 230
heat content, 131t
hydrolysis, 78-79, 97, 103, 143, 231-232
lignocellulose, 79, 104, 231-232
manufacture, 165-166, 192-193, 228
newsprint and newspaper, 88t, 228, 230, 232
packaging, 116-118
protein production, 232-235
recycling, 228-229
recycling rate, 31, 228
repulping, 10, 23, 25, 228-229
sacrifical recycling, 17, 229
from sawmill wastes, 229
secondary fiber classification, 229
secondary fiber degradation, 229
segration, 121, 189, 229
separation from contaminants, 23, 228
separation from plastic film, 63-64, 219, 229-230
synthetic fiber, 193
in urban waste, 32, 110, 113-115, 117, 121-123, 146, 228
see also Cellulose
Paperboard, *see* Paper
Paper drives, 28
Papermill wastes, 29, 57, 58, 69, 82, 84, 165-166, 182, 184, 192-193, 228, 236
Paraffins, *see* Hydrocarbons
Paraformaldehyde, 226, 229
Paramagnetism, 33, 295g
Parasites, 295g
in composting, 102
Parking barriers, plastic, 63
Parking lots, *see* Paving
Partial oxidation, 87, 91, 139, 144t
urban waste, 91, 143, 146, 158, 168-169, 171-172
Particle board, plastic, 221
wood, 225-226, 236
Pasture, 106
Pathogens, 295g
in anaerobic digestion, 104, 130
in composting, 101-102
Paving, 17, 62, 63, 156-157, 181-182, 201. *See also* Roads
PCBs, 23, 229
Peanut wastes, activated carbon, 236
as metal ion scavengers, 69, 189
Peat, empirical composition, 88t
Pectin, 295g
from citrus peelings, 227
Pelletization, aggregate manufacture, 184
Pentoses, 97, 234, 295g
Periodate, in cellulose oxidation, 79
Peroxide radical, 78
Pesticides, controlled release, 227
in crop wastes, 226
in water, 60, 207, 209
Pet food, *see* Animal feed
Petrochemicals, 238
food base, 239
plastics, 216-217
sulfuric acid waste, 193
Petroleum, 238-239
cracking, 238
microbiological decomposition, 99, 193, 239
plastics feedstock, 216-217, 237
pollution, 239
refining, 14, 238
residuals management, 238
resources, 16f, 17, 216-217, 240
sulfur content, 238
see also Hydrocarbons

Phase difference, separation by, 39-52, 55, 60
Phenols, 54, 187-188, 236
Phenylpropane, 75
Philips Research Laboratories, 11t
Philosophy of recycling, 1, 10
Phoenix, Arizona, 17
Phosphate, coating of metals, 196
plant nutrient, 211
recovery from sludge, 51, 196
removal from wastewater, 189, 196, 211
Phosphogypsum, 190-191, 295g
Phosphoric acid, manufacture, 190-191, 193
Phosphorus, microbiological recycling, 107
plant nutrient, 211
recycle literature, 262
Photo cell sensors, 37-39
Photochemical degradation, cellulose, 232
polymer, 80
Photochemical smog, 206, 214
Photographic bleach process, 66
Photographic wastes, iron cyanides, 66
silver, *see* Silver
Photodegradation, polymer, 80
Photolysis, 295g
cellulose, 232
polymer, 80
Photo-sensitive copy paper, 24
Photosynthesis, 95-96, 105-107, 251, 295g
Physical properties, 17, 30-64
Physical separation, 17, 30-64
Pickling solutions, 296g
acid recovery, 29, 193
ferrous, 51, 57, 194-196
non-ferrous, 70, 196-199
for phosphate removal, 189
Pig feed, 130-131, 148, 233-235
Pigment, 296g
dispersion, 52
iron oxide, 195
paint wastes, 57
titanium dioxide, 18
Pine bark, in wastewater treatment, 69
Pinene, 236
Pinus radiata, 236
Pipe, from waste, 182, 220-221
Pipelines, 18, 122
Planning, "backwards," 147, 186
extended life, 22, 28
Plant nutrients, 211
Plants, *see* Cellulose; Photosynthesis
Plasma, 94, 296g
Plaster-of-Paris, 191
Plastic, 296g
Plastic bottles, *see* Beverage containers
Plastic fibers, 81
Plastic film, laminated, 221
separation from paper, 63-64, 219, 229, 231
thermal recycling, 220
Plastic-impregnated paper, 119
Plasticizer, 296g
biodegradation, 108
solvent, 221
Plastics, 216-224
from amino acids, 236
airline waste, 219
in automobiles, 17-18, 153, 222
biodegradation, 80, 108-109, 128, 217, 251
bonding with cellulose, 81, 221
building industry, 216
from carbohydrates, 236-237
collection, 17
compatabilizer, 221
density, 41-43
effluents from manufacture, 187-188
embrittlement, 81, 221
empirical composition, 88t
fabrication methods, 220
feedstock, 237
fiber, 81, 221
film spools, 220
graft copolymerization, 81, 222
heat content, 131t, 214, 224
incineration, 83-84, 118, 133, 217, 223-224
industry, 217-218
laminated film, 221
in lightweight aggregate, 184
litter, 108-109
milling, 221
molding, 27, 63, 75, 77, 220
packaging, 75, 117-118, 217-218
paving, 63
photodegradation, 80, 108, 217
plumbing, 75, 216, 220-221
pyramid, 221-222
pyrolysis, 75-77, 87-88, 93, 139, 223

re-extrusion, 63, 81, 217, 221
reuse, 217-220
remolding, 27, 63, 75, 217, 220-221
segregation, 219
separation, 219
by cryogenic embrittlement, 62-63, 219
by density methods, 41, 219
by electrostatics, 37
by solvent extraction, 51-52, 219, 221
by surface properties, 46-48
by thermal properties, 63-64, 219, 231
from paper, 18, 63-64, 219, 229-230
softening, 221
solubility, 51-52, 219, 221
specifications, 221
spinning, 81
stress-cracking, 81, 221
in urban waste, 63, 114-118, 133, 219, 223
waste disposal, 217
see also Polymers
Plating, *see* Surface coatings
Plating wastes, *see* Electroplating wastes
Plating metals, 252-261. *See also specific metals*
Platinum, recycle, 8t
recycle literature, 261
resources, 16f
Plumbing, plastics, 75, 216, 221
Plywood adhesive, 226, 236
Pneumatic separation methods, 48-50, 53, 124-125
Pneumatic pinched sluice, 49-50
Polarization, concentration, 69
Polishing materials, from waste, 227
Pollution, 10, 14, 22-24, 110, 176, 214, 238, 243
air, 24, 133, 201-207
dispersal, 22
thermal, 248
visual, 110
water, 82, 187, 207-209, 211, 227, 229
see also Environmental protection; Litter
Polyacrylic, 66
Poly(acrylonitrile), from cellulose, 237t
formation, 74
hydrolysis, 80
Polyalcohols, 97
Polyamides, 74
hydrolysis, 222
membranes, 78
nylon in tires, 155
nylon solvent, 221
Polybutadiene, from cellulose, 237t
Polybutylene, from cellulose, 237t
Polycarbonate, hydrolysis, 222
membranes, 78
Polychlorobiphenyls, 23, 229
Polyelectrolytes, 149, 296g
Polyesters, 74, 108, 236. *See also* Poly(methylmethacrylate); Polyurethane
Polyethenimine, 78
Polyethylene, airline waste, 219
biodegradation, 108-109
from cellulose, 237
containers, 218-219
copolymer, 182
density, 42t
depolymerization, 76-77
heat content, 131t
packaging, 118
photodegradation, 108
recycle, 63
remolding, 63
separation by sink-float, 41-42
solvent, 51-52, 221
structure, 75
thermal treatment, 220
uses, 75
Poly(ethylene terephthalate), depolymerization, 80
hydrolysis, 222
lamination, 221
Polymer, 296g
Polymer beads, adsorbent, 55
magnetic, 18, 22, 32
Polymerization, 73-75
with glass, 182
with ionizing radiation, 240
Polymer powders, 220-221
Polymers, addition, 74-75
from amino acids, 236
biodegradation, 81, 108-109
condensation, 74-75, 222
conversion to chemical products, 65, 75-81
depolymerization, 75-77, 222
glass transition temperature, 62-63
hydrolysis, 78-80, 222

incineration, 83-84, 118, 133, 217, 223
inorganic, 81-82
irradiation, 81
mastication, 81
molecular mass, 76-77
natural, 75, 99. *See also* Cellulose; Chitin; Lignin; Protein
oxidation, 27, 77-78, 80-81, 220
photolysis, 80
pyrolysis, 75-77, 87-88, 93, 139, 223
repolymerization, 222
solvent extraction, 51-52, 219, 221
solvolysis, 78-80
structure, 73
synthesis, 74-75
synthetic, *see* Plastics; Textiles
thermal depolymerization, 75-77
ultrasonic degradation, 81
see also Plastics; Resin; Thermoplastics
Poly(methyl methacrylate), depolymerization, 75-76
Poly(methyl styrene), depolymerization, 76
Polyolefins, *see* Polyethylene; Polypropylene
Polypropylene, density, 42t
from cellulose, 237t
heat content, 131t
remolding, 63
separation by sink-float, 41-42
structure, 74-75
uses, 75
Polypropylene oxide, 80
Polysaccharides, 97, 296g
Polystyrene, airline waste, 219
density, 42t
depolymerization, 76-77
heat content, 131t
hydrolysis, 222
remolding, 63
segregation, 219
separation by water classification, 41-42
solvent, 51-52, 221
structure, 74-75
uses, 75
Polystyrene sulfonate, 66
Polysulfone, 58
Poly(tetrafluoroethylene) degradation, 81, 108
recycling, 218
separation, 38, 218
Polyurethane, automobile waste, 222
biodegradation, 108
heat content, 131t
hydrolysis, 80, 222
incineration, 133
manufacture, 205
solvent, 221
solvolysis, 80
Poly(vinyl acetate), copolymer, 182
from cellulose, 237t
Poly(vinyl chloride), biodegradation, 108
cryogenic embrittlement, 63
density, 42t
from cellulose, 237t
heat content, 131t
incineration, 83, 118, 133
pyrolysis, 139, 223
remolding, 63
separation, by water classification, 41-42
by surface properties, 46
solvent, 51-52, 221
structure, 74-75
uses, 75
Poly(vinyl pyridine), 66
Portland cement, *see* Cement
Postconsumer waste, 6, 110-174, 215, 217
Potable water recycle, *see* Water recycle
Potassium carbonate, 205
Potassium chloride, 71, 187, 190
Potassium oxide, 181, 185, 190
Potato chip wastes, 175, 212, 227, 233
Pottery wastes, 181
Poultry manure, 233
Poultry-packing wastes, 46, 227
Powder metallurgy, 70-71
Power stations, clinker, *see* Clinker
effluents, *see* Furnace effluent gases
fly ash, *see* Fly ash
solid waste fuel, 124, 133, 146, 160, 172-173
Pozzolans, 185-186, 296g
fly ash, 184
in waste treatment, 82
Precious metals, 215, 261
Precipitate flotation, 66
Precipitation, 66, 187-191, 296g
with alkali, 188
coprecipitation, 34
with ferrous hydroxide, 34

of metal ions, 34, 66
of silver, 38
see also Cementation
Pressure, in membrane processes, 57-58
Pressure vessel, for partial oxidation, 91
Producer gas, 296g
from waste cellulose, 91
Producers, microbiological, 106
Prompt scrap, 6, 296g
plastic, 217
Propane, in oil refining, 239
Propanol, 98t
Propionic acid, 98t, 236
Propylene, 29, 237t
Protease, 230
Protein, 99, 232-235, 296g
animal, 232-233
coagulation, 68, 227
enzyme protease, 230
from fermentation, *see* Single cell protein
leaf, 227
meat, 232-233
microbiological decomposition, 98-99, 230
separation, 226-227
by centrifugation, 227
by coagulation, 68, 227
by flotation, 46, 227
by ion exchange, 68, 227
by precipitation, 68, 227
by reverse osmosis, 58, 226
by ultrafiltration, 57, 226
Prototype urban waste systems, 157-174, 248-249
Psychology in recycling, 27-28, 121, 201
PTFE, *see* Poly(tetrafluorethylene)
Public, participation in recycling, 121
Publications, 10, 250-269, 300-377
element recycling, 250-263
reviews, 264-269
water recycle, 213
Public relations, 28, 119
Pulp and paper mill wastes:
B.O.D., 69
chemicals, 29, 60, 84, 192-193, 228, 236-237
construction materials, 182, 184, 192
incineration, 166, 192
lignin and lignosulfonates, 57
sludges, 69, 82
water recycle, 60, 69
see also Paper
Pulverization, *see* Shredding
Pulverized fuel ash, 183. *See also* Fly ash
Pump cell, 70
"Purox" process, 136, 167
Putrefaction, *see* Anaerobic digestion
Pyrolysis, 84 88, 93, 296g
active carbon, 55
enthalpy changes, 86-87
of plastics, 75-77, 87-88, 93, 139-140, 223
of rubber tires, 88, 92, 93, 139-140, 156
of urban waste, *see* Urban waste
Pyrolysis-incineration, 86-87, 91, 139, 144t
Pyrometallurgy, 61, 93-94, 189-190, 202, 216, 297g
Pyruvic acid, 97-98

Quality of recycled material, 26-27, 151
Quarry wastes, 178

Radiation, ionizing, 36, 106, 191, 240
sources, 240
Radiators, automobile, 49, 150
Radical, *see* Free radical
Radioactivity, *see* Radiation, ionizing
Radium, in phosphogypsum, 191
Rags, *see* Textiles
Rare earths, metallurgy, 51
Rayon manufacture, zinc recovery, 51, 189
Raytheon, 125
RC separation, 43
Reactions, chemical, 30, 65-94
exchange, 66
Reactivation, active charcoal, 54, 72
ion exchange resin, 67-69
Reactors, 92-93, 297g
anaerobic digestion, 104-105
fixed bed, 91, 92
fluidized bed, *see* Fluidized beds
gasification, 91
high temperature, *see* Electric arc furnace; Electric induction furnace; Furnaces; Molten metals; Molten salts; Smelting
incineration, 132-133
partial oxidation, 91
pyrolysis, 139
rotary kiln, 92
Re-Chem International, 176

Reclamat International, 221
Reclamation, definition, 6. *See also specific substances*
Recovery, definition, 6. *See also specific substances*
Recreation, 3, 10
Recrystallization, 41, 66
Recycle ratio, 8
Recycling, categories, 5t, 7t
  centers, 28, 121
  energy costs, 245-248
  factors, 13-29
  index, 29
  philosophy, 1, 10-12
  see also specific substances
Red mud, 181
Reducers, microbiological, 96, 106
Reduction, 82-92, 297g
  carbothermic, 94, 190
  metal hydroxides, 188
Re-extrusion, plastics, *see* Plastics
Reflected light, in separation, 37
Refractories, 17, 60, 72-73, 136, 216, 297g
Refrigerants, 205
Refrigeration, *see* Cryogenic
Refuse, 5, 297g. *See also* Urban waste
Refuse derived fuel, 124, 134-136, 146, 160, 172-173
Refuse Energy Systems Co., 171
Regeneration, active charcoal, 54, 72
  ion exchange resin, 67-69
Remolding, plastics, 27, 63, 75, 217, 220-221
Reservoirs, water, 210
Residential waste, 5, 110. *See also* Domestic waste
Residuals, 5, 14, 297g
  management, 14, 297g
Resin, from nutshells, 227
  from sulfite waste liquor, 193
  ion exchange, 66-69
  solvent extraction, 51-52
  synpulp, 193
  see also Plastics; Polymers
Resource Recovery Act, 6, 112
Resources, 3, 10-11, 14-17, 241, 243
  "crisis," 241
Returnable bottles, *see* Beverage containers
Reverse osmosis, 58-60, 297g
  carbohydrate separation, 226-227
  oil, 239
  protein separation, 58, 227
  water desalination, 213
  water treatment, 209, 211, 212
Reversible process, 241-242
Revert scrap, 6
"Reverzer" plastics recycle process, 220
Reynolds, Metal Co., 11t
Rhodotorula, 234
Rhodotorula gracilis, 234
Ribonuclease, 234
Rice hulls, ash, 227
Rising current separation, 43
Rivers, 207-208, 248
Roads, construction material from wastes, 178-179, 201, 221
  surface material, 17, 156, 157, 181-182, 192, 201, 223, 239
Roadside litter, *see* Litter
Rock phosphate, 190-191, 193
Roofing felt, 225, 229
Rotary kiln reactors, 92
Rubber, crumb, 62, 157, 182
  in asphalt, 17, 156, 157
  depolymerization, 73, 77, 81
  mastication, 81
  natural, 73, 75
  reclamation, 155-156
  remolding, 77
  softening, 221
  synthetic, 75
  in urban waste, 115t
Rubber tires, cryogenic embrittlement, 62, 157
  discarded, 154-157
  heat content, 131t, 156t
  hydrogenation, 156
  incineration, 155-156
  pyrolysis, 88, 92, 93, 139, 156
  retreading, 155
  uses of waste, 55, 155
Rubbish, 5, 297g
Ruminants, 297g
  in urban farming, 130
Ruthenium, 251
Rutile, 16, 25t
Rye, grass, 106

Saccharomyces cerevisae, 96, 102, 236

Saccharomyces fragilis, 234
Sacrificial dissolution, 72
Sacrificial recycling, 4, 201
Salt, *see* Brine; Desalination; Seawater; Sodium chloride
Salts, ionic properties, 50
Salvage, 6, 30-31, 297g
Sand, *see* Silica
Sand filters, 55
San Diego, California, 169-171
Sanitary landfill, *see* Landfill
Santee, California, 11t, 150
Saprophytes, 297g
  in composting, 102
Saugas, Massachusetts, 14, 133, 171
Sawdust, in bricks, 184, 227
  long fiber, 229
  oil absorber, 226
  pyrolysis, 88
Science fiction, 1, 2
SCP, *see* Single cell protein
Scrap dealers, 30
Scrap metal, 30, 199, 215
Screening, 32, 39-41, 113t, 125, 230. *See also* Wet screening
Scrubbing, 201-202
Sealers, microbiological decomposition, 99
Seattle, Washington, 167
Seawater, desalination, 56-58, 66, 133, 213
  magnesium source, 16, 25
  see also Brine
Secondary fiber, *see* Paper
Second Law of Thermodynamics, 241
Sedimentation, 41, 227
Seebeck effect, 38
Seed wastes, 227, 233
Selective dissolution, 71
Selective melting, 60
Selenium, environmental effect, 23
  in plants, 23, 226
Semipermeable membrane, 56, 58, 297g
Semiprecious metals, 261. *See also* Silver
Semi-wet pulverization, 124, 165
Sensors, 31-32, 126
Separation, 18-22, 30-64
  air classification, 48-50, 53, 124-125
  ballistic, 52-54
  bounce, 52-54
  chemical 18-21, 30, 65-94
  density-drag, 42-50
  density, 42-50, 125, 200, 218, 219
  drying, 30, 64, 187
  dry screening, 32, 39-40, 125, 231
  electrochemical, 69-72
  electromagnetic, 32-39
  electrostatic, 35-37, 219
  energy requirements, 25
  filtration, 39-41, 54-55, 66, 188, 209
  fluidized bed, 48-50
  frictional, 52-54
  froth flotation, 46, 66, 180, 200, 227
  hydraulic, 41-48, 123, 125, 138, 160, 165, 230
  ion exchange, 66-69, 150
  magnetic, 18, 31, 32-35
  manual, 31, 219
  mechanical, 31-32
  membrane, 55-60. *See also* Dialysis; Electrodialysis; Reverse Osmosis; Ultrafiltration
  phase difference, 39-52, 55, 60
  physical, 17, 18, 30-64
  pneumatic, 47-50, 124-125
  precipitation, *see* Precipitation
  size, 32, 39-41, 125, 230
  solvent extraction, 50-52, 68, 189, 214, 219, 230
  spectroscopic, 32-39
  surface properties, 46, 52-55, 219
  thermal, 60-64, 219, 231
  vaporization, 61-62, 64
  viscous drag, 42-50
  water classification, 43
  wet screening, 40, 123, 125, 138, 160, 165, 230
Settling tank, 187
Sewage, 147-150
  animal feed products, 148
  chemical treatment, 149-150
  domestic treatment, 148-149
  fly ash treatment, 149
  ion exchange treatment, 68, 150
  methane production, 105, 148
  microbiological oxidation, 147-149
  phosphate removal, 189, 196
  water recycle, 11t, 150, 209-211, 248
Sewage canals, industrial, 208
Sewage sludge, algae, 148, 235
  anaerobic digestion, 129, 148
  as fertilizer, 148-149

composting, 106, 128, 148
hydroponics, 106-107, 148
liquefaction, 90, 149
spray irrigation, 102, 148
treatment with silicates, 82, 149
wet combustion, 91, 149
Sewer pipe, 182
Shale wastes, 134, 180-182, 184-185
Sheep, 130
Shredding, 31, 123-124
automobiles, *see* Automobile, shredder scrap
cryogenic embrittlement, 18, 62-63, 151, 157, 216, 219
plastics, 221
rubber tires, 62, 81, 157
urban waste, *see* Urban waste
woody wastes, 227
Silica, ash, 227
in blast furnace slag, 180t
in cement manufacture, 185, 227
fluidized bed, 49t
in glass manufacture, 199-200
resources, 3, 199
Silicates, in fly ash, 183t
in slag, 179
soluble, 60, 82, 149, 182-183, 228
synthetic, 82
in waste treatment, 82, 149, 182-183, 193, 228
see also Slag
Silicon carbide, 73
Silicon dioxide, *see* Silica
Silicones, 205-206, 214
Silos, fumigant recovery, 55
Silver, recycle, 8t
recycle literature, 261
resources, 16f
separation by cementation, 72
separation, by eddy current methods, 35t
by electrochemical processes, 71
by ion exchange, 67
from photographic wastes, 38, 72
Single cell protein, 95, 103, 232-235, 297g
actinomyces, 234
algae, *see* Algae
as animal feed, 106, 233-234
bacteria, *see* Bacteria
from cellulose, 233-235
from feed-lot wastes, 103, 232
food value, 232-233
fungi, *see* Fungi
nucleic acid content, 234
palatability, 234-235
from petroleum, 239
from plastics, 108
protein-lipid ratio, 234
from sewage, 148
from urban waste, 131, 143
from wood sugars, 192
yeasts, *see* Yeasts
Sink-float separation, 41-43
Sintered glass, 187
Sintered pumice, 184
Sirotherm®, 67
"Sitall," 180, 186, 298g
Size characteristics, in separation, 32, 39-41, 125, 230
Skinned membranes, 57-58
Slag, 187, 298g
aluminum, 190
blast furnace, *see* Blast furnace slag
in construction, 179-180, 183
in electroslag melting, 61
in glass manufacture, 175
high temperature incineration, 136
in high temperature solvent extraction, 52
neutralization, 319
Slag cements, 202
Slagging incineration, 136-137, 181, 201, 298g
Slagging pyrolysis, 136-137, 139-140, 143, 167, 171-172
Slaked lime, *see* Calcium hydroxide
Slate, lightweight aggregate, 134, 184
Slaughterhouse wastes, 46, 227
Slow release fertilizer, 193, 236
Sludges, 178, 187-192
dewatering, 187
dissolution, 187
lime, 181, 188
see also Sewage sludge
Slurry, 178, 187
Smelter flue dusts, 70, 190
Smelting, 61, 94, 154, 190, 202. *See also* Blast furnace; Furnace effluent gases
Smog, photochemical, 206, 214
Smokeless incinerator, 151
Snow-melting, 249

Soapmaking, 227
Social attitudes to recycling, 10, 27-28
Social service, *see* Volunteer recycling
Social factors in recycling, 10, 27-28, 120-122, 153-154
Society, 10, 12
Soda ash, *see* Sodium carbonate
Sodium bisulfite, 203t
Sodium carbonate, in glass manufacture, 200
  in hydrometallurgy, 19f-20f
  manufacture, 200
  molten, 205
  sulfur oxides recovery, 203t, 205
  in wood pulp liquor, 192-193
Sodium chloride, aluminum flux, 187, 190
  in electrochemical machining, 199
  see also Brine; Desalination
Sodium dichromate, 197
Sodium hydroxide, 188, 192, 203-205
Sodium lignosulfonate, 57, 184, 192, 227
Sodium nitrate, 199
Sodium nitrite, 232
Sodium oxide, 181, 185, 190
Sodium silicate, in fly ash brick, 184, 186
  in lightweight aggregate, 184
  in metal treatment, 60
  in paper fiber cleaning, 228
  in waste treatment, 82, 149, 182-183
Sodium sulfate, 21f, 29, 60, 192
Sodium sulfide, 192
Sodium sulfite, 203-205
Soft technology, 11t, 91, 103, 147
Soil conditioner, bark, 69, 226
  compost, *see* Compost
  glass, 201
  rubber crumb, 155, 157
  tire fiber, 156
  urban waste, 128-129
  wood pulping wastes, 193
Soil stabilization with waste materials, 17, 179
Solar energy, for drying, 64
  for heating, 10-12
  photosynthesis, 95-96
  photosynthetic storage, 105-107, 237-238
  research, 249
  thermodynamic potential, 245
Solder, scrap, 70, 93, 216
Solid-liquid solubility, 50-52
  temperature dependence, 188, 195
Solid waste, classification, 5-6, 9. *See also* Agricultural waste; Urban waste
Solid Waste Management Act, 6, 112
Solubility, *see* Gas-liquid solubility; Liquid-liquid solubility; Plastics solubility; Solid-liquid solubility
Solubility parameter, 50, 221
Solution process, 66
Solutions, for recovery, 187-199. *See also specific materials*
Solvay process, 312
Solvent extraction, 50-52, 189
  cellulose, 229
  liquid ion exchangers, 68
  polymers, 51-52, 219, 221
  solvents, 214
  see also Hydrometallurgy
Solvents, evaporation, 40-41
  from carbohydrate fermentation, 236
  recovery, 214, 223
Solvolysis, 78-80, 298g
  of polymers, 78-80
"Sortex" optical separation system, 160
Sorting, *see* Separation
Source reduction, 120-122
Southampton Water, 248
South Yorkshire County Council, 174
Soviet Union, glass-ceramic, 180, 186
Space, 1, 10
Space heating, 133, 249
Spark beam, 38
Specific ion exchange, 67
Specifications, of recycled material, 27, 63, 179
Spectroscopic separation, 32-39
Spinning techniques, for polymers, 81
Spirulina maxima, 235
Spontaneous process, 244
Sports surfaces, 62, 155-156
Spray irrigation, 102, 106-107, 148
Stack gases, 65, 181, 193, 201-207
Stainless steel, 60, 61, 215
Standard electrode potential, 69-72
Standford Research Institute, 125
Stannate, 68
Starch, 298g
  grafted onto fiber polymer, 81
  wastes, 57

see also Carbohydrates
Starvation, 233-234
Steam distillation, 61
Steam drying, 64
Steel, in automobiles, 150-154
effect of contaminant metals, 61, 151
industry, 14, 94, 151-154, 167, 190
metallurgy, 18-21, 61, 145, 151, 190
packaging, 117
pickling, 51, 57, 194-196
recycled, 27, 190
recycle literature, 252-254, 256-258
Steel scrap, *see* Ferrous metal
Steel Can Group, 164
Steel cans, 118-120
Steel can scrap, 27, 94
Sterilization, of animal feed, 130
incineration, 132
ionizing radiation, 240
of microbiological media, 102
St. Louis, Missouri, 124, 134-135, 168, 172-173
Stock feed, *see* Animal feed
Stoneware, 186
Stress-cracking, plastics, 81, 221
Strontium, removal by fly ash, 189
Structural materials from wastes, 17, 178-187
Structural metals, 252-261
Styrene, 75, 182, 222
Sublimation, 62
Submerged combustion, 41, 193
Subsidies, 14
Substitute natural gas, 89, 91, 140
Substitution of materials, 22
Succinic acid, 98t
Sucrose, 97, 227
Sugar, 97, 226
from cellulose, 78-79, 97, 102-103, 143, 231-232
in microbiological processes, 96-98, 102-103, 226
refining, 45, 84
wood, 192
see also Lactose
Sugar acids, 97
Sugar industry wastes, 14, 58, 212, 227, 229, 232, 234
Suint grease, 227
Sulfide copper ore, 25t
Sulfite woodpulping liquor, 192-193, 233-234, 236
Sulfur, in blast furnace slag, 180t
in coal, 134t, 192
in construction, 182, 202
in fly ash, 183t
microbiology, 108
in paving, 182
recycle, 65, 107, 192, 202-205
recycle literature, 262-263
resources, 192, 202
in steel, 151
in tires, 155, 156t
in urban waste, 181, 134t
Sulfur dioxide, catalytic oxidation, 65, 191-193, 203-205
recycle, 202-205
see also Sulfur oxides
Sulfuric acid, cellulose hydrolysis, 78-79, 143, 231
coagulant, 227
dissolution of sludge, 188
from furnace gases, 65
hydrofluoric acid manufacture, 191
petrochemical industry, 193
phosphoric acid manufacture, 190-191, 193
protein recovery, 68
recovery, 193
Sulfurous acid, 192
Sulfur oxides, 201-205
from furnace gases, 65, 181
from sulfuric acid recovery, 193
Sulfur trioxide, in fly ash, 183t
from sulfur dioxide, 65
Sun, *see* Solar energy
Supperalloy waste, 18-21, 298g
Superheated steam, for drying, 64
Surface-active agents, 55, 298g
Surface coatings, decorative, 186
facing bricks, 186
from nutshells, 227
glass, 201
metals, 215
phosphate, 196
plastic, 216
recovery, 18
removal, 93
solvents, 214
wastes, 214

Surface separation processes, 46, 52-55, 219
Surface properties, of plastics, 46, 55
of precipitates, 66
Surfaces, cleaning of metal, *see* Pickling
Survival function, 245
Suspension firing, 93. *See also* Fluidized beds
Sweating (selective melting), 60
Sweden, Clivus Multrum, 11, 148-149
energy recovery incineration, 249
solvent extraction metal recovery, 199
Sydney University, 11t
Symbiotic cultures, 103
Synpulps, 193
Synthesis gas, 87-90, 141, 298g
Synthetic adsorbents, 55
Synthetic papermaking fiber, 193
Synthetic polymers, *see* Plastics; Polymers; Resin; Textiles; Thermoplastics
Synthetic rubber, 75. *See also* Rubber
Syrups, purification, 45. *See also* Sugar
Systems, hierachy, 2-3, 241, 246, 250

Taconite, 25t
upgrading, 84, 145
Tailings, 298g. *See also* Mining wastes
Tall oil, 236
Tallow, 227. *See also* Fats
Tannin, 69, 226, 236
Tanning, recovery of brines, 64
recovery, of chrome liquors, 187, 227
of protein, 227
of unhairing liquors, 187
synthetic agents, 236
Tantalum, 8t
Tap water supply, *see* Water recycle
Tarmacadam, *see* Asphalt
Tartaric acid, from cellulose, 79, 236
recovery by ion exchange, 68
Taxation, 14, 112, 239
Tea leaves, animal feed, 233
wax extraction, 227
Technology, 17, 24. *See also* Soft technology
Teflon, *see* Poly(tetrafluoroethylene)
Temperature range, climate, 17, 184
stoneware, 186
Temperature scale, 241
Terminology, 3-6, 8-9, 287g-299g
Terrazzo, 186, 298g
Tetrabromethane, 41
Tetraethyl lead, 215
Textiles, 224-225
dyeing, 58
heat content, 131t
pyrolysis, 55
recycling rate, 224
in urban waste, 224
Thallium formate-thallium malonate mixtures, 41
Thermal conductivity, glasses, 181-182
Thermal decomposition, *see* Pyrolysis
Thermal efficiency, 248-249
Thermal energy utilization, 67, 248-249
Thermal insulation, *see* Insulation
Thermally sensitive materials, 62, 214
Thermal pollution, 248
Thermal separation methods, 60-64, 219, 231
Thermodynamic potential, 245, 248
Thermodynamics, 10, 241-245, 298g
Second Law, 241
Thermoelectric effect, 38
Thermophilic organisms, 101, 104-105, 234, 298g
Thermoplastics, 63-64, 75, 231, 298g
incineration, 83-84, 118, 133, 217, 224
molding, 27, 63-64, 75, 77-78, 217, 220-221
pyrolysis, 87
separation, 231
Thermosetting plastic, 27, 63, 299g
Thiobacillus ferrooxidans, 107
Tiles, from waste, 63, 181, 186. *See also* Surface coatings
Timber, *see* Cellulose; Sawdust; Wood
Time, 3, 10
Tin, 260
in automobiles, 61t, 150
from incinerator flue dusts, 190
recycle, 8, 215
recycle literature, 252-254, 261
removal from steel, 61
resources, 3, 16f, 216
separation, by carbothermic reduction, 190
by eddy current methods, 35t
by electrochemical processes, 71
by ion exchange, 67-68
by pyrometallurgy, 94, 190

by selective melting, 61t
in steel, 151
Tinned steel cans, 3, 18, 118-119, 216. *See also* Beverage containers; Steel cans
Tipping, *see* Landfill
Tires, *see* Rubber tires
Titania, *see* Titanium dioxide
Titanium, 261
metallurgy, 25t, 311
recycle, 18
recycle literature, 252-254, 261
resources, 16
uses, 61
Titanium carbide, 73
Titanium dioxide, 18, 25t, 195
Titanium tetrachloride, 72
Tobacco, 227
Toluene, 221
Toluene diamines, 80
"Torrax" process, 137, 171
Torula yeast, 234
Total organic carbon, 208
Town gas, from coal, 84, 239. *See also* Substitute natural gas
Toxic material, 22-24, 214, 216
in compost, 129
incineration, 83
removal by ion exchange, 67-68
in single cell protein, 234
in water supplies, 210-211
Toxic metals, *see* Heavy metals
Toys, recycled plastic, 221
Traffic barriers and markers, plastic, 63
rubber, 155
Trajectory, in separation, 53-54
Transition temperature, 63
Transmutation, 250
Transportation, 14, 17
automobile hulks, 150-154
cullet, 200
hydrochloric acid, 223
used oil, 239
see also Freight rates
Trees, *Acacia mollisima,* 236
Pinus radiata, 236
see also Forest; Wood
Trezek Recovery System, 126, 230
Trichoderma lignorum, 235
Trichoderma viride, 103, 231-232, 235
Tripoli, Libya, 128
Tripolite, 290g. *See also* Diatomaceous earth
Trisodium phosphate, 51
Tropical island industry, 91
"Tufbord," 221
Tungsten, 261
hydrometallurgy, 51
recycle, 215
recycle literature, 252-254, 261
resources, 16f, 17
Tungsten carbide, fluidized bed, 49t
Turbine blades, 199
Turpentine, 192, 236
Two-shot injection molding, 221
Tyne and Wear County Council, U.K., 134, 174

Ultrafiltration, 41, 56-57, 299g
carbohydrate separation, 226-227
oil, 239
protein separation, 226
Ultrasonics, polymer degradation, 81
Ultraviolet radiation, 37
in cellulose hydrolysis, 232
effect on polymers, 77-78, 80
in silver precipitation, 38
Unemployment, 27, 120
Union Carbide, 223
Linde Division, 136, 143, 146
Union Electric Co., 134, 146, 172
United Kingdom, automobile shredder capacity, 151
blast furnaces, 136
blast furnace slag, 179
coal slag wastes, 180
construction materials from waste, 178-179
Department of Energy, 249
Department of Industry, 176. *See also* Warren Spring Laboratory
Department of the Environment, 174
domestic waste, 110, 114
fly ash composition, 183t
gypsum production, 191
lubricating oil recycle, 239
National Coal Board, 249
obsolete scrap recovery, 9
packaging, 117
petroleum use for plastics, 216-217
phosphoric acid production, 191

plastics wastes, 216
pulp and paper industry wastes, 193
"Reverzer" process, 220
rubber reclamation, 155
sewage gas production, 105
solvent recovery rate, 214
textile recycle, 224
urban waste composting, 128
urban waste incineration, 133
urban waste recovery, 173
Working Party on Refuse Disposal, 129
United States, agricultural wastes, 9t, 225
animal wastes, 225
Army Natick Laboratory, 232
ash utilization, 178
automobile composition, 153
automobile industry, 150
beverage container centers, 121
beverage container legislation, 120
blast furnace slag, 179
Bureau of Mines, *see* Bureau of Mines
chemicals recycling, 29
coal slag wastes, 180-181
commercial waste, 9t, 110
crop wastes, 225
Department of Agriculture, Forest Products Laboratory, 121, 125, 228
Department of Commerce, 6, 154
Department of the Interior, 6. *See also* Bureau of Mines
domestic waste, 9t, 110
electroplaters, 197
energy use, 25
Environmental Protection Agency, *see* Environmental Protection Agency
federal government, 6, 12
federal wastes, 9t
fly ash composition, 183
highway construction, 178
housewives recycling attitudes, 121
industrial waste, 9t
legislation, 6, 24, 112
lubricating oil recycle, 239
metal content of urban waste, 215-216
metal recycling rates, 215
mineral waste, 9t
municipal waste, 9t
National Research Council, Highway Research Board, 120
nuclear energy research, 249
obsolete scrap recovery, 8t
Office of Science and Technology, 6, 9
packaging, 117
paper recycle rate, 228
paper usage, 230
plastic container reuse, 218
plastics wastes, 216
pulp and paper wastes, 193
residential waste, 9t, 110
resources, 16f
rubber reclamation, 155
scrapped automobiles, 150
solid wastes, 9t, 17, 110
Standard Industrial Classification, 175
tire use, 154
urban waste, 9t, 15, 110, 114, 157
urban waste composting, 128-129
urban waste fuel source, 132
urban waste incineration, 133
urban waste metal content, 215-216
urban waste paper recycle, 226
volunteer paper segregation, 230
waste utilization, 224-228
water reuse, 210
water supply, 210
Universe, 2, 241
University courses, 12
University of California, 130
University of Illinois, 130
University of Minnesota, 11t
University of Wisconsin, 121
Uranium, hydrometallurgy, 51
microbiological leaching, 108
recovery by ion exchange from mine leach waters, 68
resources, 16f
Urban farming, 106-107, 131
Urban waste, 5, 9t, 110, 299g
aluminum, 117, 216
anaerobic digestion, 14, 129-130
analysis, 114, 138, 162
animal feed, 130-131
ash, 138, 185-186
baling, 110
biological treatment, 126-131
calorific value, 131-132, 134t, 140t, 144t
cellulose, 116, 125, 160, 226
chemical hydrolysis, 79, 143-144
chemical products, 145
cinder, 114t

collection, 17-18, 122, 147, 157
compaction, 122
comparison of treatment processes, 143-147
composition, 87-88, 114-116, 138, 162
composting, 52, 101, 113t, 128-129, 146t, 158, 166-167
"Dano" compost process, 129, 158
demonstration systems, 157-174
economics, 110, 145-147, 201, 231
empirical formula, 88-89, 116
energy costs, 248
energy recovery, 131-145, 248
ferrous metal, 114-115, 117, 123, 138, 162
fuel source, 23, 131, 134-144, 184-185
gasification, 91, 113t, 141-144, 146, 158, 168-171
gas turbine operation, 134, 164
glass, *see* Glass
heat content, 131-132, 134t, 140t, 144t
high temperature gasification, 91
high temperature incineration, 136-137, 182, 201
hydrogasification, 140-141, 144t
hydrolysis, 79, 113t, 143-144, 231
hydropulping, 40, 123, 125, 138, 158-162, 165, 230
incineration, 83, 132-134, 146t
incineration energy recovery, 132-133, 165, 169, 171, 248-249
incineration flue dust, 35, 133, 190
incineration reactors, 92
incineration residue, 133, 136-138, 181, 201
industry participation, 112
iron ore upgrading, 84, 145
landfill, 110, 124, 146t, 181, 183
liquefaction, 89, 113t, 141, 144t
local authorities, 112, 146, 157
magnetic separation, 124, 162-164, 166, 168
metal content, 113-115, 215-216
microbiological treatment, 126-131, 158
nonferrous metals, 114-115, 117, 138
packaging, 116-118
paper, *see* Paper; Wood fiber
partial oxidation, 91, 141-143
planning, 147
plastics, 63, 114-115, 117-118, 133, 219, 223
prototype systems, 157-174
publications, 112
pulping, 40-41, 123, 125, 138, 158-162, 230-231
pyrolysis, 88, 113t, 139-140, 144t, 146, 167-172
pyrolysis residues, 139-140
recovery methods, 112-113, 120-145
research, 112
rubber, 115t
scale of operation, 146
screening, 32, 40, 113t, 123, 125, 137-138, 158-162, 230-231
segregation, 120-122, 219, 230
separation, 122-126
  at source, 120-122, 219, 230
  by air classification, 113t, 124-125, 136
  by ballistic methods, 53-54, 125
  by centrifugal methods, 44-45
  of metals by cryopulverizing, 216
  by density methods, 125, 200
  by item identification, 31-32, 37, 126
shredding, 113t, 123-124, 129, 173-174
single cell protein production, 131
size characteristics, 32
slagging pyrolysis, 136-137, 139, 143, 167, 171-172
solid fuel, 137, 146
sulfur content, 116, 134t
supplementary power station fuel, 124, 134-136, 146, 172-173
terminal sorting, *see* Urban waste, separation
textiles, 224-225
thermodynamic potential, 245
toxic contaminants, 129
utilization, 112
voluntary separation, 120-122
water content, 116t, 134t
wet combustion, 143-144
wet screening, *see* Urban waste, screening
Urea, 198
Uric acid, 234
Uricase, 234
U.S.S.R., glass-ceramic, 180, 186

Vacuum centrifugal distillation, 62
Vacuum distillation, 62, 214
Vacuum melting, 61

Vanadium, hydrometallurgy, 51
in steel, 151
Vanillan, 236
Vaporization, in chemical transport, 72-73
in distillation, 61-62
natural gas, 63
Vapor transport, 72-73
Varro composting process, 129
Vegetable oils, from waste, 227
Vibratory casting, 182-183
Vinyl compounds, for polymerization, 74-75, 236-237
Vinyl acetate, 182
Vinyl chloride, manufacture, 193
Viruses, in composting, 102
in water, 209-210
Viscose rayon manufacture, zinc recovery, 51, 189
Viscosity, separation methods, 45
variation in ferrohydrodynamics, 34
Visible radiation, 36-37, 200
Vitamins, in yeast, 234
Vitreous slag, 181, 185
Volatile compounds, chemical vapor transport, 72-73
in incineration, 169, 181
Volunteer recycling, 120-122, 219, 225, 230
Von Roll, 171
Vortex classifier, 45
Vortex incineration, 93

Wallboard, *see* Building board
Walnut expeller meal as metal ion scavenger, 69
Warren Spring Laboratory, 49-50, 60, 123, 125, 134, 173-174, 176-177
Waste, classification, 4, 9, 175. *See also* Agricultural waste; Commercial Waste; Consumer waste; Domestic waste; Federal Waste; Industrial Waste; Municipal Waste; Urban Waste
Waste derived fuel, 124, 134-136, 146, 160, 172-173
Waste disposal, publications, 112
Waste materials exchange, 174, 176-177
Waste paper, *see* Paper
"Waste-plus-waste" methods, 13-14, 41, 189, 199, 202, 212
Wastes, adsorbent, 55
incineration, *see* Incineration
pyrolysis, *see* Pyrolysis
Wastewater, *see* Water
Water, activated carbon treatment, 72
advanced treatment, 41, 58, 65, 209-211
cooling, 188, 207-208, 212, 248
crystallization, 62
desalination, 56-60, 66-69, 133, 213, 249, *see also* Brine
heavy metal removal, *see* Heavy metals
immiscibility, 50, 61
industrial, 207-208
mixture with alcohol, 41, 221
municipal reuse, 210
natural system pollution inventory, 208
nutrient removal, 189, 196, 211-212
pollution, 82, 187, 207, 211, 228, 229
primary treatment, 209
purification, by electrodialysis, 56, 150, 209
by filtration, 54-55
by freezing, 62, 210
by high gradient magnetic separation, 210
by ion exchange, 66-69, 150, 209
by microbiological techniques, 209
by oxidation, 209
by precipitation, 189, 211
by reverse osmosis, 58, 150, 209, 212-213
secondary treatment, 209
solvent, 50, 207
storage, 210
supplementary supplies, 210
total reuse, 207-208, 212
treatment, with alum and lime, 189
with waste metal ions, 189
in urban waste, 116t, 134t
Water classification, 43
Water-gas shift, 86-89, 92, 145, 299g
Water glass, *see* Sodium silicate
Water heating, 133, 249
Water hyacinth, 106
Water recycle, 3, 207-214
energy costs, 248
industrial, 14, 23, 58-60, 192-193, 212
literature, 213
potable, 27, 209-211, 213
pulp and paper industry, 192-193, 207, 212

from sewage, 11t, 150, 210, 248
Water-solubility, 61
Water-steam mixture, 41
Water treatment, chemical reuse, 212
Wax, 101, 299g
recovery, 223, 227
thermoplastic property, 64
Wax-impregnated paper, 64, 119, 229
Wet beneficiation, 41-48, 123, 125, 158-162, 165, 230
Wet combustion, 91, 143, 144t
Wet screening, 40, 113t, 123, 125, 158-162, 165, 230
Wheat protein, 57
Wheat silos, 55
Wheelabrator-Frye, Inc., 171
Whey wastes, 57-58, 226-227, 232, 234
Windpower, 12
Wine-making, 96, 102
Winnowing, 47-50, 53, 124-125, 299g
Wiping cloths, by copolymerization, 81
Wire scrap, 53, 63
Wire-stripping wastes, 18, 44
Wiring, in automobiles, 22, 151
Wood, demolition, 179, 226
empirical composition, 88t
fuel, 237
gasification, 91
heat content, 131t
hydrolysis, 78-80, 231-232
liquefaction, 89-90
plastics feedstock, 237
pyrolysis, 84-88
see also Cellulose; Sawdust
Wood bark, 226
as metal ion scavenger, 69, 189, 229
pyrolysis, 88
tannin, 69, 226, 236
Wood fiber, in bricks, 184, 227
in controlled release pesticides, 227
in plastic board, 221
oil absorber, 226
from sawdust, 229
from urban waste, 125, 158-162, 228, 230
see also Paper
Wood pulping wastes, *see* Pulp and paper mill wastes
Wood sugars, 192, 233
Wool, *see* Textiles
Wool scouring wastes, 227
Worcester Polytechnic Institute, 90
Worthing, U.K., 110, 174
Wringers, 187

X-rays, 36-37
Xylene, 51, 221
Xylose, 97

Yeasts, 95, 233-234, 299g
Candida utilis, 234
facultative anaerobes, 96
lipid content, 234
Rhodotorula, 234
Rhodotorula gracilis, 234
Saccharomyces cerevisiae, 96, 102, 236
Saccharomyces fragilis, 234
single cell protein, 103, 148, 234
Torulopsis utilis, 234

Zeolite, 66
Zinc, 260-261
in automobiles, 61t, 150-153
in brass, 60, 190
cementation of copper, 188
cryogenic embrittlement, 63
die castings, 63
environmental effect, 24
from flue dusts, 189-190
metallurgy, 51, 94, 188-190, 202
microbiology, 108
plating wastes, 198-199
rayon manufacture wastes, 51, 189
recycle, 8t, 18, 61t, 196, 215
recycle literature, 252-254, 260-261
removal from water by fly ash, 189
resources, 16f
salts in concrete aggregate, 181
separation, by eddy current methods, 35t
by electrochemical processes, 70
by hydrometallurgy, 51, 188, 197
by ion exchange, 67-68
by precipitation, 188-189
by pyrometallurgy, 94, 190
by selective melting, 61t
by solvent extraction, 189
in steel, 151
in tires, 156t
Zinc nitrate, 203t
Zinc oxide, in brass fumes, 190
in tires, 155
Zinc phosphate, 196
Zinc sulfate, 190
Zinc sulfide, microbiological leaching, 108